Towards the Wireless Information Society

Heterogeneous Networks

Volume 2

DISCLAIMER OF WARRANTY

For a listing of recent titles in the *Artech House Universal Personal Communications Series,*
turn to the back of this book.

Towards the Wireless Information Society

Heterogeneous Networks

Volume 2

Ramjee Prasad

Editor

Library of Congress Cataloging-in-Publication Data
Prasad, Ramjee.
Towards the wireless information society: heterogeneous networks/Ramjee Prasad
p. cm.—(Artech House universal personal communications series)
Includes bibliographical references and index.
ISBN 1-58053-364-7 (alk. paper)
1. Personal communication service systems. 2. Internetworking (Telecommunication)
3. Telecommunication systems. 4. Heterogeneous computing. 5. Information society.
I. Title.

TK5103.485.P65 2006
621.384—dc22 2005056276

British Library Cataloguing in Publication Data
Prasad, Ramjee
Towards the wireless information society
Heterogeneous networks.—(Artech House universal personal communications series)
1. Wireless communication systems
I. Title
621.3'84

ISBN-10: 1-58053-364-7

Cover design by Yekaterina Ratner

International Standard Book Number: 1-58053-364-7
Library of Congress Catalog Card Number: 2005056276

10 9 8 7 6 5 4 3 2 1

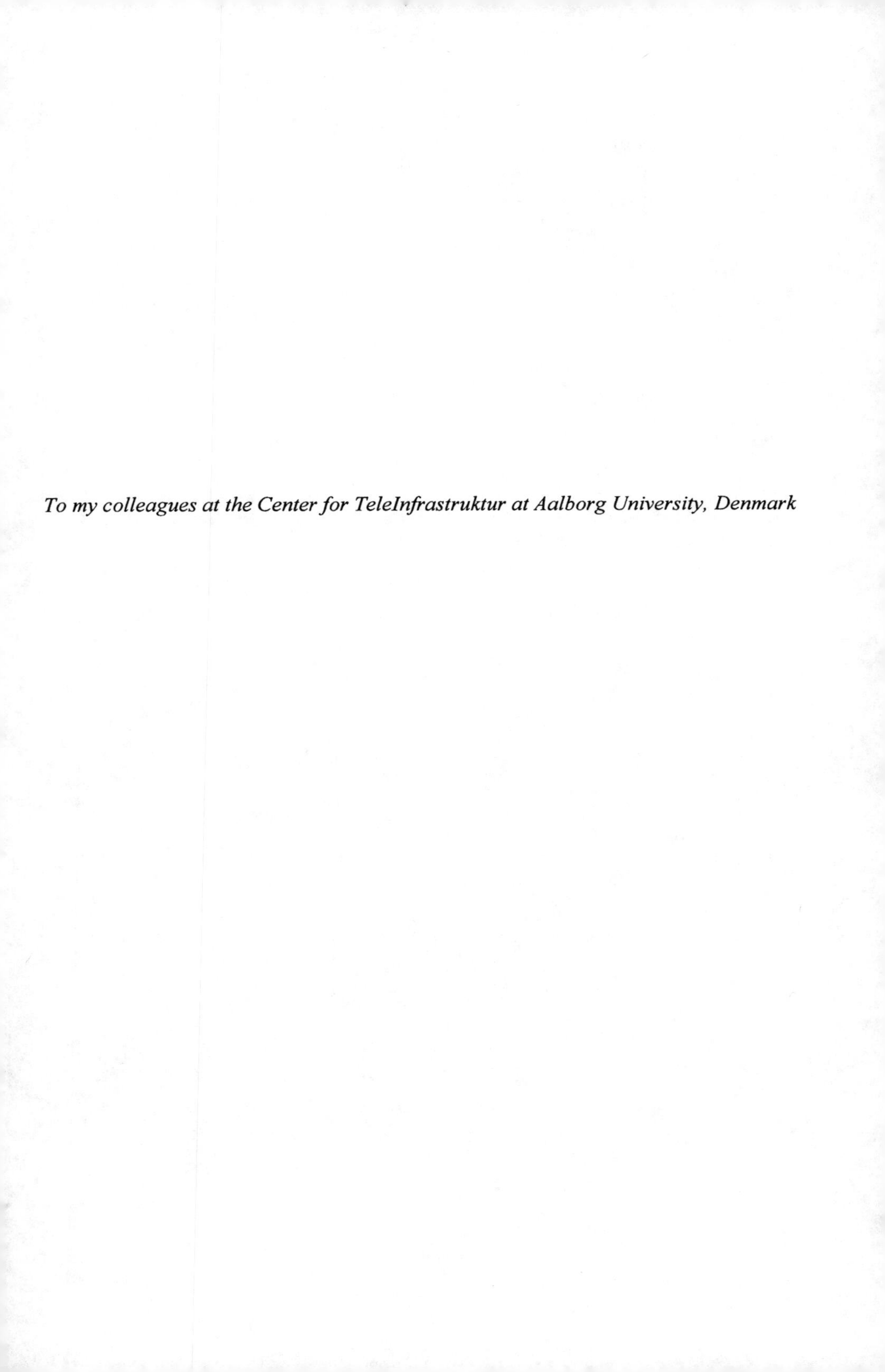

To my colleagues at the Center for TeleInfrastruktur at Aalborg University, Denmark

Contents

Preface

ज्ञानं ज्ञेयं परिज्ञाता त्रिविधा कर्मचोदना ।
करणं कर्म कर्तेति त्रिविधः कर्मसंग्रहः ॥१८॥

jñānaṁ jñeyaṁ parijñātā
tri-vidhā karma-codanā
karaṇaṁ karma karteti
tri-vidhaḥ karma-saṅgrahaḥ

jñānam—knowledge; *jñeyam*—the objective of knowledge; *parijñātā*—the knower; *tri-vidhā*—of three kinds; *karma*—of

Knowledge, the object of knowledge, and the knower are the three factors that motivate action; the senses, the work, and the doer are the three constituents of action.

—*The Bhagavad Gita (18.18)*

The Information Society Technologies (IST) research program was launched in 1999 as a successor of the Advanced Communications Technologies and Services (ACTS) research framework.

The IST research and development (R&D) initiative Mobile and Personal Communications and Systems including Satellite-Based Systems and Services launched many projects that focused on the development of advanced wireless-based technologies, systems and networks, both terrestrial and satellite-based, and on the integration of the associated services in a seamless infrastructure ensuring global personal connectivity and access to broadband wireless multimedia communications and services by anyone, from anywhere, at any time.

The enormous research effort concentrated in the various R&D project activities required a special supporting initiative that would span the whole domain of projects *and* promote, structure, and disseminate the research effort and results. This was the main idea behind the European IST project PRODEMIS (Promotion and Dissemination of the European Mobile Information Society). This book is one of the results of the project PRODEMIS. The project collected and integrated the final research results of the numerous projects in the above-mentioned R&D European initiative with the objective of creating a permanent record of their achievements that is available as a series of three books.

In particular, the work within the European IST mobile R&D initiative addressed reconfigurable radio systems and networks, the objective of which was to lay the foundation for allowing the radio network, including terminals and base stations, to adaptively/automatically adjust to traffic and user requirements. It encompassed work on terrestrial wireless systems and networks, the objective of which was to investigate, develop, test, and validate advanced terrestrial wireless systems and architectures (e.g., UMTS, MBS, WLAN, MMDS, and LMDS) and their interworking and interoperation in particular with fixed/broadcasting networks. It included the development of advanced positioning and navigation concepts and systems, their integration with communications-based systems, and their overall end-to-end management. The work on satellite systems and services developed, demonstrated, and validated novel technologies, architectures, and innovative broadband services in the context of satellite-based communication systems with regional or global coverage and integrated where appropriate with terrestrial and satellite-based navigation services. Advanced tools and technologies for wireless communications were investigated and developed to validate the conceived concepts and to facilitate a mass-market take-up of diversified wireless terminals, networks, services, and applications. Particular emphasis was placed on the integration of such technologies in future generation broadband systems and networks, from cellular to broadband fixed radio access and broadband wireless local area networks for both interactive and distributive services.

A key aspect of the work undertaken within the R&D initiative Mobile and Personal Communications and Systems including Satellite-Based Systems and Services was that of the validation and demonstration of mobile and wireless broadband multimedia technologies and generic services for novel systems and networks. Some of the demonstrations and validation platforms are available in the DVD accompanying this book.

The material collected in this book was edited to provide useful reading material to senior and junior engineers, undergraduate, graduate, and post-graduate students, and anyone else interested in the development of current and future mobile communications.

I hope that all readers will experience the benefits and power of this knowledge.

Acknowledgments

This book was completed because of the full support of Albena Mihovska, who is employed as an assistant research professor at the Center for TeleInfrastruktur (CTIF) at Aalborg University.

The material in this book was collected and structured with the support of the European Commission within the frames of the IST supporting project PRODEMIS in an effort that spanned a period of 3 years. It originates from the technical reports and documentation of the projects that were involved in the R&D initiative Mobile and Personal Communications and Systems including Satellite-Based Systems and Services of the Fifth European Research Program (FP5). The editor, therefore, would like to thank the European Commission, and in particular the leading project officers Dr. Jorge M. Pereira, and later Dr. Andrew Houghton, for their support and guidance during the lifetime of the project and toward its successful completion. Additionally, the editor thanks Dr. Joao Da Silva, Dr. Bartolome Arroyo, Dr. Francisco Guirao Moya, and Dr. Demosthenes Ikonomou, formerly with the unit Mobile, Personal, and Satellite Communications, for their indirect support of the project initiatives.

The project PRODEMIS consisted of an international consortium of seven member organizations. It was coordinated by Aalborg University, and more specifically by the CTIF, former Center for PersonKommunikation (CPK) that had the sole responsibility for this book. PRODEMIS adopted a three-pronged approach that included technology watch and concertation and dissemination activities. Other project results were the technology roadmaps and the proceedings of the First Mobile e-Conference that are available at the Web site of the project (http://www.prodemis-ist.org). The project created a Web discussion space that remains active. I would like to thank all those individuals, who supported these project activities in particular, my thanks go to Dr. John Farserotu and Dr. Fanny Platbrood (CSEM, Switzerland), Professor Rahim Tafazzoli, Dr. Woo-Lip Lim, Dr. Paul Leaves, and Angorro Widiawan (University of Surrey, the United Kingdom), Dr. Sofoklis Kyriazakos, and Dr. George Karetsos (National University of Athens, Greece), Dr. Renata Guarneri (Siemens Mobile Communications, Italy), Robert van Muijen and Philip Jackson (Vodafone Group, the Netherlands), and Albena Mihovska (CTIF, Denmark).

I would also like to thank Dr. Julie Lancashire, formerly of Artech House, for always being available for fruitful discussions related to the books and for being supportive of the overall project activities, as well as her successor Tiina Ruonamaa for her effort in the final project stage.

Chapter 1

Introduction

The demand for mobile multimedia communications has been rapidly increasing. The radio spectrum, however, is a precious and scarce resource. Therefore, novel technologies for efficient spectrum utilization to enhance the capacity of existing and future systems are keenly anticipated.

Adaptive antennas improve the spectral efficiency of a radio channel and, thus, greatly increase the capacity and coverage of most radio transmission networks. This technology uses multiple antennas, digital processing techniques, and complex algorithms to modify the transmitted and received signals at a base station and at a user terminal. In addition, multiple input multiple output (MIMO) techniques can provide significant improvements in the radio-link capacity by making positive use of the complex multipath propagation channels found in certain terrestrial mobile communications. MIMO techniques are based on establishing several parallel independent communication channels through the same space and frequency channel by using multiple antenna elements at both ends of the link. Furthermore, efficient multiple access schemes and adaptive modulation improve the bandwidth efficiency of the systems.

Advanced radio resource management (RRM) algorithms and flexible frequency sharing methods are beneficial in maximizing and optimizing the frequency resource utilization. In addition, antenna and coding technologies such as smart antennas, diversity techniques, coding techniques, space-time coding, and combined technologies improve the radio-link quality in multipath Rayleigh fading channels.

Multihop radio network technology provides means to expand the coverage per base station and allow scalability of the radio network to match offered and demanded traffic capacity. Therefore, this technology leverages fast deployment of wireless networks with low cost.

This chapter is organized as follows. Section 1.1 addresses technology topics that appear relevant to the future development of systems beyond third-generation (3G). New radio technologies and their impact on spectrum utilization, including technologies for improving spectrum efficiency, those using multiple antennas such as adaptive antennas and MIMO, and those for handling traffic asymmetry and time division duplex (TDD) are discussed. Efficient spectrum utilization is the topic of Section 1.2. Advanced RRM algorithms are discussed in Section 1.3. Adhoc networks are a new type of network generation, different from the traditional infrastructure-based networks. Adhoc networks are introduced in Section 1.4. Finally, Section 1.5 gives a preview of the book.

1.1 BEYOND 3G—THE NEXT GENERATION OF COMMUNICATIONS

With the development of core network intelligence, both voice and broadband multimedia traffic will be directed to their intended destination with reduced latency and delays, speeds will be increased, and there will be far more efficient use of network bandwidth and resources. The Universal Mobile Communication System (UMTS), as a modular concept, takes full regard of the trend towards convergence of fixed and mobile networks and services, enabling a huge number of applications to be developed. As an example, a laptop with an integrated UMTS communications module becomes a general-purpose computing and communications device for broadband Internet access, voice, video telephony, and conferencing for both mobile and residential use. As the number of Internet Protocol (IP)-based mobile applications keeps growing, UMTS becomes the most flexible broadband access technology because it allows for mobile, office, and residential use in a wide range of public and nonpublic networks.

UMTS can support both IP and non-IP traffic in a variety of modes, including packet, circuit switched, and virtual circuit, thus benefiting directly from the development and extension of IP standards for mobile communication. New developments allow parameters like quality of service (QoS) and data rate to be set by the operator and/or service provider. Developments of new domain name structures will increase the usability and flexibility of the system, providing unique, fixed addressing for each user, independent of terminal, application, or location. The question of capacity is just as important as the push towards higher data rates and advanced services. Future networks aim at allowing millions of users to access broadband mobile multimedia services, creating the need to efficiently use available radio spectrum. Increased capacity for IP traffic will be possible via wireless local area networks (WLANs) as an extension of UMTS and the further evolution of UMTS.

A wireless IP-based architecture is a heterogeneous system that integrates different access technologies. As a result, the possibility of providing a specific service using different access system technologies is introduced. What needs to be taken into account is that different services can be supported more or less efficiently by a given access network, which in turn offers increased overall spectrum efficiency as the optimal transmission technology can be chosen for a given service type [1].

All recent wireless technology developments, although pulled by different market requirements and applications such as WLANs for portable connectivity in office information technology (IT) enterprises, WPANs for cable replacement and personal sphere interconnectivity, and so on, have been characterized by a common quest for higher capacity density and link speed on the one hand, and greater reliability and scalability in the presence of varying channel conditions and/or traffic loading and QoS system requirements on the other. Guided by strong evidence of these common persistent trends, research has focused on the exploration of the suitability of higher frequency bands for next generation wireless communications.

The 17.1- to 17.3-GHz frequency range recommended by the European Telecommunications Standards Institute (ETSI) was adopted, for example, in the IST project WIND-FLEX (Wireless Indoor Flexible High Bitrate Modem Architecture) [2] in order to achieve the required capacity performance at a reasonable front-end cost. This choice allows access to wider bandwidth, providing high bit rate at the same transmitter power on smaller ranges, and exploiting propagation characteristics in smaller ranges and higher frequencies in order to achieve higher spatial capacity density and spectral efficiency. Moreover, such frequency ranges provide a higher level of capacity due to the combined effect of smaller cell size and higher frequency reuse, while at the same time easing concerns about security because of the intrinsic limited propagation through walls.

A contiguous extension of the available 200 MHz in the 17-GHz range was recognized by the ITU study group JRG 8A-9B [1], which proposed further 400-MHz extensions from 17.3 to 17.7 GHz. In the United States and Japan similar frequency bands are generically allocated for radio communications. Given the clear need for wireless systems to move to upper carrier frequency ranges in order to offer high capacity and higher spatial efficiency and the fact that the 17-GHz allocation does not provide any harmful interference to other wireless systems, it is expected that Federal Communication Commission (FCC) and global regulatory approval of this band will not be an issue, assuming industry support of such bands.

Other technologies, specifically targeting next generation WPANs are either making use of much wider bandwidth in the low-frequency regions [i.e., ultra wideband (UWB)] or at even higher frequencies (i.e., 60 GHz). As far as 60-GHz technologies are concerned, the main reason to explore this bandwidth for high bit rate (fixed) wireless access technologies was driven by the high attenuation produced by the oxygen absorption in that range, considered an enabler for easier and denser frequency reuse. However, in an indoor environment these propagation characteristics are not exploited since propagation is actually limited by obstruction (i.e., walls). Moreover, the 60-GHz propagation requiring full line of sight (LOS) is not suitable for mobile/nomadic purposes. Another obvious consideration is the much higher front-end cost that makes this technology unsuitable for low-cost/consumer applications.

The general requirements for a next generation system will mainly be derived from the types of service a user will require in the future. Therefore, communication systems will need to cope with new forms of personal communications as well as man-machine interaction. This demand will be helped by delivering IP-based, real-time, person-to-person multimedia communications, integrating real-time person-to-person with nonreal-time person-to-machine, providing the ability for different services and applications to interact, and allowing the end user to easily set up multiple services in a single session. An information transport platform must provide high speed, large volume, good quality, global coverage, and flexibility to roam. Hence, person-centered communication infrastructures with a high degree of adaptivity and high performances have to be designed. Communication, information, and entertainment needs vary from one individual to another, and the personalization of a mobile access device will enrich the end user's experience while driving increased usage and revenues for operators.

Personalization can have many aspects, ranging from the visual formatting of an individual's home page to user-selected alerts for news headlines, weather reports, stock quotes, or personal calendar appointments. As the size of wireless terminals shrink, the use of traditional keyboard or keypad input and screen output is becoming increasingly inconvenient to use. A new generation of efficient, flexible, and user friendly human interfaces for communication and data transfer must be designed, implemented, and tested against a number of network-based application services.

Global wireless networks are meant to enable mobile users to communicate regardless of their location. One of the most important issues is location management in a highly dynamic environment because mobile users may roam between different wireless systems, network operators, and geographical regions. Focusing also on the user, it is important to work on mechanisms for reducing the location tracking costs, as well as how to significantly decrease call-loss rates and average paging delays. For mobile operators, the potential for interoperability of mobile and fixed networks and a strong service creation platform are some of the factors that will play a vital role in enhancing differentiation of 3G services. Robust, high-value services that integrate multimedia activities and allow services to interact with each other will enhance the natural, intuitive process of the end user, in whatever network they are operating, whether fixed, mobile, or roaming. The resultant integration and interaction of media types opens up new possibilities for far richer services.

Network changes will be apparent in two major aspects. One is the evolution towards all-IP networks. Thus, besides the physical layer techniques to provide for burst-type communications, medium access control (MAC)/link and network layer technologies to implement wireless IP networks and to control and optimize radio and network operation a highly variable environment are much needed. The other aspect that requires major network changes involves the possible convergence of various technologies and services, such as WLAN, cellular, and fixed.

As an aspect of the radio access area, reconfigurable radio access points and technologies are needed to allow the hardware and software to adopt the best radio access technique suited to each case (spectrum adaptation, channel conditions, network adaptation). Reconfigurable networks are needed to provide access to a number of radio technologies and make a part of the adaptation process.

Finally, future wireless networks are expected to support IP-based multimedia traffic with diverse QoS requirements that will be derived from the types of service a user will require. The challenge of providing new and effective services and applications to a future wireless market requires research into future generation networks and namely into highly heterogeneous and time-varying QoS from the underlying protocol layers. Key aspects to providing the needed QoS relate to fast adaptation to the channel and traffic conditions. On the other hand, issues of great significance are the provision of high-capacity, low-cost, and intelligent radio resource management strategies to provide for reliable wireless multimedia communications networks.

1.1.1 New Access Technologies and Trends

Multimedia user interfaces and context-aware technologies must be designed such that they are able to negotiate a way to create links to the network and to organize data transfer in the most efficient way (e.g., by simultaneously taking QoS and cost-effectiveness into account).

Figure 1.1 shows the trend towards a multiaccess future telecommunications environment.

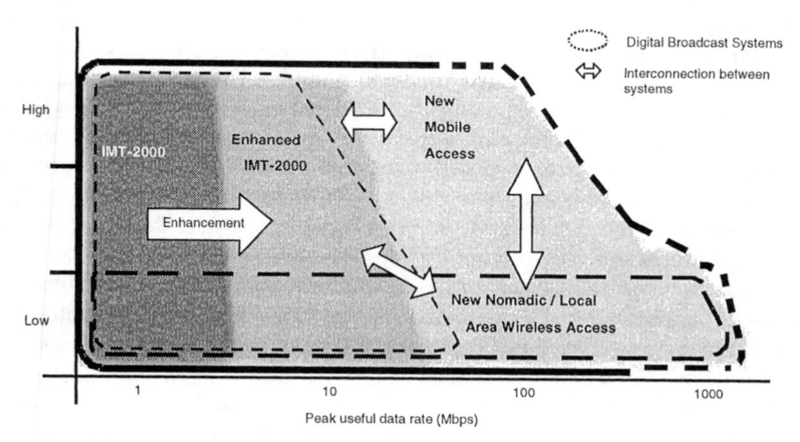

Figure 1.1 Trends for systems beyond IMT-2000—multiaccess. (*From:* [3].)

This trend implies research work in the following directions:

1. Multiaccess user interfaces;
2. Multiaccess functionality;
3. Multiaccess systems architectures;

4. Multiaccess service models;
5. Multiaccess service models;
6. Multiaccess business models.

The multiaccess functionality basically impacts the whole mobile system, so important research issues are the terminal architecture, radio access and radio network, authentication and security, subscription and policies, mobility support, services, transport, and signaling, so that such a system will be easy to use and economically viable to offer.

Throughput, defined as the data rate successfully received, is a key measure for the performance of wireless data transmission systems (e.g., in terms of QoS). Throughput is affected by the channel environment such as the distance between the transmitter and the receiver, the fading state of the channel, and the noise and interference power characteristics. It is also influenced by the choice of design parameters, for example, symbol rate, modulation and coding, constellation size, power level, multiple-access scheme, and many others [4].

Adaptive modulation has been proposed as an efficient technique to improve the data rate of such wireless systems by adapting some of those design parameters to the time-varying channel environment to maintain an acceptable bit error rate (BER) [4–7]. Most works in adaptive modulation, however, have only aimed to improve the link layer performance. As application requirements become more complex, there is a need to take into account higher layer metrics. For instance, several works in the literature have developed adaptive techniques that support different types of traffic with different QoS requirements [8].

Technologies for improving bandwidth efficiency include orthogonal frequency division multiplexing [9] (OFDM), ultra-wide band (UWB), and flexible frequency sharing. In pedestrian and indoor environments, there will be severe fluctuations in traffic demands, high user mobility and different traffic types. This highly complex environment will require advanced RRM algorithms. It could be beneficial to have a central intelligent unit that can maximize the resource utilization. This capability is provided by the so-called bunched systems. The bunched system consists of a limited number of remote antenna units (RAUs) that are connected to a functional entity named the central unit (CU). All intelligence as well as significant parts of the signal processing are located in the CU. The RAUs are simple antenna units capable of transmitting and receiving user signals. The local centralization at the CU level permits the use of near optimal algorithms for resource management because the CU has complete knowledge of all allocated resources at any time. This results in very efficient resource utilization within the bunched system. Furthermore, the bunched system can be enhanced to allow the radio access network (RAN) to detect changes, make intelligent decisions, and implement appropriate actions, either minimizing or maximizing the effect of the changes.

1.1.2 Advanced Antenna Techniques

Advanced antenna systems inherently have the capability to increase the channel and, hence, the system capacity. The information capacity of wireless communication systems increases dramatically by employing multiple transmit and receive antennas[10, 11]. An effective approach to increasing data rate over wireless channels is to employ coding techniques appropriate to multiple transmit antennas, for example, space-time coding (STC) [12]. Space codes introduce temporal and spatial correlation into signals transmitted from different antennas, in order to provide diversity at the receiver, and coding gain over an uncoded system without sacrificing the bandwidth. The spatial-temporal structure of these codes can be exploited to further increase the capacity of wireless systems with a relatively simple receiver structure [13].

Space-time block coding (STBC) was also introduced in [14], generalizing the transmission scheme discovered by Alamouti to an arbitrary number of transmit and receive antennas for a Rayleigh fading channel. These codes retain the property of having a very simple maximum likelihood-decoding algorithm based on linear processing at the receiver.

Consider a wireless communications system with M and N antennas at the transmitter and the receiver respectively at each time slot t; signals c_t^m, $m = 1, 2, \ldots, M$ are transmitted simultaneously from the M antennas. Assuming a flat fading channel with transmission matrix H and having a path gain (entry) with a variance of one, quasistatic so that the path gains are constant over a frame of length p, at time t the signal r_t^n, received at antenna n, is given by

$$r_t^n = \sum_{m=1}^{M} H(n,m) c_t^n + \eta_t^n \tag{1.1}$$

In (1.1) η_t^n are independent noise samples of a zero-mean complex Gaussian random variable with variance M/SNR, and the transmitted symbols, which are the entries of the code, C, have an average normalized energy (power) of one, so that the average power of the received signal at each receive antenna is M and the signal-to-noise ratio is SNR.

Assuming that perfect channel estimation is available at the receiver, the receiver computes the decision metric:

$$\sum_{t=1}^{p} \sum_{n=1}^{N} \left| r_t^n - \sum_{m=1}^{M} H(n,m) c_t^m \right|^2 \tag{1.2}$$

over all possible codewords $c_1^1 c_1^2 \ldots c_1^N c_2^1 c_2^2 \ldots c_2^N \ldots\ldots c_p^1 c_p^2 \ldots c_p^N$ and decides in favor of a codeword that minimizes the sum. The performance of the STBC depends on the selection of the space-time codes. At any time a block of data with *kb* bits arrives at the encoder, select the *k* constellation signals depending on the modulation type and the encoder generates a $p \times M$ transmission matrix to be transmitted in *p* time slots. Therefore, the rate of the encoder is $R=k/p$, and the overall bandwidth efficiency of such a system is *R* times the bandwidth efficiency of the used modulation type (e.g., BPSK, QPSK).

Maximum likelihood decoding of any space-time block code can be achieved using only linear processing at the receiver. For example, in a space-time block code Γ_2, which uses the matrix shown in (1.4), the transmission is carried out in two time slots. The columns denote the antennas and the rows denote the time slots. Antenna 1 (R_1), for example, transmits x_1 first and then $-x_2^*$, while antenna 2 (R_2) transmits x_2 followed by x_1^*. The signal processing at the receiver solves:

$$\begin{aligned} x_1 &= H_1^* R_1 + H_2 R_2^* \\ x_2 &= H_2^* R_1 - H_1 R_2^* \end{aligned} \tag{1.3}$$

where H_1 and H_2 are channel estimates

G_2: R=1 G_3: R=1/2 G_4: R=1/2

$$\begin{pmatrix} x_1 & x_2 \\ -x_2^* & x_1^* \end{pmatrix} \begin{pmatrix} x_1 & x_2 & x_3 \\ -x_2 & x_1 & -x_4 \\ -x_3 & x_4 & x_1 \\ -x_4 & -x_3 & x_2 \\ x_1^* & x_2^* & x_3^* \\ -x_2^* & x_1^* & -x_4^* \\ -x_3^* & x_4^* & x_1^* \\ -x_4^* & -x_3^* & x_2^* \end{pmatrix} \begin{pmatrix} x_1 & x_2 & x_3 & x_4 \\ -x_2 & x_1 & -x_4 & x_3 \\ -x_3 & x_4 & x_1 & -x_2 \\ -x_4 & -x_3 & x_2 & x_1 \\ x_1^* & x_2^* & x_3^* & x_4^* \\ -x_2^* & x_1^* & -x_4^* & x_3^* \\ -x_3^* & x_4^* & x_1^* & -x_2^* \\ -x_4^* & -x_3^* & x_2^* & x_1^* \end{pmatrix} \tag{1.4}$$

The performance of the G_2, G_3, and G_4 space-time codes (see 1.4) is discussed in [15] for a Rayleigh channel. The code G2 is the one considered by Alamouti [14]. The higher dimension of codes such as G_4 and G_3 will perform better than G_2 in a scattering-rich environment. The higher the number of scatterers considered in the simulations, the closer the result will be to the Rayleigh channel, which is the real scenario that can happen in a fully scattering-rich environment like the indoor channel.

The Bell-Labs Layered Space-Time (BLAST) [16, 17] architecture uses multielement antenna arrays at both transmitter and receiver to provide high-capacity wireless communications in a rich scattering environment. The BLAST technology takes advantage of several independent spatial channels through which different data streams can be transmitted. It has been shown that the theoretical capacity approximately increases linearly as the number of antennas is increased [16]. Two types of BLAST realizations have been widely publicized: vertical BLAST (V-BLAST) [16] and diagonal BLAST (D-BLAST) [18]. The V-BLAST is a practical algorithm that is shown to achieve a large fraction of the MIMO channel capacity [19] in the case of narrowband point-to-point communication scenarios. The V-BLAST algorithm implements a nonlinear detection technique based on Zero Forcing (ZF) combined with symbol cancellation to improve the performance. The idea is to look at the signals from all the receive antennas simultaneously, first extracting the strongest substream from the received signals, then proceeding with the remaining weaker signals, which are easier to recover once the strongest signals have been removed as a source of interference. This is called successive interference cancellation (SIC) and is somewhat analogous to decision feedback equalization. When symbol cancellation is used, the order in which the substreams are detected becomes important for the overall performance of the system. In fact, the transmitted symbol with the smallest postdetection SNR will dominate the error performance of the system. Postdetection SNR is determined by ordering. The optimal ordering is based on the result that simply choosing the best postdetection SNR at each stage of the detection process leads to the maximization of the worst SNR over all possible orderings.

1.1.3 Evolution of Radio Access Network

With the adoption of the General Packet Radio Service (GPRS), the packet switched domain was introduced to the core network (CN) of the Global System for Mobile Communications (GSM). For the users, GPRS offers the possibility of being permanently on-line but only paying for the actual retrieval of information. The radio interface provides peak data rates of up to 21.4 Kbps per time slot can be combined with the transmission of multiple time slots per user. With eight time slots a peak data rate of around 171 Kbps can be reached.

An important feature of GPRS is the presence of four coding schemes (CS) with different levels of resistance to transmission impairments [20]. [20] derives rate adaptation algorithms that maximize throughput by switching among the four coding schemes as channel conditions change. As in other data transmission systems, the most important variable that affects the choice of a coding scheme is the channel quality as measured by carrier-to-interference ratio (C/I). In addition to channel C/I, packet size and channel fading characteristics play important roles in coding scheme selection.

The throughput for GPRS users is affected by a range of factors, which have been provided by Vodafone, New Zealand:

1. *Terminal capability.* For a class-10 GPRS terminal, for example, the best capability is a 4x time slot (TSL) on the downlink (4x 13.3 Kbps at the physical layer, dropping to 4x 9.6 Kbps if the radio conditions are poor) and 2x TSL on the uplink.
2. *Radio network capacity.* Radio resource is always scarce. GPRS typically competes with voice, and usually voice has priority in a given cell site. So, during voice busy hours there may be very little resources available for GPRS.
3. *Radio network quality.* GPRS is more sensitive to a poor radio network environment than voice. Factors such as interference and lack of a single dominant cell can dramatically reduce GPRS performance. Of course, TCP reacts to this, so TCP-based applications will not work well in poor radio conditions or while mobile in urban environments. Small-packet UDP applications are currently best suited to GPRS, but round-trip times will typically vary between 600 ms and several seconds. Early adopters of GPRS, using applications like courier dispatch and mobile sales force, usually used some communications server middleware to manage UDP messages between clients and back-end systems.

So, the user experience is a factor of all this—at best, 50 Kbps on the downlink, but during the day in urban environments much less (say, 20 to 30 Kbps), and even less if there is a large number of users.

The Enhanced Data Rates for the Global Evolution (EDGE) system incorporates the 8PSK modulation together with link adaptation and incremental redundancy (IR) across the GPRS air interface to increase the spectral efficiency. The resulting peak data rate per slot is around 60 Kbps. With the support of four time slots, a peak data rate of up to 237 Kbps can be reached in the downlink. EDGE can deliver an average user throughput of around 128 Kbps when the network is loaded, and with the previous time slot configuration [21]. Such capabilities make possible services such as e-newspapers, images and sound files, tele-shopping, and IP-based video telephony. EDGE includes the same traffic classes as UMTS, and the same core network.

Wideband code division multiple access (WCDMA) is the most widely adopted air interface for 3G systems. It provides peak rates of 2 Mbps, variable data rate on demand, a 5-MHz bandwidth, and a significant reduction of the network round-trip time.

Though the spectral efficiency of WCDMA may possibly not represent a major benefit as in EDGE [21], it can provide for higher average user data rates under loaded network conditions (in the range 200 to 300 Kbps) because of the larger bandwidth. The services to be provided by WCDMA are similar to those of EDGE, with potential enhancement of the user's perceived quality for some high bit rate services.

High-speed downlink packet access (HSDPA) is a technology concept included by the Third Generation Partnership Project (3GPP) in the specifications of the Release 5 as an evolutionary step to boost the WCDMA downlink performance of packet traffic. HSDPA appears as an umbrella of technological enhancements that permits the increase in user peak data rates, reduces the service response time, and improves the spectral efficiency for downlink packet data services. The concept is based on a new downlink time-shared channel that supports 2-ms transmission time interval (TTI), adaptive modulation and coding (AMC), multicode transmission, and fast PHY-layer hybrid automatic repeat request (H-ARQ). The link adaptation and packet scheduling functionalities are executed directly from the node B. This enables the node Bs to acquire knowledge of the instantaneous radio channel quality of each user. This knowledge allows advanced packet scheduling techniques that can profit from a form of selection (multiuser) diversity.

The data rates that are contemplated by HSDPA systems are in the range of 20 Mbps (10 Mbps is expected in practice), achievable through a notion of shared channels, and a special channel called high-speed dedicated shared channel (HS-DSCH) is established to suit the bursty packet data delivery over the air interface. The HS-DSCH delivery optimization is based on adding various new physical functionalities including adaptive modulation and coding schemes (AMCS), fast cell site selection (FCSS), fast-ARQ, and very short TTIs. In the great scheme of things everything is adaptive, including channel coding, transmit diversity, and rate matching, to cope with the channel conditions. This is illustrated in Figure 1.2.

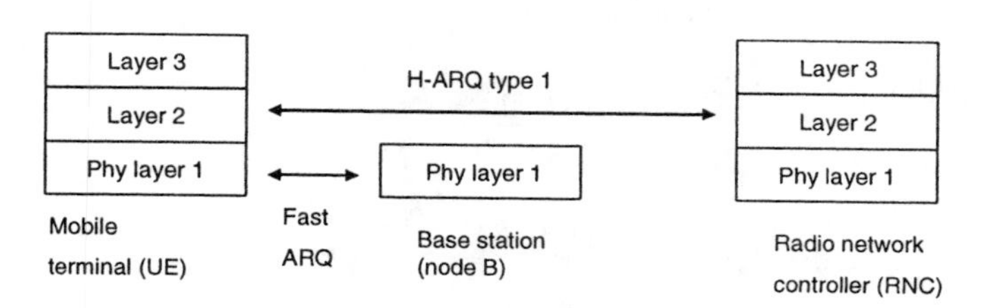

Figure 1.2 Relocation of scheduling to Layer 1 for HS-DSCH channel.

To achieve very high data rates over the air interface, the node-B functionality should be enriched with not only the adaptation at the physical layer (PHY L) but also with protocol level optimization. Figure 1.2 illustrates how a fast ARQ mechanism is introduced at the node-B in 3G systems. Here, in order to achieve very high data rates, joint optimization of the physical layer and the data link layer (DLC layer 2) is necessary. Fast ARQ through a special MAC entity is currently accepted in 3G.

At the same time, at CN level, an all-IP network philosophy has already been embraced and the various functionalities are now distributed on a server-based approach. One of the key servers is a resource manager, which absorbs a lot of slow functionalities of the radio network controller (RNC) and the remaining fast functionalities of the RNC can be moved to the node-B. This could mean that multiple node-Bs will be talking to each other as well as talking to the resource manager in the CN. In fact, this separation can be viewed as a logical separation, which means that some resource manager (RM) functionality can be colocated with certain node-Bs. The RM at the node-B would work in a client-server relationship with the network level resource manager.

Based on the building blocks of high-speed data systems approved in the 3G standards, and the maturity of software reconfigurability, it is expected that future-proof RAN system design will be very important. This will yield mechanisms to rapidly develop the service deployment architectures based on protocol optimization across all the layers.

The concept of *intelligent network adaptation* for the radio dynamics proposed in [22] is built upon the existing concepts of CN and RAN in the third generation standard. The key difference is within the RAN architecture, where the base stations (node-Bs) are divided into *core* node-Bs and *leaf* node-Bs. This novel concept is shown in Figure 1.3. For the purpose of resource control, the core node-Bs are logically associated to the core network and the leaf node-Bs are governed within the RAN. The leaf node-Bs are connected via wireless links to the core node-Bs (double-sided arrows). Mobile terminals can connect both to the core and leaf node-Bs. The small vertical lines symbolize the communication between core node-Bs and leaf node-Bs by means of the multiple antennas.

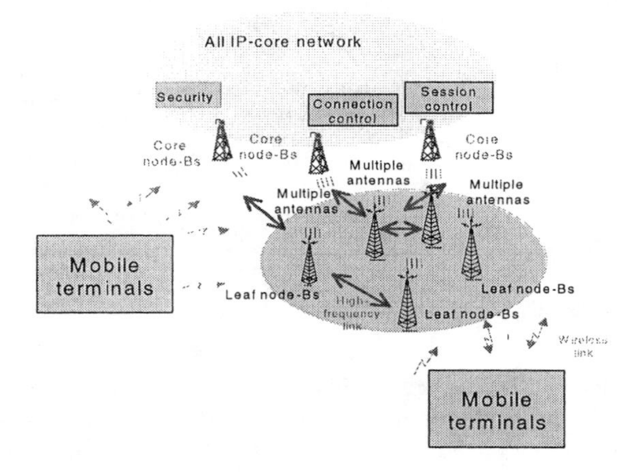

Figure 1.3 Concept of virtual node-Bs.

To overcome the backhaul leased line problems, flat type architecture can be adopted (partially decentralized). This architecture is shown in Figure 1.4.

The difference from the current 3G architecture is that each core node-B is associated with a group of leaf node-Bs with wireless modems. As the user equipment (UE) moves between the node-Bs, the custodian will be a new leaf node-B that is controlled by another core node-B.

The UE migration influences the mobility and QoS, and network adaptation mechanisms are required at each node to provide flexible and efficient management of resources. Within this evolved architecture, two mechanisms for handover (HO) may be available:

- *HO between cells.* Here a user (mobile station) moves from one cell to the other and associates itself with a new base station. Due to this the traffic originating from a cell might change.
- *HO between node-Bs.* Here the network rearranges the connections between the core and leaf node-Bs. It can assign/associate as many core node-Bs to a leaf node-B as required by the originating traffic.

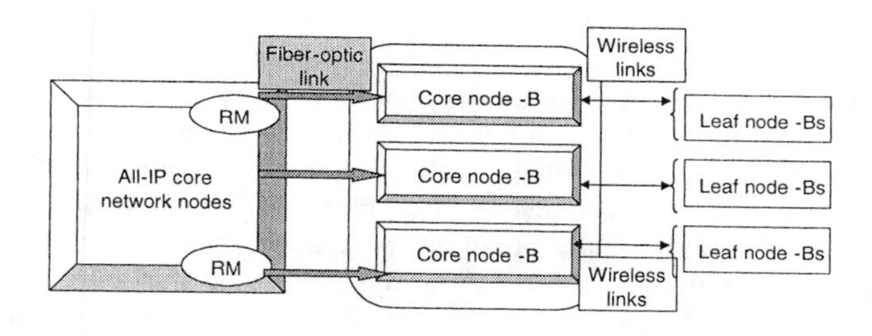

Figure 1.4 Evolved 3G architecture.

These two mechanisms offer the possibility to optimize the assigned resources to the traffic. Leaf node-Bs with a high load are connected with more capacity to the core (using the adaptive antennae). At the level of the leaf node-Bs the traditional mechanisms are applied.

1.1.4 Evolution of User Equipment

It is difficult to predict exactly how a 3G mobile device will evolve [23]. However, mobile devices will be important to the adoption and evolution of third generation services. There will be constraints originating from users, operators, service providers, and manufacturers requirements. Revenues will be based on the sharing of certain services, including the device.

New mobile user equipment (UE) are assuming characteristics of general-purpose programmable platforms [24] by:

- Containing high-power general-purpose processors that follow Moore's law of dramatically increasing price/performance;
- Providing a flexible, programmable platform that can be used for an ever-increasing variety of uses.

The convergence of wireless connectivity and a general-purpose programmable platform heightens some existing concerns and raises new ones, so that environmental factors as well as traditional technology and market drivers will influence the architecture of these devices. Some important environmental factors are the following economic, security, and privacy.

Complementing the environmental factors are the following traditional market and technology drivers: the user value pull, the security requirements pull, and the technology enablers.

1.2 SPECTRUM CONSTRAINTS

Efficient usage and availability of radio spectrum is a key prerequisite for wireless mobile services. Current methodologies to estimate spectrum requirements are limited and should be augmented to allow improved estimations, taking into account the vision for systems beyond 3G, which includes the cooperative use of different wireless systems and a broad, ever-increasing range of services. In this book spectrum requirements estimation methods used for current specified wireless systems are reviewed with respect to new spectrum requirements arising from the predictions of future services and systems beyond 2010. Current methodologies to estimate spectrum requirements are spreadsheet-based, limited to a selected example service/market scenario and restricted in the examined deployment scenarios. As the current methods do not take into account convergence scenarios and cooperative networks, a new spectrum requirement methodology estimation framework is proposed and outlined in this deliverable to accomplish the envisaged scenario of ubiquitous mobile services and ambient wireless networks foreseen in the next decades. Flexible statistical methods and Monte-Carlo simulation models are suggested for the estimation of frequency requirements. Such models allow the consideration of spatial and temporal distributions of the market requirements and network deployment scenarios in virtually unlimited flexibility. In the proposed methodology framework, relevant parameters are defined as statistical distribution functions, the results are not fixed spectrum numbers but a range of results, reflecting the broad range of assumptions used in the modeling which also reflect future scenarios of spectrum sharing and interworking of different radio access systems. This section also provides valuable information to assist future spectrum identification/assignment discussions for mobile communication within the International Telecommunication Union (ITU) and the World Radio Conferences, where suitable frequency bands for systems beyond 3G is identified to be used from 2015 onwards.

The current method of assigning spectrum to different radio systems is a fixed-spectrum allocation (FSA) scheme. With this technique, radio spectrum is allocated to a particular radio standard, and these spectrum blocks are of fixed size and are usually separated by a guard band. The spectrum then remains solely for the use of the radio license owner. This method of spectrum allocation can control interference between differing networks using the spectrum, provided adequate guard bands are maintained, with no coordination between the networks. However, most communications networks are designed to cope with a certain maximum amount of traffic called the *busy hour*, which is the time of the peak use of the network. If this network uses its allocated spectrum fully during this hour, then the rest of the time the spectrum is not fully used. In fact, almost all services, such as speech, video, Web browsing, and multicast applications, which are envisaged as future mobile services, have distinct time-varying traffic demands. The demand for different services on different networks depends on location, leading to a spatial variation in the spectrum usage. Therefore, the radio spectrum, while scarce and economically valuable, is often underused or idle at certain times or areas.

Due to the success and expected growth of mobile services, especially seen with GSM boosting national economies and international trade, the need for more spectrum for 3G services was foreseen and acted upon at the ITU World Administrative Radio Conference in 1992 (WARC '92). This conference provided a framework for agreement at the global level for spectrum allocations to support growth in cellular mobile services towards 3G, capable of supporting mobile demand for nonvoice services and global international roaming, as essential requirements.

WARC '92 identified 230 MHz of spectrum for IMT-2000 in the bands 1,885 to 2,025 MHz and 2,110 to 2,200 MHz. This included 1,980 to 2,010 MHz and 2,170 to 2,200 MHz for the satellite component; the terrestrial component thus comprised 170 MHz and is known as the Core Band. In Europe, all except 15 MHz (already allocated for the Digital Enhanced Cordless Technologies (DECT) standard) is or will be available for UMTS services. In June 2000, the ITU World Radio Communication Conference 2000 (WRC-2000) made important decisions on the identification of additional spectrum for IMT-2000. Studies of spectrum requirements led to a general recognition of a requirement for an additional 160 MHz of spectrum beyond the 2G spectrum and the 230 MHz mentioned earlier as having been identified at WARC '92. Agreement was reached on a small number of bands from which each administration would find the total spectrum requirements. The main results from WRC-2000 were:

- Protection of existing IMT-2000 core bands;
- 2,500–2,690-MHz band identified for IMT-2000 in all three ITU regions (this becomes the main expansion band in Europe and Asia);

- 1,710–1,885-MHz band was identified for IMT-2000 in all three ITU regions (this is the preferred band for the CITEL countries and will allow evolution from GSM-1800 to IMT-2000);
- First and second generation cellular bands (within 806 to 960 MHz as used, for example, by U.S. cellular and GSM) identified globally for migration to IMT-2000.

Taking these and other decisions together, the achieved results create an excellent radio spectrum environment for 3G networks to be deployed to the benefit of everyone. As a result, manufacturers know the frequency bands for which devices must be designed, and the small number of globally common bands provides them with the best opportunity to drive scale economies and global roaming for a mass market in 3G services.

Dominated by the pervasiveness of IP and the advent of reconfigurable radio [26], spectrum allocation and utilization in the context of next generation has invoked the necessity for research in adaptive spectrum and bandwidth allocation to provide for dynamic utilization of spectral resources. The idea of having a spectral resource allocated, assigned, or used in a way that is not fixed is under investigation in Europe, Asia, and the United States, and is beginning to be considered a possibility by regulators [27]. Figure 1.5 gives an overview of the current spectrum allocation for 2G, 3G, and beyond.

Some additional bands allocated by the WRC 2000 are shown in Figure 1.6.

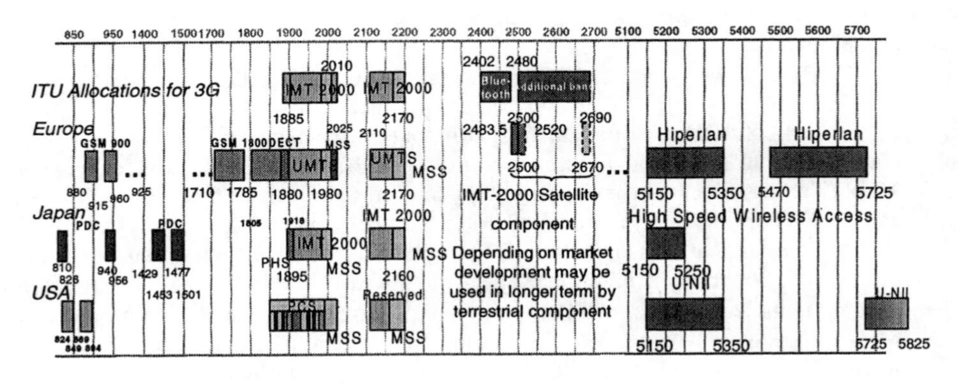

Figure 1.5 Frequency ranges for 2G, 3G, and beyond [28].

As the automotive industry advances towards the information age, telematics will play an increasingly pivotal role in the integration of mobile, in-vehicle computing and wireless technologies. This is the motivation for different spectrum efficient techniques, for example, dynamic spectrum allocation. The bandwidth limitations of 3G are becoming self-evident, and the cost of a license plus the upgrade may not be economically feasible based on return on investment.

However, with 4G it will be possible to achieve 2 Mbps or greater bandwidth per mobile user in the same spectrum in use today. While all the spectrum attention has been placed on mobile wireless, fixed wireless is also becoming a communications industry hot topic. The fixed wireless industry currently relies on expensive local multipoint distribution services technology, which uses licensed spectrum in the access WAN and IEEE 802.11 with frequency hopping to deliver speeds of 1 to 2 Mbps to the wireless LAN.

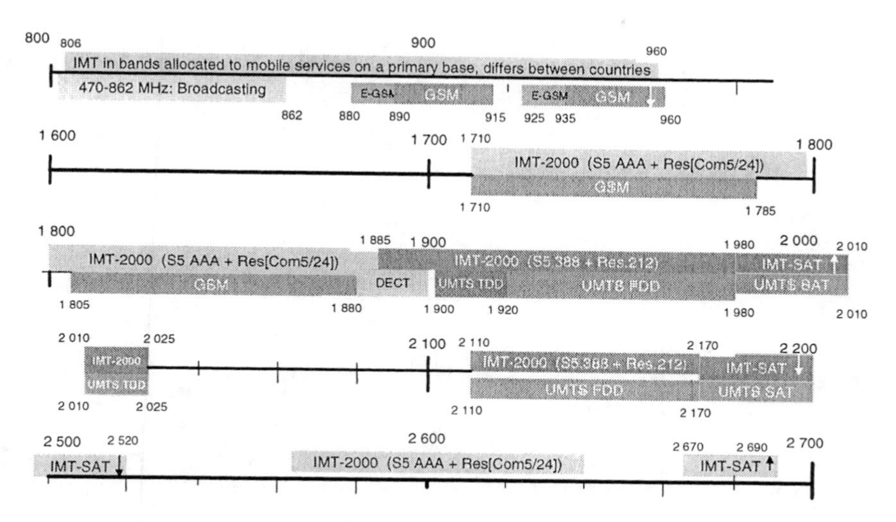

Figure 1.6 Additional bands allocated by WRC 2000 [29].

1.2.1 3G Licensing Situation

All 110 issued and 60 additionally planned 3G licenses worldwide use the globally agreed IMT-2000 core band. More than 70 granted licenses in Europe provide for 2x 15-MHz paired or 2x 15-MHz + 5-MHz unpaired per operator, as recommended by the UMTS Forum [30].

A status of 3G licenses is shown in Figure 1.7.

This is important for network planning and cross-network coordination. Adherence to the core band is crucial for global roaming and creating mass-market volumes for terminals and infrastructure for the benefit of consumers and the success of 3G/UMTS. Roaming is greatly simplified as a common band and avoids the need for multiband terminals. There is a proposal in North America to use the PCS band (1,830 to 1,910 MHz paired with 1,930 to 1,990 MHz) for 3G services, but this is not consistent with the IMT-2000 bands agreed on at WARC'92 and WRC 2000, and introduces further technical difficulties. Countries using PCS1900, mainly in North and South America, face the problem of how to clear spectrum to align their mobile communications globally. There are concerns that IMT-2000 allocated spectrum in some markets will be used for non-IMT-2000 applications, such as WLAN technology or CDMA2000 1xev-DO, which are not IMT-2000 standards. Their use would create interoperability difficulties. There are other views that new 3G data services can be embedded into the existing voice spectrum. However, studies show that this will neither be economic nor sustain growth. Other solutions such as pairing of frequency subbands within 1,710 to 1,850 MHz and with 2,110 to 2,170 MHz are considered where the frequency availability is incompatible with IMT-2000 decisions globally. Although such frequency arrangements do not support early harmonization of IMT-2000 equipment, roaming can be provided with dual-band terminals (GSM1800 and WARC '92 bands).

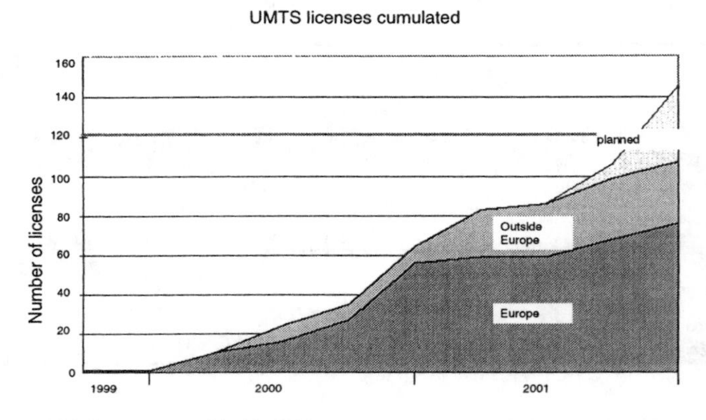

Figure 1.7 Status of 3G licensing worldwide [31].

1.3 ADVANCED NETWORK AND RESOURCE MANAGEMENT

A straightforward and simple answer to the traffic demand of emerging multimedia applications is to add resources (i.e., capacity) to the links [1]. However, IP networks are becoming more and more heterogeneous, ranging from fiber-optic in the core networks to various wireless technologies in the RAN part, each one specified by a certain transport capacity. Radio resource management (RRM) strategies will play an important role in a mature 3G scenario. It is true that some of the available resource management techniques for GSM networks can be reutilized in all-IP networks [33]. For example, the half-rate/full-rate traffic channel allocation resource management technique can be seen in a bigger QoS context in packet-switched networks. In an UMTS network, adaptive multirate codec (AMR) can be used to provide various levels of speech-call quality and resource reservation. Thus, it can provide trade-off between call quality and resource consumption in the same way as a half-rate/full-rate in GSM. Moreover, since a lot of different service types are foreseen in all-IP networks, more extensive measures for the same purpose can be seen. The QoS levels of new calls could be negotiated in a call admission phase if the network load is high and the user is willing to settle for a lower QoS level. Also, the QoS levels of existing users might be downgraded by performing QoS renegotiation in order to reduce the network load if it has been driven to too high a level.

Handovers will be more complicated in networks such as UMTS with respect to GSM, but at the same time will provide more possibilities for resource management. The soft handover used in UMTS implies that the user can be communicating with multiple base stations at the same time especially if he resides in the cell border where signals from multiple base stations are likely to be received approximately with the same signal level. Support for some handover operations like link addition, dropping, and replacement has been specified in 3GPP specifications but the handover algorithm can be implemented freely based on those operations and the measurements provided by the network. Thus, algorithms relying on the common pilot channel powers or qualities can be envisaged and even more complicated algorithms taking explicitly into account the traffic load situation and distribution between cells might be used. The subsequent actions could encompass changing soft handover threshold values in such a way that the links to some congested base stations are not added so easily.

Functionality for some of the techniques proposed for GSM such as user groups, is directly available in UMTS networks as well according to the 3GPP specifications. Also, some techniques, such as time-limited calls, can be applied in a similar manner to UMTS networks.

Due to underlying differences in the access technologies of GSM (TDMA/FDMA) and UMTS (CDMA), some techniques are different in UMTS. Cell resizing is one technique that should partially take place automatically in UMTS—there is an inherent coupling between coverage and capacity in UMTS and thus if the load of a cell increases, the radius of the cell decreases since the interference level gets higher. However, it is also possible to provide additional control on the cell sizes in UMTS by modifying the common pilot channel (CPICH) transmission power and thus affecting the cell selection/reselection process as well as handovers.

1.3.1 Influence of Radio Network Planning and Optimization

Second generation networks (GSM) were designed as noise-limited systems. However, since tight reuse patterns became necessary as the number of users tremendously increased, these systems may today be regarded as practically interference limited.

Power-aware techniques, such as mobile station (MS) and base station (BS) power control for GSM or MS power control with a slow adaptation rate, are used in existing systems to adjust the transmission power to the required level.

In contrast to GSM systems, transmit power efficiency has a much stronger importance in 3G WCDMA systems as the strength of interference ultimately determines capacity; hence, in order to optimize capacity, efficient means for controlling transmit powers have been implemented.

1.3.2 Dynamic Reconfiguration

The idea of dynamic reconfiguration in wireless networks has emerged from the need to have access to diverse wireless networks from a single terminal. The usual method for achieving this is to download configuration software on the terminal whenever there is a need to access a network with different characteristics from the current one. This approach is called software defined radio (SDR) and there are already many initiatives that are trying to extend this concept to the whole wireless networking system covering the base stations and even the way services are created and introduced into the network with encouraging results. The application of the idea of reconfigurability for resource management, where the network is made capable of dynamically assigning resources to specific users according to their demands, the availability of resources and experienced load is also part of the SDR framework.

SDR technology is one of the key technologies for systems beyond IMT-2000, promising the flexibility required for total mobility and, from the user's perspective, pervasive access to services [34]. Such technology will herald a new generation of flexible and reconfigurable user terminals, able to detect their operating environments and adapt to make optimum use of the available resources, downloading appropriate software to the terminal when necessary to achieve interoperability, increased performance, and security, or improving spectrum utilization efficiency. It is clear that there are many similarities between SDR technology and personal computers, both providing flexible platforms for execution of software that is potentially sourced from multiple vendors. The difference lies in the fact that SDR technology applies software implementation to functions normally performed in hardware (or embedded into operating system kernels), to increase the level of flexibility. This will enable lower terminal manufacturing costs through economies of scale and will open up new revenue generating possibilities (via core software download) and fulfill the demand of the users for higher service availability and innovative new services.

1.3.3 Dynamic Spectrum Allocation

In a multiradio environment comprised of disparate systems, such as cellular and broadcast networks (e.g., UMTS and DVB-T), there will be significant variations in the traffic demand (and therefore spectrum utilization) on the networks over both time and space. Dynamic spectrum allocation (DSA) attempts to improve efficiency by allocating the spectrum dynamically over space or time, as required by the networks, while managing or preventing any resulting interference. The current method of assigning spectrum to different radio systems is an FSA scheme [35, 36]. With this technique, radio spectrum is allocated to a particular radio standard, and these spectrum blocks are of fixed size and are usually separated by a guard band. The spectrum then remains solely for the use of the radio license owner. This method of spectrum allocation can control interference between differing networks using the spectrum, provided adequate guard bands are maintained, with no coordination between the networks. However, most communications networks are designed to cope with a certain maximum amount of traffic called the busy hour, which is the time of the peak use of the network. If this network uses its allocated spectrum fully during this hour, then the rest of the time the spectrum is not fully utilized. In fact, almost all services, such as speech, video, Web browsing, and multicast applications, which are envisaged as future mobile services, have distinct time-varying traffic demands. It will also be seen that the demand for different services on different networks depends on location, leading to a spatial variation in the spectrum usage. Therefore, the radio spectrum, whilst scarce and economically valuable, is often underused or idle at certain times or areas.

1.4 AD HOC NETWORKS

The emergence of a multitude of modern multimedia applications creates the demand for a new network generation, different from the traditional, infrastructure-based networks.

With adhoc networking, local communications can depend on local communications channels and transmission technologies and protocols, provided sufficient spectrum is available for local use. Nodes in an ad hoc network are often assumed to have IP addresses that are preassigned, or assigned in a way that is not directly related to their current position relative to the rest of the network topology. This differs substantially from the way that IP addresses are assigned to nodes in the global Internet. Routing within today's Internet depends on the ability to aggregate reachability information to IP nodes. This aggregation is based on the assignment of IP addresses to nodes so that all the nodes on the same network link share the same routing prefix. In practice, good network administration requires that networks that are nearby should have similar prefixes. The process of aggregation can be iterated if network administrators for nearby sites cooperate to use networks with prefixes that share a common initial bit string (i.e., common smaller prefix).

This creates a hierarchy of network prefixes—smaller prefixes that fit at higher levels of the hierarchy. Reachability to all nodes within the hierarchy can be described by advertising a single, smallest routing prefix. This drastically reduces the amount of routing information that has to be advertised and provides the necessary economy for the Internet to continue to grow. Thus, aggregating routing information is the key to the Internet scalability.

With ad hoc networks, however, such aggregation is typically not available. Some proposed methods attempt to reintroduce aggregation by controlling the IP addresses of the mobile nodes, but this requires that the IP addresses (and, subsequently, routing information relevant to the mobile node) be changed depending on the relative movement of the node. It is not at all clear that the benefit of improved aggregation is worth the cost of complicated readdressing and route table revisions. Thus, the scalability afforded by aggregation within the Internet is not likely to be available for ad hoc networks. This means that there may be major limitations on the viability of ad hoc network algorithms for extremely large populations of mobile nodes. The limits will also depend on the relative speed of movement between the mobile nodes. More movement means more maintenance so that the available routing information remains useful.

Ad hoc networks may apply either hierarchical or flat infrastructure both in logical and physical layers independently. As in some flat ad hoc networks, the connectivity is maintained directly by the nodes themselves, the network cannot rely on any kind of *centralized* services. In such networks the necessary services such as the routing of packets and key management have to be *distributed* so that all nodes have responsibility in providing the service. As there are no dedicated server nodes, any node may be able to provide the necessary service to another. Moreover, if a tolerable number of nodes in the ad hoc network crash or leave the network, this does not break the availability of the services. Finally, the protection of services against *denial of service* is in theory impossible. In ad hoc networks, *redundancies* in the communication channels can increase the possibility that each node can receive proper routing information. Such approaches, however, do produce more overhead both in computation resources and network traffic. The redundancies in the communication paths, however, may reduce the denial of service threat and allow the system to detect malicious nodes from performing malicious actions more easily than in service provisioning approaches that rely on single paths between the source and destination.

A common concern in ad hoc networks is the access control: there needs to exist a method for restricting the access of foreign nodes to the network, which requires the use of a proper authentication mechanism. Moreover, the communication between the insider nodes in the network must be protected from attacks on confidentiality. This is especially important in military applications. If the link layer does not support a valid encryption scheme, such a mechanism must be involved in the network layer also. The group membership is noted in all of the mentioned multicast protocols, but they do not suggest any specific access control or authorization policy protocols.

The security aspects related to ad hoc networks form a very complex problem. On the other hand, ad hoc networks vary from each other greatly from the viewpoint of the area of application. Some ad hoc networks may not need security solutions other than simple encryption and a username-password authentication scheme, while networks operating in highly dynamic and hostile environments demand extremely efficient and strong mechanisms. As the security requirements and their implications vary, general security architecture for ad hoc networks cannot be constructed. The development of a secure ad hoc networking framework seems to be just starting, as all the most severe security problems are not even fully solved in ad hoc networking proposals.

Within the frames of European research, ad hoc networks have been investigated from different aspects. Some of the most important research projects and their focus are summarized in Table 1.1.

While a lot of work was performed in the 1970s and 1980s, most of the recent work on ad hoc network has been carried out under the Mobile Ad Hoc Network (MANET) [37] group project with a focus on the network aspects. Relatively small networks of large density are considered, with high mobility characteristics. The project Broadway [38] focuses on the definition, development, and demonstration of the key components of a hybrid dual frequency system based on HiperLAN/2 at 5 GHz and an innovative fully ad hoc extension at 60 GHz named HIPERSPOT and equipped with a novel, more robust, modified multicarrier transmission scheme. Thus, Broadway focuses on small/medium network sizes (equivalent to the network size of HiperLAN/2), but the innovative elements of the project are applicable to high-density networks with low mobility. The project PACWOMAN [39] develops the concept of the personal area network (PAN) and tries also to solve power constraints issues. More about the rest of the projects can be found in [40].

1.5 PREVIEW OF THE BOOK

This book focuses on advanced resource and network optimization methods, investigated within the framework of the Fifth European Information Society and Technology research program, and in particular within IST projects within the Mobile and Satellite project domain.

Generally, it is expected that data services instead of pure voice services will play a predominant role, particularly due to a demand for mobile IP applications. Variable and especially high data rates (> 20 Mbps), will be requested, which should also be available at high mobility, in general, or high vehicle speeds in particular. Highly asymmetrical data services between up and downlink carried most of the traffic and needs the higher data rate compared with the uplink. However, it is important to remember that 4G will not only be about higher data rates but more about efficient use of scarce spectrum [1]. Increasing spectrum efficiency will require methods such as cell splitting and adaptive coverage, relay mechanisms, bandwidth on demand, adaptive modulation, and heterogeneous system integration. These topics have been described in this book and the resulting chapters were based on public material available from the IST projects.

Table 1.1

Summary of EU Fifth Framework Program Research Work on Ad Hoc Networks

EU Projects	Topology Characteristics (network size/ density/mobility)	Traffic	Node Capabilities
MANET	Small (tens of meters)/high (hundreds of nodes)	Error sensitive	Security, mobility support, power constraints, node complexity
BROADWAY	Small, medium/high/low	Delay and error sensitive	Link quality awareness, physical topology awareness, MAC/PHY interaction, QoS support, mobility support, logical/physical topology interaction
PACWOMAN	Small/high/low	Delay and error sensitive	Link quality awareness, power constraints, security, mobility support, battery capabilities
WIND-FLEX	Small/high/low	Error sensitive	Link quality awareness, QoS support
MIND (BRAIN)	Small/high/medium	Delay and error sensitive	QoS support, mobility support and IP multicast, QoS support
TopInd	All/all/all	Delay and error sensitive	—

Chapter 2 has as a main theme the management of heterogeneous and hybrid networks. Some major contributing projects are the IST projects Spectrum Efficient Uni- and Multicast Services Over Dynamic Multi-Radio Networks in Vehicular Environments (OVERDRIVE), Management of Networks and Services in a Diversified Radio Environment (MONASIDRE), Broadband Radio Access for IP-Based Networks (BRAIN), and Mobile IP-Based Network Developments (MIND).[1]

Chapter 3 focuses on advanced radio resource management techniques. Major contributing European IST projects to that area of research were the projects Advanced Radio Resource Management for Wireless Services (ARROWS), Capacity and Network Management Platform for Increased Utilization of Wireless

[1] The reader is referred to the appendix of this book, where a complete list of the IST projects included in this book as well as links to them are given.

Systems of Next Generation++ (CAUTION), Resource Management and Advanced Transceiver Algorithms for Multihop Networks (ROMANTIK).

Short-range technologies, such as UWB are expected to play an increasingly important role in the development of information and communication technology, particularly in utilizing different applications of the computing part of the personal pocket terminal [2]. A major project, and the only one within the IST Fifth Framework working on optimization problems for short-range networks, was the project PACWOMAN. Part of the project results are presented also in Chapter 3. Chapter 4 discusses advanced radio technologies, including UWB. The IST projects contributing to this chapter are the projects PACWOMAN, Multi-Element Transmit and Receive Antennas (METRA), Intelligent Multi-Element Transmit and Receive Antennas (I-METRA), Spectrally Efficient Fixed Wireless Network Based on Dual Standards (STRIKE), and Wireless Indoor Flexible High Bitrate Modem Architecture (WIND-FLEX). Providing highly reliable wireless link technologies seems more and more to be possible through use of multicarrier transmission techniques, such as OFDM and MC-CDMA, multiuser modulation, coding schemes, and advanced multiaccess technologies.

IST research focused on the interactions between the performance of packet-switched data transmission protocols and the use of advanced physical layer schemes, such as smart antenna, OFDM techniques, and MC-CDMA techniques, as well as network descriptors such as instantaneous traffic. A specific focus is given on the benefits and implications that the use of smart antennas technology has on protocol decisions and performances.

The wireless channel is frequently the bottleneck for wideband wireless systems in terms of power, interference, and capacity; accurate channel models for the selected scenarios are necessary for the realistic representation of the propagation conditions. Although some environments (mainly indoors) have been extensively studied and modeled, there are others for which few models exist or for which existing models have been demonstrated to be inaccurate. For these, additional measurements need to be performed. Research work within the frames of the Fifth IST research program is also reported in Chapter 4.

Chapter 5 describes advances in the development and new meaning of satellite systems as part of a global integrated communication system or 4G. Major contributing IST projects here were Satellite-UMTS IP-Based Network (SATIN), Functional UMTS Real Emulator), VIRTUOUS (Virtual Home UMTS on Satellite (FUTURE), Broadband Access for High-Speed Multimedia Via Satellite (BRAHMS). High altitude platforms (HAPS) are also described here, although contributions are solely based on monitoring work carried out within the project PRODEMIS. HAPS as an upcoming and relatively new topic has been undertaken within the FP6 program.

Chapter 6 focuses on advances made through IST projects in the area of interactive broadcasting. This chapter is based on contributions from the IST projects OVERDRIVE, Fast Internet to Fast Train Hosts (FIFTH), Ultra WideBand Audio Video Entertainment System (ULTRAWAVES), and Multi-

Segment System for Broadband Ubiquitous Access to Internet Services and Demonstrator (SUITED).

Chapter 7 describes end-to-end solutions for delivery of services and guaranteed QoS. Contributions were made by the IST projects BRAHMS, VIRTUOUS, and FUTURE.

The accompanying DVD includes software tools, detailed demonstrator descriptions, and other items relevant to the contents of the book material.

References

[1] Tönjes, R., et al., "Architecture for a Future Generation Multiaccess Wireless System with Dynamic Spectrum Allocation," *IST Summit 2000,* Galway, October 4–8, 2000.

[2] IST 1999-10025 Project WIND-FLEX, http://www.vtt.fi/ele/research/els/projects/windflex.htm.

[3] http://www.itu-org/imt-vis.

[4] Webb, W. T., and R. Steele, "Variable Rate QAM for Mobile Radio," *IEEE Transactions on Communications*, Vol. 43, July 1995, pp. 2223–2230.

[5] Goldsmith, A. J., and S.-G. Chua, "Variable-Rate Variable-Power MQAM for Fading Channels," *IEEE Transactions on Communications*, Vol. 45, October 1997, pp. 1218–1230.

[6] Goldsmith, A. J., and S.-G., Chua, "Adaptive Coded Modulation for Fading Channels," *IEEE Transactions on Communications*, Vol. 46, May 1998, pp. 595–602.

[7] Matsuoka, H., et al., "Adaptive Modulation System with Variable Coding Rate Concatenated Code for High Quality Multi-Media Communication Systems," *IEICE Transactions on Communications*, Vol. E79-B, March 1996, pp. 328–334.

[8] Pursley, M. B., and J. M. Shea, "Adaptive Nonuniform Phase-Shift-Key Modulation for Multimedia Traffic in Wireless Networks," *IEEE Journal on Selected Areas of Communication*, Vol. 18, August 2000, pp. 1394–1407.

[9] Van Nee, R., and R. Prasad, *OFDM for Multimedia Communications,* Norwood, MA: Artech House, 2000.

[10] Foschini, G. J., and M. J. Gans, "On Limits of Wireless Communications in a Fading Environment When Using Multiple Antennas," *Wireless Communication Magazines*, Vol. 6, March 1998, pp. 311–335.

[11] Telatar, E., "Capacity of Multi-Antenna Gaussian Channels," *Technical Memo,* AT&T Bell Labs, June 1995.

[12] Seshadri, N., V. Tarokh, and A. R. Calderbank, "Space-Time Codes for High Data Rate Wireless Communications: Code Construction," *Proceedings of IEEE VTC'97,* Phoenix, Arizona, 1997, Vol. 2, pp. 637–641.

[13] Naguib, A. N., et al., "Applications of Space-Time Block Codes and Interference Suppression for High Capacity and High Rate Wireless Data Systems," *Proceedings of 32 Asilomar Conference Signals, Systems, and Computers*, Pacific Grove, California, Vol. 2, November 1998, pp. 1803–1810.

[14] Alamouti, S. M., "A Simple Transmitter Diversity Scheme for Wireless Communications," *IEEE Journal on Selected Areas of Communications,* Vol. 16, October 1998, pp. 1451–1458.

[15] Tarokh V., et al., "Space-Time Block Coding for Wireless Communications: Performance Results," *IEEE Journal of Selected Areas in Communications,* Vol. 17, No. 3, March 1999.

[16] Foschini, G. J., and M. J. Gans, "On Limits of Wireless Communications in a Fading Environment When Using Multiple Antennas," *International Journal of Wireless Personal Communications*, Vol. 6, 1998, pp. 311–335.

[17] Golden, G. D., et al., "Detection Algorithm and Initial Laboratory Results Using V-BLAST Space-Time Communication Architecture," *IEEE Electronic Letters,* Vol. 35, No. 1, January 1999.

[18] Foschini, G. J., Jr., "Layered Space-Time Architecture for Wireless Communication in a Fading Environment When Using Multi-Element Antennas," *Bell Labs Technical Journal*, Autumn 1996, pp. 41–59.

[19] Foschini, G. S., et al., "Simplified Processing for Wireless Communications at High Spectral Efficiency," *IEEE Journal on Selected Areas of Communication,* Vol. 17, November 1999, pp. 1841–1852.

[20] Chen, X., and D. J. Goodman, "Theoretical Analysis of GPRS Throughput and Delay," *IEEE International Conference on Communications,* June 2004.

[21] Halonen, T., et al., *GSM, GPRS, and EDGE Performance*, New York: Wiley, 2002.

[22] Mihovska, A., et al., "A Novel Flexible Technology for Base-Station Architecture Support," *Proc. of WPMC 02*, Honolulu, Hawaii, October 2002.

[23] http://www.ist-fitness.org.

[24] ITU Report Document 8F/TEMP/31, "Technology Trends," October 27, 2003.

[25] ITU Study Groups, "Spectrum aspects of Fixed Wireless Access," *Documents 8A-9B/58-E, 28*, September 1998.

[26] Pereira, J. M., "Fourth Generation: Now It Is Personal," *PIMRC 2000*, September 2000, London, CD-ROM.

[27] Lu, W. W., and R. Berezdivin, "Technologies on Fourth-Generation Mobile Communications," *IEEE Wireless Communications*, Vol. 9, No. 2, April 2002, pp. 6–7.

[28] Mohr, W., "Vision for Systems Beyond 3G," *Sixth ITU-R WP8F Meeting*, Tokyo, Japan, October 2001.

[29] Eylert, B., "The Way from Second to Third Generation of Mobile Communications," *CEPT Conference,* Vienna, Austria, April 2002.

[30] http://www.umts-forum.org.

[31] Huber, J. F., "3G Services from the Mobile Industries View," *The 5th 3G China*, Bejing, China, May 2002.

[32] Giordano, S., et al., "Advanced QoS Provisioning in IP Networks: The European Premium IP Projects," *IEEE Communications Magazine*, Vol. 41, No. 1, January 2003.

[33] http://www.telecom.ntua.gr/caution.

[34] Clemo, G., et al., "Requirements on Service Discovery, Middleware and Flexible Protocols for Software-Defined Reconfigurable Terminals," *Proc. of Mobile IST Summit,* Aveiro, Portugal, June 2003, CD-ROM.

[35] Leaves, P., and J. Huschke, "A Summary of Dynamic Spectrum Allocation Results From DRiVE," *Proc. of IST Mobile and Wireless Telecommunications Summit,* Thessaloniki, Greece, June 2002, pp. 245–250.

[36] http://www.ist-overdrive.org.

[37] http://www.ietf.org/html.charters/manet-charter.html.

[38] http://www.ist-broadway.org.

[39] http://www.imec.be/pacwoman.

[40] http://www.cordis.lu/ist/ka4.

Chapter 2

Management of Heterogeneous Hierarchical Networks

2.1 INTRODUCTION

The historical trend of wireless systems is the introduction of a new system every decade. First generation appeared in the early 1980s while the digital systems emerged in the 1990s. Despite the delay of the commercial launches of 3G systems, investigations are ongoing for the definition of the next generation of communications. Instead of promoting one new and ideal generation, and likely compromising the commercial success of 3G systems, a new communication model is appearing, altering the 10-year regular life cycle. This vision, often named Beyond 3G (B3G), consists of an open, smooth, and flexible integration of different access technologies such as cellular, broadcast, or wireless LAN, around IP. The different radio technologies will be integrated in an expandable manner, to complement each other, providing an efficient use of the scarce radio spectrum and, above all, a wide range of services to the final user. On the terminal side, it is expected with the progress of SDR that a terminal would be able to configure itself from one radio technology to the other in a seamless manner.

A major challenge for the future generation networks is that the architecture will have to be very flexible and open, capable of supporting various types of networks, terminals, and applications. The fundamental goal is to make the heterogeneous networks transparent to users. Another goal is to design a system architecture that is independent of the wireless access technology. These considerations led to a set of requirements that are specifically relevant to heterogeneous networks. The major elements are: (1) multiservice user terminal (multimodule/SDR-based) for accessing different RANs; (2) wireless system discovery; (3) wireless system selection; (4) unified location update and paging; (5) cross-system handover; (6) simple, efficient, scalable, low-cost; (7) energy-efficient; (8) secure; (9) QoS support; and (10) personal mobility/universal ID [1].

Future wireless networks will therefore provide added value by allowing the deployment of a large variety of services over them. A shift from a technology-and transport-oriented view towards a service-oriented one is happening, bearing in mind that these services should be provided over heterogeneous infrastructures allowing coexistence of cellular (UMTS), broadcast [digital video broadcasting (DVB)], short-range (802.11x-based), and upcoming technologies, connected to an all-IP core network. In order to make such services commercial, an additional research effort must be made to provide to the terminal equipment (but also to the radio access points, on the network side) a real flexibility. The capability to self-adapt to the radio conditions and to the requested services requires the capability of dynamically adapting the involved radio access layers: mainly the physical layer (modulation, coding) and the link layer (retransmission). Cross-layer optimization, therefore, has gained momentum in recent years. More and more, transmission, modulation, coding, and retransmissions operations are merged or jointly operated. It is the case, for instance, of EGPRS where the link adaptation scheme involves different modulation and coding schemes; H-ARQ combines retransmissions and forward error coding (FEC) (also used in high-speed downlink packet access—HSDPA and 1xEV-DO systems). While using wired networks, reliability is obtained using repetition with the application of the standard ARQ protocol. In the case of wireless transmissions, because of the poor quality of the link, packets should be protected against the channel noise, the fading caused by mobility, and the interference created by other users. Protection is mainly given by FEC, (e.g., transmitting additional bits to the packet). However, to provide the same quality as in a wired system, FEC overhead could lead to very inefficient transmissions. As a result, hybrid schemes combining FEC and ARQ have been defined. The last decade witnessed the emergence of new forward error codes, mimicking random codes such as turbo codes or low density parity codes, which showed near Shannon limit performance when associated with an iterative decoder. H-ARQ schemes, for example, were investigated in the context of low-density parity-check codes in the IST project OVERDRIVE [2] and a performance criteria based on the analysis of the density evolution was introduced.

Emerging Internet applications, such as distribution of newsclips, music, sports information, widely used multiplayer games, application sharing, and virtual worlds, require one-to-many or many-to-many communication combining several information media. There are many services in which the same multimedia information must be distributed to a large group of users. This distribution can be done either by using unicast or multicast. By using multicast we save network resources and serve the entire group instead of individual users. Multicast is an enabling technology that has been a study item in fixed network environments for a long time.

Multiaccess is the system capability of combining and using a number of radio accesses. Figure 2.1 shows that heterogeneous access networks proliferate as well as the variety of existing access technologies and devices for the provision of multimedia services.

Considering that several mobile users, with different terminal capabilities, have several accesses available at a given time, the optimal transmission of data can be different at any time.

There has to be an entity responsible for managing the delivery of mobile multimedia services to the mobile users in a resource-efficient way. This can be done either using unicast or multicast.

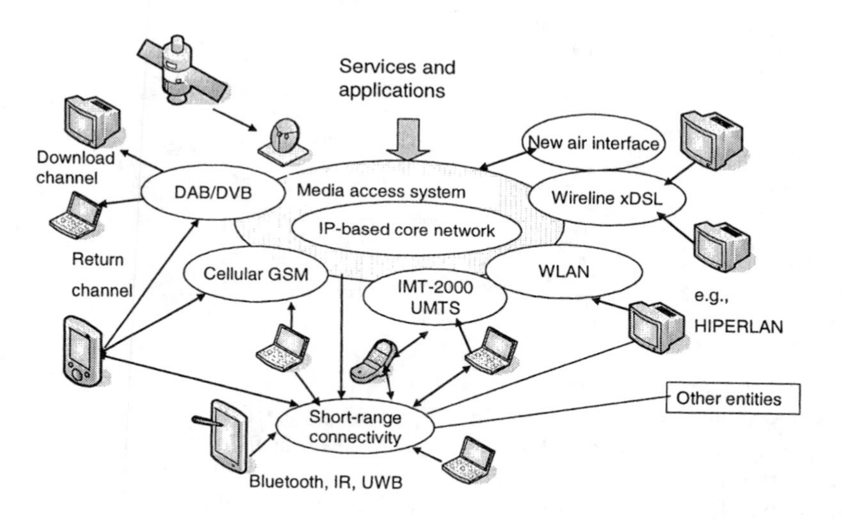

Figure 2.1 Heterogeneous access networks proliferate.

Multicast transmission, in fact, can be enabled by means of enhancements at different levels of the network with different impacts on the network architecture and standardization. The goal of physical and protocol improvements is to increase multicast transmission reliability while maintaining high spectrum efficiency. The management of multicast radio resources means exploiting base station resources in the most efficient way, also taking into account unicast connections.

At the physical layer the advantage of a high-speed OFDM downlink can be used, one should note that OFDM is a technique already used by broadcast technologies such as digital video broadcasting-terrestiral (DVB-T) or digital audio broadcasting (DAB), as well as adopted by IEEE802.11a/g. Power control constraints are highly reduced because there is no intracell interference.

Furthermore, OFDM enables single frequency network deployment where the same information can be sent by several base stations on the same set of subcarriers, thus performing a sort of soft handover.

Multicast over the UMTS terrestrial radio access network (UTRAN) is a very challenging task since UMTS was not initially designed to bear this type of transmission. The air interface complicates the scenario being a highly unreliable medium: the issue of managing packet errors, their retransmission, and the management of radio resources to transmit a multicast channel in the most efficient way are issues investigated in this chapter.

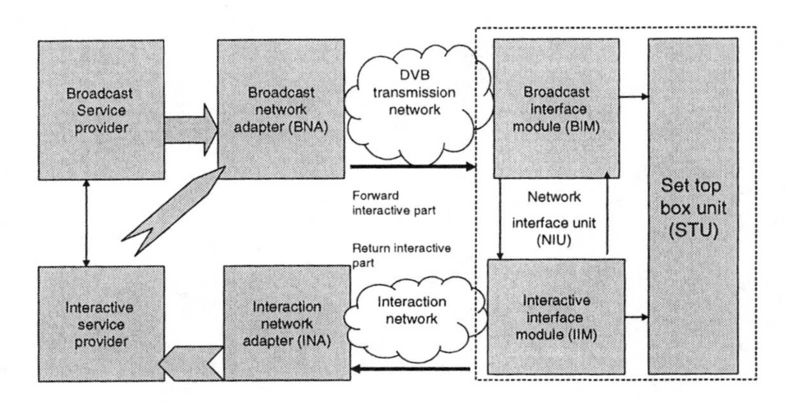

Figure 2.2 DVB-RCT generic model.

At protocol level a deep investigation is needed on hybrid ARQ transmissions, an advanced technique that increases transmission efficiency. Then there is the issue of power control in the multicast scenario, where users experience different propagation conditions while receiving the same information. Other investigations can relate to the switching point between multiple dedicated connections and a single multicast transmission, (i.e., the level of offered traffic at which there is an advantage in leaving dedicated channel features to set up a fixed power channel).

Next generation wireless networks will be a conglomeration of different networking technologies and will support multiple broadband wireless access technologies. This will allow global roaming across systems constructed by individual access technologies (see Figure 2.3).

This chapter is organized as follows. Section 2.2 describes the concept of hybrid wireless networks. This concept was a focus for a number of IST projects. Therefore, the material for this section was based on input from the projects OVERDRIVE [2], for describing different mobile multicast techniques, and MONASIDRE [3], for describing the cooperation among heterogeneous radio access technologies, such as WLANs based on HiperLAN/2, UMTS, and DVB. Some of the outlined issues are related to radio resource management, QoS provisioning, and the delivery of services through such technologies. Section 2.2 further describes the process of radio network planning and network optimization. Reference architectures are also described as an example.

Section 2.3 describes research activities related to multiradio access. Focus is on the integration of different air interface types of widely different characteristics and how these can cooperate to deliver a range of applications. The input to this section was from the IST project BRAIN [4] and its continuation with the IST project MIND [4].

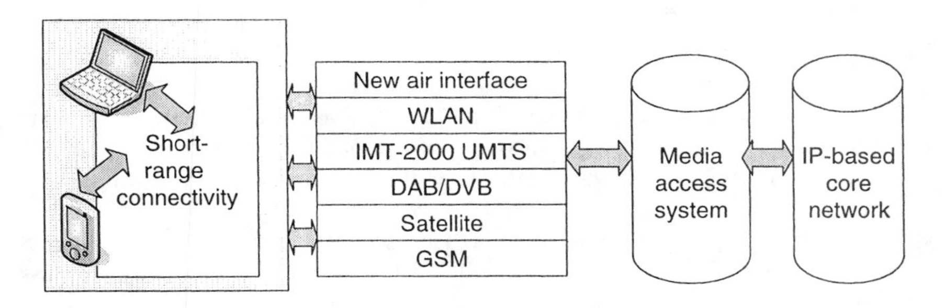

Figure 2.3 Summary of mobile and wireless systems in the concept of a heterogeneous radio environment.

Section 2.4 continues the theme of the chapter and describes approaches to secure heterogeneous access. The described approaches were investigated within the IST project BRAIN [4] and Secure Heterogeneous Access for Mobile Applications and Networks (SHAMAN) [5]. Security requirements, possible threats, and different protection mechanisms are also described here. Section 2.5 concludes the chapter by summarizing the most important points related to the management of heterogeneous networks.

2.2 HYBRID WIRELESS NETWORKS

User demand for accessing the Internet and other services while on the move is a driver for the research in the direction of cooperation of different types of services, and cost-efficient access to information, entertainment, and communication, and capacity of frequency spectrum and networks. The usage drivers for the proper management of hybrid wireless networks can be summarized as follows [1]:

- Multiplicity of usage environments (various locations, radio accesses, data rates, fixed/mobile);
- Simple and intuitive usage;
- Provision of appropriate usage context, giving the people what they want;
- Information system harmonization;
- Seamless handovers over heterogeneous technologies (WLAN, GPRS, UMTS, and upcoming ones);
- Reachability of users.

These drivers imply complementarity of the access networks rather than their competition, service interoperability, and a generalized management of mobility. The following sections show some of the approaches that can be undertaken to ensure the management of hybrid wireless networks.

2.2.1 System Enhancements Beyond 3G

The explosive growth in the number of mobile users and their increasing demand for flexible access to diverse services has motivated significant research, standardization, and development efforts in the area of future wireless access systems such as UMTS, mobile broadband systems (MBS), such as HiperLAN, and digital broadcasting systems (DBS), such as DVB. In future wireless access systems involving UMTS, HiperLAN, and DVB-T, as well as upcoming wireless systems that comprise technologically different access networks on one hand and a number of technologically different mobile devices on the other hand, the provision of wireless services becomes duly complex and requires the implementation of a network and service management system capable of interconnecting/interfacing the service provider to the network operator(s).

The IST project MONASIDRE [3] investigated the cooperation of UMTS, MBS, and DBS as three cooperating and complementary components. Through such infrastructure, operators are enabled to provide users with efficient (in terms of cost and QoS) wireless access, and service providers with the means for offering sophisticated services. The investigation was done under the assumption that there exist multiple service providers and network operators (owning one or more of the managed systems, but also potentially cooperating).

One of the main support actions to ensure such cooperation is to find a means for improving statistical performance indicators. Figure 2.4 shows the high-level structure of the UMTS, MBS, and DBS network and service management system.

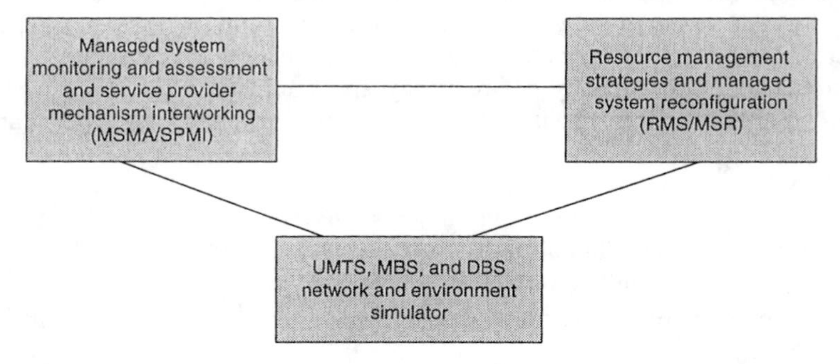

Figure 2.4 High-level description of the UMTS, MBS, and DBS network and service management system. (*From:* [3].)

The first component is meant to manage system monitoring and assessment through a service provider mechanism for ensuring interworking (MSMA/SPMI).

The following issues need to be addressed to ensure proper functioning of this mechanism:

- *Definition of interfaces with the service provider mechanisms.* Through this interface service providers can request the reservation of resources (establishment of virtual networks) from the managed network. Appropriate application programming interfaces (API) must be developed to allow service providers to create their own services with a minimum interaction with networks carriers.
- *Contracts with service providers processing, establishment, and maintenance.* One of the aspects framed in this context is the accurate and efficient—in terms of the (potentially combined UMTS, MBS, and DBS) network resources required—translation of the service provider view to a corresponding network level view. Pertinent problems can be resolved in conjunction with the other components of the management system.
- *Definition of interfaces with the (highly heterogeneous) elements or segments of the managed network.* The anticipated heterogeneity of typical UMTS, MBS, DBS, as well as upcoming network configurations may be addressed by common standards on the interfaces that the managed elements, or network segments, should provide.
- *Integration of sophisticated managed system monitoring methodologies and procedures.* Traffic, mobility, and propagation conditions in a heterogeneous environment (service area) are highly complex. Sophisticated procedures for (readily and accurately) obtaining, or detecting changes in, the probability distribution of parameters are means for coping with this aspect. Pertinent work from probability theory and data mining forms the basis in this context and, hence, must be considered and integrated.

The second component in Figure 2.4 supports resource management strategies (RMS) development and integration, and managed system reconfiguration (MSR). In essence, the role is to apply resource management strategies that dynamically find the appropriate system reconfigurations, through which the service provider requests and/or handles the (new) service area conditions in the most cost-efficient manner. The optimization processes envisaged in this component should take into account both cost and QoS criteria. The use of efficient, in terms of processing time and search-space, algorithms and tools is required and should be developed and integrated in the RMS/MSR component. The scope of the resource management strategies in the context of this example is as follows:

- *Multiple air interface network planning.* Operators will be allowed to operate or coordinate operations of heterogeneous wireless systems.

Implementing the best suited wireless technology in terms of service demand and QoS will definitively optimize system deployment costs, hence enabling lower cost wireless access.

- *Dynamic air interface configuration.* This requires the development of software components that may dynamically find the most cost-effective radio resource configuration pattern that satisfies service providers' requests or the current environmental conditions. Dynamic spectrum assignment aspects and dynamic configuration of radio access point air interface are also the focus of such software components. Activation (cessation) of the appropriate radio resource control strategies (e.g., call admission control, handover threshold) is also included in this context. The target air-interfaces (for the example given here) are those of UMTS (namely, the FDD and TDD components of UTRAN), MBS (namely, the BRAIN architecture—BRAN—[4]), and DBS (namely, DVB-T or DAB). The coexistence of other legacy systems, (e.g., GSM/EDGE) is also important. Another investigation step is the definition of service classes.
- *Access network and/or interface with fixed network.* Handling service provider requests or new environment conditions may require reconfigurations in the access network or in the interface with the core network. Bridges to existing solutions of the access or the core network (e.g., IP control concepts like differentiated or integrated services) should be implemented to address transition from existing solutions. In essence, traffic that originates from/to the radio network should be mapped into appropriate IP classes. In the example of Figure 2.4 the core infrastructure is all-IP and it considers all known Internet services (e.g., streaming) in the characterization of the service classes (parallel-streaming services will also be considered).

The third component is the UMTS, MBS, and DBS network and environment simulator (UMDB-NES) component (see Figure 2.4). This component is motivated from the need of validating a management decision prior to its application in the real network. Therefore, the tool will enable off-line testing, verification, validation, and demonstration of the management software in a wide range of test cases. The next sections describe the reference architectures of the UMTS, MBS, and DBS networks.

2.2.1.1 UMTS Reference Architecture

Functional Elements in the UMTS Reference Architecture

For the all-IP solution for UMTS [6], all data in the core network is transported on IP, including traditional circuit-switched voice. Real-time services supported by Releases 4 and 5 are circuit-switched voice service and IP-based multimedia service. The mobile switching center (MSC) in Release 4 was split into an MSC server for the control part and into media gateways (MGWs) for the transport part.

The evolution towards the all-IP core network brings new elements into the architecture. These are as follows [7]:

- *MSC server:* The MSC server controls all calls coming from circuit-switched mobile terminals and mobile terminated calls from the Public Switched Telephone Network (PSTN)/Integrated Services Digital Network (ISDN)/GSM to a circuit-switched terminal. The MSC server interacts with the media gateway control function (MGCF) for calls to and from the PSTN. In the functional split introduced by Release 5, the MSC server handles the call control and services part. The MGW, an IP router, replaces the switch. This functional split reduces deployment cost while supporting all existing services.
- *The call state control function (CSCF):* The CSCF is a Session Initiation Protocol (SIP) server (SIP has been adopted as the call control protocol between terminals and the mobile network) providing and controlling multimedia services for packet-switched (IP) terminals.
- *MGW on PSTN side:* Calls coming from the PSTN are translated to voice over IP (VoIP) calls for transport in the UMTS IP-based core network.
- *MGW on the UTRAN side:* The MGW transforms VoIP packets into UMTS radio frames. This enables the support of 3G circuit-switched terminals in a full IP UMTS core network and provides backward compatibility with Release 99 terminals. The MGW is controlled by the MGCF using H.248 media gateway control protocol. The MGW can be located at the UTRAN side of the Iu interface (when Iu is IP-based) or at the core network side (to keep the Iu interface unchanged from release 99).
- *Signaling gateway (SG):* The SG relays all call-related signaling to and from the PSTN and the UTRAN on IP-bearers and sends the signaling data to the MGCF.
- *MGCF:* Controls the MGWs via H.248. It also performs translation at the call control signaling level between the ISDN User Part (ISUP) and IP signaling.
- *Home subscriber server (HSS):* The HSS is an extension of the Home Location Register (HLR) database integrating the multimedia profile of the subscribers. The HSS is an HLR with IP-specific functions such as authentication, authorization, and accounting (AAA); domain name system (DNS); and dynamic host configuration protocol (DHCP).

UMTS QoS Classes, Bearer Services, and Parameters

UMTS is able to offer high user bit rates to ease the introduction of new services such as video telephony and quick downloading of data. Rates of 384 Kbps on circuit-switched connections and 2 Mbps on packet-switched connections can be reached. This makes possible the quick access to information and its filtering according to the location of the user. The requested information, however, is typically on the Internet and this calls for efficient handling of the transport control protocol (TCP)/user datagram protocol (UDP)/IP traffic flows in the UMTS network. At the beginning of the UMTS introduction, delay-critical applications such as speech and video were planned to be carried on circuit-switched bearers. UMTS, however, implements additional QoS functions to support delay-critical services as packet data. The most important new feature of UMTS is allowing negotiation of the radio bearers.

During the transfer throughput, transfer delay and data error rate can be controlled and negotiated over various bearers. To optimize UMTS for many applications, UMTS bearers have to be generic by nature to enable support of existing applications and facilitate the evolution of new applications. The service platform should follow the same rules to facilitate application developments.

UMTS Bearer Service

User and applications can negotiate bearer characteristics that best fit the information and can change the bearer properties in the course of an active connection or session. Bearer negotiation is initiated by the application while renegotiations may be triggered either by the application or the network. The application requests a bearer depending on needs; the network checks the availability of resources and the user's subscription and responds. The user may accept or reject the offer. The properties of the bearers directly affect the quality and charging of the service. For example, the IST project MONASIDRE [3] extended the notion of bearers to complete networks in a diversified radio environment by introducing a MSMA/SPMI mechanism called a MASPI mechanism. Here again, a selection of a given technology or network bearer service has a direct influence on the service price and QoS. The UMTS bearer service architecture is shown in Figure 2.5.

The bearer class, bearer parameters, and parameter values are directly related to the application and the networks that lie between the sender (source) and receiver (destination).

The parameters should be selected so that negotiations are simple and unambiguous and so that easy policing and monitoring can take place. For the UMTS bearer service layered architecture shown in Figure 2.5, each bearer service on a specific layer offers its own services while relying on those provided by the layers below. The UMTS bearer service plays a major role in the end-to-end service provision as highlighted in Figure 2.5.

To provide end-to-end QoS, all bearers must satisfy the required QoS at their own level. The UMTS bearer service guarantees QoS only if the RAN and the CN bearer services can achieve the desired quality. This can only be provided if all lower layer bearers fulfill their own requirements. In particular, radio technologies and transport technology planning, control, and management must be accomplished with the end-to-end target QoS in mind. This overall quality must be mapped into lower layer QoS to feed the overall UMTS bearer service architecture and the service management platform. The classification of services can be given based on the 3GPP service classification:

- Conversational class;
- Streaming class;
- Interactive class;
- Background class.

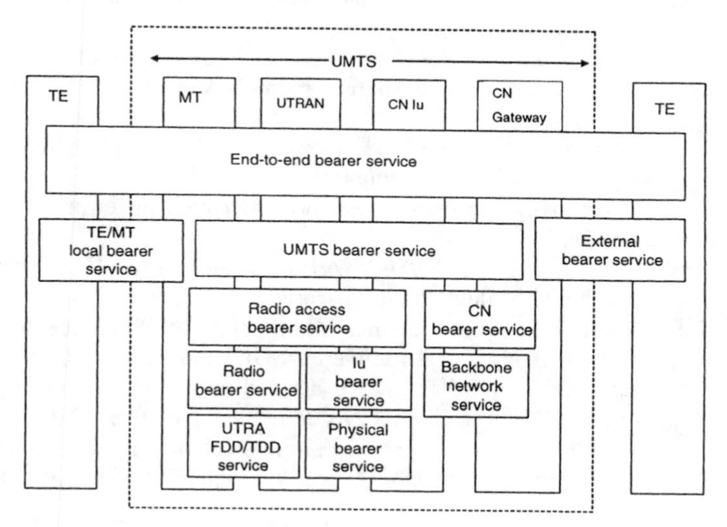

Figure 2.5 UMTS bearer service architecture.

QoS Parameters

The UMTS bearer service parameters describe the services provided by the UMTS network to the user of the bearer service. The service is characterized by a set of parameters that constitutes what is referred to as the QoS profile. Each bearer service is characterized by a set of attributes.

The list of attributes for the UMTS bearer service can be described as follows:

- *Traffic class:* type of application for which the UMTS bearer service is optimized (i.e., conversational, streaming, interactive, and background).
- *Maximum bit rate (Kbps):* the maximum number of bits delivered by UMTS at a service access point (SAP) within a period of time divided by the time period. Traffic conforms to the maximum bit rate as long as a required token bucket algorithm controls the rate. Token rate equals

maximum bit rate and bucket size equals maximum service data unit (SDU) size.

- *Guaranteed bit rate (Kbps):* guaranteed number of bits delivered at an SAP within a time period normalized by the duration of the period. A required token bucket algorithm, with a token rate that equals the guaranteed bit rate and a bucket size that equals the maximum SDU size, ensures conformance.
- *Delivery order (y/n):* indicator of in-sequence delivery of SDU requirement by the UMTS bearer.
- *Maximum SDU size (octets):* maximum allowed SDU size. Used for admission control and policing.
- *SDU format information (bits):* list of possible exact SDU sizes. Information needed for transparent radio link control (RLC) protocol operation mode.
- *SDU error ratio:* indicates the fraction of SDUs lost or detected as erroneous. Applies to conforming traffic.
- *Residual bit error ratio:* indicates the undetected bit error ratio in the delivered SDUs.
- *Delivery of erroneous SDUs (y/n):* indicates if SDUs detected as erroneous should be delivered or discarded.
- *Transfer delay (ms):* indicates maximum delay for 95th percentile of the distribution of delay for all delivered SDUs during the lifetime of the bearer service. Delay for an SDU is defined as the time from a request to transfer of an SDU at one SAP to its delivery at the other SAP.
- *Traffic handling priority:* specifies the relative importance for handling all SDUs belonging to the UMTS bearer compared to the SDUs of other bearers.
- *Allocation/Retention priority:* specifies the relative importance compared to other UMTS bearers for allocation and retention of the UMTS bearer. This allocation/retention priority attribute is not negotiated from the mobile terminal.

Part or all of these attributes apply to each traffic class (conversational, streaming, interactive, and background). Observe that a traffic class is also an attribute itself.

The radio access bearer service is applied to both circuit-switched (CS) and packet-switched (PS) domains and consists of the very same attributes (those used for the UMTS bearer service level). However, the values taken by these attributes are different for obvious reasons (e.g., header compression, resource management, delay contribution to overall delay).

The radio access bearer service requires an additional attribute to describe the flows for transport over radio access bearers (RAB): the source statistics descriptor (SSD).

The SSD specifies characteristics of the submitted SDUs source. The SSD is set if the RAB carries a speech service or streaming speech so that UTRAN calculates a statistical multiplexing gain on the radio and Iu interfaces and uses it for admission control. The RAB attribute settings in the interactive and background classes are identical to those of UMTS bearer services.

The Iu bearer service relying on the physical layer service provides the transport between UTRAN and CN. The operator selects the QoS capabilities in the IP layer or asynchronous transfer mode (ATM) layer depending on transport technology selection.

IP-based Iu bearer services operate on an Internet Engineering Task Force (IETF) differentiated services basis. When the operator selects ATM-SVC as the internal dedicated transport bearer, interoperation with the IP-based networks relies on differentiated services. Operators use service level agreements (SLA), an integral part of the DiffServ architecture, for interoperability.

The UMTS CN bearer service supports different backbone bearer services for a variety of QoS on the very same basis.

The parameter value ranges for UMTS bearer service attributes are summarized in Tables 2.1 and 2.2. The value ranges for the UMTS bearer service provided in Table 2.1 reflect the capability of the UMTS network.

Table 2.2 provides the capability of UTRAN and describes the value ranges for the radio access bearer service attributes.

In order to provide end-to-end QoS, the parameters in different QoS levels (bearer service level) must be mapped. That means that mapping from application parameters to UMTS bearer service parameters and subsequently to radio access bearer service parameters must be achieved [7].

In addition the working assumptions for the management functions in Figure 2.5 should adhere and come as close as possible to the QoS management functions envisaged and envisioned within 3GPP as described below.

The QoS management functions establish, modify, and maintain a UMTS bearer service with a specific QoS. The QoS management components of all the UMTS network entities together ensure the desired/contracted QoS.

The functions that are expected within the QoS management framework are in the control plane. These can be summarized as follows:

- *Service manager:* coordinates the functions of the control plane for establishing, modifying, and maintaining the service for which it is responsible.
- *Translation function:* converts between the internal service primitives for UMTS bearer service and the various protocols for service control of interfacing external networks.
- *Admission capability control:* maintains information about all available resources of a network entity and about all resources allocated to a UMTS bearer service.

- *Subscription control:* checks the administrative rights of the UMTS bearer service user to use the requested service with the specified QoS attributes.

Table 2.1

Value Ranges for UMTS Bearer Service Attributes

	Traffic Class	Traffic Class	Traffic Class	Traffic Class
Attributes	Conversational	Streaming	Interactive	Background
Maximum bit rate	< 2,048	< 2,048	< 2,048-overhead	< 2,048-overhead (Kbps)
Delivery order	Yes/No	Yes/No	Yes/No	Yes/No
Maximum SDU size	≤ 1,500 or 1,502	≤ 1,500 or 1,502	≤ 1,500 or 1,502	≤ 1,500 or 1,502 (octets)
SDU format information	TBD by 3GPP RAN WG3 (RLC transparent)	TBD by 3GPP RAN WG3 (RLC transparent)	—	—
SDU error ratio	10^{-2}, $7*10^{-3}$, 10^{-3}, 10^{-4}, 10^{-5}	10^{-1}, 10^{-2}, $7*10^{-3}$, 10^{-3}, 10^{-4}·10^{-5}	10^{-3}, 10^{-4}, 10^{-6}	10^{-3}, 10^{-4}, 10^{-6}
Residual bit error rate	$5*10^{-2}$, 10^{-2} $5*10^{-3}$, 10^{-3} 10^{-4}, 10^{-6}	$5*10^{-2}$, 10^{-2} $5*10^{-3}$, 10^{-3} 10^{-4}, 10^{-5}, 10^{-6}	$4*10^{-3}$, 10^{-5}, $6*10^{-8}$	$4*10^{-3}$, 10^{-5} $6*10^{-8}$
Delivery of erroneous SDUs	Yes/No	Yes/No	Yes/No	Yes/No
Transfer delay (ms)	100 - maximum value	250 - maximum value		
Guaranteed bit rate (Kbps)	< 2,048	< 2,048		
Traffic handling priority			1, 2, 3	
Allocation/retention priority	1, 2, 3	1, 2, 3	1, 2, 3	1, 2, 3

In the user plane the QoS management functions maintain signaling and user traffic data within limits defined by the QoS attributes. These functions that ensure provision of the QoS negotiated for a UMTS bearer service are as follows:

- *Mapping function:* provides each data unit with specific marking required to receive the intended QoS at the transfer by a bearer service.

- *Classification function:* assigns data units to the established services of a mobile terminal (MT) on the basis of QoS attributes if multiple service bearers are established.
- *Resource manager:* distributes the available resources over all services sharing the same resource.
- *Traffic conditioner:* provides conformance between the negotiated QoS for a service and the data unit traffic. This is achieved by traffic shaping or policing.

Table 2.2

Value Ranges for Radio Access Bearer Service Attributes

	Traffic Class	Traffic Class	Traffic Class	Traffic Class
Attributes	Conversational	Streaming	Interactive	Background
Maximum bit rate (Kbps)	< 2,048	< 2,048	< 2,048-overhead	< 2,048-overhead
Delivery order	Yes/No	Yes/No	Yes/No	Yes/No
Maximum SDU size (octets)	≤ 1,500 or 1,502	≤ 1,500 or 1,502	≤ 1,500 or 1,502	≤ 1,500 or 1,502
SDU format information	TBD by 3GPP RAN WG3 (RLC transparent)	TBD by 3GPP RAN WG3 (RLC transparent)	—	—
SDU error ratio	10^{-2}, $7*10^{-3}$, 10^{-3}, 10^{-4}, 10^{-5}	10^{-1}, 10^{-2}, $7*10^{-3}$, 10^{-3}, 10^{-4}, 10^{-5}	10^{-3}, 10^{-4}, 10^{-6}	10^{-3}, 10^{-4}, 10^{-6}
Residual bit error rate	$5*10^{-2}$, 10^{-2} $5*10^{-3}$, 10^{-3} 10^{-4}, 10^{-6}	$5*10^{-2}$, 10^{-2} $5*10^{-3}$, 10^{-3} 10^{-4}, 10^{-5}, 10^{-6}	$4*10^{-3}$, 10^{-5}, $6*10^{-8}$	$4*10^{-3}$, 10^{-5} $6*10^{-8}$
Delivery of erroneous SDUs	Yes/No	Yes/No	Yes/No	Yes/No
Transfer delay (ms)	80–maximum value	250–maximum value	—	—
Guaranteed bit rate (Kbps)	< 2,048	< 2,048	—	—
Traffic handling priority	—	—	1, 2, 3	
Allocation/retention priority	1, 2, 3	1, 2, 3	1, 2, 3	1, 2, 3
Source statistic descriptor	Speech/ unknown	Speech/unknown	—	—

A management architecture, such as the one shown in Figure 2.6, should classify its management functions onto these high-level and generic QoS management functions as defined in the UMTS specifications in order to provide a common ground and facilitate dissemination of information to standardization working groups (ETSI and 3GPP) whenever appropriate. In the architecture described in Figure 2.6, the service provider (SP) interacts with the MASPI component of the whole network and service management system, through interface I1. Similarly, the rest of the defined interfaces support the interaction of the MASPI component with the different radio access networks, the resource management strategies and managed system reconfiguration (RMS) component, and with the UMD-NES component.

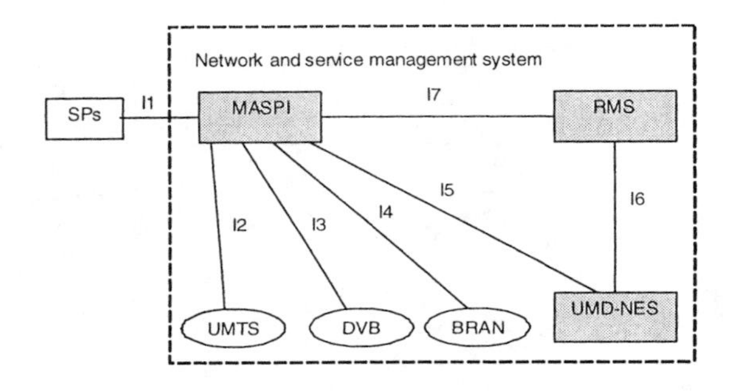

Figure 2.6 Example of a management architecture—project MONASIDRE. (*From:* [3].)

The result of the SP network resources request is the reservation and allocation of network resources managed by the network operator (if possible). The resources reserved or in other words, the virtual networks established over the managed infrastructure are the result of a network monitoring functionality and a resource management strategies definition (as performed by the MASPI and RMS components, respectively). The specific infrastructure involved by the established virtual network is hidden to the SP.

The interaction between the SP and the MASPI component requires that the MASPI component takes into account the SLA established between the SP and the network operator (NO).

Moreover, in order for the NO to allow the user to access the services provided by the SP, the MASPI component must be also aware of the users allowed and their profiles. This is shown in Figure 2.7.

Scenarios providing end-to-end QoS from the UE to the UMTS network depend on whether the Public Land Mobile Network (PLMN) portion of the overall architecture is IP-based or Go Text Protocol (GTP) tunnel-based. The all-IP architecture will in the long run extend all the way to the RNC and even to the base stations [1]. These network entities are expected to include IP capabilities and possess an IP address in future releases of the UMTS network. Mobility management will be Mobile IP based.

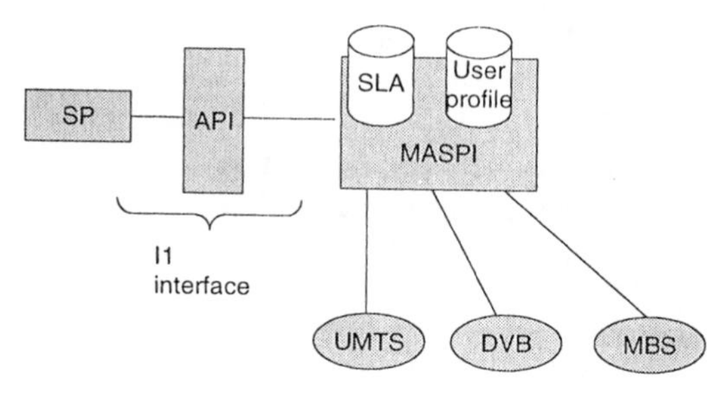

Figure 2.7 SP – MASPI interworking mechanism. (*From:* [7].)

The end-to-end IP QoS bearer service is controlled by the gateway GPRS service node (GGSN) that may support DiffServ edge functions. The CN is not mandated to use DiffServ exclusively but can use mechanisms such as RSVP, over-provisioning, and aggregate RSVP. All these conceptual models designed to handle end-to-end QoS were part of the network infrastructure analyzed by the project MONASIDRE [3]. Figure 2.8 indicates the portion where various QoS mechanisms may be used depending upon the architecture in place between the UE and the remote host.

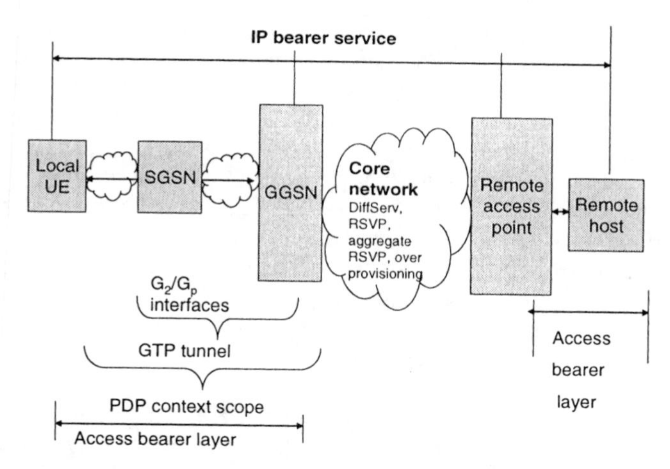

Figure 2.8 Potential network architectures. (*From:* [7].)

There can be a variety of potential scenarios, but in Figure 2.8 two main classes are retained.

One case is the scenario when the UE does not provide an IP bearer service manager. The end-to-end QoS towards the remote terminal is then controlled from the GGSN. The other case corresponds to a UE performing IP bearer service functions, which enable end-to-end QoS. This can happen with or without IP layer signaling towards the IP bearer service function in the GGSN or the remote terminal. In addition the UE and GGSN may support DiffServ edge function or support RSVP signaling instead.

Transport Technologies

The radio network layer (RNL) can be assumed independent from the transport network layer (TNL). This assumption is legitimate since it is a basic requirement within 3GPP [6] and necessary to maintain backward compatibility with previous releases of the UMTS network.

This separation of RNL and TNL except for minor addressing issues allows the introduction of any desired transport technology. Transport can be based on ATM, STM, or IP depending upon operator choices. In addition, a broker network can provide transport for a RAN provider or a CN provider. The interface between the different providers is operated on a user-to-network interface (UNI) or a network-to-network interface (NNI) depending upon the contracts that can be set up between the various providers. Typically the transport layer will act as a broker providing and respecting a QoS agreement and SLA, thus providing a clear and clean-cut separation of the providers from a management and network control standpoint.

The transport protocol across the Iu interface (between the radio network controller—RNC and the core network) for UTRAN can be based on ATM transport or IP transport. The coexistence of these two technologies is also provisioned for within the 3GPP recommendations. The transport options rely on the same functional split between network elements as well. The interface is also assumed to support all services such as classical telephony, Internet-based services, and multimedia services. The Iu interface is assumed to support efficiently dedicated circuits for voice connections, best-effort packet services, and real-time multimedia services requiring high QoS. The current specification for Iur and Iub uses AAL2/ATM to provide the services to the RNL.

The introduction of IP transport requires that the transport bearer establishment does not induce any change to the RNL functional split. Further, there shall be no impact on the serving RNC (SRNC) (Iu/Iur)/controlling RNC (CRNC) (Iub) TNL and RNL transactions during bidirectional bearer establishment. This indicates clearly that for the example MONASIDRE architecture, the RNL-TNL interactions can be based on a unique functional split regardless of the transport technology option selected to respond to a service provider request through the MASPI and the management platform.

The core network transport technology can be based on ATM, STM, or IP transport. The ATM-SVC technology is well established and in use in many private networks and public services. It is also well defined by the ITU Telecommunications Standardization Sector (ITU-T) [8] and the ATM Forum.

By using this technology, bandwidth is allocated on demand and network efficiency is greatly improved. ATM-SVC can be used for public networks and it supports stringent QoS and offers scalability. Further, STM-SVC has gained recognition in the IETF world as well. An ATM-PVC–based approach cannot realize on demand based QoS control or efficient resource allocation.

If ATM-SVC is the most appropriate technology to support various QoS requirements and to use network resources efficiently, the UMTS CN should include ATM-SVC capability as one potential and realistic solution. Even if in the long term an IP based transport technology may be advocated to achieve an all-IP-based UMTS network, the coexistence of alternate transport technologies remains a reality and such solutions shall not be overlooked or precluded.

Figures 2.9 to 2.14 show the diversity of transport technologies within the UMTS framework and how they evolve.

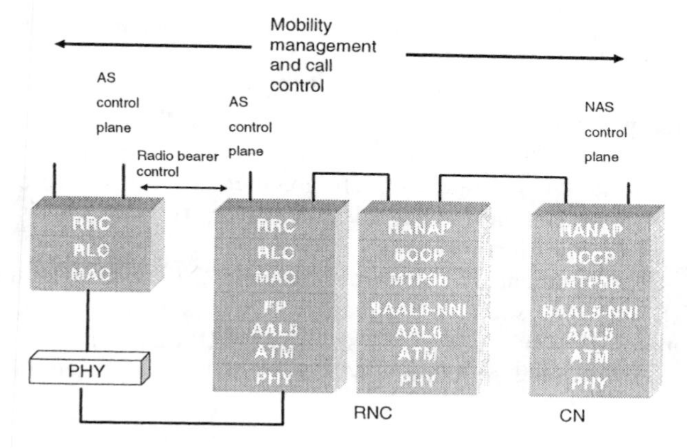

Figure 2.9 Control plane.

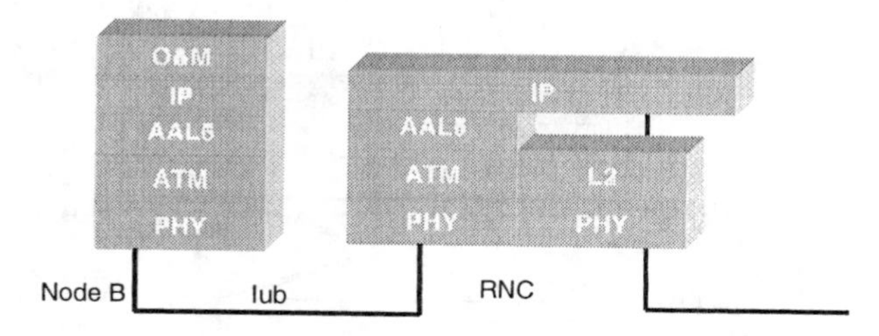

Figure 2.10 Use of IP for operation and maintenance over the Iub interface.

Over the RAN both ATM and IP transport are available over the control and user plane. Differentiation of services over the UTRAN is not guaranteed over every portion of the end-to-end path. For instance, referring to Figure 2.11, the node-B is transparent and cannot differentiate between the various UMTS QoS classes. From the UE to the RNC, it is not possible to handle classes according to their respective QoS and priority cannot be introduced within the base station.

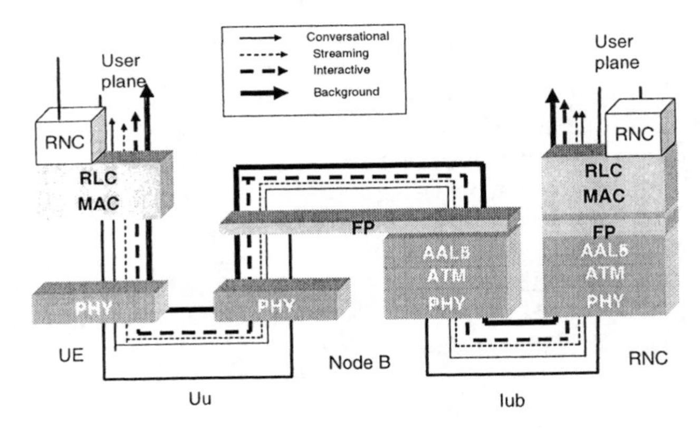

Figure 2.11 The node-B is a service class transparent terminal to network data transfer. User plane protocol stack over UTRAN.

For the downlink or forward link the RNC can intervene by treating classes differently and providing the needed priorities through the packet scheduling unit (PSU), which is part of the RNC. The only part where IP transport has penetrated the RAN at this stage is the operation and maintenance leg whose corresponding protocol stack is shown in Figure 2.10. However, IP transport is expected to colonize the interconnection networks.

This is clearly indicated in Figures 2.12 and 2.13 where IP transport is a viable and lower cost solution for carrying SS7 signaling from the circuit-switched domain. Finally, Figure 2.14 shows how IP transport over optical links using multiprotocol label switching (MPLS) will be introduced in the core IP domain.

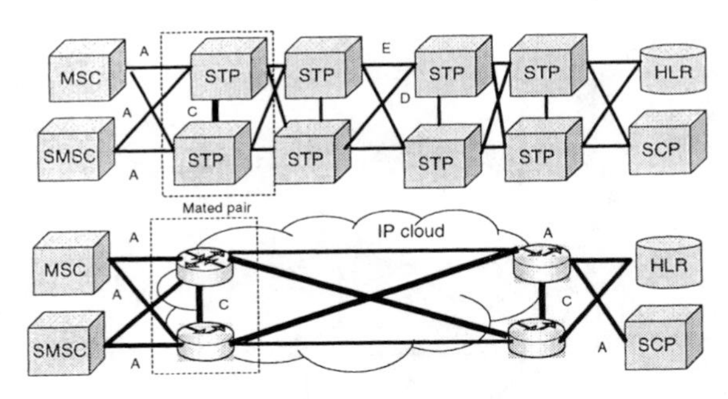

Figure 2.12 Replacing the circuit-switched-based signaling network by IP clouds.

As far as the IP transport technology is concerned there, are no restrictions on the type of solution selected for the user/control plane and the user/control plane transport signaling solutions. Solutions such as MPLS and AAL2, and some others, are possible and remain fairly transparent to the type of system in Figure 2.8.

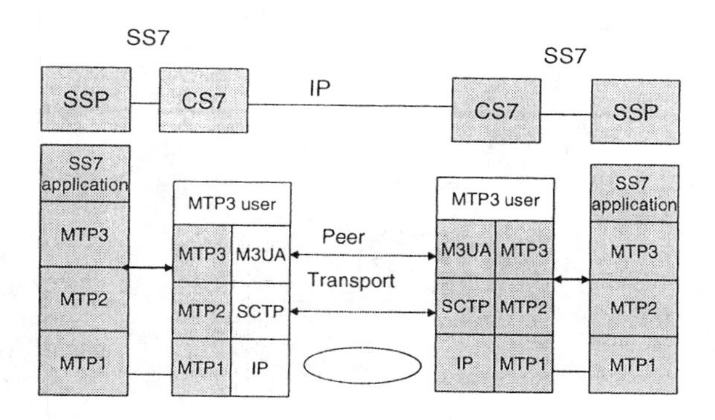

Figure 2.13 Protocol pile for IP-based SS7 signaling transport.

For the described example in Figure 2.6 with the MONASIDRE platform [7], only the service provided by the transport layer matters as long as all flows are served at the required level of QoS. In other words, focus could be given only to the capabilities of the transport network and the management actions that could be triggered to control (reconfigure) the transport network. Hence, effort should be put towards analyzing each transport network solution and the associated dynamic control and configuration flexibility and possibilities, as it is, for instance, the ability of controlling each individual flow versus controlling aggregate flows within UTRAN and the CN. The number of handled simultaneous connections and response times via these transport networks are other important characteristics.

Service Architecture

An important component of the Release 4 architecture [6] is the service architecture. The service architecture sets the stage in terms of interfaces and API towards the service provider.

The management capabilities of the service platform depend upon the service servers, functionality, and capability provided by the UMTS service infrastructure to external entities. The potential options available to network providers and brokers determine the services that the service and management platform (see Figure 2.6) should provide to service providers via the MASPI service provider mechanism interface. These aspects need to be investigated and jointly analyzed with the UMTS role model to identify scenarios where the cooperative network architecture (in the discussed example UMTS/MBS/DBS) and service management system is integrated. This action does not preclude ownership of the management system by any third party.

In order to achieve a sufficient degree of service differentiation, the UMTS in Release 4 provides the following fundamental improvements:

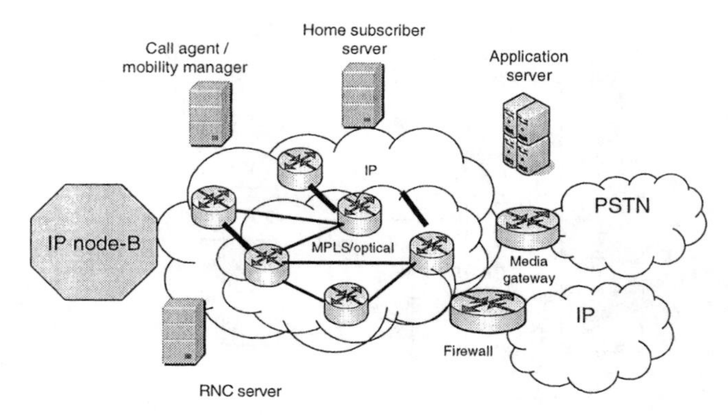

Figure 2.14 Overwhelming presence of IP transport over the long term.

- Wideband access via the RAN for multimedia service provision over the air;
- Mobile-fixed Internet convergence via the virtual home environment (VHE) concept to offer users in a unique way cross-domain services;
- Flexible service architecture by standardizing the building block that makes up services.

This enhances creativity and flexibility when inventing new services. The virtual home environment (VHE) concept promotes the view of a layered service architecture, which enables the development of services independently from the underlying networks (see Figure 2.15).

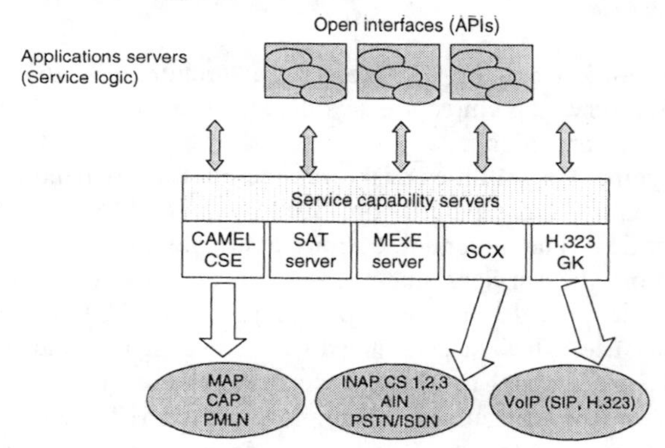

Figure 2.15 VHE service architecture in UMTS.

This concept of service portability allows service providers to develop UMTS applications that can run on several networks and terminals. The idea is to standardize service capabilities instead of services. This trend in the design of the UMTS service architecture must be examined with the trend to move toward an all-IP UMTS network architecture.

A VHE enables end users to access the services of their home network/service provider even when roaming into the domain of other network providers.

Independence from the underlying networks is achieved by standardizing the interfaces between the network layer (network elements under the operator's control) and the service layer (third party servers running service logic). Service capability servers (SCSs) are all the servers in the network that provide functionality used to build/construct services. From a software standpoint, the open service access (OSA) interface is an object-oriented API. All the functionality provided by the SCSs is grouped into logically coherent software interface classes. The MSC is an example of a SCS where call control is a class consisting of several call control related functions. The CSCF (or SIP server) in the all-IP architecture is also an example of SCS. The classes of the OSA interface are called service capability features (SCFs) and can be seen in Figure 2.16 for the release 2000 all-IP UMTS architecture.

The SCFs are just added as a software layer of interface classes on top of the existing network elements. The network elements equipped with these SCFs are called SCSs. By providing services in the service layer access to the SCFs, OSA offers an open standardized interface for service providers toward underlying networks. The service logic resides in the application servers in the service layer. The SCSs and application servers can be interconnected via an IP-based network to allow distributed deployment.

The purpose of the SCFs/SCSs is to raise the abstraction level of the network interfaces toward the service providers and thereby ease (and free/liberate) application development. In addition, OSA hides network-specific protocols, offers connectivity to both circuit switched and IP networks and protects core networks from misuse or intrusion via AAA and management interfaces towards SCSs.

As depicted in Figure 2.16, the offered functionality to the service layer is represented by the SCFs, which are implemented by the underlying UMTS transport network protocols.

Examples of such protocols are Camel (CAP), Mobile Application Part (MAP), and the Wireless Application Protocol (WAP). The SCSs within the UMTS architecture are as follows:

- UMTS call control servers such as MSC and CSCF;
- Home subscriber server (HSS) handling subscribers location and information;
- Mobile execution environment (MExE) server offering WAP or Java based value-added services through the MExE client in the terminal;
- SIM Application Toolkit (SAT) server;

- CAMEL server extending Internet service provisioning to the mobile environment.

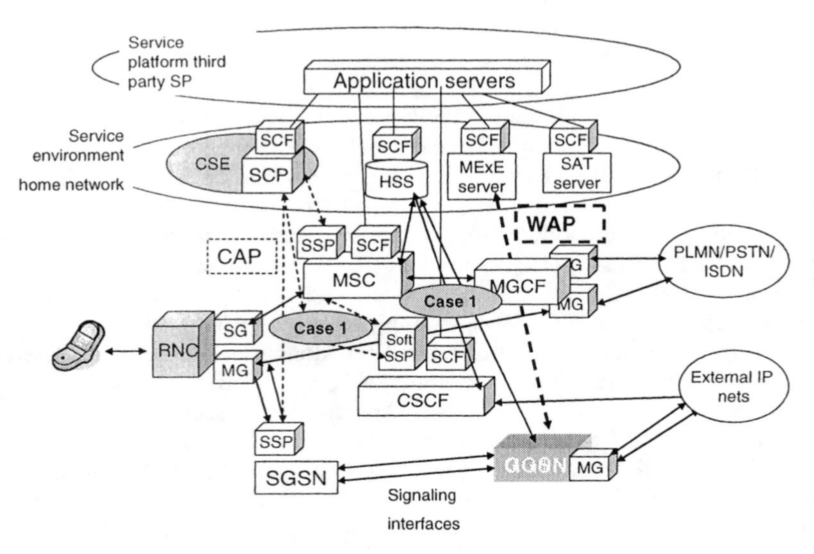

Figure 2.16 Release 5 service architecture providing an open standardized architecture. (*From:* [6].)

Certain services only require a UMTS bearer, while for other services like WAP a server like the MExE is essential. Traditional network operators keep complete control from inside their private network/service environment by providing services via their servers (MSC, Secure CoPy—SCP, HSS, MExE server, and SAT server) and the associated protocols (MAP, CAP, WAP, and SAT). This is depicted as case 1 in Figures 2.16 and 2.17.

The strength of OSA is to offer, via the OSA interfaces towards the SCSs, the opportunity for service providers to run their service logic on the application servers in their own domain while using the capabilities of the underlying networks via the operator's SCSs. This scenario is identified as case 2 in Figure 2.17. It enables flexible deployment of future innovative multimedia applications and services by third party providers.

In release 4 there are two call control elements: the MSC server for circuit-switched telephony services, and the CSCF or SIP server for VoIP and multimedia over IP (MMoIP) services delivery. The MSC server has been split into two parts; the server itself and the media gateway, thus separating control and transport. Since this split has no major functional impact, circuit-switched services are offered in the same way as in release 99 via a CAMEL platform (see Figure 2.16). The CSCF on the other hand introduces totally new multimedia capabilities and several solutions are possible. The services can be offered via the classic

IN/CAMEL service control via the network provider SCP. In the case of third party call control, the open standardized interface on top of the CSCF is used.

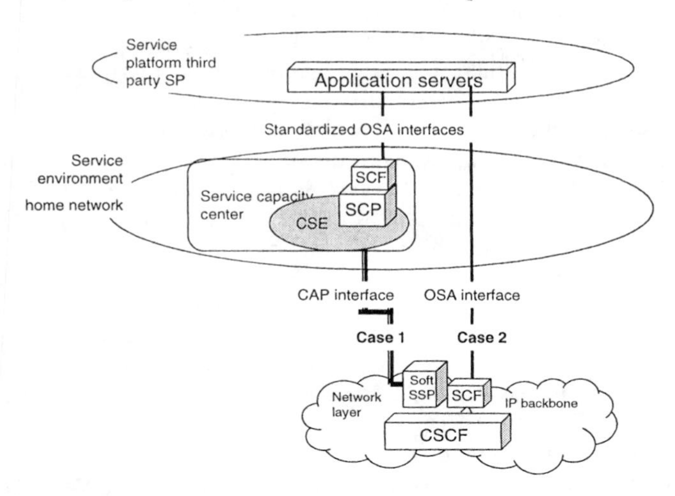

Figure 2.17 OSA options for network control by third-party service provider.

In the soft secure system provisioning (SSP) scenario, where the third party service providers can get access to the operator's network only through the central access point SCP, all applications for legacy as well as new IP services can be created via the proven CAMEL service creation environment (the CSE). The soft SSP maps the SIP call state model and the state model of the CAMEL service logic. In this soft SSP solution, the network operator and the actual capabilities of the CAMEL version as well as the actual implementation unfortunately limit third-party service providers. However, this solution allows operators to reuse their CAMEL investments from Release 99. Separation of the service logic from the SIP server is possible because SIP allows a third party to instruct a network entity to create and terminate calls to other network entities. The common gateway interface (CGI) and call processing language (CPL) can be used as well to provide such control. CGIs are meant for trusted users such as administrators, while CPL provides limited access to subscribers and third parties. All these servers running specific service logic can be interconnected via a distributed service platform such as the Common Object Request Broker Architecture (CORBA). In the example of Figure 2.6 the call control, the management, and communications between objects is achieved via a CORBA communications plane.

For research purposes, sometimes, it is best to consider a hybrid architecture for the service platform as depicted in Figures 2.16 and 2.17. Network providers are likely to possess both options as traditional operators will migrate from Release 99 to an all-IP network solution. New UMTS operators, who do not own legacy circuit-switched networks, adopt directly Release 4 all-IP architecture, and will use the third party call control approach. To analyze these solutions, from a network management standpoint, however, emphasis must be put on the impact these service provision solutions have on service and network management in the architecture components (in the case of Figure 2.6, these are MASPI, the resource management strategies and RMS, and UMD-NES).

For roaming users the capabilities supported by the terminal and the involved networks determine how personalized services are rendered. Home networks should compare the differences in the supported capabilities in the visited networks. Based on the comparison, the home network selects the most suitable environment and interfaces for service delivery. The concept of VHE is clearly nontransparent in terms of data flows between terminals, visited and home networks. The impact on service and network management must be assessed and integrated in the platform developments. For example, if the service required by the roaming terminal is a legacy service (telephony services such as prepaid) perfectly supported by CAMEL and if the visited network supports the right version of CAMEL, the home network may decide to delegate call control to the visited network. In the case of a third party service provider, the service logic (provided via the OSA interface on top of the CAMEL SCP in the home network) communicates with the call control through CAP, the home and visited networks. For new VoIP and MMoIP services that cannot be supported by the visited network CAMEL capabilities, the home network may decide to handle call control from the home network itself. In this case the service logic of a third-party service provider communicates with the call control in the home network by means of the OSA interface directly on top of the CSCF in the home network.

To provide integrated legacy and new VoIP/MMoIP services, operators must implement OSA interfaces on top of SCPs and OSA interfaces directly on top of CSCFs to ensure optimal delivery on the basis of the VHE. In addition this solution enables both network providers and third party service providers to offer both mobile and fixed subscription. Consequently, the project must consider the VHE concept as an enabler of wireless multimedia and personalized service portability across UMTS networks and terminal boundaries. For systems not embracing the VHE concept, interworking units and functions will be required to offer service continuity with or without service portability.

UMTS Management System

The UMTS reference architecture presented in Figure 2.18 is based on a complete IP solution deployed in the core network. This architecture corresponds to a mainstream IP network capable of offering Mobile IP+ service. In this architecture MIP+ handles intra-CN mobility and inter-PLMN mobility, as well as intersystem mobility in the packet domain. The functionality of the serving GPRS support node (SGSN) and GGSN are combined into one node, the Internet GPRS support node (IGSN). Functionality is added in the IGSN to use Mobile IP for inter-IGSN mobility. The IGSN/FA represents the end and entry points of the UMTS specific part of the PLMN. During a transition period, the IGSN should be able to act as an SGSN and a GGSN to allow compatibility with UMTS/GPRS networks.

Figure 2.18 shows the resource management policies, systems, and QoS strategies proposed respectively in the access and core network segments.

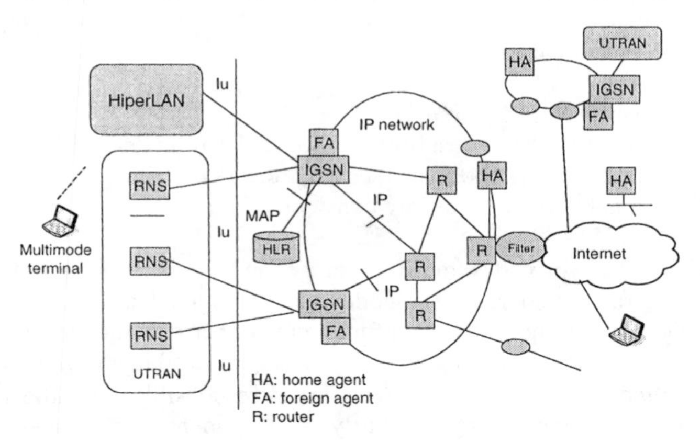

Figure 2.18 UMTS reference architecture: CN architecture with Mobile IP MM within the CN and between different systems and GPRS PLMNs.

In general, the following high-level split applies:

- *Access network.* Resource management policies affect the radio interface and the fixed infrastructure between nodes-B and SGSN1 routers in the core network. Call admission techniques must be used on the radio interface to comply with the specific constraints exhibited by a WCDMA interface. Resources available between nodes-B and SGSN (thus including RNCs) must be handled in a different manner. The following paragraph will outline some issues to be considered.
- *Core network.* In an all-IP core network infrastructure resources are handled using policies and techniques derived from IETF recommendations duly adapted to the mobile context. 3GPP specifications provide some guidelines, but at this level many initiatives are left to manufacturers' inventiveness.

The following sections analyze the UMTS access and core segments from the perspective of introducing the network elements subject to management actions as well as policies and protocols recommended by standardization bodies.

It is worthwhile mentioning that, especially at the core network level, manufacturers have the autonomy to propose and implement proprietary solutions to optimize resources. In this area, research can contribute to algorithms, protocol stacks, and QoS policies.

UMTS Access Network

This section introduces the nodes and the interfaces included in the access network segment. It is not meant to provide accurate insights into resource management policies or network nodes control, but rather give a high-level description of these elements.

Access Network Nodes and Interfaces

The access network includes the radio interface and any node and link between nodes-B and the core network interface (see Figure 2.19).

The following network elements are considered here:

- *Node-B.* This provides radio coverage and cooperates with RNCs in handling radio resources. The node-B converts the data flow between the Iub and Uu interfaces: its main function is to perform the air interface L1 processing (channel coding and interleaving, rate adaptation, spreading
- *Radio network controller (RNC).* This is responsible for radio resource management and users' mobility management (soft handover). It interfaces the CN and also terminates the Radio Resource Control (RRC) protocol that defines the messages and procedures between the mobile and the access network. The interfaces considered are Iub (between node-B and RNC), Iur (between RNCs), and Iu (between RNC and core network). RNC is a complex node especially in a mature UMTS scenario where IP will be used as the unique layer 3 protocol. RNC also hosts a functionality essential to guarantee services like VoIP. Figure 2.20 shows an example of RNC functional architecture.

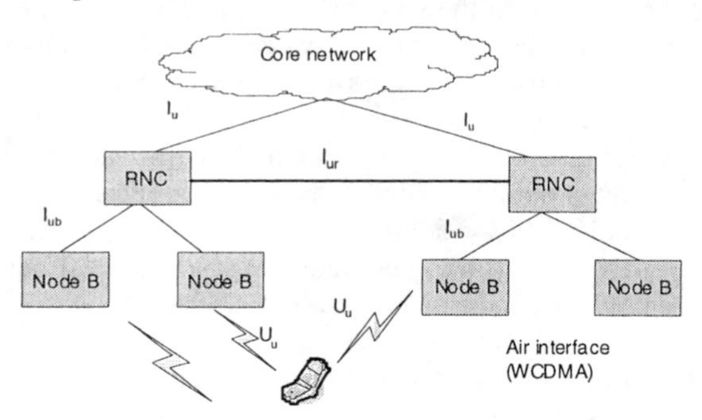

Figure 2.19 UMTS access network.

The radio network control server (RNCS) manages signaling issues between the mobile terminals and the core network. Handheld terminals will not have H.323 functionality and therefore it is necessary to handle UMTS-specific signaling streams towards proxy servers located in the core network border. RNCS will therefore act as a signaling interworking unit between the access and the core network.

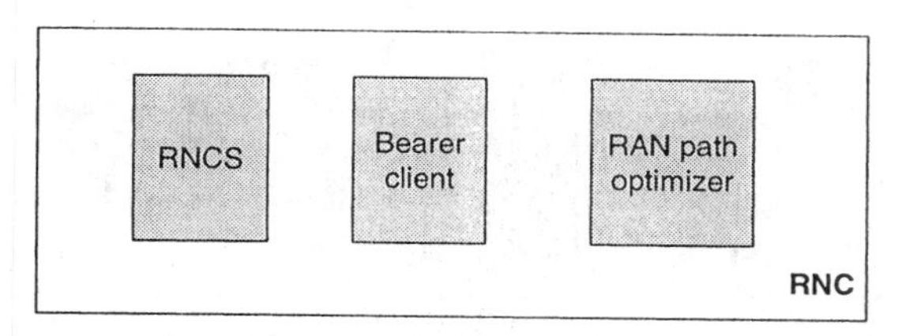

Figure 2.20 RNC functional architecture.

The bearer client plays a proxy role between the access and the core network as far as the bearer stream is concerned. Note that this function is responsible for running Real-Time Transport Protocols (RTPs) when users invoke VoIP services. It is, therefore, a bearer interworking unit between the access and the core segment.

Finally, the RAN path optimizer cooperates with a similar node in the core network segment to handle soft handover issues. Path optimizers at both the access and core levels, are essential to reconfigure the signaling and bearer path between RNCs and SGSN whenever a mobile terminal leaves the serving RNC and moves towards a new RNC.

IP functionality is managed introducing control protocols widely defined by ITU and IETF. H.323, in particular, is used in any gateway control transaction. At the border between the access and the core network, the SGSN plays a role of proxy. The SGSN is also responsible for managing QoS policies at the border between the access and core network. These policies were described when the core network management system was introduced.

Radio Resource Management

UMTS offers a wide variety of services ranging from messaging to speech, data and video communications, and high bit rate communication up to 2 Mbps. UTRA is based on two different radio access schemes: WCDMA and time-division code division multiple access (TD-CDMA). In the paired bands (with FDD duplexing) the system adopts the WCDMA access scheme, while in the unpaired bands (with TDD duplexing) the system adopts the TD-CDMA access scheme. In CDMA systems the topic of radio resource management is more complex when related to the case of TDMA/FDMA system, as with GSM. In fact, in a system like GSM channels available to users are fixed, and when these channels are all allocated there is no way to give service to another user (hard capacity limitation).

In CDMA systems all users share the same bandwidth at the same time, and the multiplexing is carried out using orthogonal codes. Consequently, power control is very important because all the communications in the same bandwidth must arrive at the receiver with the same power level, and so users are not admitted to transmit at a too high power level. In power-controlled systems such as CDMA systems, the number of users that each cell can support is limited by the total interference figure received at each base station and it varies with time. In fact, CDMA systems are interference-limited systems by definition.

Consequently, the estimate of the cell capacity is based on S/I ratio and both the uplink and downlink cases should be considered. When a system is congested, admitting a new call can only make the link quality worse for ongoing calls and may result in call dropping. Thus, a CDMA system needs a call admission policy for new call requests to maintain acceptable connections for existing users. In other words, in CDMA networks, the "soft capacity" concept applies: each new call increases the interference level of all other ongoing calls, affecting their quality. Therefore it is very important to control the access to the network in a suitable way [e.g., by means of call admission control (CAC)].

If the air interface loading is allowed to increase excessively, the coverage area of the cell is reduced below the planned value, and the quality of service of the existing connections cannot be guaranteed. Before admitting a new connection, admission control needs to check that the admittance will not sacrifice the planned coverage area or the quality of the existing connections. Admission control accepts or rejects a request to establish a radio access bearer in the radio access network. The admission control algorithm is executed when a bearer is set up or modified. The admission control functionality is located in the RNC, where the load information from several cells can be obtained. The admission control algorithm estimates the load increase that the establishment of the bearer would cause in the radio network. This has to be estimated separately for the uplink and downlink directions. The requesting bearer can be admitted only if both uplink and downlink admission control admit this, otherwise it is rejected because of the excessive interference that it would produce in the network. The radio network planning sets the limits for admission control.

It is possible to identify two phases into the call admission control process:

- **Phase 1:** Admission control is performed according to the type of required QoS. "Type of service" is to be understood as an implementation-specific category derived from standardized QoS parameters (3GPP TS 23.107—QoS Concept and Architecture). So, the first step is to map QoS parameters in "Type of service." Table 2.3 illustrates this concept.

The meaning of type of service classes is as follows:

  - Premium Service: service with low delay (high priority).
  - Assured Service: service with a minimum rate below the mean rate guaranteed and that may use more bandwidth if available (medium priority).
  - Best Effort: service with no guaranteed QoS (low priority).

- **Phase 2:** Admission control is performed according to the current system load and the required service. The call should be blocked if none of the suitable cells can efficiently provide the service required by the UE at call setup (i.e., considering the current load of the suitable cells, the required service is likely to increase the interference level to an unacceptable value). This would ensure that the UE avoids wasting power affecting the quality of other communications. In this case, the network can initiate a renegotiation of resources of the ongoing calls in order to reduce the traffic load.

Table 2.3

Type of Service and QoS Parameters

Service	Domain	Transport Channel	Type of Service	CAC Performed
Voice	CS	DCH	Premium	Yes
	IP	DCH	Premium	Yes
Web	IP	DSCH	Assured service	Yes
	IP	DSCH	Best effort	No

It is worthwhile considering the meaning of a common terminology used in UMTS, regarding the different roles that an RNC (or an RNS) can assume in UMTS. In particular, the following three different roles are identified.

The controlling RNC term identifies a role that an RNC can take with respect to a specific set of node-Bs. There is only one controlling RNC for any node-B. The controlling RNC has the overall control of the logical resources of its node-B's.

The serving RNC term identifies a role that an RNC can take with respect to a specific connection between a UE and UTRAN. There is one serving RNC for each UE that has a connection to UTRAN. The serving RNC is in charge of the radio connection between a UE and the UTRAN. The serving RNC terminates the Iu for this UE.

The drift RNC term identifies a role that an RNC can take with respect to a specific connection between a UE and UTRAN. An RNC that supports the serving RNC with radio resources when the connection between the UTRAN and the UE needs to use cell(s) controlled by this RNC is referred to as drift RNC. In particular, the role of the drift RNC is related to the concepts of soft handover and macro diversity.

In the following, some simple CAC scenarios can be described, starting from the assumption that admission control is performed by the CRNC controlling under request from the SRNC (serving). Figure 2.21 is a possible example.

It describes the general scheme that involves admission control when no Iur is used and the CRNC takes the role of SRNC.

The various steps that constitute the communication related to the mechanism of CAC are described here:

- The CN requests SRNC to establish a RAB indicating QoS parameters.

- The RRM entity assigns a type of service to the requested service according to QoS parameters.
- CAC is performed according to the type of service.
- The resources are allocated according to the result of CAC.
- The acknowledgment is sent back to CN according to the result of CAC. The sublayers are configured according to the decision taken before.

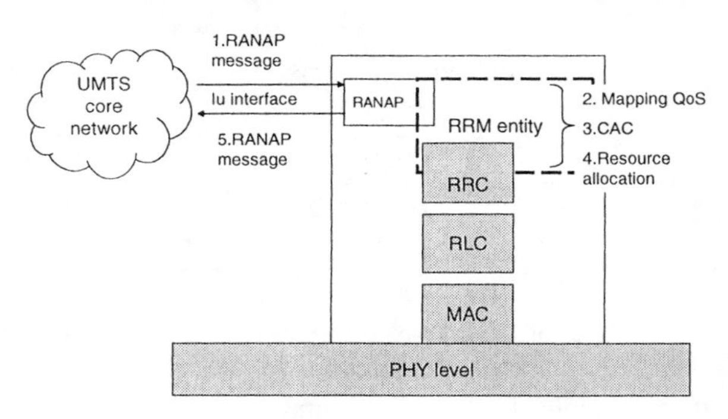

Figure 2.21 Functionality involved in a CAC mechanism.

It is useful to note that steps 2 to 4 may also be triggered by the SRNC for reconfiguration purposes within the SRNC (intra-RNC handovers, channel reconfigurations, location updates). Other scenarios may involve a drift radio network controller (DRNC): if a radio link is to be set up in a node-B controlled by another RNC than the SRNC, a request to establish the radio link is sent from the SRNC to the DRNC. CAC is always performed in the CRNC, and in this case the Iur interface (interface between different RNCs) must be used, so this time the DRNC takes the role of CRNC.

UMTS Core Network QoS Architecture

As described in the previous paragraphs, the UMTS core network is assumed to be IP-based.

IP networks do not have a native mechanism to guarantee consistent, predictable data delivery service. IP provides what is called a best-effort service: it can make no guarantees about when data will arrive, or how much it can deliver. This limitation has not been a problem for traditional Internet applications like Web browsing, e-mail, and file transfer. But the new breed of applications, including audio and video streaming, demand high data throughput capacity (bandwidth) and have low latency requirements when used in two-way data communications (real-time applications like conferencing and telephony). In other words, it is necessary to implement a mechanism to provide QoS.

QoS is the ability of a network element (e.g., an application, host, or router) to have some level of assurance that its traffic and service requirements can be satisfied. QoS with a guaranteed service level requires resource allocation to individual data streams. There are essentially two types of available QoS mechanisms [9], namely:

- *Integrated services:* network resources are reserved according to an application's QoS request, which can be executed on a per flow basis, and subject to bandwidth management policy. RSVP, for example, provides the mechanisms to do IntServ.
- *Differentiated services:* network traffic is classified—according to bandwidth management policy criteria—into several classes of service. The network elements will be in charge of providing a different traffic shaping according to the class of service. To enable QoS classification gives preferential treatment to applications identified as having higher priority. DiffServ provides this service.

These QoS protocols and algorithms are not competitive or mutually exclusive; on the contrary, they are complementary. As a result, they are designed for use in combination to accommodate the varying operational requirements in different network contexts. Table 2.4 shows a complete picture of how the QoS technologies (IntServ and DiffServ) can work together to provide "end-to-end QoS" [10].

Table 2.4

IntServ Versus DiffServ Techniques

Features/Network Technique	Advantages	Disadvantages
IntServ	Standardized end to end (E2E) signaling protocol (RSVP)	Not scalable and complex because of the need of core routers to maintain control state for each flow
DiffServ	Scalable (focuses on traffic aggregates with similar service requirements) Requested services offer availability of resources E2E	No signaling protocol to check the availability of resources E2E Not suitable for E2E QoS
IntServ in the access and DiffServ in the core	Scalable (DiffServ) Standarized IntServ/RSVP interdomain signaling protocol	Common agreement needed between ISPs on the use of IntServ/RSVP for interdomain signaling

IntServ provides resources for network traffic, whereas DiffServ simply marks and prioritizes traffic. IntServ is more complex and demanding than DiffServ in terms of interrouter signaling overhead requirements, so it can negatively impact backbone routers. This is why the best common practice says to limit IntServ's (e.g., RSVP) use on the backbone, and why DiffServ can exist there.

DiffServ is a perfect complement to IntServ as the combination can enable end-to-end QoS. End hosts may use IntServ requests with high granularity (e.g., bandwidth, jitter threshold). Border routers at backbone ingress points can then map those IntServ reservations to a class of service. At the backbone egress point, the IntServ provisioning may be honored again, to the final destination.

Ingress points perform traffic conditioning to assure that SLAs are satisfied. The architecture represented in Figure 2.22 has momentum and support.

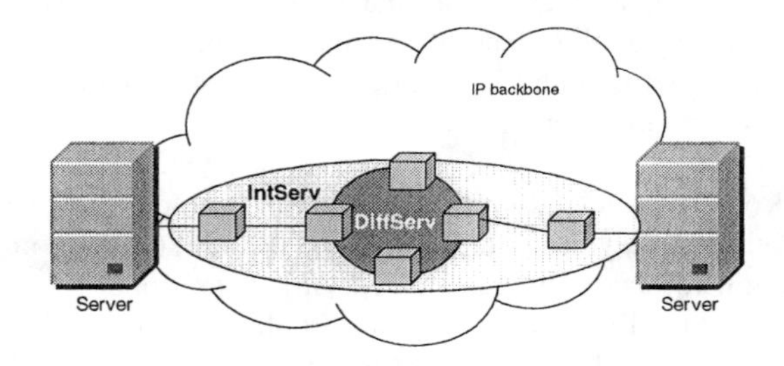

Figure 2.22 IntServ in the access/edge and DiffServ in the core.

The accomplishment of complex sections involving various media—possibly requiring different QoS degrees—implies the cooperation between session management mechanisms and QoS mechanisms. Moreover, the introduction of better-than-best-effort transport services calls for some policy mechanism able to monitor and verify the access to these services by allowed users only and to ensure fair and accurate accounting and billing.

It could be possible to envisage a three-layer QoS aware architecture depicted in Figure 2.23.

The policy and billing layer decides how to configure network resources to realize QoS according to policies and user profile information (at this level AAA mechanisms are realized, e.g., using RADIUS); moreover it provides a mechanism to manage billing, policy, and user profile information.

The session control layer takes care of the establishment and the maintenance of the end-to- end QoS data flow.

The QoS-aware network layer" enforces the QoS establishment on the network devices on the basis of the decisions taken by the policy and billing layer.

UMTS Radio Network Planning

This section presents some aspects relevant to UMTS radio network planning, including dimensioning, capacity and coverage planning, and network optimization. Such a radio network planning process is shown in Figure 2.24 [11].

The first step of the process is to roughly estimate the number of base station sites and their configurations needed to fulfill the operator's requirements in terms of coverage, capacity, and QoS. Capacity and coverage are strictly related in WCDMA networks and, therefore, both must be considered simultaneously in the dimensioning process.

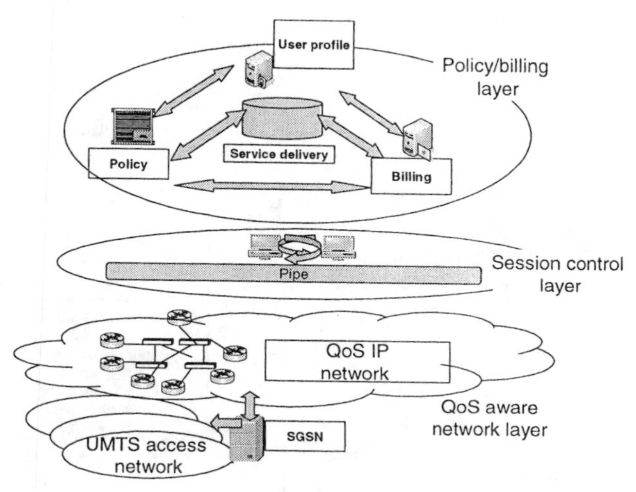

Figure 2.23 Layer QoS-aware architecture.

When the network is in operation, its performance can be observed by measurements, and the results can be used to monitor and optimize network performance; in this context it is clear that planning and management (optimization) processes are interdependent. In particular, since the planning phase requires traffic data estimates in each area, and this data is often unstable and not precise enough, the resource management strategy plays a fundamental role, especially for controlling the network load and avoiding a congestion situation.

Network Dimensioning

Radio network dimensioning is a process through which possible configurations and the amount of network equipment are estimated on the basis of operator's requirements related to:

- Coverage;
- Coverage regions;
- Environment information;
- Propagation conditions;
- Capacity;

- Available spectrum;
- User's growth forecast;
- Traffic density information;
- QoS;
- Coverage probability;
- Blocking probability;
- End user throughput.

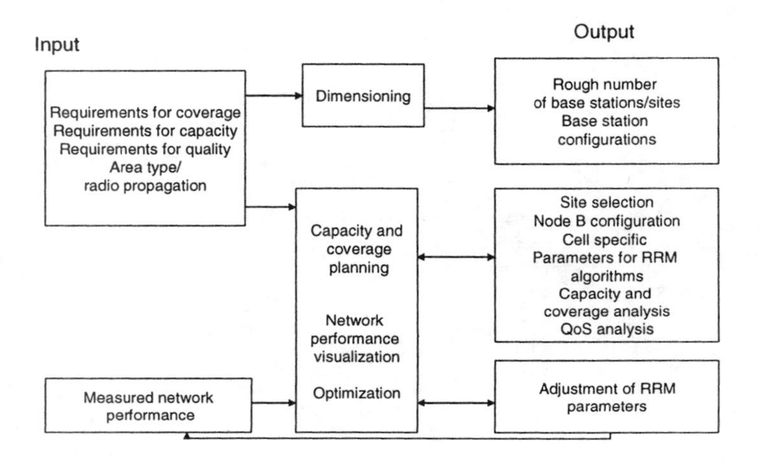

Figure 2.24 UMTS radio planning process.

Dimensioning activities include radio link budget calculations and coverage analysis, capacity estimation, and, estimations of the amount of sites, RNCs, and core network elements (i.e., circuit-switched and packet-switched domain core network).

The radio link budget is used to determine the maximum allowed path loss: this could be defined as the maximum loss allowed between the MS and the BS while maintaining a connection satisfying the required QoS for the user in question and all the other users within the maximum cell range. The radio link budget should be considered for both the uplink and the downlink: assuming an uplink link budget at the receiver end, the base station should receive the user's signal with specific E_b/N_0 requirements in order to maintain the QoS.

The link budget takes into account many different parameters, some of them more related to the coverage aspect, others more related to the capacity analysis (see Table 2.5).

Table 2.5

Link Budget Input Parameters

Coverage Analysis	Capacity Analysis
E_b/N_0 requirement/service	E_b/N_0 requirement/service
Geographic environment	Bearer rate/service
Coverage reliability	f-factor
Uniform cell loading	Service activity factor
Penetration losses	Target interference margin
System/MS/BS parameters (chip rate, soft handover gain, power control, body losses)	Processing gain per service
Propagation model	Outage probability

Besides, there are some technology specific parameters. These are described here for WCDMA:

- *Interference margin (or noise rise):* This takes into account the cell loading, which affects the coverage. The more loading is allowed in the system, the larger is the interference margin needed in the uplink, and the smaller is the coverage area.
- *Fast fading margin (power control headroom):* Some margin is needed in the mobile station transmission power for maintaining adequate closed-loop fast power control. This applies especially to slow-moving pedestrian mobiles, where fast power control is able to effectively compensate the fast fading.
- *Soft handover gain:* Handovers, soft and hard, give a gain against slow fading (log-normal fading) by reducing the required log-normal fading margin. This is because the slow fading is partly uncorrelated between base stations, and by making handover the MS can select a better BS. Soft handover gives an additional macro diversity gain against fast fading by reducing the required E_b/N_0 relevant to a single radio link, due to the effect of macro diversity combining.

For further details on the typical parameters used in link budgets, please refer to [12].

Following the link budget approach, the cell range R can be easily calculated for a given propagation model, for example, Okumura-Hata. The propagation model describes the average signal propagation in a specific environment, converting the maximum allowed propagation loss in decibels to the maximum cell range in kilometers. Once the cell range is determined, the site area, which is also a function of the base station sectorization configuration, can be derived.

For example, for a cell with hexagonal shape covered by an omni-directional antenna, the coverage area can be approximated as $2.6R^2$.

Capacity

The radio link budget that is used to calculate the maximum allowed path loss puts constraints on the maximum cell radius. The link budget includes parameters,such as the interference margin (noise rise), related to the capacity of the system: for this reason the second step of dimensioning is estimating the amount of supported traffic per base station site, in other words, the capacity of the system. In this section the relationship between noise rise and capacity will be derived.

Uplink Loading Factor

From the uplink point of view, the BS should receive the MS signal with a given strength, usually quantified by the required E_b/N_0. This value is affected by the path loss, thus the transmission power needs to be accommodated. In WCDMA, the noise rise represents the amount of interference generated by users in a cell. Furthermore, interference from out-of-cell must be taken into account. The noise rise can be derived from the loading factor of the cell, usually called h [11].

The first step in this derivation is to define the E_b/N_0, the energy per user bit divided by the noise spectral density:

$$\frac{E_b}{N_{0j}} = P_G \; of \; user \; j \times \frac{Signal \; of \; userj}{Total \; RX \; power \, (excluding \; own \; signal)} \tag{2.1}$$

This can be written (P_G is the system processing gain):

$$\frac{E_b}{N_{0j}} = \frac{W}{SAF_j \times R_j} \times \frac{P_j}{(I_{total} - P_j)} \tag{2.2}$$

where W is the chip rate, P_j is the received signal power of user j, SAF_j is the activity factor of user j; R_j is the bit rate of user j, and I_{total} is the total received wideband power including thermal noise power in the base station (indicated as P_N). Solving equation (2.2) for P_j and defining the noise rise as the ratio of the total received wideband power to the noise power

$$NR_{UL} = \frac{I_{total}}{P_N} \tag{2.3}$$

it is possible to obtain

$$NR_{UL} = \frac{I_{total}}{P_N} = \frac{1}{1 - \sum_{j=1}^{N} \frac{1}{1 + \frac{W}{\left(\frac{E_b}{N_0}\right)_j \times R_j \times SAF_j}}} = \frac{1}{1 - h_{UL}} \tag{2.4}$$

with N being the users in the same cell and the uplink loading factor h_{UL} defined as:

$$h_{UL} = \sum_{j=1}^{N} \frac{1}{1 + \frac{W}{\left(\frac{E_b}{N_0}\right)_j \times R_j \times SAF_j}} \tag{2.5}$$

When h_{UL} becomes close to 1, the corresponding noise rise approaches infinity and the system reaches its pole capacity; this means that the MS should transmit infinite power to be correctly received by the BS.

Additionally, in the loading factor, the interference from the other cells must be taken into account by the ratio of other cells to own cell interference, f:

$$f = \frac{\text{other cells inteference}}{\text{own cells inteference}} \tag{2.6}$$

The uplink loading factor can be then be written as

$$h_{UL} = (1 + f)\sum_{j=1}^{N} \frac{1}{1 + \frac{W}{\left(\frac{E_b}{N_0}\right)_j \times R_j \times SAF_j}} \tag{2.7}$$

The noise rise is equal to $10\log10(1 - h_{UL})$: the interference margin in the link budget computation must be equal to the maximum planned noise rise.

The required E_b/N_0 can be derived from link level simulations and from measurements. It includes the effect of the closed loop power control and soft handover. The f-factor is a function of cell environment and cell isolation (e.g., macro/micro, urban/suburban) and antenna pattern (e.g., omni, 3-sector, or 6-sector). The parameters are further explained in Table 2.6.

Table 2.6

Parameters Used in Uplink Loading Factor Calculation

Symbol	Definitions	Recommended Values
N	Number of users per cell	—
SAF	Service activity factor	0.6 for speech, assuming 50% voice activity and DPCCH overhead during DTX 1.0 for data
E_b/N_0	Signal energy per bit divided by noise spectral density (required to meet a predefined QoS). Noise includes both thermal noise and interference	Depends on service, bit rate, multipath fading channel, receive antenna diversity, mobile speed
W	WCDMA chip rate	3.84 Mcps
R	Service bit rate	Depends on service
F	Other cells to own cell interference ratio seen by the base station receiver	Macro cell with omni-directional antennas: 55%

The load equation is commonly used to make a semi-analytical prediction of the average capacity of a WCDMA cell. For a classical all-voice service network, where all N users in the cell have a bit rate R, it is possible to note that:

$$\frac{W}{\left(\frac{E_b}{N_0}\right)R \times SAF} >> 1 \tag{2.8}$$

and the previous uplink load equation can be approximated and simplified to

$$h_{UL} = N \times (1+f) \times SAF \times \frac{R}{W} \times \frac{E_b}{N_0} \tag{2.9}$$

It can be easily verified that a noise rise of 3 dB corresponds to a 50% loading factor, and a noise rise of 6 dB corresponds to a 75% loading factor. Typical planning values for the loading factor are in the range of 60% to 70%.

In a mixed service scenario, (2.2) becomes more complex, as M different services need to be taken into account. The following equation shows the general form for the uplink loading factor for a mix of services:

$$h_{UL} = \sum_{m=1}^{M} N_{01} \times (1+f) \times SAF_m \frac{R_{01}}{W} \times \frac{E_b}{N_{0m}} \quad (2.10)$$

To find the M solutions of this equation it is necessary to use other $M-1$ equations; these equations could be, for example, the relationships between the number of users of each service with respect to a reference service (usually speech):

$$\frac{N_{01}}{N_1} = k_m \quad (2.11)$$

Downlink Loading Factor

The downlink loading factor, h_{DL}, can be defined by following the same principle as for the uplink, although the parameters are slightly different (see Table 2.7) [13]:

$$h_{DL} = \sum_{j=1}^{N} Nn[(1-N_j)+f_j]SAF_j \frac{R_j}{W} \frac{E_b}{N_{0j}} \quad (2.12)$$

where $-10\log10\ (1-h_{DL})$ is equal to the noise rise over thermal noise due to multiple access interference. The other parameters are explained in Table 2.7.

Compared to the uplink load equation, the most important new parameter is α_j, which represents the orthogonality factor in the downlink. WCDMA employs orthogonal codes in the downlink to separate users, and, without any multipath propagation, the orthogonality remains when the mobile receives the BS signal.

For a sufficient delay spread in the radio channel, the mobile will see part of the BS signal as multiple access interference. The orthogonality of 1 corresponds to perfectly orthogonal users. Typically, the orthogonality is between 0.4 and 0.9 in multipath channels.

In the downlink, the ratio of other cells to the own cell interference, f, depends on the user location and is therefore different for each user j. The effect of soft handover transmission can be modeled as having additional connections in the cell. The soft handover overhead is defined as the total number of connections divided by the total number of users minus one. At the same time the soft handover gain relevant to the single E_b/N_0 is taken into account. The gain, called macro diversity combining gain, can be derived from simulation analysis and is measured as the reduction in the required E_b/N_0 for each user.

The downlink loading factor, h_{DL}, presents very similar behavior to the uplink loading factor, h_{UL}, in the sense that when approaching unity, the system reaches its pole capacity and the noise rise over thermal goes to infinity.

Table 2.7

Parameters Used in Downlink Loading Factor Calculation

Symbol	Definitions	Recommended Values
N	Number of connections per cell=number of users per cell (1+soft handover overhead)	
SAF	Service activity factor	0.6 for speech, assuming 50% voice activity and DPCCH overhead during DTX 1.0 for data
E_b/N_0	Signal energy per bit divided by noise spectral density (required to meet a predefined QoS). Noise includes both thermal noise and interference	Depends on service, bit rate, multipath fading channel, receive antenna diversity, mobile speed
W	WCDMA chip rate	3.84 Mcps
α_j	Channel orthogonality of user *j*	0.4 and 0.9 in multipath environment
R_j	Bit rate of user *j*	Depends on service
f_j	Ratio of other cells to own cell interference received by the user *j*	Each user sees a different f_j depending on its location in the cell and log-normal shadowing

In both the uplink and the downlink, the air interface load affects the coverage, but the effect is not exactly the same: in the downlink the coverage depends more on the load than in the uplink.

The reason is that, in the downlink, the maximum transmission power is the same regardless of the number of users and is shared between the downlink users, while in the uplink each additional user has its own power amplifier. Therefore, even with low load in the downlink, the coverage decreases as a function of the number of users. In general, the coverage is limited by the uplink, while capacity is limited by the downlink loading factor.

Soft Capacity

Another important characteristic of interference-limited networks is the soft capacity concept. A CDMA system suffers from two types of limitations, soft and hard. On one hand, the capacity is hard blocked by the amount of hardware; on the other hand, the maximum capacity is limited by the amount of interference on the air interface: this is the definition of soft capacity, since there is no single fixed value for the maximum capacity. For a soft capacity limited system, the classical Erlang capacity cannot be calculated from the Erlang B formula, since it would give too pessimistic results. The total channel pool is larger than just the average number of channels per cell, since the adjacent cells share part of the same interference, and therefore more traffic can be served with the same blocking probability. The soft capacity can be explained as follows. The less interference is coming from the adjacent cells, the more channels are available in the middle cell; with a low number of channels per cell, the average loading must be quite low to guarantee low blocking probability. Since the average loading is low, there is typically extra capacity available in the neighboring cells. This capacity can be borrowed from the adjacent cells; therefore, the interference sharing gives soft capacity.

Referring to these concepts, WCDMA soft capacity is defined as the increase of Erlang capacity with soft blocking compared to that with hard blocking with the same maximum number of channels per cell on average with both soft and hard blocking:

$$\textit{soft capacity} = \frac{\textit{Erlang capacity with soft blocking}}{\textit{Erlang capacity with hard blocking}} - 1 \tag{2.13}$$

The wideband power-based admission control strategy gives soft blocking and soft capacity.

Uplink soft capacity can be approximated based on the total interference at the base station; this includes both own cell and other cell interference. Therefore, the total channel pool can be obtained by multiplying the number of channels per cell in the equally loaded case by $1+f$.

The procedure for estimating the soft capacity is summarized as follows:

1. Calculate the number of channels per cell, N, in the equally loaded case based on the uplink loading factor.
2. Multiply that number of channels for $1+f$ to obtain the total channel pool in the soft blocking case.
3. Calculate the maximum offered traffic from the Erlang B formula.
4. Divide the Erlang capacity by $1+f$.

To better explain this approach, a few numerical examples are given in Table 2.8, with the assumptions shown in Table 2.9.

Table 2.8

Assumptions in Soft Capacity Calculations

Parameter	Value
Bit rate	Speech: 12.2 Kbps Data: 64.144 kbps
SAF	Speech: 67% Data: 100%
E_b/N_0	Speech: 4 dB Data: 2 dB (64 Kbps), 1.5 dB
Noise rise	3 dB (=50% loading factor)
Blocking probability	2%
f	0.55

Table 2.9

Soft Capacity Calculation in the Uplink

	Channels Per Cell	Hard Blocked Capacity	Soft Blocked Capacity	
12.2	60.5	50.8 Erl	53.5 Erl	5%
64	12.5	7.0 Erl	8.2 Erl	17%
144	6.4	2.5 Erl	3.2 Erl	28%

Network Optimization

Network optimization is a process to improve the overall network quality as experienced by the mobile users and to ensure that network resources are used efficiently. Optimization includes the analysis of the network and improvements in the network configuration and performance. Statistics of the key performance indicators for the operational network can be used to feed a network status analysis tool, and the radio resource management parameters can be tuned for better performance. An example of an optimization problem is to optimize the soft handover areas. The first phase of the optimization process is to define the key performance indicators. These consist of measurements in the network management system and of field measurement data, or any other information that can be used to determine the QoS of the network. With the help of the network management system, it is possible to analyze the past, present, and predicted future network performance.

The performance of the radio resource management algorithms and their parameters can be analyzed using their key performance indicator results. The RRM includes handovers, power control, packet scheduling, admission, and load control.

Automatic optimization is important for 3G mobile networks, since there are many more services and bit rates than in second generation networks and manual optimization would be too time-consuming; in particular, automatic adjustment should provide a fast response to the changing traffic conditions.

UMTS Radio Resource Management

Interference-Based Radio Resource Management

RRM is responsible for utilization of the air interface, in order to guarantee QoS, to maintain the planned coverage area and to offer high capacity. RRM can be divided into:

- Handover;
- Power control;
- Admission control;
- Load control.

Power control is needed to keep the interference levels at a minimum in the air interface and to provide the required quality of service. Handovers are needed in cellular systems to handle the users' mobility when moving from the coverage area of one cell to another.

In 3G networks, other RRM algorithms (admission and load control) are required to guarantee the quality of service and to maximize the system throughput with a mix of different services, bit rates, and quality requirements.

The RRM algorithms can be based on the amount of available hardware in the network (hard blocking) or on the interference levels in the air interface (soft blocking). Hard blocking is defined as the case where the hardware limits the capacity before the air interface load is estimated to be above the planned limit; if soft blocking RRM is applied, the air interface load needs to be measured. It has been shown previously that this second method gives a higher capacity.

Typical locations of the RRM algorithms in a WCDMA network are shown in Figure 2.25.

Measurement of the Air Interface Load

If the RRM is based on the interference levels in the air interface, the air interface load needs to be measured. In this section two load measures are presented: load estimation based on wideband received/transmitted power and load estimation based on throughput. These are both examples that could be used in WCDMA networks.

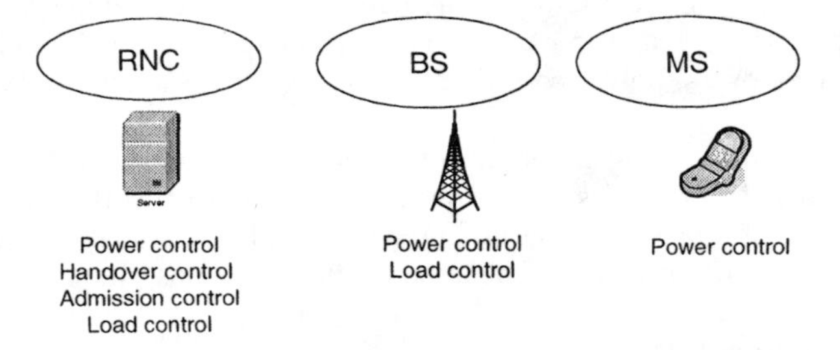

Figure 2.25 Typical locations of RRM algorithms in a WCDMA network.

Power-Based Load Estimation

The wideband received power level can be used to estimate the uplink load; such power level can be measured at the BS: starting from these measurements, the uplink load factor can be obtained as shown next.

The received wideband interference power, I_{total}, can be split into the power of own cell users (I_{intra}), other cells (I_{inter}) and background noise P_N:

$$I_{total} = I_{intra} + I_{inter} + P_N \tag{2.14}$$

The uplink noise rise is defined as the ratio of the total power and the noise power:

$$Noise\ rise = \frac{I_{total}}{P_N} = \frac{1}{1 - h_{UL}} \tag{2.15}$$

This equation can be rearranged to give the uplink load factor:

$$h_{UL} = 1 - \frac{P_N}{I_{total}} = \frac{Noise\ rise}{Noise\ rise - 1} \tag{2.16}$$

where I_{total} can be measured by the BS and P_N is known beforehand. The uplink load factor h_{UL} is normally used as the uplink load indicator. In a similar way, the downlink load factor of a cell can be determined by the total downlink transmission power, P_{total}. The downlink load factor h_{DL} is defined as the ratio of the current total transmission power divided by the maximum BS transmission power P_{max}:

$$h_{DL} = \frac{P_{total}}{P_{max}} \tag{2.17}$$

Throughput-Based Load Estimation

As shown before in Equation (2.7), the uplink load factor can be calculated as the sum of the load factors of the users connected to the BS:

$$h_{UL} = (1+f)\sum_{j=1}^{N} \frac{1}{1+\dfrac{W}{\left(\dfrac{E_b}{N_0}\right)_j \times R_j \times SAF_j}}$$

In dimensioning, where the average number of users N of a cell needs to be estimated, while the others are input parameters; these values are typical for that environment and can be based on measurements and simulations. In load estimation the instantaneous measured value for $\frac{E_b}{N_0}$, I, SAF, and N are used to estimate the instantaneous air interface load.

In the downlink, throughput-based load estimation can be affected by using the sum of the downlink allocated bit rates as the downlink load factor, as follows:

$$h_{DL} = \frac{\sum_{j=1}^{N} R_j}{R_{max}} \tag{2.18}$$

where N is the number of downlink connections, including common channels, R_j the bit rate of the user j, and R_{max} is the maximum allowed throughput of the cell.

Comparison of Uplink Load Estimation Methods

In the wideband power-based method, interference from the adjacent cells is directly included in the load estimation because the measured wideband power value includes all interference that is received in that carrier by the BS. If the loading in the adjacent cells is low, it can be detected with this approach and a higher load can be allowed in the cell (i.e., soft capacity can be obtained). This method keeps the coverage within the planned limits, and the delivered capacity depends on the loading factor in the adjacent cells: such an approach effectively prevents cell breathing that would exceed the planned values. The problem with wideband-based load estimation is that the measured power can include interference from external sources, thus overestimating the load. Throughput-based load estimation does not take interference from other cells directly into account. If soft capacity is required, information about the adjacent cell loading can be retrieved from the RNC. The throughput-based RRM keeps the throughput of the cell at the planned level. If the loading in the adjacent cells is high, this affects the coverage area of the cell.

Admission Control

If the air interface loading is allowed to increase excessively, the coverage area of the cell is reduced below the planned value (cell breathing phenomenon), and the QoS of the existing connections cannot be guaranteed. Before admitting a new connection, admission control needs to check that the admittance will not sacrifice the planned coverage area or the quality of the existing connections. Admission control accepts or rejects a request to establish a radio access bearer in the radio access network; it is executed when a bearer is set up or modified. The admission control functionality is located in the RNC, where the load information from several cells is available. The admission control algorithm estimates the load increase that the establishment of the bearer would cause in the radio network. This has to be estimated separately for the uplink and downlink paths: the requested bearer can be admitted only if both uplink and downlink admission control admit it, otherwise it is rejected because of the excessive interference that it would produce in the network. The radio network planning sets the limits for admission control. Several call admission control algorithms have been proposed in the literature [14–18], mainly based on the total power received and transmitted by the BS.

Power-Based Admission Control Strategy

In the interference-based admission control strategy, the new user is not admitted by the uplink admission control algorithm if the new resulting interference level is higher than a predefined threshold value:

$$I_{total\_old} + \Delta I > I_{threshold} \tag{2.19}$$

The threshold value $I_{threshold}$ is the same as the maximum uplink noise rise and can be set by the radio network planning. This noise rise must be included in the link budget calculations as the interference margin. The wideband power-based admission control is shown in Figure 2.26, where the algorithm estimates the load increase due to a new user. The idea is to estimate the increase ΔI of the uplink received wideband interference power I_{total} due to a new user; the admission of the new user and the power increase estimation are handled by the admission control functionality. The proposed power increase estimation method is based on an integral approach: the derivative of the interference with respect to the load factor is integrated from the old value of the loading factor (h_{old}) to the new value ($h_{new} = h + \Delta h$), as follows:

$$\Delta I = \int^{+\Delta} dI_{total} \tag{2.20}$$

which gives

$$\Delta I = \frac{I_{total}}{1 - h_{new} - \Delta h} \Delta h \tag{2.21}$$

In this last formula, the loading factor Δh of the new user is the estimated load factor of the new connection and can be obtained as

$$\Delta = \frac{1}{1 + \dfrac{W}{SAF \times \dfrac{E_b}{N_0} \times R}} \tag{2.22}$$

The downlink admission control strategy is the same as in the uplink, (i.e., the user is admitted if the new total downlink transmission power does not exceed the predefined target value):

$$P_{total\_old} + \Delta P_{total} > P_{threshold} \tag{2.23}$$

The threshold power threshold P is set by the radio planning; the load increase total P_D in the downlink can be estimated based on the initial power, which depends on the distance from the BS.

Throughput-Based Admission Control Strategy

In the throughput-based admission control strategy the new requesting user is not admitted into the radio access network if:

$$h_{UL} + \Delta h > h_{UL\_threshold} \tag{2.24}$$

The same rule applies for the downlink:

$$h_{DL} + \Delta h > h_{DL\_threshold} \tag{2.25}$$

where the uplink and downlink loading factors are evaluated with the formulas presented earlier, while the loading factor of the new user is calculated with $h_{new} = h + \Delta h$.

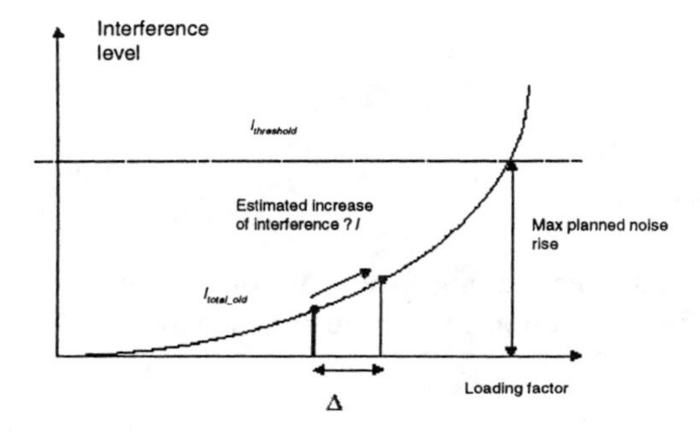

Figure 2.26 Uplink load curve and estimation of the load increase due to a new user.

Load Control (Congestion Control)

One important task or the RRM functionality is to ensure that the system is not overloaded and remains stable. If the system is properly planned, and the admission control works well, overloading situations should be exceptional. If overloading is encountered, however, the load control functionality returns the system controllably back to the targeted load, which is defined by the radio network planning.

The possible load control actions to reduce the load are listed here:

- Downlink fast load control: deny downlink power-up commands received from the MS;
- Uplink fast load control: reduce the uplink E_b/N_0 target used by the uplink fast power control;
- Reduce the throughput of packet data traffic;
- Handover to another WCDMA carrier;

- Handover to GSM;
- Decrease bit rates of real-time users;
- Drop calls in a controlled fashion.

The first two in this list are fast actions that are carried out within a base station and provide fast prioritization of the different services. The instantaneous frame error rate of the nondelay-sensitive connections can be allowed to increase in order to maintain the quality of those services that cannot tolerate retransmission. These actions only cause increased delay of packet data services while the quality of the conversational services, such as speech and video telephony, is maintained.

One example of a real-time connection, whose bit rate can be decreased by the radio access network, is adaptive multirate (AMR) speech codec (refer to [11] for further details).

Interfrequency and intersystem handovers can also be used as load balancing and load control algorithms. The final load control action is to drop real-time users (i.e., speech or circuit-switched data users) in order to reduce the load in the system. This action is taken only if the load on the system remains very high even after other load control actions have been performed in order to reduce the overload.

2.2.1.2 MBS Reference Architecture

In the process of standardization and development of the MBS, some initiatives of ETSI evolved, such as the Broadband Radio Access Network (BRAN) [4], which is a radio access system supporting broadband services with limited mobility.

The BRAN family of standards includes: HIPERLAN Type 1 (high-speed wireless LANs), HIPERLAN Type 2 (short-range wireless access to IP, ATM, and UMTS networks)—both operating in the 5-GHz band—HIPERACCESS (fixed wireless broadband point-to-multipoint), and HIPERLINK (wireless broadband interconnection) operating in the 17-GHz band. This is represented in Figure 2.27 together with the operating frequencies and indicative data transfer rates on the air interface.

HIgh PErformance Radio Local Area Network 1 (HIPERLAN Type 1)

HIPERLAN/1 is a wireless local area network. It is intended to allow high-performance wireless networks to be created, without existing wired infrastructure. Multiple HIPERLANs can coexist in the same geographical area with equitable bandwidth sharing without coordination between them. In addition, HIPERLAN/1 can be used as an extension of a wired local area network. HIPERLAN/1 offers unconstrained connectivity based on directed one-to-one communications as well as one-to-many broadcasts. The channel provides both self-configurability and flexibility of use thanks to a distributed channel access and standardized forwarding feature.

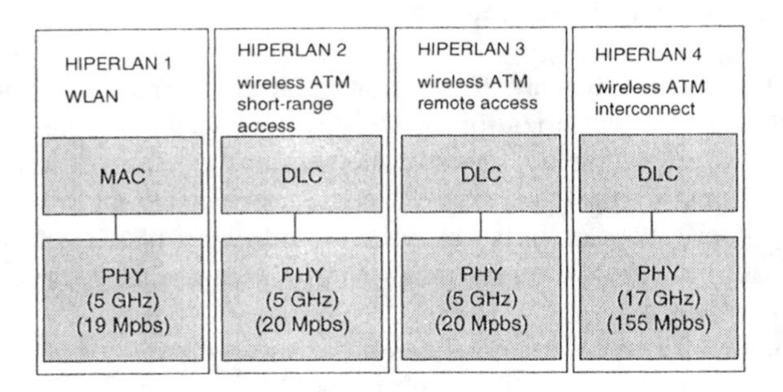

Figure 2.27 Overview of HIPERLAN types, HIPERACCESS, and HIPERLINK.

HIgh PErformance Radio Local Area Network 2 (HIPERLAN Type 2)

HIPERLAN/2 is a complement to HIPERLAN/1, ETSI's high-speed wireless LAN, and provides high-speed (25 Mbps is a typical data rate) communication between portable computing devices and broadband ATM and IP networks, aimed at telecommunications access and capable of supporting the multimedia applications of the future. The typical operating environment is indoors.

User mobility is supported within the local service area; wide area mobility (e.g., roaming) is supported by standards outside the scope of the BRAN project.

HIgh PErformance Radio ACCESS Network (HIPERACCESS)

HIPERACCESS is an outdoor, high-speed (25 Mbps is a typical data rate) radio access network, providing fixed radio connections to customer premises and capable of supporting multimedia applications (other technologies such as HIPERLAN/2 might be used for distribution within the premises). HIPERACCESS will allow an operator to rapidly roll out a wide area broadband access network to provide connections to residential households and small businesses. However, HIPERACCESS may also be of interest to large organizations wishing to serve a campus and its surroundings and operators of large physical facilities such as airports, universities, and harbors HIPERACCESS will have no mobility or this will be very limited.

HIgh PErformance Radio LINK (HIPERLINK)

HIPERLINK is a very-high-speed (up to 155-Mbps data rate) radio network for static connections and is capable of multimedia applications. A typical use is the interconnection of HIPERACCESS networks and/or HIPERLAN access points (APs) into a fully wireless network.

HIPERLAN/2 Network

A HIPERLAN/2 network typically has a topology where the mobile terminals communicate with the APs over an air interface as defined by the HIPERLAN/2 standard. There is also a direct mode of communication between two MTs. The user of the MT may move around freely in the HIPERLAN/2 network, which will ensure that the user and the MT get the best possible transmission performance. An MT, after an association has been performed (which can be viewed as a login), communicates with only one AP in each point in time. Each AP takes care of automatic radio network configuration, taking into account changes in radio network topology (i.e., there is no need for manual frequency planning).

A HIPERLAN/2 network for a business environment consists typically of a number of APs, each of them covering a certain geographic area. Together they form a radio access network with full or partial coverage of an area of almost any size. The coverage areas may or may not overlap each other, thus simplifying roaming of terminals inside the radio access network.

Each AP serves a number of MTs that have to be associated to it. In the case where the quality of the radio link degrades to an unacceptable level, the terminal may move to another AP by performing a handover.

For a home environment, the HIPERLAN/2 network is operated as an ad hoc LAN, which can be put into operation in a plug-and-play manner. The HIPERLAN/2 home system shares the same basic features with the HIPERLAN/2 business system by defining the following equivalence between both systems:

- A subnet in the ad hoc LAN configuration is equivalent to a cell in the cellular access network configuration.
- A central controller in the ad hoc LAN configuration is equivalent to the access point in the cellular access network configuration. However, the central controller is dynamically selected from HIPERLAN/2 portable devices and can be handed over to another portable device, if the old one leaves the network.
- Multiple subnets in a home are made possible by having multiple central controllers (CCs) operating at different frequencies.

HIPERLAN/2 supports two basic modes of operation:

- Centralized mode: In this mode, an AP is connected to a core network that serves the MTs associated with it. All traffic has to pass the AP, regardless of whether the data exchange is between an MT and a terminal elsewhere in the core network or between MTs belonging to this AP. The basic assumption is that a major share of the traffic is exchanged with terminals elsewhere in the network. This feature is mandatory for all MTs and APs.

- Direct mode: In this mode, the medium access is still managed in a centralized manner by a CC. However, user data traffic is exchanged between terminals without going through the CC. It is expected that in some applications (especially in the home environment), a large portion of user data traffic will be exchanged between terminals associated with a single CC. This feature is intended for use within the home environment and, hence, is mandatory in the data link control (DLC)-home extensions.

The general features of the HIPERLAN/2 technology can be summarized as follows:

- High-speed transmission;
- Connection-oriented;
- QoS support;
- Automatic frequency allocation;
- Security support;
- Mobility support;
- Network and application independent;
- Power save.

The HIPERLAN/2 reference model is shown in Figure 2.28.

HIPERLAN/2 has a very high transmission rate, which at the physical layer extends up to 54 Mbps and on layer 3 up to 25 Mbps. To achieve this, HIPERLAN/2 makes use of OFDM, in order to transmit the analog signals. OFDM is very efficient in time-disperse environments (e.g., within offices) where the transmitted radio signals are reflected from many points, leading to different propagation times before they eventually reach the receiver. Above the physical layer, a brand new MAC protocol implements a form of dynamic time division duplex to allow a more efficient utilization of radio resources.

In a HIPERLAN/2 network, data is transmitted on connections between the MT and the AP, established prior to the transmission using signaling functions of the HIPERLAN/2 control plane. Connections are time division multiplexed over the air interface. There are two types of connections, point-to-point and point-to-multipoint. Point-to-point connections are bidirectional whereas point-to-multipoint are unidirectional in the direction towards the MT. In addition, there is also a dedicated broadcast channel through which traffic reaches all terminals transmitted from one AP.

The connection-oriented nature of HIPERLAN/2 makes it straightforward to implement support for QoS. Each connection can be assigned a specific QoS in terms of bandwidth, delay, jitter, bit error rate, and so forth. It is also possible to use a more simplistic approach, where each connection can be assigned a priority level relative to other connections. This QoS support in combination with the high transmission rate enables the simultaneous transmission of many different types of data streams (e.g., video, voice, and data).

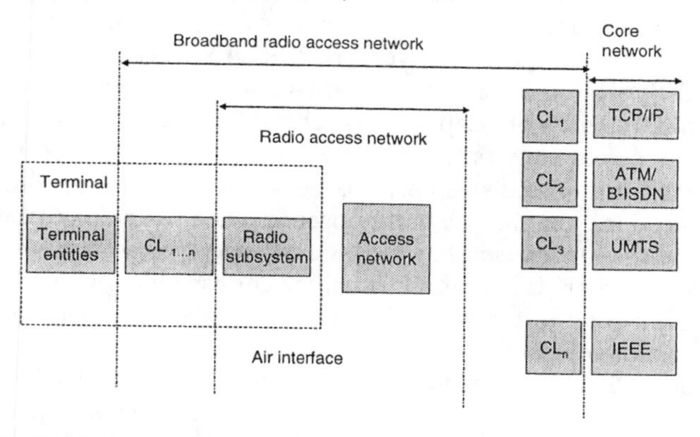

Figure 2.28 HIPERLAN/2 reference model.

In a HIPERLAN/2 network, there is no need for manual frequency planning as in cellular networks like GSM. The radio base stations, which are called access points in HIPERLAN/2, have a built-in support for automatically selecting an appropriate radio channel for transmission within each AP's coverage area. An AP listens to neighboring APs as well as to other radio sources in the environment and selects an appropriate radio channel based on which radio channels are already in use by those other APs and to minimize interference with the environment.

The HIPERLAN/2 network supports both authentication and encryption. With authentication, both the AP and the MT can authenticate each other to ensure authorized access to the network (from the AP's point of view) or to ensure access to a valid network operator (from the MT's point of view). Authentication relies on the existence of a supporting function, such as a directory service, which is outside the scope of HIPERLAN/2. The user traffic on established connections can be encrypted to protect against eavesdropping and man-in-the-middle attacks.

The MT will check that it transmits and receives data to/from the nearest AP; the MT uses the AP with the best radio signal as measured by the signal-to-noise ratio.

Thus, as the user and the MT move around, the MT may detect that there is an alternative AP with better radio transmission performance than the AP to which the MT is currently associated.

The MT will then order a handover to this AP. All established connections will be moved to this new AP resulting in the MT staying associated to the HIPERLAN/2 network and continuing its communication. During handover, some packet loss may occur. If an MT moves out of radio coverage for a certain time, the MT may loose its association to the HIPERLAN/2 network resulting in the release of all connections.

The HIPERLAN/2 protocol stack has a flexible architecture for easy adaptation and integration with a variety of fixed networks. A HIPERLAN/2 network, can for instance, be used as the last hop wireless segment of a switched ethernet, but it may also be used in other configurations (e.g., as an access network to third generation cellular networks). All applications today that run over a fixed infrastructure can also run over a HIPERLAN/2 network.

In HIPERLAN/2, the mechanism allowing an MT to save power is based on MT-initiated negotiation of sleep periods. The MT may at any time request the AP to enter a low power state (specific per MT) for a specific sleep period. At the expiration of the negotiated sleep period, the MT searches for the presence of any wake-up indication from the AP. In the absence of the wake-up indication, the MT reverts back to its low power state for the next sleep period, and so forth. An AP will defer any pending data to an MT until the corresponding sleep period expires. Different sleep periods are supported to allow for either short latency requirement or low power requirement.

In Figure 2.29, the protocol reference model for the HIPERLAN/2 radio interface is depicted. The protocol stack is divided into a control plane part and a user plane part following the semantics of ISDN functional partitioning (i.e., user plane includes functions for transmission of traffic over established connections, and the control plane includes functions for the control of connection establishment, release, and supervision).

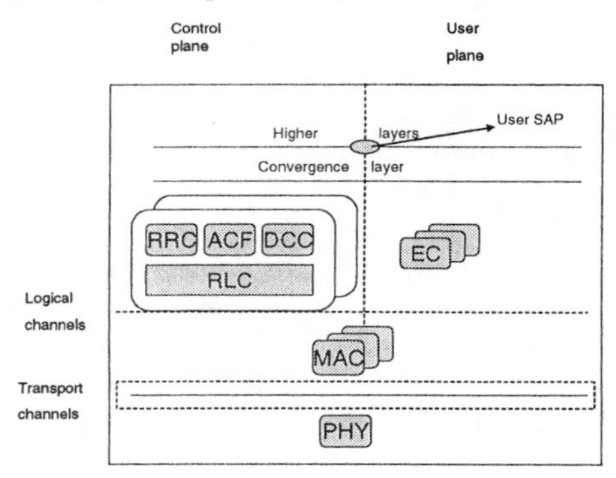

Figure 2.29 HIPERLAN/2 protocol reference model.

The HIPERLAN/2 protocol has three basic layers: physical layer (PHY), data link control layer (DLC), and convergence layer (CL). At the moment, there is only control plane functionality defined within DLC. The PHY, DLC, and the CL are detailed further in the following sections.

The transmission format on the physical layer is a burst, which consists of a preamble part and a data part, where the latter could originate from each of the transport channels within DLC. OFDM has been chosen due to its excellent performance on highly disperse channels. The channel spacing is 20 MHz, which allows high bit rates per channel but still has a reasonable number of channels in the allocated spectrum (e.g., 19 channels in Europe). There are 52 subcarriers per channel, where 48 subcarriers carry actual data and 4 subcarriers are pilots that facilitate phase tracking for coherent demodulation. The duration of the guard interval is equal to 800 ns, which is sufficient to enable good performance on channels with a delay spread of up to 250 ns. An optional shorter guard interval of 400 ns may be used in small indoor environments.

The DLC layer constitutes the logical link between an AP and the MTs. The DLC includes functions for medium access and transmission (user plane) as well as terminal/user and connection handling (control plane). Thus, the DLC layer consists of a set of sublayers as follows:

- MAC protocol;
- Error Control (EC) protocol;
- Radio Link Control (RLC) protocol, with the associated signaling entities DLC connection control (DCC), RRC, and association control function (ACF).

MAC Protocol

The MAC protocol is the protocol used for access to the medium (the radio link) with the resulting transmission of data onto that medium. The control is centralized to the AP which informs the MTs of the point in time in the MAC frame that they are allowed to transmit their data, which adapts according to the request for resources from each of the MTs.

The air interface is based on TDD and dynamic time division multiple access (TDMA). For example, the time-slotted structure of the medium allows for simultaneous communication in both downlink and uplink within the same time frame, called MAC frame in HIPERLAN/2. Time slots for downlink and uplink communication are allocated dynamically depending on the need for transmission resources. The basic MAC frame structure on the air interface has a fixed duration of 2 ms and comprises transport channels for broadcast control, frame control, access control, downlink (DL) and uplink (UL) data transmission, and random access. All data from both AP and MTs is transmitted in dedicated time slots, except for the random access channel where contention for the same time slot is allowed. The duration of broadcast control is fixed whereas the duration of other fields is dynamically adapted to the current traffic situation. The MAC frame and the transport channels form the interface between DLC and the physical layer.

The Error Control Protocol

Selective repeat (SR) ARQ is the EC mechanism that is used to increase the reliability over the radio link. EC in this context means detection of bit errors and the resulting retransmission of user-protocol data units (U-PDUs) if such errors occur. EC also ensures that the U-PDUs are delivered in sequence to the convergence layer. The method for controlling this is by giving each transmitted U-PDU a sequence number per connection. The ARQ ACK/NACK messages are signaled in a logical channel (LCCH). An erroneous U-PDU can be retransmitted a number of times (configurable). To support QoS for delay-sensitive applications (such as voice) in an efficient manner, a U-PDU discard mechanism is defined. If data becomes obsolete (e.g., beyond the playback point), the sender entity in the EC protocol can initiate a discard of all U-PDUs with lower sequence number and which have not been acknowledged. The result is the retransmission of DLC missing data, while retaining the DLC active connection. It is up to higher layers, if there is a need, to recover missing data.

The RLC protocol gives a transport service for the signaling entities ACF, RRC, and the DCC function. These entities comprise the DLC control plane for the exchange of signaling messages between the AP and the MT.

Association Control Function

The MT listens to different APs and selects the AP with the best radio link quality. The MT then continues listening to the broadcast of a globally unique network operator ID, in order to avoid association to a network that is not able or allowed to offer services to the user of the MT. If the MT decides to continue the association, it will request and be given a MAC-ID from the AP. This is followed by an exchange of link capabilities starting with the MT providing information about (not exhaustive):

- Supported PHY modes;
- Supported convergence layers;
- Supported authentication and encryption procedures and algorithms.

The AP will respond with a subset of supported PHY modes, a selected convergence layer (only one), and a selected authentication and encryption procedure (where one alternative is to not use encryption and/or authentication).

If encryption has been negotiated, the MT will start the Diffie-Hellman key exchange to negotiate the secret session key for all unicast traffic between the MT and the AP. In this way, the following authentication procedure is protected by encryption. HIPERLAN/2 supports both the use of the DES and the 3-DES algorithms for strong encryption.

Broadcast and multicast traffic can also be protected by encryption through the use of common keys (all MTs associated to the same AP use the same key). Common keys are distributed encrypted through the use of the unicast encryption key. All encryption keys must be periodically refreshed to avoid flaws in the security. There are two alternatives for authentication: one is to use a preshared key and the other is to use a public key.

When using a public key, HIPERLAN/2 supports a public key infrastructure (PKI) by means of generating a digital signature. Authentication algorithms supported are Message-Digest 5 (MD5), Keyed-Hashing for Message Authentication (HMAC), and RSA encryption. Bidirectional authentication is supported for authentication of both the AP and the MT. HIPERLAN/2 supports a variety of identifiers for identification of the user and/or the MT (e.g., network access identifier (NAI), IEEE address, and X.509 certificate). After association, the MT can request a dedicated control channel that it uses to set up radio bearers (within the HIPERLAN/2 community, a radio bearer is referred to as a DLC user connection). The MT can request multiple DLC user connections where each connection has a unique support for QoS.

An MT may disassociate explicitly or implicitly. When disassociating explicitly, the MT will notify the AP that it no longer wants to communicate via the HIPERLAN/2 network. Implicitly means that the MT has been unreachable for the AP for a certain time period (see more details under MT alive below). In either case, the AP will release all resources allocated for that MT.

DLC User Connection Control

The MT (as well as the AP) requests DLC user connections by transmitting signaling messages over the DCCH logical channel. The DCCH controls the resources for one specific MAC entity (identified through the MAC-ID). No traffic in the user plane can be transmitted until there is at least one DLC user connection between the AP and the MT. The signaling is quite simple: a request is followed by an acknowledgment when the connection can be established. For each request, the connection characteristics are given. The established connection is identified with a DLC connection identifier, allocated by the AP. A connection might subsequently be released using a procedure analogue to the establishment.

HIPERLAN/2 also supports modification of the connection characteristics for an established connection.

Radio Resource Control (RRC)

The handover starts from radio link quality measurements. This may then result in a request for a handover initiated and requested by the MT. HIPERLAN/2 supports two forms of handover—reassociation and handover via the support of signaling across the fixed network. Reassociation basically means to start over again with an association as described above, which may take some time, especially in relation to ongoing traffic. The alternative means that the new AP, to which the MT has requested a handover, will retrieve association and connection information from the old AP by transfer of information across the fixed network. The MT provides the new AP with a fixed network address (e.g., an IP address) to enable communication between the old and new AP. This alternative results in a fast handover minimizing loss of user plane traffic during the handover phase.

RRC supports dynamic frequency selection (DFS) by letting the AP have the possibility of instructing the associated MTs to perform measurements on radio signals received from neighboring APs. Due to changes in environment and network topology, RRC also includes signaling for informing associated MTs that the AP will change frequency.

The AP supervises inactive MTs that do not transmit any traffic in the uplink by sending an alive message to the MT for the MT to respond to. As an alternative, the AP may set a timer for how long an MT may be inactive. If there is no response from the alive messages, or alternatively if the timer expires, the MT will be disassociated.

The power save function is responsible for entering or leaving low consumption modes and for controlling the power of the transmitter. This function is MT initiated. After a negotiation on the sleeping time, the MT goes to sleep. After N frames, there are four possible scenarios:

- The AP wakes up the MT (cause: e.g., data pending in AP);
- The MT wakes up (cause: e.g., data pending in MT);
- The AP tells the MT to continue to sleep (again for N frames);
- The MT misses the wake-up messages from the AP. It will then execute the MT alive sequence.

Convergence Layer

The CL has two main functions: to adapt the service request from the higher layers to the service offered by the DLC and to convert the higher layer packets (SDUs) with a variable or a possibly fixed size into a fixed size that is used within the DLC. The padding, segmentation, and reassembly function of the fixed-size DLC SDUs is one key issue that makes it possible to standardize and implement a DLC and PHY that is independent of the fixed network to which the HIPERLAN/2 network is connected. The generic architecture of the CL makes HIPERLAN/2 suitable as a radio access network for a diversity of fixed networks (e.g., Ethernet, IP, ATM, UMTS). There are currently two different types of CLs defined: cell-based and packet-based. The former is intended for interconnection to ATM networks; whereas the latter can be used in a variety of configurations depending on fixed network type and how the interworking is specified.

The structure of the packet-based CL with a common and service-specific part allows for easy adaptation to different configurations and fixed networks. From the beginning though, the HIPERLAN/2 standard specifies the common part and a service-specific part for interworking with a fixed ethernet network.

Radio Network Functions

The HIPERLAN/2 standard defines measurements and the signaling to support a number of radio network functions. These are currently defined as dynamic frequency selection, link adaptation, radio cell handover, multibeam antennas, and power control. All algorithms are vendor specific. The following sections outline the contents of the radio network functions.

Dynamic Frequency Selection
The HIPERLAN/2 radio network shall automatically allocate frequencies to each AP for communication. This is performed by the DFS function, which allows several operators to share the available frequency spectrum and can be used to avoid the use of interfered frequencies. The frequency selection made by each AP is based on filtered interference measurements performed by the AP and its associated MTs.

Link Adaptation
To cope with the varying radio quality, in terms of the signal-to-interference ratio (S/I), a link adaptation scheme is used. The range of S/I levels varies depending on the location where the system is deployed, and it also changes over time depending on the traffic in surrounding radio cells. The link adaptation scheme adapts the PHY robustness based on link quality measurements.

Antennas
Multibeam antennas are supported in HIPERLAN/2 as a means to improve the link budget and increase the S/I ratio in the radio network. The MAC protocol and the frame structure in HIPERLAN/2 allow up to seven beams to be used.

Handover
The handover scheme is MT initiated (i.e., an MT performs the necessary measurements on surrounding APs and selects appropriate APs for communication). The handover measurements are not defined in the standard (i.e., a vendor can choose to base the handover on signal strength or some other quality measurements). The standard defines necessary signaling to perform the handover.

Power Control
Transmitter power control is supported in both the MT (uplink) and the AP (downlink). MT power control is mainly used to simplify the design of the AP receiver. AP power control is part of the standard out from regulatory reasons (i.e., to decrease the interference to satellite systems).

2.2.1.3 DBS Reference Architecture

DBS refers to the transmission systems that were initially designed for the delivery of television content. Such systems include satellite, cable, and terrestrial transmission. More recently, such systems were enhanced with features enabling encapsulation of IP traffic and return path from users to broadcasters.

Two international bodies have been very active in the specification of DBS systems:

- Digital Video Broadcasting (DVB) [19] is the historical European specification group of digital television broadcasting systems. Its approach has mainly consisted of specifying and standardizing the user equipment and the interface to the network, letting broadcasters freely design the network part. Its scope is mainly the specification of physical layers, mapping of MPEG content, signaling, and encapsulation of IP traffic.
- Digital Audio-Visual Council (DAVIC) [20] was an international specification body whose role consisted of specifying end-to-end architectures for any system able to deliver broadcast content [21]. Its approach mainly consisted of defining a framework and interfaces between the elements of the end-to-end system, without reinventing the wheel; in other words, reusing existing standards as much as possible.

Other broadcast systems have been or are being developed but they either do not support mobility or are still in development, such as the terrestrial integrated services digital broadcasting (ISDB-T).

DVB-T is the terrestrial component specified in DVB and standardized by ETSI [22, 23], and it is already deployed in Europe and elsewhere in the world. It enables a very local coverage compared to DVB-S, and it is able to support mobile reception [24]. Furthermore, guidelines are provided to encapsulate IP traffic [25] and support interactive services [26, 27]. It is therefore an interesting complement to cellular systems, which other IST projects have already identified [28, 29].

Broadcast architectures are usually proprietary and so far do not require extensive standardization because the users have little or no interaction with the network. In this context, only the air interface needs to be standardized. MONASIDRE used DAVIC as a reference because it gave some insights on the key services and components of a broadcasting system. Therefore, the reference system architecture was derived from the DAVIC specifications and adapted to the DVB-T context [3].

DAVIC Reference Architecture

The general model that was defined in DAVIC is reproduced in Figure 2.30.

The model distinguishes four domains, interconnected by means of standardized interfaces or reference points:

- The content provider system (CPS) has the role of producing the content.
- The service provider system (SPS) has the role of formatting and aggregating contents from different sources.
- The service consumer system (SCS) makes reference to the user terminal or network.

- The delivery systems (DS) are the networks interconnecting the previous domains.

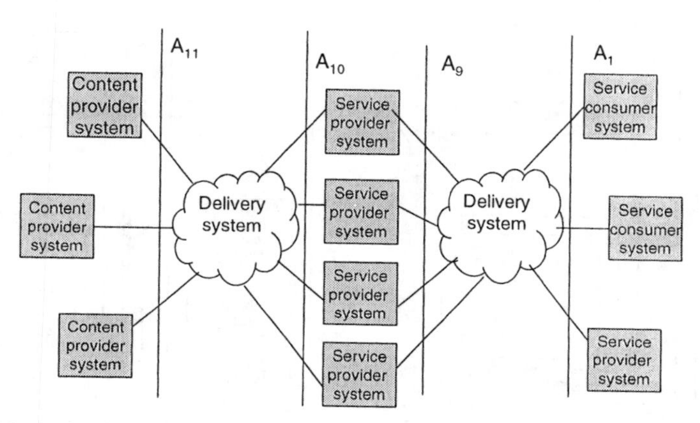

Figure 2.30 DAVIC general model.

Each domain can match a specific actor in the value chain. The left-hand part of the general model up to interface A10 can be seen as the studio part (e.g., TV channel), while the rest (excluding the SCS) would be the broadcaster subsystem.

The model is further detailed in Figure 2.31. It decomposes the service provider system and the delivery system. It also highlights some of the logical flows. For complete details the reader is referred to [7].

The SPS includes functions applicable to the system as a whole such as the network interface, network control, session control, and transport control. The content provider domains can be seen as mirrors of the CPSs; the service provider makes resources available to the content providers for performance enhancement (e.g., local storage). Each host in the content provider domains implements service elements like application, content, stream, directory, and download services.

These domains are interconnected to the native domains via the A10 interface. Note that an interworking unit is required to adapt access to the Internet. The service gateway is the entry point for a client into the SPS.

There are five standardized information flows between the systems (see Table 2.10).

S2 flow makes use of the user-to-user part of the digital storage media command and control (DSM-CC), while S3 uses the user-to-network part of the same protocol. S4 implements network signaling protocols for call connection and control. On this topic, it is interesting to note that this architecture is strongly inspired by telecommunications concepts in the sense that there are user-to-user and user-to-network protocols in order to perform resource reservation. In the Internet world, explicit user-to-network protocols are almost nonexistent.

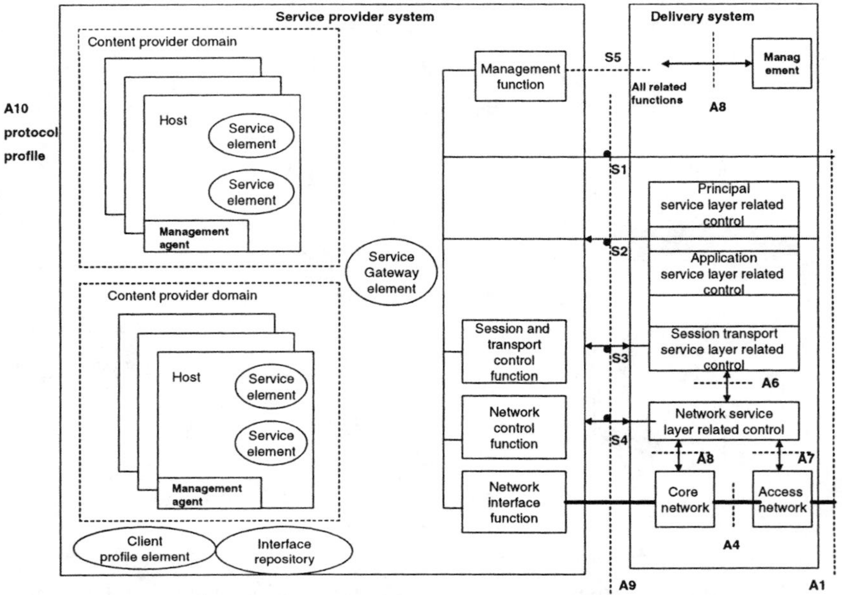

Figure 2.31 Detailed broadcaster reference model.

S5 deals with the management plane. DAVIC specifications only indicate that the Common Management Information Protocol (CMIP) or Simple Network Management Protocol (SNMP) agents may be used in the managed domains. However, they only address the management of access networks. In particular, they propose a list of generic classes of objects that can be found in access networks.

These flows are logical flows transported over the DVB transport stream for S1 and over TCP/IP for the others.

DAVIC Reference Architecture Applied to DVB-T

The model described earlier needs a few amendments to reflect the precise situation of DVB-T networks. First, it has to take into account that DVB-T networks may not provide a return path to the network, and therefore have to rely on alternative networks such as cellular or WLANs.

Second, current regulation context does not allow considering DVB-T as a pipe for delivering unicast IP traffic only to mobile users.

Most of the DVB-T bandwidth is to be used to deliver MPEG2 broadcast content to fixed or nomadic users. However, in the context of MONASIDRE, it will be assumed that, in the future, regulation will allocate part of the DVB-T bandwidth to be available for mobile users' IP traffic.

The block diagram in Figure 2.32, is a retailored view of the DAVIC reference architecture applicable to DVB-T.

Table 2.10
Information Flows Applicable to a Broadcasting System

Flow Name	Plane	Type	Description
S1	User	Content information	Flow between a source object and a destination object Transparent to any intermediate object Does not alter the behavior of source and destination objects
S2	User	Content information	Flow between an application service layer source object and a peer destination object Transparent to any intermediate object Behavior of source or destination objects may change as result of a flow Examples of messages are play, stop, fast-forward
S3	Control	Transport control information	Flow between a session and transport service layer source object and a peer destination object Example of messages are establish, release connections, communicate addresses, port information, and routing data
S4	Control	Network control information	Flow between a network service layer source object and a peer destination object Examples of messages are establish, release connections, communicate addresses, port information, and routing data
S5	Management	Management information	Flow between a source object to a peer destination on the management plane

The interaction network needs to be able to interconnect with the broadcaster SPS, as the latter is responsible for the encapsulation of the IP traffic into the broadcast stream. In the scenario where the interaction network is part of the delivery system (e.g., the case of terrestrial return path provided by the broadcaster, as specified in [30]), the terminal should fully support the information flows as defined in DAVIC. In case of an interaction network being a cellular network or a WLAN, S3 and S4 will be encapsulated and forwarded to the SPS and will not have direct influence on the interaction network. Equivalent protocols are needed, however, so DAVIC call control and interaction network call control would coexist in this scenario. In both cases, it is foreseen that an interworking unit will be required in the SPS in order to apply relevant QoS policies to the IP traffic that is requested to pass via the unidirectional network.

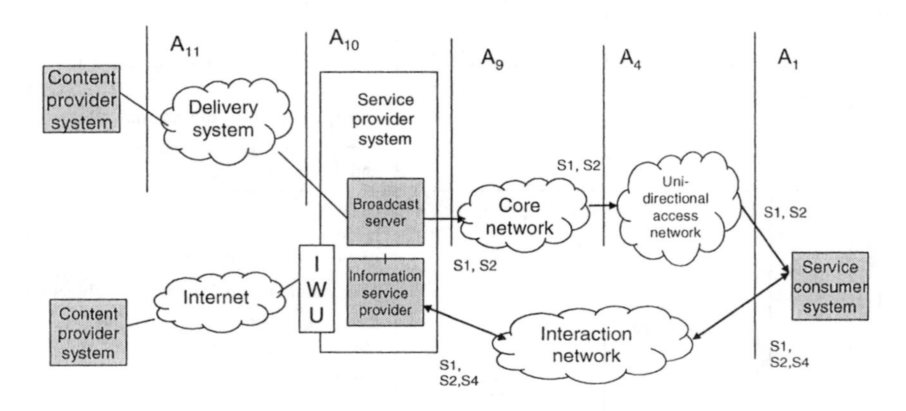

Figure 2.32 DAVIC compliant reference model for DVB-T.

From a service standpoint, it is desirable for the DBS service provider to implement the interworking unit in order to enable DBS/Internet integrated or at least cooperative services.

On the management plane, the situation is more complex when the broadcast system and the interaction network are independent while the management systems are not coordinated. One added value of the coordinated system will be the coordination of the management decisions.

A typical broadcast network's physical architecture developed from [31] is depicted in Figure 2.33.

The broadcast programs (audio/video) are created in studios, encoded in MPEG2, and distributed to regional platforms via PDH or SDH networks. The distributed streams are properly encapsulated and synchronized by means of network adapters, as specified in [32, 33]. On each regional platform, the audio/video streams are multiplexed in order to create the MPEG2 transport stream that is broadcast. Regional content can be inserted at that point.

Once the programs are multiplexed, they are distributed to each DVB-T transmitter either directly or by means of a network.

Such architecture is not well suited for transmission of data to individual users as it could only be inserted at a regional level, and therefore broadcast over a complete region (millions of potential users over several thousands of square kilometers, covered by a few tens of transmitters). For this reason, enhancements of broadcast networks are likely to happen in order to support the delivery of localized content. This change can be achieved by associating an MPEG2 multiplexer to each DVB-T transmitter, as shown in Figure 2.34.

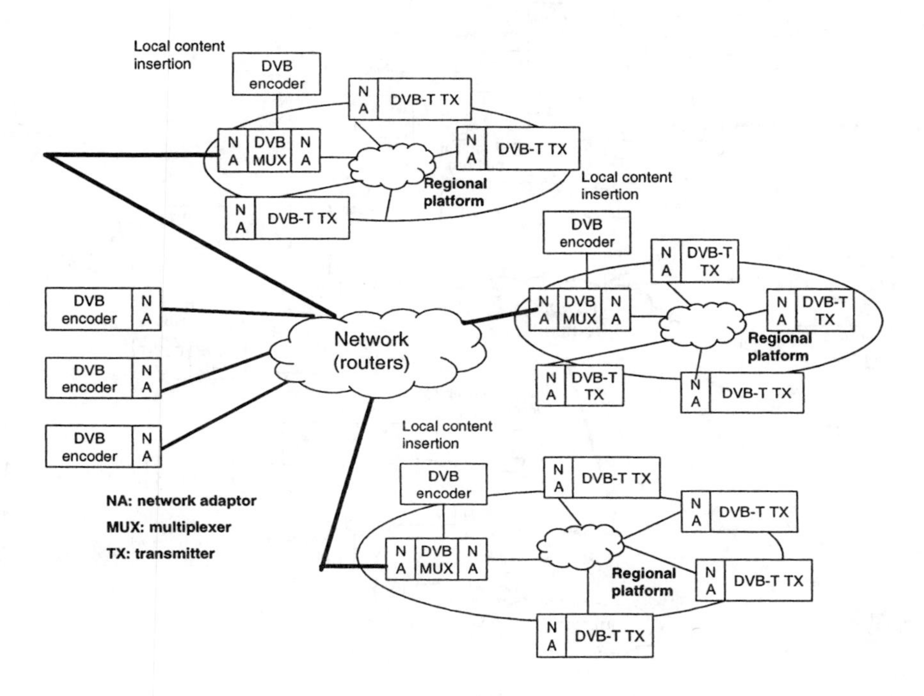

Figure 2.33 Typical broadcast distribution network.

Local insertion of Internet content can be specific to each transmitter. From a spectral efficiency standpoint, this solution is more satisfactory than the previous one, although the area covered by each transmitter is still large (a few hundreds of square kilometers). It seems clear that the capacity made available in DVB-T transmitters should be devoted to multicast traffic, as it would efficiently reach a significant number of users. However, to optimally route multicast sessions to the correct transmitter, there must be mobility management mechanisms that enable the interactive service provider to determine when and where to route multicast sessions. If the DVB-T network offers a return path, such mobility mechanisms are internal to the access network. Otherwise, the mobility management protocol messages need to travel via the return path technology available on the terminal.

The local multiplexers can either be collocated with the transmitters or centralized with the input multiplexer in the region. Both scenarios are depicted in the figure. The choice for one or the other solution is an operator choice. In any case, there is a need for one multiplexer per transmitter as the Internet content to each transmitter is likely to be different. Note also that the Internet traffic needs to be shaped in order to prioritize flows in such a way that they fit the available bandwidth. The bandwidth devoted to Internet traffic can either be fixed or variable, depending on the encoder and multiplexing technologies used.

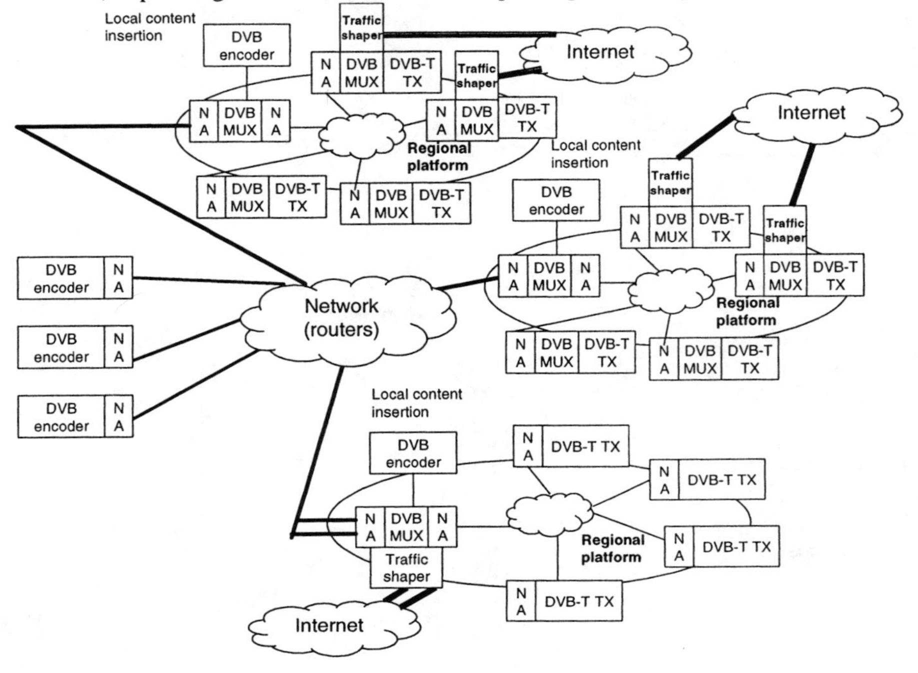

Figure 2.34 Broadcast network enhanced for local insertion of Internet traffic.

In the early days of MPEG2, encoders were designed to provide a fixed bit rate, in order to simplify the multiplexing of programs coming from different sources. It appears nowadays that video quality is improved when encoded video bit rate is not kept constant, and multiplexers are capable of statistical multiplexing. A consequence is that the bandwidth made available to Internet traffic is not constant and can only be statistically guaranteed. Another consequence is that the Internet traffic fed to a multiplexer needs to be shaped based on QoS policies. A network management platform will have a role here to select the appropriate QoS policies, based on the assumption that the available bandwidth in a DVB-T multiplex will be variable and in the minority compared to MPEG2 content.

The regulator may direct the number and nature of the economical actors involved in the DVB-T deployment [34]. There are four economical actors in the value chain that together make possible digital terrestrial broadcasting. Their names and functions are in Table 2.11.

Table 2.11

Actors and Functions in the DVB-T Value Chain

Economical Actor	Function	Mapping to DAVIC Model
Program editor	Gets an authorization for a certain capacity and manages it	Content provider system, and/or service provider system
Commercial distributor	Contact with the consumer (distinct from the editor, e.g., pay TV) May provide the terminals	
Multiplex operator	Gets a frequency channel Manages capacity and multiplexes programs	Delivery system
Network operator	Provides the broadcasting infrastructure	Access network in DS

Since these actors may be independent companies, there can potentially be as many different management domains.

DBS Management System

DAVIC recommends that a typical telecommunications management network (TMN) architecture should be used for the management of broadcaster networks. The elementary element is an operations support function (OSF). An OSF reference model is composed of four management layers and two reference points, as depicted in Figure 2.35 (a).

The management layers are as follows:

- Element management layer (control and coordination of network elements, maintenance of statistical data and logs);
- Network management layer (control and coordination of the network view, provision/cessation/modification of network capabilities, maintenance of statistical data on performance, and usage).
- Service management layer (interaction with service providers and between services, maintenance of statistical data on QoS);
- Business management layer (operations and maintenance budgets, business models for operations, others).

Reference point q3 [in Figure 2.35, (a)] is used between entities of the same management system, while reference point x is used when management systems need to cooperate.

Figure 2.35(b) gives a deployment example of a management application. Such application gets data from the managed systems by means of agents distributed in the system and communicating with standardized protocols such as CMIP, and SNMP.

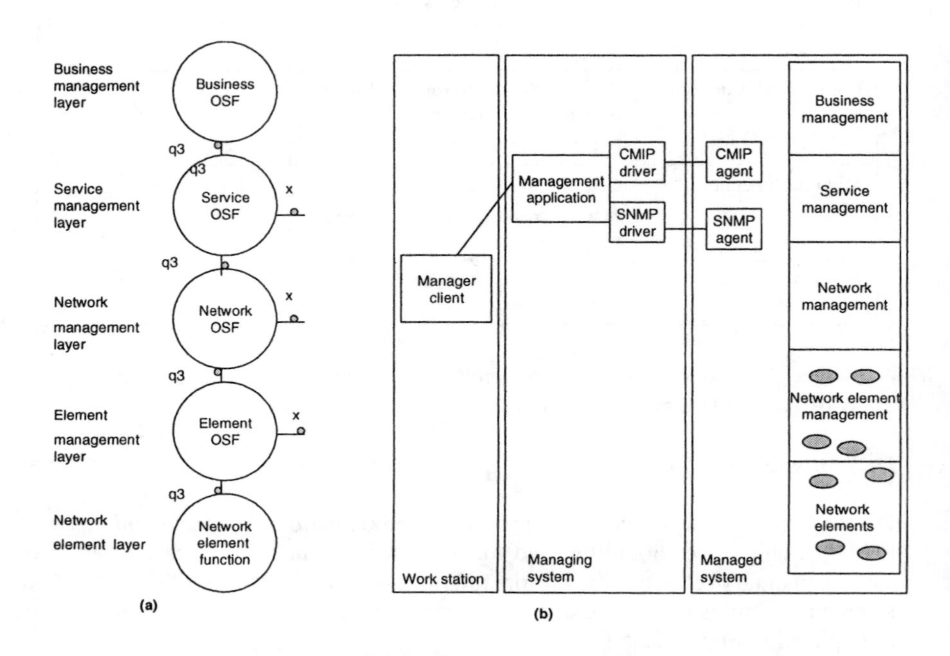

Figure 2.35 (a) OSF architecture and (b) partitioning example. (*From*: [7].)

The work consists of defining the object classes in the access network information model and their relationships.

The identified network elements to be controlled are listed in Table 2.12, with the controllable and measurable parameters.

It must be noted that not all parameters may be accessed by the management platform, depending on its ownership in the value chain. This kind of scenario has been anticipated in a pure DAVIC context and listed in [35].

Table 2.12
Sample List of Network Elements and
Corresponding Controllable/Measurable Parameters

Network Element	Domain	Controllable Parameters	Measurable Parameters
Power amplifies	Network operator	Transmitted power	—
DVB-T modulator	Network operator	Modulation scheme FFT size, guard interval Coding scheme	—
DVB multiplexer	Multiplex operator	Multiplexing algorithms MPEG2 signaling	Multiplexer load
IP traffic shaper	Multiplex operator	Address filters Service filter Classification rules QoS policies	Dropped packets per address, per service, per class Buffer status
DVB IP encapsulator	Multiplex operator	IP address/MPEG2 mapping rule	Load
Network adapter	All domains		Frame error rate Load
Routers	All domains	Buffer limits	Packet error rates Buffer status Load per input/output
Terminal	Commercial distributor		Received power Error rates
Network servers	Several domains		Signaling load

2.2.1.4 RMS Requirements

In this section, the functional requirements that should be taken into account during the design of the RMS module of Figure 2.6 are listed. The RMS environment and possible deployments that it could have to support are described. Such deployments have consequences on the information flow between actors, as well as on the overall architecture of the management system. The requirements that should be fulfillled by the RMS module as depicted in Figure 2.7 are identified next.

Evolutionary Platform

UMTS, DVB-T, and HIPERLAN/2 systems have to exist first, before joint use is possible. The architectural platform should be made of elementary components belonging to each system, which will ultimately be able to cooperate with an entity at a higher level of abstraction (i.e., RMS driving other RMSs rather than network elements). The structure could be chosen as hierarchical as well as peer to peer.

Multiservice Management

Within each system, each management platform must be able to make provision for multiple services. This is rather new to evolved cellular systems, such as 2.5G and 3G. It can be understood that inputs from service providers can help, for the reason that they know which types of services the subscribers may want to use. The reason for this new situation is the potential separation between the network operator role and the service provider role. The type of agreement between service providers and network operators is also made more complex because they now have to include the mixture of different service classes that the network needs to support.

Multiple Service Provider Management

A network operator will not necessarily offer services to end users. Rather, the network operator offers access to services sold by service providers. As a result, multiple service providers will be accessed through the same radio network. There is, therefore, the need for the network operator to use a management platform able to manage provisioning for multiple service providers, in particular arbitrating the access to resources when they are temporarily restricted.

Independent/Complementary User Management

In traditional systems, the network operator also performs user management, and users only access one type of transport service. In future systems, for example, priority is given to enable the user to access its service provider. When using a specific network, the network operator can indeed manage the users. However, the user will request information to help him or her select the best network. This information must be made available by an operator guaranteeing equitable and nondiscriminatory access to service providers. This is a reason why actors other than radio network operators are better positioned to provide this information. The technical consequence is that an information channel between users and management platform (across multiple systems) must exist. Note that this requirement is not to have a one-to-one connection between users and the management platform (due to scalability reasons).

RMS Environment Assumptions

The RMS environment is anything that interfaces with the RMS component. The first basic assumption is that the RMS interfaces with the UMD-NES on one side and the MASPI on the other side. The UMD-NES is a component that when used in an operational environment will help in validating RMS decisions. The MASPI component is the communication resource that provides the managed network status to the RMS and sends actions. The example platform considered here has a distributed nature [7]; therefore, the MASPI component also is assumed to provide communications means with other RMS components. Figure 2.36 shows the immediate environment of the RMS component.

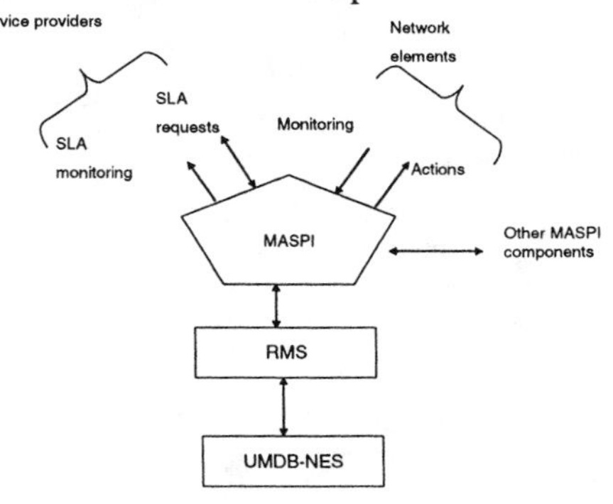

Figure 2.36 Assumed immediate RMS environment.

In the requirements described above it was indicated that distribution of the management platform can be based on a hierarchical approach or on a peer-to-peer approach. In the first approach there is the idea that one platform has at the end some form of control over the others; for instance, the service provider platform can drive the resource management of the networks used by the service provider subscribers. In the peer-to-peer approach, there is no such dependency. Both approaches are shown in Figure 2.37.

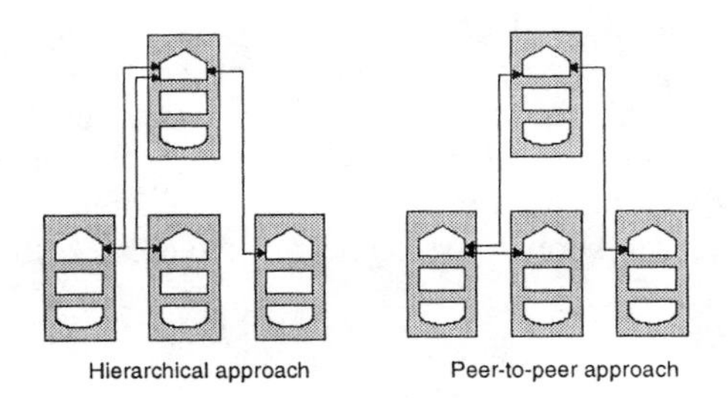

Figure 2.37 Communications approach for a management platform.

The consequences brought by the second approach have an impact on the information exchange protocols between entities and as a result on the type of information the RMS component can receive.

RMS Conceptual Model

The RMS is a network management function, meaning that it generates actions whose effects are visible on large time scales (a few minutes/hours, rather than seconds). Consequently, the RMS primary correspondent within a given network is a radio resource control module that will apply the RMS decisions. This is depicted in Figure 2.38.

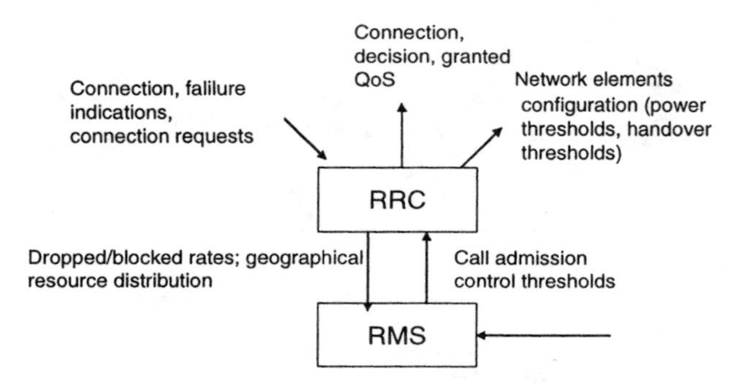

Figure 2.38 RMS environment functional model. (*From*: [7].)

Assuming that the network radio planning is appropriately done for normal network usage conditions, the RMS should only perform dynamic management of applicable SLAs in a multiservice context. Main actions would probably consist of dynamically tuning the call admission control thresholds per service, so that the available resources are not exhausted. To some extent, it could also consist of controlling some network elements. Adjustments are driven by performance measurements brought by the RRC module such as dropped call rates, actual resource consumption, and so on.

The SLA applicable to a given RMS domain can change over time if it is assumed that a service provider may be provisioned across multiple radio networks. There is then the need for an entity that processes the service provider SLA into SLAs applicable to all networks in which the service provider has subscribers. Another role of this entity could be to generate the information that subscribers use, in order to know to which network to connect to.

2.2.1.5 UMD-NES

The objective of the UMD-NES is twofold:

- To palliate the current lack of large-scale experimental test-beds;
- To allow network operators to simulate the effect of network configuration changes prior to their application to the real network.

In an operational management system, the network is not part of the system itself. The network can also be categorized in the other types of components: elements that are not part of the system but that interact with it. As a consequence, the network is an actor (in the sense of UML) of the application, as can be seen in Figure 2.39.

The objective of the UMD-NES component is to replace (to simulate) the network. Therefore, the simulator can be seen as an actor with relation to the MASPI and RMS components. The simulator can be designed independently from the rest of the application, provided that the interaction of the application with any network is clearly specified.

Figure 2.39 shows the use cases that involve the application and the network. The other existing actors in a management system (i.e., users, service providers, network operators) also interact with the system but these interactions do not concern the network. Nevertheless, these actors also interact with the network. In the case of a real network, these interactions do not need to be modeled and implemented because they are part of the "life" of the overall actors and a result of their internal behavior (NO, US, SP, and the network). When the network is no longer a real network but a simulator, there is a need to design the internal behavior of the network as well as its interactions with the other actors of the overall system. In other words, although the user as an actor also needs to be simulated, he or she in fact becomes a part of the simulator. Let us now consider the simulator as the system to be designed and the components that interact with it as the actors.

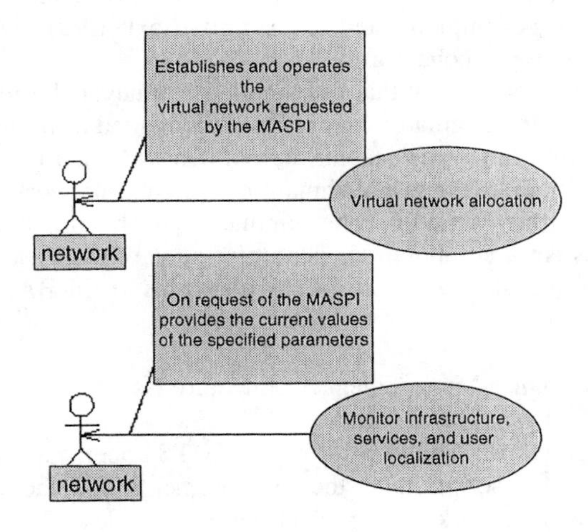

Figure 2.39 Use cases involving the network actor.

The two use cases resulting from this point of view are shown in Figure 2.40. It has to be noted that Figures 2.39 and 2.40 both give a general list of use cases involving the simulator.

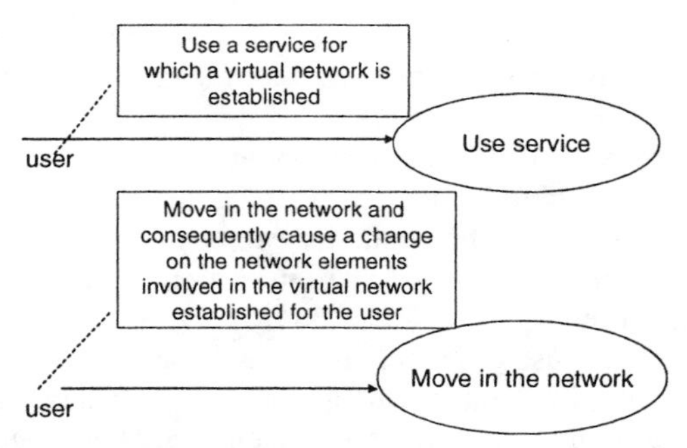

Figure 2.40 Use cases involving the network actor.

2.2.1.6 Implementation Issues

The use cases presented in Figure 2.39 and Figure 2.40 define how the application can interact with the network and therefore define the main functionality of the UMD-NES component.

They also define the interface, which needs to be implemented over any real network in order to allow this network to work with the specific application. The use cases presented here are artificial use cases that help simulate the behavior of actors of the application that interacts with the network (namely, the user). These use cases could be implemented as scripts (text files), which contain the description of the user's behavior.

The four use cases show that the simulator is never the initiator of any use case. Nevertheless the simulator cannot be implemented as a fully reactive object. Indeed, the network has a permanent internal behavior, and requests from outside of the network (such as the establishment of a virtual network) could have long-term impacts on this behavior. The simulator can be implemented as one (or several) process (in a computer science point of view) and can interact with the application through standard distributed platforms (CORBA, MPI, PVM, and others).

2.2.1.7 Performance Management Architecture Issues

This section gives an analysis of the UMTS open network management architecture and a description of the measurements and the collected data for UMTS service and network management purposes as described in the IST project MONASIDRE [7].

The network management actions that can be conducted on the UMTS RAN and the core network will rely on the operation and maintenance (O&M), management and network monitoring, and fault management architecture expected within the UMTS services and management framework. Figure 2.41 shows the management infrastructure and the interfaces used for both O&M and fault management.

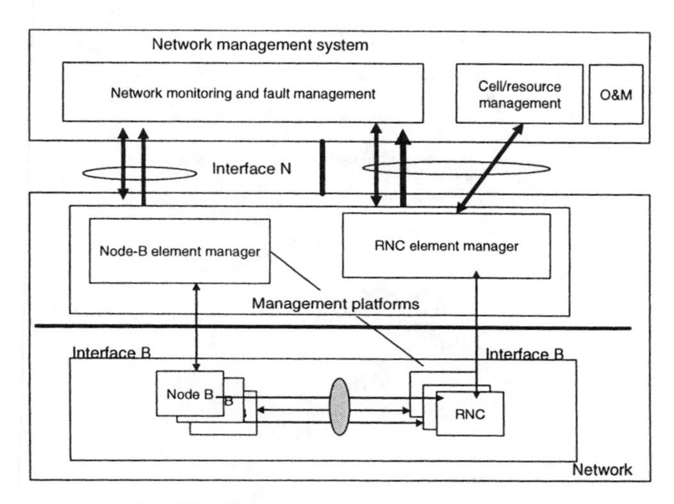

Figure 2.41 UTRAN management interfaces.

Through the interfaces and the node-B and RNC element managers, it is possible to configure logical resources owned and controlled by the RNC (but physically supported by the node-B). The RNC element manager sends the new cell configuration parameters to the associated RNC. The RNC is the traffic controlling entity within UTRAN. The network monitoring and fault management subsystem handles event and alarm notifications and customer complaints.

The purpose of any performance management is to collect data that can be used to verify the physical and logical configuration of the network in order to locate potential problems as early as possible. The measurement data requirements include in general:

- Traffic load on the radio interface in terms of signaling and user traffic;
- Usage of resources within network nodes;
- User activation and use of supplementary services;
- Number of pages per location area over some management platform defined time scale;
- Number of busy hour call attempts per RNC per MSC;
- Handover per RNC over a given observation time.

In the example of coexisting UMTS, BRAN, and DVB access networks, network management can be achieved by means of the telecommunications management architecture envisaged for UMTS and enhanced accordingly. The UMTS management architecture is shown in Figure 2.42.

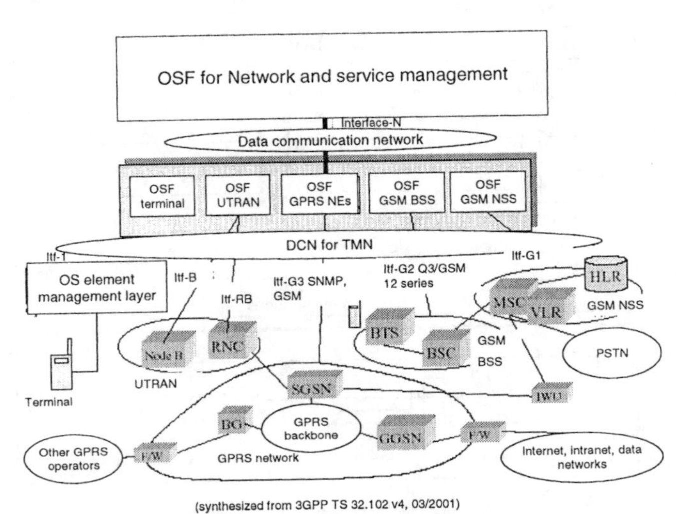

Figure 2.42 Telecom management domains and interfaces for UMTS.

The TMN provides an architecture made of operating systems (OS) and network elements (NEs) along with the required interfaces among them (Q interfaces within an operator domain and X interface between different operators). The UMTS management architecture can be mostly based on TMN. However, UMTS management must explore new management paradigms widely accepted and deployed and integrate new ones. UMTS TMN implies a focus change from net element management towards information management. Basic information such as alarms is essential for localizing faults, for instance, but it may also be the key information to allow setting up a service with a service level agreement. Enabling interoperability between network operators (providers) and service providers for the exchange of management/billing (charging) information sets new requirements on the management and physical architecture of UMTS.

The billing system (BS) is a key component for this open service and network management architecture. The wealth of information and measurements reported to the BS for profile, subscription, and service management should be sufficient to achieve this required interoperability. The UMTS charging and billing system for circuit-switched and packet-switched services as in [7, 9].

The management domain of UMTS consists of many TMNs related to the various technologies and components making up the UMTS network. One architectural requirement for UMTS management is therefore to support distributed TMNs and TMN interoperability on a peer-to-peer basis. This requirement is inherent in the architecture presented in Figure 2.42 where the interface Itf-N between the network management/service management (NM/SM) OSFs and the NE OSFs is expected to be open with standardized information models. This interface could be used by the network and the service management systems to transfer management messages, notifications, and service management requests via the NE OSFs to the NEs. The interfaces between the NEs and the NE managers are also provided in Figure 2.42.

An important aspect of the overall management architecture is the special case of the network element management architecture where one type of network element needs information coordination of a subnetwork. This is the case of the RNC and the node-B where the management information shared between these nodes will not reach the operators and is not considered as part of the UMTS TMN. This concern relates particularly to the resource management functions for the node-B located in the RNC and the resource management functions within the node-B. All other management information related to the node-B will be transparently transferred by the RNC to the UMTS TMN.

Call and event data from the circuit-switched and packet-switched UMTS domains are required for a number of network management activities. These can be summarized as follows:

- Billing of home subscribers, either directly by the mobile network providers or via the service providers, for network utilization charges;
- Settlement of accounts for traffic carried or services performed by the fixed network or other operators;
- Settlement of accounts with other PLMNs for roaming traffic;
- Statistical analysis of service usage;
- Customer billing and service complaints.

Besides the collection of this information, performed by the network element functions (NEFs), network management functions are required for the administration of on-line charging data.

This data is used to drive the charge display in the MS to achieve Advice of Charge (AoC) service. The data collected by the NEFs are sent to the appropriate OSF blocks for storage and further processing. The data required by the NEFs to provide on-line charging information is also distributed by the appropriate OSF. This is depicted in Figure 2.43.

The next section will describe each of the radio access networks adopted for the MONASIDRE management system.

2.2.1.8 Management Actions in the Radio Access Networks

UMTS RAN

The management actions in the RAN concern mostly the control and the efficient use of radio resources. RAN management consists essentially of network element configuration and reconfiguration, where key parameter values are modified to manage and optimize performance, parameter control, and triggering of network operation options (algorithms, antenna control, power control, and so on).

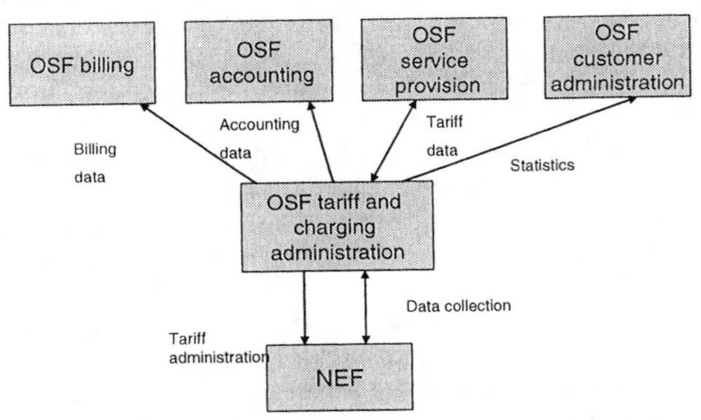

Figure 2.43 Tariff and charging administration architecture.

The SHO parameters are used mostly to control interference and benefit from macro diversity. Aspects that can be handled or modified by the management system could be:

- Active set update method;
- Hysteresis thresholds (add, drop and replacement thresholds);
- Load control in node-B is also important to deal with inner and outer cell interference in CDMA systems;
- Pilot power;
- BS power budget;
- CAC threshold;
- Amount of power reserved for control channels (DPCCH);
- Congestion control policy changes;
- Target E_b/N_0 per service as a function of traffic load assessments conducted by RNC (intra- and intercell interference);
- Maximum acceptable interference above the noise level modification due to environment changes (noise floor varies with area and environment type).

Code tree management control as a function of traffic load and ongoing services proportions is also an important aspect of traffic and interference control and management in the UMTS RAN. A nonexhaustive set of management actions that can be exercised, in terms of physical channel resources, via the code tree consists of:

- Fixed rate or rate adaptation;
- Proportion of code tree branches reserved for DSCH and reserved for DCH (depends highly on service classes being served—real-time (RT) such as conversation and streaming or nonreal time (NRT) such as interactive and background traffic).

Channel management, where reconfiguration of node-B in terms of physical and logical channel arrangements are modified, includes:

- RACH channel codes list (random access codes);
- Assigned scrambling codes to node-B or cell Ids:
- Number of carriers in each node-B.

More generally stated, it includes the establishment of logical resources in one or several cells of the node-Bs.

At the UTRAN and RNC levels, the management actions include network expansion (of existing elements or integration of new elements) and reconfiguration of network elements.

Reconfiguration concerns mostly the modification of parameters related to the network elements or entities, but it can entail a lot more important changes.

The RNC controls the traffic carrying entities (the cells in node-B) and their associated parameters. The generic cell model is held at the RNC that has real time traffic conditions and can therefore initiate a cell reconfiguration. Evidently, the management system can also initiate the reconfigurations. Both entities support the ongoing reconfiguration of cell parameters.

Among the key changes that can be initiated by the network management layer, the network self-healing actions are of particular interest. These include the following:

- Addition (or removal) of carrier/frequency from a node-B to handle traffic fluctuations and temporary hot spots.
- Transmit and receive diversity control.
- Transmit and receive antenna beam control (beam steering or switch diversity) to cover the appropriate areas (hot spots) in order to carry traffic and control interference.
- Swapping a node-B from one RNC control to another RNC (in order to handle traffic load on RNCs and redistribute the load). Node-B expansion management procedures are triggered in all affected neighboring node-Bs.
- Upon a new node-B installation or an existing node-B parameter modification, the management platform or layer must reset/modify

parameters in existing node-Bs. An automatic configuration of all parameters that are not vendor specific should at least be handled.

- The management actions include remote software updates using either a pull or push technology to update the network entities and particularly the RNCs (where much of the resource management strategies are expected to evolve gradually as UMTS networks become operational and experience is gained over time with respect to service usage and user behavior).

Note that the management actions can be initiated by the system monitoring (MASPI in the system architecture of Figure 2.7), which triggers the network optimization procedures (located and related to the RMS). Once the RMS module has identified the appropriate strategy, the MASPI is instructed to conduct the actual management actions via the management platform on the network elements. The information and measurements that can be used to monitor the system are typically the following:

- Percentage of blocked requests;
- Traffic load per application;
- Lost packets;
- Proportion of connections in poor QoS conditions (for instance, the proportion of mobile users operating close to their maximum power).

In order to identify the possible modifications that bring an improvement to the overall network performance, a collection of measurement data within a cell and in node-B must be performed. On the basis of these reported measurements, the network optimization procedures can trigger network reconfiguration or configuration or even expansion. Even if focus is more on the reconfiguration aspects, the management procedures are fundamentally the same. The measurement reports include the following information:

- Number of active channels in the cell along with the data rates/service used on these channels. The measurements occur over given time periods (which can also be modified by the management system).
- Additional important information that could be supplied by the RNC to the management system is an estimation of QoS via the number of flows (connections) that are lagging, synchronized, and lead their required data rates.
- Sector call setup attempts. This number has to be measured along with the number of setup failures indicating whether this has occurred in node-B or in the RNC.
- Number of RF losses per cell must be counted and correlated with a failure indicator.

- Node-B total transmit power has to be reported to the management system. This is a way to evaluate interference to surrounding cells.
- Received signal strength indicator (RSSI) for the FDD mode must be reported to compute the uplink interference at the node-B.
- The received and transmitted signal code power at the antenna connector must be measured.
- Within node-B a resource usage (channel number, code number) report must be indicated.
- Data rates used and the multicode usage.

For handover performance purposes, a number of successful/unsuccessful handover attempts are necessary for optimization of the UTRAN and the core network (this can enable RA and LA area resizing or control). The handover report shall contain intra and intersystem/domain handovers, namely:

- Softer HO;
- Intra/inter-RNC soft HO;
- Intra/intercell hard HO;
- Internode-B hard HO;
- Inter-RNC hard HO;
- HO UMTS to GSM and GSM to UMTS;
- HO from UMTS to UMTS FDD and TDD;
- HO UMTS to BRAN, DVB and vice versa.

HIPERLAN RAN

The management of HIPERLAN networks is essentially based on TMN and the use of SNMP. This is also true of IEEE 802.11 WLAN, which provides a set of parameters via its MIBs. The important parameters are the power levels used by the APs and report on traffic load experience over the AP interfaces (uplink and downlink load on the radio Interface, ethernet interface). HIPERLAN Type 1 provides essentially the same information upon which MONASIDRE can rely to achieve coordinated management in a diversified radio environment. The number of packets over these interfaces is counted and reported, and the number of lost packets is also made available. Each packet is time stamped and its source is identified (TCP or UDP packet). The actual signal strength and the QoS are passed to the management plane as well. Since the AP does not conduct any service differentiation, the reported link quality measures can be used by the access routers into intranets and the Internet to achieve service differentiation. The load on each AP can also be used to control traffic and eventually transfer a number of connections on a geographically collocated radio access technology provided it can handle the flows and respect the required bit rate and QoS. This interRAN handover or connection transfer can be achieved with the help of the MONASIDRE management system. For instance, a handover from HIPERLAN to

UMTS FDD can be achieved via interaction with the management planes of these two technologies and the intersystem handover (ISHF) function in the UMTS core network that handles intersystem hand off. This kind of intersystem handover can also be motivated by service cost minimization for the user and bandwidth availability in each technology. The allocation to a given RAN is a function of time (period of the day, week) and user subscription profile (flat rate fee or time dependent connection cost). The preferential allocation of users to a given technology may change over time if tariff policies are changed by the network or service providers. These changes may occur for strategy reasons resulting from the analysis network service usage statistics or simply for economic and SLA modification reasons. The management system should enforce the policy changes.

For example, the management information bases (MIBs) defined in the BRAN specification documents [38] concern mostly the provision of a common view to the network manager (human or system) of HIPERLAN devices and particularly APs. Basic performance and fault monitoring and a basic set of configuration parameters is defined for network monitoring and control, respectively. Systems management, like device setup and software upgrade, are vendor specific and are placed in vendor-specific MIBs. Each HIPERLAN Type 2 (H/2) device with SNMP support contains at least part of the standard Internet MIB (MIB II) in addition to its own H/2 MIB. The SNMP network management requires an IP core network. Structured of Management Information (SMIv2) is used for the definition of the H/2 MIB. SNMPv3 is recommended for security reasons but is not mandatory. It is important to note that the H/2 radio interface characteristics are mapped on an interface entry of the interface table of MIB II, making a H/2 device appear like any network device to network management. The utilization of the H/2 interface can be calculated through some of the key objects (ifSpeed, ifInOctets and ifOutOctets) in the interface table entry of which an excerpt of interest is reported in Table 2.13. The ifInOctets and ifOutOctets measures are provided as normalized octets. A normalized octet represents a constant fraction/portion of the MAC frame. Counting only the data octets does not tell the use of the interface because of link adaptation. The actual utilization depends on MT location; a user using a data rate of 5 Mbps close to the AP will end with a many times higher (five times more) rate than the AP H/2 interface. The objects ifInOctets and ifOutOctets are not showing the data traffic through the AP, but the traffic can be estimated using other attributes such as ifOutUcastPkts, ifInNUcastPkts, ifInNUcastPkts, and ifOutNUcastPkts.

The actual interfaces for an H/2 AP connected to an ATM and Ethernet network are shown in Figure 2.44, which also depicts the management infrastructure. Each radio transceiver handles the traffic originating from its cell.

Table 2.13

Excerpt from the Mapping of H/2 Interface Entry

Object	Description
If Speed	An estimate of the current bandwidth in bits per seconds for the interfaces
ifInOctets	Total number of octets received on the interface (from media)
ifOutOctets	Total number of octets transmitted out to the interface (to media)
ifOutUcastPkts	Number of packets that higher level protocols requested
ifInNUcastPkts	Number of nonunicast (subnetwork-multicast or subnetwork-broadcast) packets delivered to a higher layer protocol
ifInUcastPkts	Number of subnetwork-unicast packets delivered to a higher layer protocol
ifOutNUcastPkts	Number of packets that higher level protocols request to be transmitted to a nonunicast (subnetwork-multicast or subnetwork-broadcast) address, including those that have been discarded or not sent

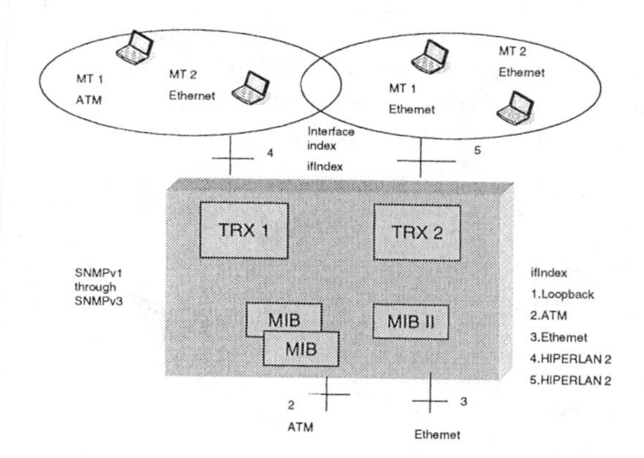

Figure 2.44 HIPERLAN/2 management infrastructure.

The AP contains a set of vendor-specific MIBs but typically supports at least part of the Internet MIB II. The interface table of MIB II contains at least one entry for each physical network interface including the AP of the H/2 radio interface. In the H/2 MIB a table shows the handovers from neighboring cells. The table indicates how MTs are moving to this cell from the neighboring cells.

2.2.1.9 Management Actions in a Heterogeneous Radio Environment

Service and network management actions that can be conducted on network entities can be listed as follows:

- Routing area and location area resizing (change in MSC area and SGSN routing area by BTS and node-B relocation or swap).
- Bandwidth provisioning for a given user or a group of users over a given (scheduled) time period according to the user service upgrade request. For instance, providing extra bandwidth or throughput to an expected audiovisual conference.
- Radio access technology selection according to service type and subscription/profile/cost policies. This can be achieved through COPS to connect a terminal to a given radio technology according to rights, subscription, profile, cost, service type, and requested bit rate.
- Traffic load management through load balancing among available technologies.
- RNC relocation (falls in the realm of network configuration management or reconfiguration management).
- SGSN relocation and GGSN relocation.
- Management in the SIP servers through call and connection control (management is achieved via the CSCF upon call establishment and call control based on state information, provided for instance through COPS. By enforcing policies that are modified in a static or dynamic way, calls are connected to the appropriate network entities, e.g., those that are not loaded).

Example of Network Management Through Dynamic State Modification

In this example, network management usage is demonstrated by network reconfiguration through dynamic state modification.

Several cases where this could happen exist. These are the following:

- Dynamic access network relocation;
- Dynamic GGSN relocation;
- Dynamic node-B relocation;
- Dynamic routing update.

These examples are not exhaustive but they represent different layers of the complete network structure.

In the example of dynamic access network relocation, a connection has already been set up through a UMTS network, but for certain reasons it is necessary to switch to another technology like a WLAN (HIPERLAN or IEEE

802.11). The actions required in every layer to achieve service provision in the new domain are identified.

First, the IP layer has to be reconfigured. It is assumed that DHCP is the way addresses are exchanged along with any pertinent information. Information is received through management procedures that a new node is accessible via the WLAN. A DHCP process provides the node temporary addresses to start AAA procedures.

Once this is achieved, the COPS server is informed of the presence of the new IP node. It obtains the policy information base (PIB) containing authorization information (IMSI, for example). An access to the COPS policy server verifies that this particular user is authorized to use the WLAN access. If so, the COPS server will initiate a deallocation procedure to the HSS stating that the user is no longer attached to the service and that it is necessary to free the used resources.

On the other side, the COPS server considers the required resources for the WLAN access. If the connection is to use best effort, then COPS can just configure the access router to be able to achieve DIAMETER accounting procedures based on time or bytes sent from the user terminal. If the connection is to be set using a certain QoS like DiffServ, then the COPS server will update the access router (AR) with the necessary marking and shaping SLAs to enforce the required QoS.

In the example of dynamic GSN relocation, it is considered that a connection has already been set up through a GPRS call procedure (ATTACH, ACTIVATE). For a reason that originates from the management, an alarm shows that the load on the GSN is high. This state information is relayed to the TMN management and COPS server that will divert the call to an alternate GSN to enforce a policy on traffic load limits, for instance. In this manner the management system can achieve load balancing and the dynamic relocation of a GSN. In fact the same can also apply to the CSCF functions themselves where the proxy CSCF, the interrogating CSCF, and the serving CSCF roles can be modified through TMN and COPS. TMN would manage network entities through its OSF management planes while COPS would handle QoS according to the state changes. COPS would reallocate bandwidth in network elements and links to adapt to the load balancing actions. This is depicted in Figure 2.45.

Dynamic node-B relocation means that a node-B relating SNMPv3 information indicates that the cell resources are running out and there is very little capacity left. In this case COPS is not necessary to handle the unavailability of resources. Simple management procedures that will look for all backlogged connections on the RNC and on the node-B should be sufficient. The new node may be identified, however, in a COPS PIB. When this is done, all radio resource parameters of ongoing connections should be mapped on to the new node-B. The COPS server should update the available resources after the new cell is activated. The old node-B should then be removed from the available COPS PIB wireless node-B resource. The necessary information per node-B, kept within a COPS policy server, should typically be the list given in the GPRS framework.

The procedure of dynamic routing update is the same as that described in the COPS policy examples. The same steps apply to change border gateway protocol (BGP) weights on the edge routers. The objective here is to control multimedia calls and sessions routing within a network provider domain and select the edge router towards external domains that would lead to the most cost-effective solution for the user and the network provider, thus reducing costs related to the provision of service by a third party.

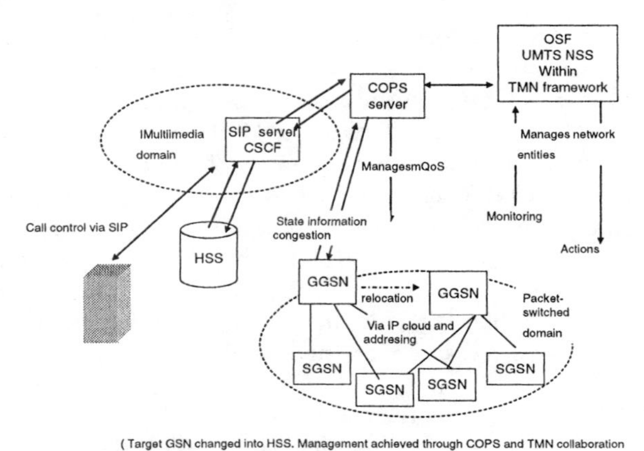

Figure 2.45 GSN relocation via COPS policy enforcement.

TMN management would act on the network element through SNMPv3 (changing weights), and COPS would be in charge of achieving QoS dimensioning bandwidth allocation and DiffServ control in the edge routers. The interconnection network needs reconfiguration as well for traffic flow and rate changes in the links. This is shown in Figure 2.46.

2.2.2 Summary

Management of heterogeneous networks is a complex problem involving definition, evaluation, and optimization, and it has influence on the achieved systems performance in terms of delays (BER, PER), QoS, capacity, throughput, power consumption, and so forth. In the context of next generation systems, because of the expected traffic asymmetry, RRM strategies should be devised from the perspective of both uplink and downlink requirements. PHY layer performance can be used to feed in simulations related to optimization of RRM algorithms. The results obtained from such input can be valuable in the context of system reconfigurability [41] and QoS management.

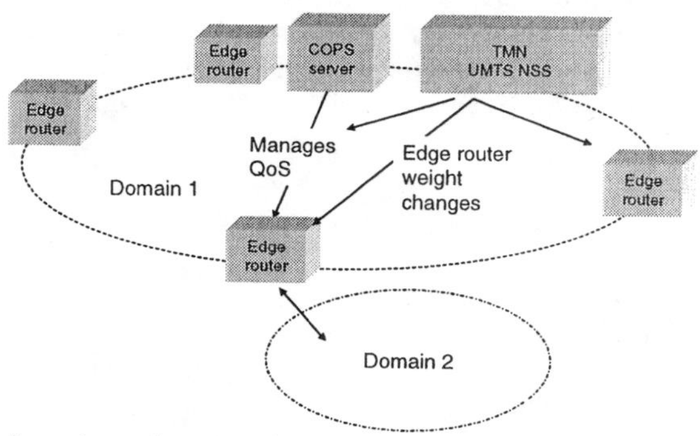

Figure 2.46 Dynamic routing update example.

2.2.3 Spectrum Efficient Interworking of Mobile and Digital Broadcast Systems

It is envisaged that for some applications, multiple users can receive the same data at the same time [42]. The benefit of multicast and broadcast in the network is that the data is sent once on each link. For example, an SGSN will send data once to an RNC regardless of the number of node-Bs and UEs that wish to receive it. The benefit of multicast and broadcast on the air interface is that many users, located in the same cell, can receive the same data over a common channel, thus not clogging up the air interface with multiple transmissions of the same data.

With the increasing use of high-bandwidth applications in 3G and next generation mobile systems, especially with a large number of users receiving the same high data rate services, efficient information distribution is essential. Thus, broadcast and multicast are techniques to decrease the amount of data within the network and to use resources more efficiently. The concept of mobile broadcast multicast systems (MBMS) is clearly defined by [43] as point-to-multipoint downlink bearer service for IP packets in the PS domain (not supported in CS Domain). Goals of MBMS are to provide an efficient use of radio/spectrum resources. Data (possibly) is transmitted over a common radio channel (broadcast or multicast); to provide video-centered content (e.g., streaming services at 8 to 128 Kbps). These services would be charged to the client or to a third party (which would allow, for example, TV-like advertisement-based business models). There would be a limit to the maximum extent the impact would have on already-defined UMTS standards and infrastructure by the use of existing common channels (FACH, PCH, BCH, RACH, CPCH) and associated protocols (BMC).

Difficult points lie mainly in handover, roaming, and mobility management. QoS issues include service continuity (service area), charging, handline UE RRC states (idle), power control, and service announcement.

Until not very long ago, no real radio broadcast bearer was considered by UMTS for multicast and broadcast services (only some dedicated broadcast radio channels were described for paging, synchronizations, and SMSCB9, and even though multicast support is standardized by 3GPP10 [6], radio delivery of IP-multicast is based on preliminary duplication of IP packets sent individually to each user).

2.2.3.1 Summary of MBMS UTRAN Requirements

MBMS services and technical requirements interpreted and appropriated by UTRAN [44, 45] can be summarized as follows.

The overall ambition of MBMS is to support multicast and broadcast services efficiently on the radio interface. Therefore, 3GPP differentiate between two point-to-multipoint UMTS modes:

- *Broadcast mode*: data is transmitted to all users in a broadcast area.
- *Multicast mode*: data is transmitted to a group of users that have subscribed to be a member of that group.

As in a classical IP multicast implementation, both of the modes will support service announcements in order to discover active MBMS services. Then users will be able to activate the desired service. 3GPP definitions regarding service notification and activation are as follows [45]:

- *MBMS notification:* A mechanism that informs users about the availability or coming availability of a specific MBMS RAB data in one given cell.
- *Service announcements/discovery:* Mechanisms that should allow users to request or be informed about the range of MBMS services available. Operators/service providers may consider several service discovery mechanisms. This could include standard mechanisms such as SMS, or depending on the capability of the terminal, applications that encourage user interrogation. Users who have not already subscribed to a MBMS service should also be able to discover MBMS services.
- *MBMS broadcast activation:* A process that enables the data reception from a specific broadcast mode MBMS on a UE. Thereby the user enables the reception locally on the UE.
- *MBMS multicast activation (joining):* Explicit point-to-point UE-to-network signaling, which enables a UE to become a member of a multicast group and thus start receiving data from a specific MBMS multicast service (when data becomes available).

To improve the basic requirements as defined in [45], two new functionalities can be introduced for MBMS, namely:

- *Counting:* A function that UTRAN performs when it wishes to identify the number of multicast subscribers (all joined subscribers, or just above a "threshold" in a particular cell that wish to receive a multicast session for a particular service.
- *Tracking:* A function that allows UTRAN to follow the mobility of multicast subscribers.

Inherently it can be used as a means of counting multicast subscribers. Considering specifically the requirements on RAN for the introduction of MBMS, a number of requirements have been identified and are currently agreed upon [45]. This RAN work is driven by the wish that in order for MBMS solutions to be adopted, the impact on the RAN physical layer should be minimized and the reuse of the existing physical layer and other RAN functionalities should be maximized. Except for this basic requirement and without giving an exhaustive list of the others, it can be noted that MBMS data shall be delivered only in cells that contain MBMS UEs joined to the multicast group.

Moreover, MBMS multicast mode transmissions should use dedicated resources (point-to-point) or common resources (point-to-multipoint). The selection of the connection type (point-to-point or point-to-multipoint) is operator dependent, typically based on downlink radio resource environment, such as radio resource efficiency. RAN proposes that a "threshold" related to the number of users may be used.

Some of the requirements for MBMS transmission are strongly connected to the introduction of the new counting and tracking functions. The other RAN requirements that have to be highlighted are as follows:

- Simultaneous reception of MBMS and non-MBMS services as well as simultaneous reception of more than one MBMS services are possible, but shall depend upon UE capabilities.
- A notification procedure shall be used to indicate the start of MBMS data transmission. This procedure shall contain MBMS radio bearer information.
- A mechanism to enable the network to move MBMS subscribers between interfrequency cells is required.
- Reception of MBMS shall not be guaranteed at RAN levels. MBMS does not support individual retransmissions at the radio link layer, nor does it support retransmissions based on feedback from individual subscribers at the radio level. This does not preclude the periodic repetitions of the MBMS content based on operator or content provider scheduling or retransmissions based on feedback at the application level.

The 3GPP also identifies a set of functions relating to service establishment and radio resource management and control for MBMS services, to mobility, and to transmission of MBMS data. The work regarding the definition of these functions is ongoing [45].

The MBMS UTRAN architecture principles are as follows: one context per MBMS service, one Iu flow per RNC and per MBMS service. The MBMS Iu bearer can be mapped to multiple DTCH and/or point-to-multipoint traffic channels, depending on the decision of the MBMS control function. Regarding the bearer channel, two logical channels are dedicated to MBMS:

- MBMS control channel (MCCH), defined as a point-to-multipoint downlink channel for transfer of control plane information between network and UEs in RRC connected or idle mode;
- MBMS traffic channel (MTCH), defined as a point-to-multipoint downlink channel for transfer of user plane information between network and UEs in RRC connected or idle mode.

A multicast functionality is also added in the MAC architecture to support MBMS. This functionality, entitled MAC-m, deals with mapping between physical channels and MBMS-related logical channels (see [44]). However, MBMS channel (logical, physical, and transport channel) structure is not yet defined. The UTRAN MAC architecture is shown in Figure 2.47.

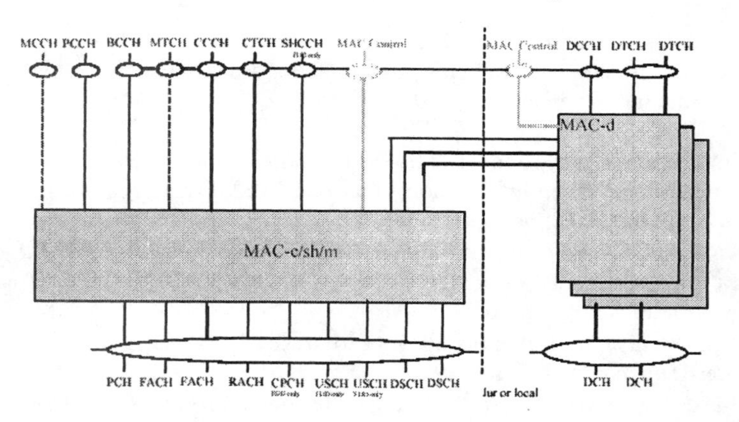

Figure 2.47 UTRAN MAC architecture.

2.2.4 Techniques for Mobile Multicast

Nowadays there are many services in which the same multimedia information is of interest to a large group of users, who want to receive it in a limited time frame.

In such cases, by using multicast, network resources can be saved and can serve the entire group instead of just individual users [42]. There is also an increased usage of mobile communication, and, in order to be resource efficient, it is necessary to merge the concepts of mobility and multicast. Moreover, when considering a hybrid multiaccess system and a large group of mobile users joining the same multicast group through these different access systems, we also see that a management entity for dealing with such multiaccess scenarios is required. This section describes the concepts for mobile multicast support in hybrid multiaccess systems. It suggests some enhancements to the existing mobile multicast approaches and defines the management entities responsible for optimizing the multicast transmission in such hybrid multiaccess systems. First, an example management model is given, which is responsible for optimizing the delivery of mobile multimedia services to a large group of mobile users in a multiaccess environment.

A group management model has a decision part, which looks at how to achieve the optimization, and an execution part, which takes care of the realization (i.e., performing the required handovers in order to achieve the optimal transmission).

This second part of the model makes use of one of the suggested enhanced approaches.

For a hybrid multiaccess system, it is also important to analyze how to couple DVB-T and UMTS systems, as well as the implications of this on the group management model.

Another issue to consider is that mobile multicast services cover both individual users as well as mobile networks. Therefore, multicast support for mobile networks should be analyzed as well as its interaction with the group management model.

2.2.4.1 Multicast Concepts

In this section several proposals for dealing with multicast over UMTS are analyzed. A general overview about multicast services issues and the state of the art of 3GPP standardization is given. Input to this section was based on work performed within the scope of the IST project OVERDRIVE [2].

At the physical layer the implementation of an OFDM-based bearer is suggested, mainly based on the fact that broadcast systems, such DVB-T already exploit it [45]. Then, H-ARQ is analyzed to increase transmission reliability, with support of investigation results. The problem of power control for a multicast group is presented by comparing three different strategies. Furthermore, some issues about the transport channel suitable for multicast are addressed and some results about fixed power allocation for a multicast channel are provided. These issues are separately investigated, but there is obviously an advantage in combining them.

This section is basically composed of two parts. The first part describes the concepts for UMTS enhancements; the second part presents preliminary results of investigations.

OFDM-Based Multicast Bearers

OFDM is one of the technologies that are proving themselves well suited to mobile radio access for high rate and multimedia services (i.e., DAB, DVB-T, 802.11a). Given the availability of this radio technology, its applicability is studied for enhancing UTRAN.

OFDM for enhancing UTRAN was first proposed at 3GPP RAN1 in February 2001 [43]. It was considered as being relevant for UTRAN Release 6, and the corresponding study item was called "Analysis of OFDM for UTRAN Evolution" [46]. The objective was to consider the performance and feasibility of OFDM in the mobile environment and to develop scenarios in which OFDM may be introduced in UTRAN in Release 6.

The following list provides examples of areas considered:

- Support for peak bit rates in the order of 10 Mbps and above;
- Support for MIMO and other advanced antenna array techniques;
- Support for personal, multimedia, and broadcast services;
- Deployment scenarios, including frequency reuse aspects, within diverse spectrum allocations.

The study of multicast is thus included in the scope of the OFDM study item. Particular investigations for multicast have not yet begun. High-speed OFDM downlink has many advantages for multicast on the physical layer. These can be summarized as follows:

- OFDM has been chosen for the physical layer of digital broadcast applications such as DVB-T, DAB, and DRM, mainly because equalization is easy with OFDM. This means that OFDM would be very well suited for multicast on the physical layer for UMTS.
- Multiple access on the downlink is easy in OFDM: this modulation transmits in parallel several subcarriers at different frequencies. There is no intracell interference between these subcarriers. A multicast channel could be allocated as a subset of those carriers. There is no intracell interference between subsets.
- On an OFDM downlink channel, there would be limited power control, and possibly no power control, mainly because of the orthogonality of users. This is an advantage over WCDMA. This means there is not much difficulty on the physical layer level to convert a downlink OFDM unicast link inside a cell into a multicast one.

- OFDM allows single frequency networks deployments (SFN): the same information is sent by several base stations, at the same frequencies (the same subcarrier sets). In OFDM, as long as the channel echoes are absorbed by the so-called cyclic prefix (or guard interval), they do not create interference. Those echoes also raise path diversity, which is frequency diversity in OFDM. The coding (FEC) operation then takes benefit from this diversity. Thus, OFDM naturally performs soft handover in a SFN case, as long as the cyclic prefix duration is well chosen.
- Thanks to SFN, channels could be broadcast on an OFDM downlink over several cells. The subcarriers sets, and cyclic frequency shifts, frequency hopping sequences, and so on, should be the same for this particular channel.

Multicast Using H-ARQ

H-ARQ-based schemes efficiently handle point-to-multipoint communication over a wireless interface. Low density parity check codes (LDPC), as a typical example of a modern coding technique suitable to iterative decoding, have been used for performance derivation. The motivation for the study is as follows.

First, H-ARQ is increasingly envisaged in mobile communication standards. For example, it has been chosen for the HSDPA bearers in UMTS. The introduction of turbo codes in 1993 raised the interest of iterative decoding techniques, and several coding techniques have been introduced (or reintroduced) with performances close to the theoretical Shannon's limits. LDPCs are an example of such codes and are currently under investigation for being part of the new release of the DVB standard.

Packet-oriented data transmission is gaining more and more importance in wireless communication systems. Typically, data transmission is not strictly delay-sensitive but requires a virtually error-free link. In order to provide such level of reliability over wireless channels affected by propagation impairments such as fading, ARQ schemes can be combined with channel coding (H-ARQ). In brief, when fading varies slowly over the duration of a codeword, coding takes care of the channel noise, while retransmissions take care of bad channel conditions (deep fades).

In [42], analysis of the throughput achievable by H-ARQ schemes based on an incremental redundancy over a block-fading channel is given. An information-theoretic analysis assuming binary random coding and typical-set decoding is provided. The performance of LPDC ensembles with iterative belief-propagation decoding is investigated, showing that, assuming infinite block length, LDPC codes yield almost optimal performance. Unfortunately, practical finite-length LDPC codes incur a considerable performance loss with respect to their infinite-length counterpart. In order to reduce this performance loss, two very effective methods are proposed: (1) using special LDPC ensembles designed to provide good frame-error rate (rather than just good iterative decoding threshold); (2) using an outer selective-repeat protocol acting on smaller packets of information bits. Surprisingly, these two apparently very different methods yield almost the same performance gain and recover a considerable fraction of the optimal throughput, thus making practical finite length LDPC codes very attractive for data wireless communications based on incremental redundancy H-ARQ schemes.

It is possible to analyze the throughput performances in terms of the information outage probability based on an existing theorem that links the information outage probability and probability of error decoding [47].

2.3 MULTIRADIO ACCESS

There is broad consensus that post-3G networks will require the integration of different air interface types of widely different characteristics, to cover a wide range of application scenarios within a common framework. To optimize the integration it is desirable to maximize the network parts that are common between the different physical and link layers, which leads to the integration at the edge paradigm [1]. At the same time, adoption of a native IP approach while making best use of air-interface capabilities means that we must consider the IP-link layer interactions in more detail.

Traditionally, IP protocols do not assume any specific functionality from the underlying link layer. For example, IP QoS architectures are based purely on IP-layer decision-making, packet buffering, and scheduling through a single link layer service access point. The recent evolution of wireless communication has driven the design of wireless link layers including more functionality than what is required for simple first input first output (FIFO) queuing. For example, the HIPERLAN/2 link layer provides priority-based packet scheduling and support for guaranteed bandwidth reservations, as described in [49].

Unfortunately, these new functionalities are likely to introduce duplications and layering violations when the IP layer functions are taken into account. The IST project BRAIN [4] studied a generic service interface called IP-to-Wireless (IP2W) between the IP layer and the link layer. This service interface is supported by an access technology specific convergence layer, which is meant to coordinate the IP QoS management and packet scheduling with the link layer and also to enable better support for address management and handovers with different link layers, for example. The resulting protocol stack functional model is shown in Figure 2.48. This model was extended in the IST project MIND [4] to include support for ad hoc behavior, specifically issues such as address handling and routing support, and also support for unified resource management mechanisms.

Effectively the integration of access technologies at the edge of the access network allows for the normal mobility support to be equally applicable to handover between different access technologies. This creates the ability to have a seamless handover between access technologies: changes in access technology do not dictate a change in access network, as the latter remains common. Therefore, the network performance is comparable to a handover in a traditional RAN (e.g., only local routing updates). In the same way, a common QoS framework can be applied across the access technologies. This means that the QoS information from the lower layers can be used in the process of handover to ensure the application is aware of the QoS being offered on the link and then an adaptive application can adjust. Equally an application can provide QoS requirements that request the use of the best access technology for the level of desired QoS.

The project MIND investigated extensions to the IP2W for its support. RRM functionalities needed to be both above and below the IP2W as routing occurs above but resource-specific functionality occurs below. Extending the approach of the IP2W to enable the provision of generic approaches above the IP2W meant that the possibility of a deployed RRM system working across a number of access technologies forming a complex communications network would be greatly increased. This would then provide significant input to the ability of the network to adjust and allow for access technology switching depending on demands on the network and would allow for the more efficient use of all resources available within the system especially related to the case of bandwidth limitations.

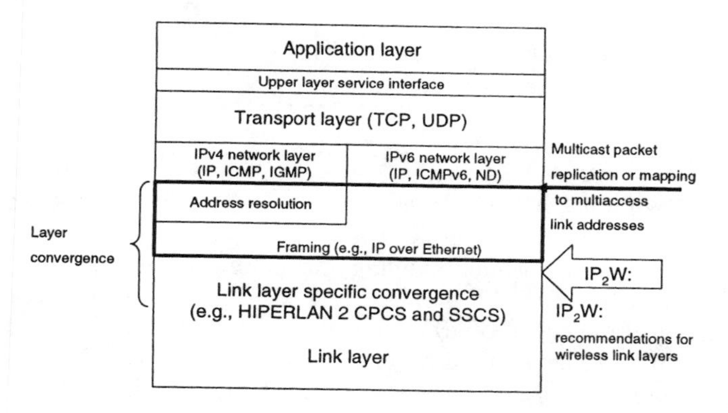

Figure 2.48 The IP2W BRAIN protocol model. (*From*: [4].)

This discussion of multiradio access networks leads to the consideration of multihoming. By the extension of IP through the RAN to the edge multihomed terminals become a realistic possibility. A terminal may, for example, have two or more coexisting connections via a singular access technology or a number of different access technologies. This raises a number of new issues that were investigated in the frames of some FP5 IST projects. One example of this would be whether a terminal with two connections should advertise link availability across the two as this could cause a hole to be produced in the routing table as routes on the b-link are advertised on the a-link. This could be advantageous in terms of connectivity but equally could mean that careful bounds on routing tables may have to be considered [1]. Another issue would be that of addressing: does a multihomed terminal have a separate address for each link or does it retain a single address advertised on both links.

2.3.1 Multicast Support

The support of IP multicast has been defined as optional in the UMTS packet domain (see Section 2.2.1.1). The GGSN will support the Internet Group Management Protocol (IGMP) towards mobile terminals and multicast routing protocols towards the Internet. Downlink multicast packets are sent point-to-point to each of the terminals that has joined that multicast group. The corresponding protocol stack is shown in Figure 2.49.

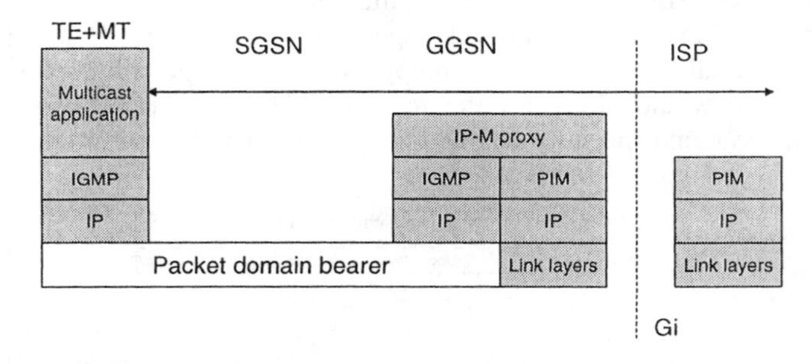

Figure 2.49 UMTS configuration for IP multicast managed access network. (*From*: [4].)

The main goal of the multicast is to offer resource-efficient multicast support in the ad hoc fringe while retaining compatibility with native IP multicast especially in the end nodes. Several different configurations and mechanisms have been studied in the scope of IST projects, and the conclusion was that the best choice would be to have two different multicast routing protocols for the ad hoc extension and the access network, as depicted in Figure 2.50.

Here, all IGMP packets coming from the end nodes towards the first hop multicast router (FHMR) and vice versa, instead of being tunneled or unicast, are forwarded using an ad hoc multicast routing protocol. This approach allows us to support many nodes with the benefits of using native IP multicast. The single point of failure of the GGSN is avoided, and the amount-stored state towards the core of the network is much reduced (only groups joined, rather than the terminals in each group). This leads to the reduction in air interface use towards the edge of the network where air interface resources are most constrained [51].

In addition to the problems of addressing, scaling, and security, the network layer for any wireless access network must clearly have some interaction with the radio resource itself. Handover will almost invariably be triggered by radio conditions, and a decision to admit a session can only be made by taking into account the detailed flow characteristics of all the traffic in that cell. Thus, the two central problems of the access network shown in Figure 2.51—mobility and QoS—are closely linked with the problem of RRM.

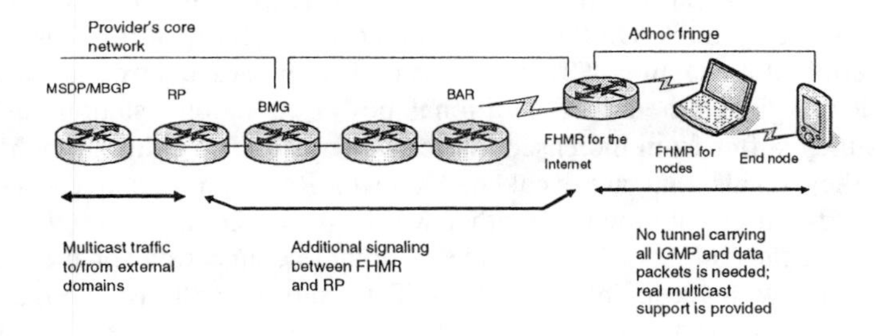

Figure 2.50 MIND multicast architecture [50].

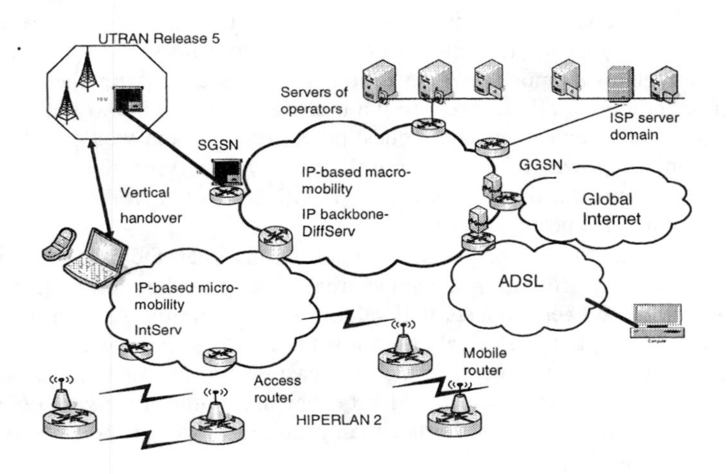

Figure 2.51 Architecture developed by the projects BRAIN/MIND. (*After*: [4; 50].)

With the aim of being independent of any specific air interface, radio resource algorithms should be studied in themselves. However, it is possible to adopt a generic approach, whereby the radio specific aspects are handled using message flows whose content is opaque to the network layer. Only the logical distribution of these messages to RRM algorithms, the triggers that these algorithms generate (such as recommend handover to cell X) are then considered. More explanation of this architecture is contained in [51] and it is related to the functionality of the IP2W interface described in the beginning of Section 2.3.1.

2.3.2 Power Control in Multicast Transmission

Tight and fast power control is one of the most important elements in a wireless system. It is an efficient technique to control interference and to increase capacity in cellular radio systems. On the unicast (point-to-point) downlink (base station to mobile terminal link) in a CDMA system, fast power control is desirable to provide a marginal amount of additional power to mobile stations at the cell edges, as they suffer from increased intercell interference. Additionally, it helps to enhance the signals that are weakened due to Rayleigh fading caused by the mobility of the user equipment. In other words, power control is used to maintain the quality of the link and thus provides a reliable connection for the receiver to obtain the data with acceptable error rates. Transmitting with just enough power to maintain the required quality for the link also ensures there is minimum interference affecting the neighboring cells on the downlink [2, 52].

In multicast (point-to-multipoint) transmissions, all the benefits described above apply when power control is employed. In a simplistic typical case of power control in physical multicasting, a base station transmits at a power level that is high enough to support the connection to the receiver with the highest power requirement among all receivers in the multicast group. This would still be efficient because the receiver with the highest power requirement would still need the same amount of power in a unicast link, and by satisfying that particular receiver's requirement, the transmission power will be enough for all the other receivers in the multicast group.

However, due to the uncorrelated channel fading that the receivers will experience, even if they are within the same area (as long as they are 6 cm away from each other), the power requests will seldom be unanimous. As a result, the transmitted power is kept at a relatively high level most of the time, which in turn, increases the quality of each receiver in the multicast group. This is because when a receiver experiences a deep fade, it is likely that most other receivers do not, hence they would benefit from the higher than required power at their receiving end.

With such a power control algorithm for multicasting, the average transmit power as well as the average received quality will be much higher than required with increased multicast users in the group. The important benefit of power control in minimizing the transmitted power (hence, intercell interference, which relates to the system capacity) is negated. Therefore, it is necessary to develop a more efficient power control algorithm for physical multicasting. The feedback commands from the MTs are essential in any closed-loop power control algorithms, and these uplink feedback power control commands will be affected by the near-far problem associated with asynchronous CDMA systems. Therefore, any power control algorithms, however efficient or perfect, are likely to be useless without overcoming or avoiding the near-far problem in the first place.

2.3.3 Radio Resource Management with Multicast Channels

Another key issue in multicast transmission is the management of radio resources towards their efficient use.

2.3.3.1 Transmitting Data over a Common Channel

No special transport channel for the purpose of multicast has yet been specified, but some proposals and preliminary studies have been provided. In [53] an investigation on power setting for a FACH multicast channel is presented, proposing a double streaming transmission where basic information for basic QoS is transmitted with the power level needed to cover the whole cell, and a second stream conveys additional information to users near the BS. In this way BS power can be saved by trading off with QoS of MTs at the cell border. In [54] the approach of combined use of FACH and DCH is followed, while in [55] a comparison between FACH and DSCH is carried out.

Therefore the driving concept to support multicast on the UTRAN is to use the existing transport channels, with minor modifications. The transport channels that are available in the downlink direction and could be used for multicast transmission are the dedicated channel (DCH), the forward access channel (FACH), and the downlink shared channel (DSCH) or its enhanced high speed DSCH (HS-DSCH). Each channel has different characteristics in terms of power control, addressing techniques, and automatic repeat request mechanisms. Furthermore, special functions like beam forming and soft handover are not available to all transport channels.

The question to be answered is under which circumstances a transport channel (with necessary modifications) is best suited for the delivery of multicast content [52].

A channel is eligible if it allows the transmission of the multicast data with the needed QoS in a spectrum-efficient manner. In a WCDMA system like UMTS, which is usually in an interference limited state, the reduction of the interference is the main target for increasing spectral efficiency, meaning that power allocation for multicast transmission should be minimized. Influencing factors for the decision of a configuration are the applied power control schemes (weakest link, for example), the way different multicast groups are addressed (e.g., one FACH per MC group or addressing using additional MAC headers), the kind of multicast traffic (e.g., traffic asymmetry), the needed QoS in terms of BER and bit loss error rate (BLER) (reliable coding and/or repetition because of missing ARQ), and special features like beam forming.

So far, some requirements have been assessed for the MBMS transport channel and the two main candidates are FACH and DSCH. Investigations are currently being carried out to enhance these two channels for MBMS service support, minimizing the impact on current channel architecture. New channel concepts could/should be investigated but the feeling is that the use of already standardized channels would speed up the provision of the MBMS services.

For an efficient overall usage of the radio resources a common channel will allow many users to access the same resource at the same time, but this depends also on the number of users belonging to the multicast group and on the type of service provided and the QoS that it can guarantee. For this reason, when only a few users are receiving the same MBMS service, it could be more efficient to use DCH for each user to minimize the transmitted power and to reduce interference. One of the key points is to determine the switching point between dedicated connection and multicast connection. The information on the number of users may not be sufficient for this purpose: a set of MTs greater than this switching point in a dedicated connection would require less power than the multicast channel would, if they are, for example, near the BS.

The main characteristics of FACH, DCH, and HS-DSCH are now listed.

FACH

This is a point-to-multipoint channel introducing MBMS group addressing in the MAC header. The following characteristics can be summarized:

- No uplink, therefore no fast power control can be performed;
- Only open loop power control is supported;
- Soft handover is not supported;
- No uplink, therefore RACH could be used for charging (overload should be avoided);
- No uplink means that modifications are required for ARQ;
- More than one FACH is needed in the cell;
- Reception of two SCCPCHs is required to allow the users to be paged while receiving the MBMS service.

DCH

- Point-to-point channel;
- Uplink channel is present;
- Fast power control is supported;
- Soft handover is supported;
- Charging is possible on a received data and/or quality basis;
- Reliable channel (ARQ);
- Connected mode only: the network knows how many users are receiving the MBMS service;
- If many users are in the multicast group, many codes and much power is used for the same information (not efficient).

HS-DSCH

- Shared channel, point-to-point connections (each TTI transmission is for one user only);
- It provides enhancements for interactive, background, and, to some extent, streaming radio access bearers;
- Uplink channel is present;
- Charging is possible on a received data and/or quality basis;
- Reliable channel;
- No inner loop power control: link adaptation is obtained with fast link adaptation (AMC+H-ARQ);
- No soft handover: scheduling and transport format selection are handled by node-B and not by RNC. Channel dependent scheduling exploits user diversity to obtain similar gains;
- Connected mode only.

The most straightforward way to implement a multicast functionality in UMTS is the use of the FACH without power control. A fixed power that has to be dimensioned taking into account the deployment, the service scenario, and the quality of service is transmitted over the cell. MTs belonging to the multicast group receive information data on this channel and cannot send uplink packets. The transmitted power should be dimensioned to cover the whole cell but it is clear that it is a trade-off between quality of service of MTs at the cell border and resource expenditure and interference caused to other MTs in the cell. When the multicast channel power is determined, it has to be found which is the offered traffic level that optimizes spectrum efficiency compared to the case in which MTs are served with dedicated channels. For example, if the operator sets the multicast channel power at 4W, this channel should be activated when the offered traffic requires a base station power over dedicated channels that is greater than 4W.

The traffic offer for level of activation is the switching point between dedicated and multicast channels.

2.4 SECURE HETEROGENEOUS ACCESS

This section specifies security architecture for Beyond 3G mobile networks. It addresses security issues related to mobility, QoS, and the protection of the first hop. The first hop is the (partly) wireless link between a mobile node and the first router in the access network. The security architecture is designed in the context of heterogeneous access whereby different radio technologies may be used to access the network [56].

There are many open issues and uncertainties concerning the specifics of post 3G network configurations. On the other hand, a security architecture must be based on a specific network architecture since it defines the functionality and interaction of concrete entities. The network reference architecture used to develop the security architecture discussed in this section is based on the architecture developed by the IST-project BRAIN [4] (see Figure 2.51), which dealt with (mostly nonsecurity related) aspects of post-3G networks. The security considerations outlined here, however, do not depend strictly on particularities of the BRAIN architecture. Indeed, they can easily be generalized to other types of all-IP mobile systems.

Addressing the various security issues of post-3G systems, a distinction must be made between security aspects of mobility and QoS. Although it is expected that there will be a considerable interaction between these two, the broad range and the complexity of the issues involved justifies a separate investigation. Activities to unify the concepts developed for both are outside of the scope of this section. Details, however, are outlined in [9].

With a focus on the use of wireless access technologies, only the security issues of the wireless transmission link are given special attention here. This topic is referred to as security for the first hop, alluding to the IP-connectivity provided to the mobile device.

With regard to the first hop, the question is at which layer to provide the protection of the first hop, taking into account that it may comprise both a wireless segment, between the mobile node and an access point, and wired segments, connecting the access point with the first router in the access network. This essentially leads to three options to protect the first hop:

1. Protection at the link layer, using the built-in mechanisms of the wireless access technologies and some additional means to protect the segment between the wireless access points and the first router;
2. Protection at the network layer, which is assumed to be implemented using IPsec;
3. A combination of both (1) and (2).

Based on UTRAN and Bluetooth, a detailed study was made [56] of the threats that remain if protection at the link layer, using the built-in mechanisms of the wireless access technologies, is not used. Optimal performance of the wireless communication cannot be guaranteed without protecting the basic signaling messages at the wireless link.

2.4.1 Security for the First Hop

The first hop is the link between the mobile node and the first router that packets encounter on the way into the network. The first hop may comprise both wireless and wired segments (e.g., in the case where wireless access points (base stations) are linked by an Ethernet backbone to which the first hop router is attached). The security issues here can be summarized as follows:

- At which layer(s) to provide integrity and encryption;
- Use or disable existing layer 2 mechanisms;
- What security context to use.

The security context for the first hop security will be derived from the security association between the mobile node and its home network. The security association between the mobile node and the visited network is established using the methods involving authentication, key establishment, and payment schemes [56, 57].

In a wireless access network the first hop always contains link layer access to the underlying wireless service (e.g., GSM, UTRAN, HIPERLAN/2, Wireless LAN 802.11b, or Bluetooth), and establishment of a link layer connection to the access point. Here the reuse of existing radio technologies does not imply that their authentication schemes apply. The main problem when developing an initial security architecture for the first hop is whether it is possible to handle the heterogeneous access procedures and security mechanisms of various wireless access systems within a unified framework. This problem was considered within the context of the model developed by the IST project SHAMAN [5], which presents one approach for its solution.

2.4.1.1 Reference Model for the First Hop

The reference model presented here is a further development of the wireless link in the BRAIN architecture [4] (see Figure 2.51) and is intended to serve as a basis for the development of the security solutions to the wireless link.

As defined by BRAIN [4], the first hop comprises the link from the mobile terminal to the first router on the network side.

In BRAIN the endpoints of the wireless link were physically identical to the endpoints of the first hop. However, this is not the usual case in access networks. More typically, the wireless link extends it from the MN to the AP or base station, and the data packets encounter the AR only after passing through a fixed connection from AP to AR.

The AP is defined as the endpoint of the wireless link. It contains the wireless base station functionality, and the required link layer interface to communicate data and control information to the AR.

The model is depicted in Figure 2.52.

It allows separation of the network routing functionality from the radio access technology. In principle, the same general-purpose AR can serve IP packets originating from terminals using different radio technologies.

In the case when AP and AR functional modules are combined physically in one network node, the model developed in BRAIN is applicable. In the case where AP and AR functionalities are separated, no previous general reference model is available. Hence, a simplified first-hop model was created by developing an extension to the BRAIN architecture, and outlining the situation of separated AP and AR is sketched in only as much detail as necessary to describe generic security solutions.

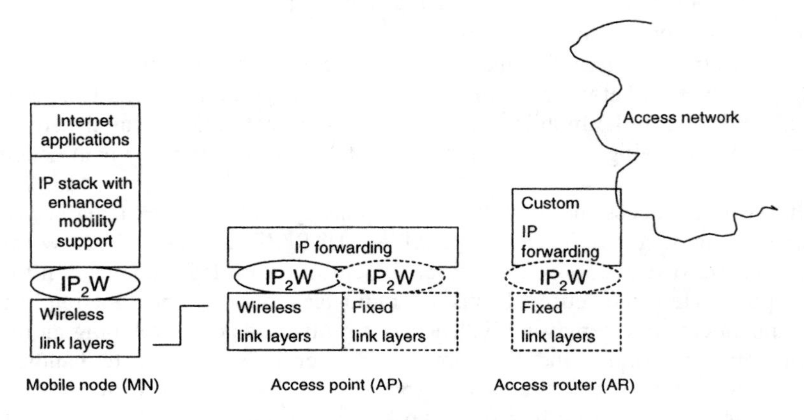

Figure 2.52 First hop reference model.

In Figure 2.53 the generalized first hop is depicted within the BRAIN framework.

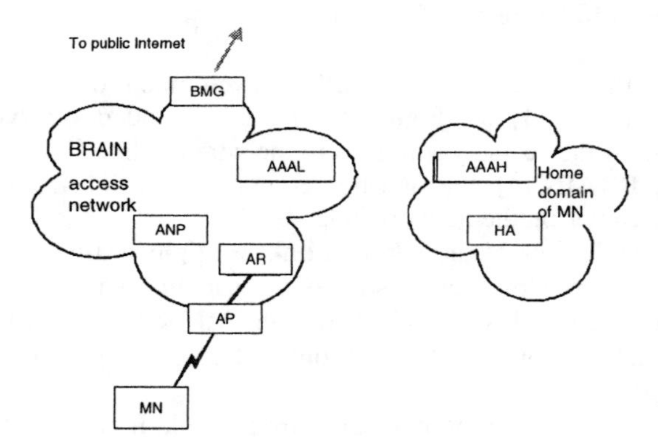

Figure 2.53 First hop in the extended BRAIN model with separated AP and AR.

2.4.1.2 First Hop Interfaces

The BRAIN architecture (see Figure 2.48, and Section 2.3.1) contains a model for IP2W and a specification of the IP2W interface. This interface is divided into two parts: the IP2W control interface and the IP2W data interface.

For a wireless link layer technology to be IP2W compatible, it is required in BRAIN that the functions performed at the link layer can be divided into two categories according to these interfaces, namely, wireless link control functions and basic data control functions.

The first category contains all signaling functions, including security management; while the second category contains functions related to the transport of user data (see Figure 2.54). The purpose of the IP2W interface is to provide a well-defined interface between the IP layer and the various technical solutions for the wireless link layer. The IP2W provides a unified framework for controlling various capabilities of the wireless link. Useful information from the link layer can also be made available to the higher layers. This interface also allows for link layer-specific optimizations of various IP-specific features in a manner that is transparent to upper layers.

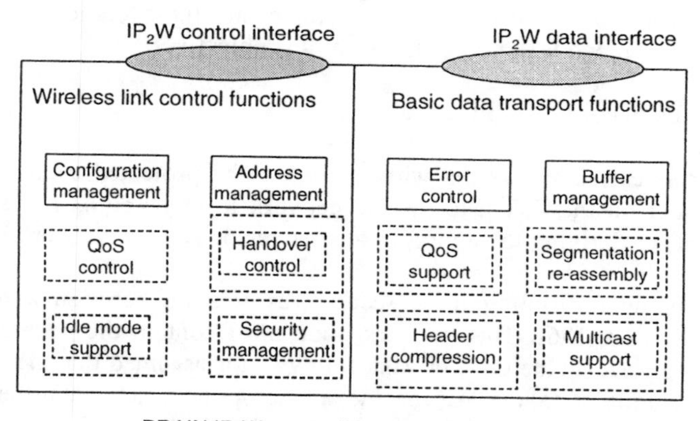

Figure 2.54 Wireless link layer functions and the BRAIN IP2W interface.

In the BRAIN architecture, the IP2W interface was implemented both in the BRAIN AR and in the MN.

In case of AP and AR functionalities separation, the link layer functionalities need to be divided into two sets, possibly duplicating or modifying some of the functions. This also implies that part of the functionality of the IP2W interface may need to be implemented in the AR as well.

The first hop IP2W interfaces were depicted in gray in Figure 2.52. In the case when AP and AR are functionally in one unit, the fixed link can be omitted, and the two interfaces are combined in one IP2W interface according to the BRAIN architecture.

The division of the IP2W interface into a control interface and a data interface induces an analogous division to the interface in the AP and to the interface in the AR.

For the purposes of defining first hop security functions, it is required that in the AP link layer the control functions include functions for security management. The IP2W interface can be used to transport key information from higher layers to link layers to be used in protecting the communication over the wireless link. On the network side the key material is first transferred to the AR, which then conveys it further to the appropriate AP to be used to protect the wireless communication between the AP and the MN. In order to perform this security task, the possibly fixed connection between the AP and AR needs to be protected.

There are basically two possibilities to implement the protection of the link between AP and AR. First, the protection can be implemented at the link layer. The second approach is to implement sufficient layer 3 functionality in AP to make it possible to use network layer protection (IPsec).

One of the main objectives in separating the access point and router functionalities is to make access points as simple and lightweight as possible. Clearly, the first approach is more in support of this objective.

2.4.1.3 BRAIN Security Architecture

The BRAIN architecture allows the wireless link to be protected using security mechanisms either on the link layer or the network layer or both. The security association used for deriving the appropriate keys resides on the network layer or higher.

Various options for setting up a security association between a network element and MN exist [56]. This network node must contain the necessary IP-layer functions to perform the task. In what follows, it is assumed that the MN has a security association with the AR or the anchor point (ANP). Further, it is assumed that this security association supports authentication and derivation of subsequent keying material for various algorithms to be used for protection of the signaling data and user data transport communication over the first hop between AR and MN.

In BRAIN, the IP2W control interface offers means and appropriate commands for establishing keys and other necessary security parameters between the BAR and the MN on the link layer using the security association on the network layer.

The BRAIN architecture does not define in which layer authentication between the access network and the mobile node is performed. The BRAIN architecture allows for separation between signaling data and user data, and also for the use of different security mechanisms at different layers.

Existing examples of first hop security solutions differ at least with respect to the following two aspects:

- Layer where MN and AP are authenticated;
- Endpoint of the link layer encryption mechanism on the network side.

For example, in UMTS the authentication between the endpoints of the wireless link is performed at the network layer. After a successful authentication, the derived encryption keys CK and AK and the integrity key IK are provided to the link layer to be used in performing the link layer security functions. In UMTS, the endpoint of the link encryption is extended beyond the base station to the RNC. Another somewhat different approach is offered by Bluetooth.

Bluetooth allows only the link authentication key, or even less, the Bluetooth PIN, to be imported directly from a higher layer. Bluetooth encryption is always preceded by link layer authentication, where the encryption key is derived. Bluetooth is also different from UMTS in that encryption terminates strictly at the endpoint of the wireless link, which is the Bluetooth module in the AP. A third example is offered by IEEE 802.11 WLAN standard, which allows link encryption to be performed independently of link authentication.

2.4.1.4 Link Layer Security

The purpose of this section is to discuss and agree on the assumptions about link layer security mechanisms. Since the goal is to provide access in heterogeneous access networks, it is necessary to make the architecture as modular as possible. The modules are connected using interfaces. For the security of the first hop, as modeled in Figure 2.52, basically two alternative approaches exist; their usefulness needs to be considered from the point of view of heterogeneous access.

The first approach considers that the terminating points of the security mechanisms are identical to the endpoints of the first hop, that is MN and AR.

The second one means that the terminating points of the security mechanisms are identical to the endpoints of the wireless link, that is, MN and AP.

Assume now that the functionalities of AP and AR are geographically separated into two different network nodes. Then these two approaches are different and mutually exclusive and then the native wireless link encryption is not usable for the first approach. Instead, the security of the first hop must rely on Layer 3 security mechanisms.

In the second approach, the native link layer mechanism can be used between MN and AP. However, as observed earlier, the fixed link between AR and AP needs to be protected. This link is used to transport secret link layer key material from AR to AP, on the one hand, and other sensitive data from AP to AR, on the other hand. The communication can be done either on Layer 3 or 2.

The second aspect that needs to be considered is what kind of communication is to be secured over the first hop. We may want to make a distinction between at least the following types of communication:

- Initial communication for authorized access;
- Subsequent network layer signaling messages, such as IP package headers (including IP addresses), QoS, and resource allocation messages;
- Communication payload (user data);
- Link layer signaling messages.

It is clear that the lower in the protocol stack the protection takes place, the more communication can be protected. In particular, network layer signaling can be protected only by means of security mechanisms implemented at lower layers.

The third aspect is protection of the access network resources in wireless access. If the security context to protect the integrity of the wireless link is derived from the AAA procedures, then the hijacking of the wireless link is protected against. This principle is used in GSM and UMTS systems.

The drawbacks caused by disabling link layer security are discussed in more detail in [57].

When discussing the pros and cons of enabling wireless link layer security mechanisms, an undeniable disadvantage is the fact that the existing link layer security systems vary a lot with respect to what kind of protection is provided, but also with respect to the strength of security. This may be partly caused by the fact that the goals and requirements for link layer security are not uniform. For some wireless systems, the security mechanisms have been provided just in order to have some security as cable replacement, while for some use cases, strong security solutions at the network or application layer are recommended.

2.4.1.5 Establishment of Security Associations for the First Hop

The security association for the first hop is derived in the authentication and key agreement procedure, which is performed between the MN and authentication, authorization and accounting home (AAAH) server of the home network of the mobile node (see section 2). This is the only way to protect the first hop from being hijacked.

The security association between the MN and the AAAH can be based either on a shared symmetric key between the MN and AAAH, or on some asymmetric cryptographic system, or both. In this context the main requirement is that it makes it possible for the parties to establish shared secret key material, what we call SHAMAN First Hop (Hop-1) Key, for securing the first hop link between the MN and the AR.

From the architectural point of view, it is desirable that the Hop-1 key as developed by the IST project SHAMAN is independent of the wireless link security technology. Then at some point the keys that are specific to the used wireless security mechanisms need to be derived from the Hop-1 key. This functionality needs to be added on top of the wireless link layer stack. The BRAIN model requires the wireless link layer to be compatible with the IP2W interface. To achieve compatibility, a specific convergence layer may need to be specified and implemented on top of the wireless link layers. Then the key conversion functions can be implemented within the convergence layer. The IP2W convergence layer is implemented in the AP (or AR), on the one hand, and in the MN, on the other hand (see Figure 2.52, where the gray shading indicates the convergence layer).

According to Figure 2.55 the Hop-1 key is established between MN and the AAAH using communication over the access network as described earlier. Then the Hop-1 key is transported over to the AP and secured using the security associations between the network nodes it passes. In the AP the Hop-1 key is taken to the IP2W convergence layer where the appropriate wireless link keys are derived from it. The same functions are performed on the MN side.

This means that the security management control functions in the IP2W compatible wireless link layer (see Figure 2.54) are mandatory and must contain mechanisms for link key derivation. The actual cryptographic mechanism to be used for this purpose remains to be specified. Ideally it takes a fixed length Hop-1 key as input and produces the required amount of cryptographic key material.

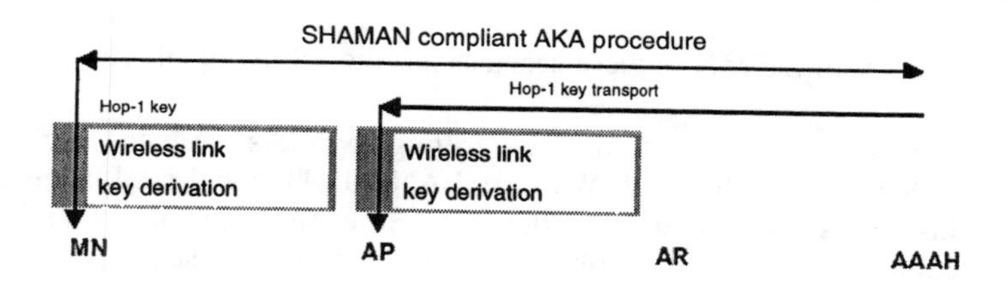

Figure 2.55 Derivation of wireless link keys from Hop-1 key.

For example, any SIM or USIM-based authentication procedure can be used to derive the SHAMAN Hop-1 key (see [58, 59]). At the IP2W convergence layer, the SHAMAN Hop-1 key can then be transformed to suitable keying material to be used to protect any wireless link using its native security mechanisms. In such a manner, keys can be derived for Bluetooth authentication (and subsequent encryption) as well as for air interface encryption, integrity, and anonymity in UMTS.

2.4.2 Secure Initial Access for Establishing Security Associations

This section discusses the security aspects of the initial communication over the first hop between the MN and the first packet router AR. In the BRAIN context the first packet router is the BAR. In the first hop reference model of Figure 2.52, a wireless link connects the MN with an AP, which is connected to the AR via a fixed net link.

2.4.2.1 Security Requirements

The security requirements for the initial access are two-fold and partially conflict with each other.

First, there is the basic requirement from the network side of giving only the minimum amount of resources to an unauthenticated, potentially unauthorized mobile node. The objective of access authorization is in correct accounting but also in the prevention of denial-of-service attacks. One approach to this problem is discussed in this section.

The second set of requirements comes from the side of the mobile node. Protection of anonymity and location privacy has become a standard requirement, for example, in the case where the purpose is to provide access over the Internet to a corporate network. Also the threat of false access networks, or false base stations, as it became known in the context of GSM networks, requires preventive countermeasures. Systems for protecting access privacy will also be discussed this section.

2.4.2.2 Phases of Initial Access

The initial access of the MN to these services has the following phases:

- Link layer access to the underlying wireless service (e.g., GSM, UTRAN, HIPERLAN/2, Wireless LAN 802.11b) and establishment of a link layer connection to the AP. Here the use of existing radio technologies does not imply that their authentication schemas apply. The security context for protecting link layer communication during initial access must be available without on-line communication between the MN and the home authentication server.
- Establishing a network layer connection to the AR and access to AAA infrastructure to carry out authentication, authorization and accounting procedures [60].
- Setup of network layer security (i.e., establishing an IPsec ESP connection between MN and AR (or ANP), which ensures integrity and confidentiality of user data on the first hop).

If implemented, phase 1 always takes place first. Phases 2 and 3 follow next. Their security requirements and functions are described in [56]. One possible approach would be to disable link layer protection. The question of whether link layer protection can be safely disabled is discussed later.

2.4.2.3 Port-Based Access Control

The IEEE 802.1X standard [61] specifies port-based network access control to prevent unauthorized Layer 2 access to IEEE 802 LAN infrastructure. There, one distinguishes between uncontrolled ports (only for authentication communication) and controlled ports (access only to authenticated ports). An authenticator handles the access request. Hence the approach is similar to [60] but with access control implemented on Layer 2 and not on Layer 3.

It may be possible to implement port-based access control in other radio technologies as well.

The IEEE 802.1X authentication framework [61] utilizes Extensible Authentication Protocol (EAP) for the transfer of authentication information. EAP was standardized by the IETF as a protocol to support multiple authentication mechanisms by encapsulating the messages used by the different authentication methods.

The term "port" is used for point of attachment to a LAN. IEEE 802.1X defines a mechanism for port-based network access. It hereby provides a means of authenticating and authorizing devices attached to an IEEE 802 LAN port that has point-to-point connection characteristics, and of preventing access to that port in cases where the authentication and authorization process fails.

The mechanism makes use of the following roles:

- *Authenticator:* the port that wishes to enforce authentication before allowing access to services that are accessible via that port adopts the role of the authenticator.
- *Supplicant:* the port that wishes to access the services offered by the authenticator's system adopts the role of the supplicant.
- *Authentication server:* the authentication server performs the authentication function necessary to check the credentials of the supplicant on behalf of the authenticator and indicates whether or not the supplicant is authorized to access the authenticator's services.

Note that the authenticator and the authentication server can be co-located, thus eliminating the need for communication with an external server. Similarly, a port can adopt the role of the supplicant in some authentication exchanges, and the role of the authenticator in others.

The authenticator makes use of two ports, an uncontrolled port for authentication messages, and a controlled port for the subsequent connection. The controlled port does not let traffic through until the authentication messages over the uncontrolled port have been successfully finished. For authentication, IEEE 802.1X uses EAP (i.e., defines the encapsulation of EAP packets). These are transferred between the supplicant and the authentication server. During this phase, the authenticator merely acts as a pass-through, looking for messages of type EAP-SUCCESS or EAP-FAILURE, depending on the outcome of the authentication.

2.4.2.4 Protecting Access Privacy

Two solutions for protecting the privacy of the MN in the initial access are described below. Both solutions are based on the security association between the access network and the home network of the MN. This security association is described in [56]. It is based on an asymmetric cryptographic mechanism, and assumes that the MN has an integrity-protected copy of the public key of its home network or AAAH, also called the home root key.

In both solutions the MN has to reveal the identity of its home network, that is, the domain that owns the root key the MN is using. The basic difference between the two solutions is what the termination point of the identity encryption is: is it some appropriate network node in the access network or is it in the home network?

The solutions presented here should be used only when no previous security association exists between the MN and the access network. After the AAA procedures have been performed and the MN is authorized, a scheme for temporary identities is recommended for use between the MN and the AAAL.

Encryption of MN Identity Terminates in the Access Network

When trying to make the first initial access to the network, the MN pages for network APs that advertise services for its home operator. The MN announces the identity of its home root key. The access network has a number of public key certificates, at least one signed by each home network with which it has a roaming agreement. The AP (AR or ANP) selects the right public key certificate and sends it to the MN. The MN verifies the certificate, extracts the public key of the access network, and uses it to encrypt its identity to the access network. The identity is decrypted in the access network and sent further to the home network of the MN. In this case, encryption mechanisms either at the link layer or at the network layer can be used. The security association is based on the public key pair of the access network. From this security association a shared secret key between the MN and AR is established. We call this key the initial key for the first hop (Hop-1-Init key) security. The Hop-1-Init key is used only for securing the initial access. It is replaced by the Hop-1 key (see Section 3.1.6) as soon as it is available.

To derive a key for securing the communication of the wireless link, the same IP2W mechanisms as in Section 2.4.1.1 can now be used. The procedure is depicted in Figure 2.56.

Some link layer encryption protocols do not hide identities of the parties in the link layer security protocols. A typical example is Bluetooth where the authentication is based on the true Bluetooth device address of the claimant. For the purposes of the anonymity protection of the MN it suffices to perform only unilateral link layer authentication between the MN and AP; that is, the MN verifies the authenticity of the AP. If implemented in this way, the link layer security protocol does not reveal the identity of the MN.

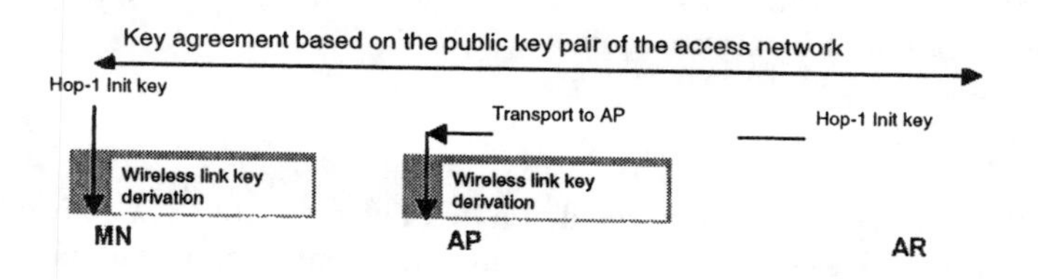

Figure 2.56 Derivation of the link layer encryption key for initial access.

Encryption of MN Identity Terminates in the Home Network

In this solution, the MN uses its home root key to encrypt its identity all the way to its home network.

In this case, only a network layer (or higher) encryption mechanism can be used. This solution offers the advantage of hiding the true identity of the MN from the access network, if so wanted. The computational requirements of the presented solutions do not differ significantly. In both cases the MN needs to use the home root key in one public key operation. The protection of the identity of the mobile node in the initial access requires identities to be sent encrypted to the access network. The decryption takes place in a node in the access network or in the home network. In both cases, protection of user identity may open the network to a denial of service attack performed by sending garbage data in place of encrypted identities.

2.4.3 Protection at Link Layer

2.4.3.1 GSM

In GSM 02.09, some signaling information elements are considered sensitive and may be protected.

To ensure identity confidentiality, the Temporary Subscriber Identity must be transferred in a protected mode at allocation time and at other times when the signaling procedures permit it. The confidentiality of user information over physical connections concerns the information transmitted on a traffic channel on the MS-BSS interface (e.g., for speech). It is not an end-to-end confidentiality service.

These needs for a protected mode of transmission are fulfillled with the same mechanism where the confidentiality function is an OSI layer 1 function. The GSM encryption scheme assumes that the main part of the signaling information elements is transmitted on DCCH, and that the CCCH is only used for the allocation of a DCCH.

The layer 1 data flow (transmitted on DCCH or TCH) is ciphered by a bit per bit or stream cipher (i.e., the data flow on the radio path is obtained by the bit per bit binary addition of the user data flow and a ciphering bit stream, generated by an algorithm). For more information, see [62].

2.4.3.2 GPRS

In GPRS the encryption is performed at a link layer control (LLC) level [62]. The plaintext at the LLC level encryption is the data consisting of the payload of the LLC frame (i.e., the information field) and the frame check sequence (FCS). The FCS is a cyclic redundancy check (CRC). This means that the header part of the LLC frame is not ciphered. The maximum length of the payload is 1,600 octets.

2.4.3.3 UMTS

In UMTS, see [63], two security mechanisms for the radio link are provided. An integrity protection mechanism is provided at the RRC level to protect the integrity of the signaling. The confidentiality protection mechanism is performed at a lower level, either at RLC or MAC layer. The nodes and protocol layers for UTRAN radio interface are depicted in Figure 2.57.

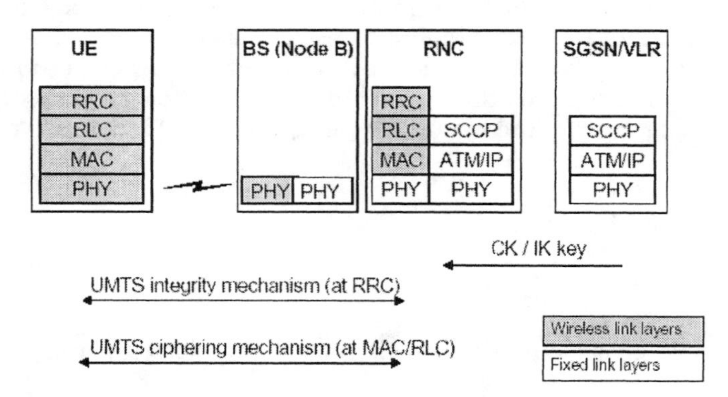

Figure 2.57 The first hop in UTRAN.

Integrity Protection

Most control signaling information elements that are sent between the ME and the network are considered sensitive and must be integrity protected. A message authentication function shall be applied on these signaling information elements transmitted between the ME and the RNC.

After the RRC connection establishment and execution of the security mode setup procedure, all dedicated ME ↔ network control signaling messages (e.g., RRC, MM, CC, GMM, and SM messages) shall be integrity protected. The mobility management layer in the ME supervises that the integrity protection is started.

The UMTS Integrity Algorithm (UIA) shall be implemented in the ME and in the RNC. Integrity protection shall be applied at the RRC layer.

Confidentiality Protection

User data and some signaling information elements are considered sensitive and may be confidentiality protected. To ensure identity confidentiality, the temporary user identity (P-)TMSI must be transferred in a protected mode at allocation time and at other times when the signaling procedures permit it.

These needs for a protected mode of transmission are fulfilled by a confidentiality function, which is applied on dedicated channels between the ME and the RNC.

The ciphering function is performed either in the RLC sublayer or in the MAC sublayer, according to the following rules:

- If a radio bearer is using a nontransparent RLC mode (data), ciphering is performed in the RLC sublayer.
- If a radio bearer is using the transparent RLC mode (voice), ciphering is performed in the MAC sublayer.
- Ciphering when applied is performed in the S-RNC and the ME, and the context needed for ciphering (CK, HFN, etc.) is only known in the S-RNC and the ME.

2.4.3.4 Bluetooth Security

The Bluetooth wireless link layer communication system [64] provides an authentication mechanism for peer device authentication. For the authentication mechanism, a key for an encryption mechanism is derived.

The Bluetooth encryption mechanism is applied to the payload of baseband packets. Signaling and user information are transmitted in the packet payload, which can be encrypted, while the access code and the packet header are never encrypted. The packets are checked for errors or wrong delivery using the channel access code, the HEC in the header, and the CRC in the payload.

Encryption is performed at link layer and controlled by the Link Manager Protocol. Authentication and encryption are based on a bond between two Bluetooth devices, which is created by establishing a common link key between the devices. The Bluetooth specifications provide a method, called the pairing procedure, whereby the two devices can establish a common link key from a pass key entered by the users to the devices. It is also possible to import a common link key to the devices from a higher layer.

The Bluetooth PAN profile specifies IP networking over Bluetooth (see also Figure 2.57). A PAN user (PANU) is able to communicate with other devices in the group ad hoc network (GN) or network access point (NAP) roles using IP on a PAN. Both the NAP/GN and a PANU may require a certain level of security as part of the PAN service establishment.

When a PANU enters the NAP/GN coverage area and detects NAP/GN presence, a part of the connection establishment procedure can be that either the NAP/GN or PANU requests Bluetooth security procedures. Three security modes are defined as part of the Generic Access Profile:

- Security mode 1: In this mode, neither the mobile node nor the NAP requires security procedures invoked so this case needs no further consideration.
- Security mode 2: This mode does not mandate the use of any security procedures in the link layer before the link setup is completed but can be invoked by the service layer. Thus, it is possible to set up a connection and retrieve information about the discovered device if it is a NAP, more about the system that it is connected to. Using security mode 2 it is possible for the devices to run an authentication and key agreement procedure at a higher layer, and in this manner establish a common Bluetooth link key.
- Security mode 3: In this mode security is enforced by the link layer and requires the security procedures to be executed before link setup is completed. If no link key exists, pairing is initiated using the common passkey. The passkey must be known and distributed to both units in advance, or entered to the devices manually with appropriate user interface. In closed environments and systems with a limited number of access units, this approach may be appropriate. In systems with multiple access points, which may be changed without notice of the mobile node, it is not feasible to use this approach.

2.4.3.5 IEEE 802.11WLAN Security

Two services are provided to bring the IEEE 802.11 functionality in line with the wired LAN assumptions: authentication and privacy [65]. Authentication is used instead of the wired media physical connection. Privacy is used to provide the confidentiality aspects of closed wired media.

IEEE 802.11 provides link level authentication between stations (STA). IEEE 802.11 provides neither end-to-end (message origin to message destination) or user-to-user authentication. IEEE 802.11 authentication is used simply to bring the wireless link up to the assumed physical standards of a wired link. This use of authentication is independent of any authentication process that may be used in higher levels of a network protocol stack.

To bring the functionality of the wireless LAN up to the level implicit in wired LAN design, IEEE 802.11 provides the ability to encrypt the contents of messages. This functionality is provided by the privacy service.

IEEE 802.11 specifies an optional privacy algorithm, Wired Equivalent Privacy (WEP). The algorithm is not designed for ultimate security but rather to be at least as secure as a wire. IEEE 802.11 uses the WEP mechanism to perform the actual encryption of messages.

The privacy may only be invoked for data frames and some authentication management frames. The {IV, frame body, ICV} triplet forms the actual data to be sent in the data frame. The WEP algorithm is applied to the frame body and integrity check value (ICV) of an MPDU.

The WEP encipherment protocol also includes computation of an ICV for each MPDU to be encrypted. The WEP ICV shall be a 32-bit field containing the 32-bit CRC calculated over the data (PDU) field.

The MAC header, which is the first part of the data frame, is neither encrypted nor integrity protected.

All STAs initially start "in the clear" in order to set up the authentication and privacy services. The default privacy state for all STAs is "in the clear." If this default is not acceptable to one party or the other, data frames shall not be successfully communicated between the parties. The privacy service depends on an external key management service to distribute data enciphering/deciphering keys. The IEEE 802.11 standards committee specifically recommends against running an IEEE 802.11 LAN with privacy but without authentication. While this combination is possible, it leaves the system open to significant threats.

After the publication of the WEP algorithm, it was demonstrated to have fatal vulnerabilities. First, severe concerns were raised about usage of the WEP stream cipher for provision of security services. In particular, the WEP authentication protocol is not secure. Later also the WEP algorithm itself was shown to have weaknesses in its key schedule. Furthermore, integrity is protected only by means of a CRC checksum, which can be manipulated. The IEEE 802 group has withdrawn the WEP algorithm and initiated work for a replacement.

2.4.3.6 HIPERLAN/2

The main purpose of the security in HIPERLAN/2 is to provide the possibility to secure the air interface [66]. The system provides authentication and encryption mechanisms.

In HIPERLAN/2 authentication has two options. It is based on either a preshared secret key or on the public keys of the terminals. In the case of public key authentication, both parties may have to trust a third party in order to verify the other part.

In HIPERLAN/2 encryption is activated before authentication. The shared secret encryption key is established in the MN and the AP using anonymous Diffie-Hellman protocol. The encryption is used to protect the identity information of the MN when it is sent to the AP at the initial access phase.

Clearly, this protection is only against passive attacks. It does not offer any protection at all against active attacks launched by false malicious access points or malicious terminals.

Only after successful authentication, in which the public Diffie-Hellman tokens are authenticated, is secure encryption possible.

2.4.3.7 Summary

All radio link layer security systems discussed above include mechanisms for peer entity authentication. Also, a mechanism for confidentiality protection has been specified for all systems, and the key for the encryption mechanism is derived as a byproduct of the authentication procedure for all systems except for IEEE 802.11 and HIPERLAN/2. In HIPERLAN/2 the encryption key can be later authenticated using the authentication mechanism. A mechanism for message authentication (integrity) is provided only in UMTS, with a clear purpose of providing integrity protection to the radio control signaling messages.

2.4.4 Protection at Network Layer

Most of the radio technologies provide their proper link layer protection. The security mechanisms vary a lot and the problems created by disabling the link layer protection are discussed in the following section. It is then part of the security architecture to employ the IPsec ESP protocol to secure the first hop on the network layer. This section contains some arguments why the IPsec ESP protocol was chosen and how it is employed in the context of the architecture developed by the IST project SHAMAN [57]. The different aspects of authentication, key agreement and negotiation of an IPsec security association are extensively discussed in [9, 56]. The focus here is only on the first hop network security.

2.4.4.1 Layer 2 Tunneling

Alternatives to straightforward application of a security protocol on the network connection exist. Namely, there is the possibility of first establishing a layer 2 tunnel on top of the network connection. There are several options:

- Using Point-to-Point Protocol (PPP) over the Ethernet [67];
- Using an enhanced PPP protocol with encryption on the L2TP layer as proposed by Microsoft in [68];

- Encryption on the network layer on top of the layer 2 tunnel (see [69] which again employs IPsec).

The advantage of these approaches is that authentication information can be incorporated in the PPP protocol and need not be transmitted separately. The disadvantages result from additional protocol overhead for the layer 2 tunneling. The PPP protocol has severe security limitations, and the latter two protocols, providing security extensions of PPP and layer 2 tunneling, are not yet standardized.

Furthermore, these approaches seem more adequate for dial-in connections (remote access). For this reason this approach is not further discussed here.

2.4.4.2 IPsec Protection on the First Hop

The only standardized protocol that provides network layer security for the IP stack is IPsec [70]. Hence the choice of IPsec is obvious.

IPsec defines two protocols, AH and ESP, and has two modes of operation, transport mode and tunnel mode. This results in a number of different combinations. Since encryption is required to protect the first hop, the ESP protocol is needed. Transport mode must be used in this setting because packets transmitted by the mobile node are not addressed to the AR directly. Instead, the inner IP packet header contains a destination address, which is different from the IP address of the AR, which demands the use of tunnel mode.
In general, it is strongly recommended to deploy ESP with encryption and integrity protection. Only the new IP header with the IPsec gateway (AR or ANP) address as destination is visible and unprotected. In [71], the authors generally recommend the exclusive use of the ESP tunnel protocol. Additional aspects arise when nested IPsec associations are required.

Before a secure network layer communication can begin, an IPsec security association (SA) needs to be set up which is identified by a triple consisting of a security parameter index (SPI), an IP destination address, and a security protocol (AH or ESP) identifier. The IPsec security association is usually negotiated by an Internet Key Exchange (IKE) (or successor) protocol. The SAs and their parameters are listed in the security association database (SAD).

The MN may now want to set up additional network security (e.g., an IPsec ESP connection to his home network, which would result in double encryption). From the access network point of view, authentication of uplink packets would suffice (e.g., by IPsec AH or ESP null encryption). The MN is then responsible for encrypting the packets to the destination network. For example, the MN's SPD can be configured to ensure the following:

- Outbound IP packets to a home network are IPsec ESP encrypted (SA with the home network) and then treated by IPsec AH (SA with the AR).

- Outbound IP packets with a destination in the access network are IPsec ESP encrypted (SA with the AR). This protects the link between MN and AR and, for example, payment-related communication with the access network.
- Outbound IP packets with all other destinations are also IPsec ESP encrypted (SA with the AR).
- Inbound IP ESP packets from the home network are decrypted (SA with home network). Optionally, there may be an outer IPsec AH wrapping (SA with the AR).
- Inbound IP ESP packets from all other sources are decrypted (SA with AR).

Note that nested IPsec is supported by the standard, although the configuration effort may be considerable or certain IPsec implementations may not support it at all. However, two separate IPsec instances on the same MN might result in difficulties (order of application, not harmonized changes of the IP stack) so that the nested IPsec applications should be configured in the SPD of a single IPsec instance.

2.4.5 Problems Created by Disabling Link Layer Protection

According to a general belief, the use of wireless link layers creates new threats. For example, unless care is taken at the link layer, it may be hard for a MN to make sure that it is actually communicating with the very access router that it thinks it is communicating with. However, these threats are mostly independent of mobile IPv6, and it is not expected that mobile IPv6 security would necessarily bring any remedy to them [72]. In this section this issue is studied in more detail.

First, the vulnerabilities caused by disabling protection at the link layer are identified for the UMTS radio interface and the Bluetooth radio link. Then link layer security requirements are investigated for the first hop of the access network within the BRAIN access network model and some approaches to first hop security are discussed.

The investigations are limited to those security services that will be provided after the access authentication at the network layer has been successfully performed. It is assumed that security associations for the first hop between the mobile node and the first access router have been derived at the access authentication procedure. It remains to investigate how to exploit these security associations in an optimal way.

The question to be studied is at which layer should protection of the first hop link be performed. The types of security mechanisms provided by the existing wireless technologies vary a lot and also, at least in some cases, lack a well-defined purpose and architectural foundation. One of the main goals of this section is to contribute to the general understanding of the role of the link layer signaling in the general network security architecture.

It is clear that signaling messages related to the control and management of the wireless communication link itself cannot be protected at a higher layer, but that potential vulnerabilities must be protected using security mechanisms at the link layer. Therefore, when studying the importance of protection at the link layer, the vulnerabilities and threats for link layer signaling messages must be assessed. Two systems can be selected (the Bluetooth wireless communication technology and the UTRAN radio access network) for a detailed analysis of threats to wireless link layer signaling messages.

Anonymity protection is deliberately left out of the discussion here because if anonymity is to be protected, the protection should start before the first authentication over the network is performed. One approach for solving this problem is presented in Section 2.4.2. Also, user identity consists of several pieces of information, such as MAC addresses, IP headers, and channel access codes that are distributed not only over the whole time axis of the access event, but also between the different communication layers.

Another question for further study is what the optimal selection for the cryptographic mechanisms to be used in the link layer should be. In some of the existing systems (e.g., GSM and Bluetooth) message authentication is provided by encryption using a key that is derived in the authentication procedure. Because the encryption mechanism in both cases is a stream cipher, such a solution for message authentication is questionable, at least in theory.

2.4.5.1 Threats for Wireless Access Links

Very straightforward attacks can be launched on the physical radio layer against communication channels over the wireless link. Brute force jamming can cause DoS for the entire link. Also, targeted jamming might be possible if the individual channel access codes are available. Physical attacks cannot, however, be launched without being detected, and they do not release the denied service for use by the attacker. Also the frequency hopping techniques and CDMA channel coding methods make targeted attacks on the physical layer very difficult to launch.

It seems that an attacker who operates on the wireless link layer and tampers with the layer 2 signaling messages can cause much greater losses in quality of service and in availability of the access service. The purpose of this section is to investigate in more detail the layer 2 signaling messages and the problems caused if link layer protection mechanisms are disabled. Two typical examples of link layer systems are selected for detailed study, the UTRAN system and the Bluetooth wireless communication technology.

Finally, conclusions are presented within the BRAIN framework by considering three different basic approaches for providing security services for the first hop.

Threats for UTRAN

The UTRAN radio link layer is divided into three sublayers performing various link layer functions. A simplified picture of the UTRAN link layers and the network nodes involved is depicted in Figure 2.57.

The main signaling flows take place between the mobile UE and the RNC. Signaling messages are exchanged between all three sublayers: RRC, RLC, and MAC. The most important and sensitive ones are those on the RRC sublayer, and they are protected by the UIA. In addition, these are also protected by the UMTS Encryption Algorithm (UEA) performed at the RLC layer for nontransparent RLC mode (data), or at the MAC sublayer for transparent RLC mode (voice).

The functions performed at the MAC sublayer are listed in [57]. Most of the MAC functions are controlled by upper layers. The following functions generate independent signaling messages at the MAC layer, or retrieve information directly from the signaling messages.

Identification of UEs on Common Transport Channels
When a particular UE is addressed on a common downlink channel, or when a UE is using the random access channel (RACH), there is a need for in-band identification of the UE. Since the MAC layer handles the access to, and multiplexing onto, the transport channels, the identification functionality is placed in MAC. The identification information sent by UE is vulnerable to tampering.

Access Service Class Selection for RACH and CPCH Transmission
The RACH resources (i.e., access slots and preamble signatures for FDD, time-slot and channelization code for TDD) and CPCH resources (i.e., access slots and preamble signatures for FDD only) may be divided between different access service classes (ASC) in order to provide different priorities of RACH and CPCH usage. In addition it is possible for more than one ASC or for all ASCs to be assigned to the same access slot/signature space. Each access service class will also have a set of back-off parameters associated with it, some or all of which may be broadcast by the network. The MAC function applies the appropriate back-off and indicates to the PHY layer the RACH and CPCH partition associated to a given MAC PDU transfer. An immediate vulnerability for ASC selection is that the service class back-off parameters can be modified by the attacker.

The functions performed at the RLC sublayer are listed in [73]. Some of the RLC functions are controlled by RRC. The following functions generate independent signaling messages at the RLC layer or retrieve information directly from the signaling messages.

- *Duplicate detection:* This function detects duplicated received RLC PDUs and ensures that the resultant upper layer PDU is delivered only once to the upper layer. If RLC PDU headers are tampered with, then duplicate detection is prevented.

- *Flow control:* This function allows an RLC receiver to control the rate at which the peer RLC transmitting entity may send information. An attacker can tamper with the flow control messages and in this manner disturb traffic flow and deteriorate the QoS of the victim UE.
- *Initial cell selection and reselection in idle mode:* Selection of the most suitable cell based on idle mode measurements and cell selection criteria. The idle mode measurements sent by the victim UE can be tampered with thus forcing it to a nonsuitable cell.
- *Configuration for CBS discontinuous reception:* This function configures the lower layers (L1, L2) of the UE when it shall listen to the resources allocated for CBS based on scheduling information received from BMC. Nonoptimal performance can be caused by tampering with the configuration messages sent by RRC to the victim UE.

2.4.5.2 Threats for Bluetooth

The link layer security mechanisms in Bluetooth can be initiated during the connection setup (security mode 3) or after the link level connection setup enforced by the service (security mode 2).

The use of security mode 3 is only possible if a common link key is already available in both Bluetooth devices, or if it is possible to use the manual pairing procedure using one-time passkeys. If network access authentication is based on service level authentication, then security mode 2 is the only possible mode of operation.

In the Bluetooth PAN profile specification security mode 2 is expanded to the PAN service-level enforced security mode, including both Bluetooth and higher layer (802.1X, IPsec) security mechanisms. The Bluetooth wireless link between the NAP and the PANU in the PAN profile is depicted in Figure 2.58.

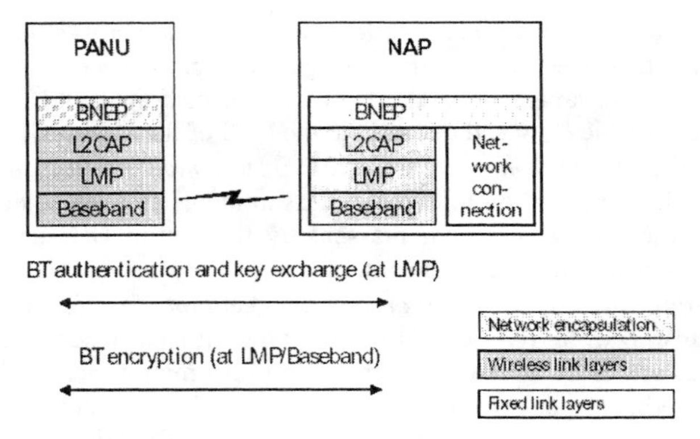

Figure 2.58 Wireless link layers in Bluetooth PAN profile.

The traffic over the Bluetooth radio link is managed by the Link Manager Protocol (LMP). The logical channels are managed and controlled by the Logical Link Control and Adaptation Protocol (L2CAP).

These protocols are common to all Bluetooth devices. The PAN profile makes use of an additional protocol layer BNEP to provide networking capabilities for Bluetooth devices.

LMP messages are used for link setup, security, and control. The LMP messages use the link manager (LM) channel, which is mapped to the payload of the baseband packet. The LMP packets are sent in data manager (DM) packets, which have error correction capability. The LMP messages are filtered out and interpreted by the LM on the receiving side and are not propagated to higher layers.

When the paging device wishes to create a connection involving layers above the LM, it sends a connection request message to the other device. If the remote host accepts the connection request, LMP security procedures, pairing, authentication, and encryption can be invoked. When both devices have sent LMP_setup_complete, the first packet on a logical channel different from the LM channel can then be transmitted.

When authentication and encryption is invoked, payloads of all subsequent baseband packets are encrypted. If at least one of the devices is in security mode 3, then authentication and encryption is invoked before the link setup is complete.

If encryption is enabled before the link setup is complete, then the following LMP control procedures will always be protected by encryption:

- Change link key;
- Change the current link key;
- Channel quality-driven change between DM and DH;
- QoS;
- Control of multislot packets;
- SCO (voice) links setup and control;

- Link supervision.

The LMP messages related to these procedures are protected only when sent after the link setup is complete.

- Clock offset request;
- Slot offset information;
- Timing accuracy information request;
- Supported features information request;
- Switch of master and slave;
- Name request;
- Detach;
- Power control.

The LMP messages are not protected by baseband encryption if the devices operate in security mode 1 or if they are in security mode 2 but have not yet established a service level enforced secure connection.

The vulnerabilities caused by tampering with the LMP signaling messages are of a fatal nature. The mobile device, the PAN user, should not accept any unauthenticated control messages from the NAP.

Tampering with the signaling messages may cause the device to disconnect or at least to try to retransmit its packets, which will be detected as deteriorated connection quality. Undetected malicious activities are also possible. In some cases the malicious device can gain better QoS or even hijack the entire channel for its own use. The most severe threats are considered below in more detail.

Channel Quality-Driven Change Between DM and DH

DM stands for data medium rate (rate 2/3 FEC which adds 5 parity bits to every 10-bit segment) and DH stands for data high rate (no FEC). If the device is forced to make changes to its data transmission rates, which are inappropriate with respect to the channel quality, then its connection will not be optimal and may disconnect.

Quality of Service

Tampering with the QoS control messages may weaken the victim device's service and cause the malicious device to get ahead of its QoS class.

Control of Multislot Packets and Slot Offset Information

These control functions are related to fragmentation and reassembly of long packets. The malicious device may make deletions and insertions to the packets undetected by tampering with the reassembly information.

Link Supervision

Each Bluetooth link has a timer that is used for link supervision. This timer is used to detect link loss caused by devices moving out of range, a device's power down, or other similar failure cases. The piconet master sends this message to the slave. A false message can cause undue detachment by the victim device.

Name Request

Using this request a malicious device can find the victim device's user-friendly name. This may be a threat if anonymity or location privacy needs to be protected.

Detach

The connection between two Bluetooth devices can be closed anytime by the master or the slave. The malicious device can use this message to hijack a communication channel. It tells the victim device to detach and then takes its place on the channel.

Power Control

If the received signal strength indication differs too much from the preferred value of a Bluetooth device, it can request an increase or decrease of the other device's transmission power. Using this signaling message, the malicious device can tell the victim device to decrease its transmission power and cause link loss and detachment.

Logical Link Control and Adaptation Protocol

The L2CAP protocol is based around the concept of channels. Each one of the end points of an L2CAP channel is referred to by a channel identifier (CID). Identifiers from 0x0001 to 0x003F are reserved for signaling messages of L2CAP functions. Other CIDs are dynamically allocated for the transmission of user data.

All L2CAP channels are mapped onto the payload of baseband packets and are protected by encryption as soon as encryption is invoked. In security mode 3, authentication and encryption are invoked before the LMP setup is complete. The first packet on a logical channel different from the LM is sent only after the LMP link setup is complete. Hence, in security mode 3, all L2CAP signaling messages, as well as user data are protected by encryption.

The functions of L2CAP include protocol multiplexing, segmentation, and reassembly, and group management.

L2CAP must support protocol multiplexing because the baseband protocol does not support any type field identifying the higher layer protocol that is being multiplexed above it. L2CAP must be able to distinguish between upper layer protocols such as SDP, RFCOMM, and Telephony Control.

The data packets defined by the baseband protocol are limited in size. Therefore, large L2CAP packets must be segmented into smaller baseband packets prior to their transmission over the air. Similarly, multiple received baseband packets must be reassembled into a larger L2CAP packet following a simple integrity check.

The L2CAP connection establishment process allows the exchange of information regarding the QoS expected between two Bluetooth units. Each L2CAP implementation must monitor the resources used by the protocol and ensure that QoS contracts are honored.

Many protocols include the concept of a group address. The baseband protocol supports the concept of a piconet, a group of devices synchronously hopping together using the same clock. The L2CAP group abstraction permits implementations to efficiently map protocol groups onto piconets. Without a group abstraction, higher level protocols would need to be exposed to the baseband protocol and link manager functionality in order to manage groups efficiently.

The L2CAP messages are not protected by baseband encryption, if the devices operate in security mode 1, or if they are in security mode 2 but have not yet established a service level enforced secure connection. Then the exchange of QoS information will be exposed to tampering by a malicious device. When security mode 2 is used in Bluetooth networking, some initial QoS negotiations are performed at the connection setup. After service level authentication, a secure connection is established and the baseband encryption is invoked. Since the initial QoS contracts may be tampered with, they must be updated using the secure connection. At this point service level requirements to the link level QoS can be taken into account.

To avoid tampering with the unprotected QoS negotiations, it would be best to offer equal QoS contracts to all devices at the initial stage and allow negotiations only over the authenticated and encrypted link.

2.4.5.3 Threats for the BRAIN First Hop

Threats for the Link Layer Signaling

From the detailed investigations of the threats to UTRAN and Bluetooth wireless link layers, it can be concluded that an unprotected wireless link can be tampered with in different ways causing undesirable damage to the victim mobile devices and victim networks. At least the following threats are opened by disabling security mechanisms for the signaling and control messages on the link layer:

- False network access points;
- Mobile station gaining unjustified high QoS level;
- Link hijacking by a mobile station;
- Link hijacking by a false network (e.g., garbage advertising and spamming);
- Tampering with signaling data causing channel deterioration.

Providing adequate message authentication mechanisms on the signaling channel can prevent these threats. The keys for these security mechanisms are derived at the network level authentication procedure when the network access connection is set up. Hence, the protection of the link level signaling can be provided only after the initial authentication.

The BRAIN First Hop

The model for the first hop developed in Section 2.4.1.1 is used here. The wireless link extends from the MN to the AP, and the data packets encounter the AR only after passing through a fixed connection from AP to AR. The AP is the endpoint of the wireless link. It contains the wireless base station functionality and the required link layer interface to communicate data and control information to the AR. This model allows the separation of network routing functionality from the radio access technology. In principle, the same general-purpose AR can serve IP packets originating from terminals using different radio technologies.

It is assumed that the keys for link layer mechanisms are securely transported to the access router at the network level by a specifically designed generic protocol. Further, it is assumed that there is a secure connection between the access router and the access point.

This connection can be either at the network level or at the link level. This secure connection is used to transport the keys for wireless link protection. Hence, it must provide message authentication and confidentiality. In case the access points are fixed nodes in the network, it would be most natural to protect them at the network level in a similar way as the other network nodes. Special attention must be taken to protect the access points in case they are only temporarily connected to the network, as in the case of a separate access point service for a special occasion.

The UTRAN wireless link layers were depicted in Figure 2.57. When the UMTS authentication and key agreement (AKA) procedure is replaced by a general type access authentication procedure, it remains to specify in which node the keys for the link layer protection are derived to be transferred to RNC. From the BRAIN point of view, the three UTRAN nodes form the network access point: node-B, RNC, and SGSN.

From the BRAIN architecture point of view, there are several alternatives about how UTRAN nodes can be mapped into BRAIN access network (BAN) nodes. There are two pure UTRAN nodes: node-B and RNC, and SGSN is the UMTS core network element to which RNC connects over the Iu interface. Note that, in the BRAIN project, even the tightest coupling between BAN and UTRAN was defined by connecting BAN to the SGSN. Hence, UTRAN and BAN were seen as parallel entities, which can be connected to the same core network. Here we assume that the whole UTRAN is embedded into the BAN.

At least the following cases can be identified:

- Node-B and RNC collude to form the AP. This is connected to the AR, which supports part of the Iu interface functionality (e.g., the key transfer). Hence, SGSN does not exist or it is combined with AR.
- Node-B, RNC, and SGSN all collude to the AP. This case is not very realistic since a major part of SGSN functionality would become unnecessary.
- Node-B is mapped into AP. The RNC exists between AP and AR as a separate physical entity. Since security functions inside UTRAN are run between the terminal and the RNC, this case is equivalent to the one first described from a security point of view.
- Node-B is mapped into AP and RNC is mapped into AR. Now AR includes part of the Iu interface functionality as an internal interface. From the security point of view this would imply that the L2 security terminates in the same point as the BRAIN first hop terminates.

2.4.6 First Hop Security Options

The purpose of this section is to investigate various approaches for first hop security. Particular attention is to be paid to the role of link layer security protection mechanisms on the wireless connection between the mobile station and the access point. The benefits and the drawbacks of protection at link layer are discussed as well as the specific requirements for the link layer security.

For the purposes of this study it is useful to make distinction between the following three types of data transmitted on the wireless link:

- User data;
- Network layer signaling messages (e.g., IP headers, which are not protected by network layer security mechanisms);
- Link layer signaling messages that can be protected at the link layer.

It should be noted that some link layer signaling, such as packet headers and device access codes, would always need to be transmitted without protection.

User data can be protected over the wireless link at the network layer or at the link layer. The protection of network and link layer signaling is possible only at the link layer. Assuming that at least the user data is protected, then three different options remain to be studied.

Option 1: Protection only at network layer (see Figure 2.59).

In this option user data and part of the network and higher layer signaling is protected using IPsec security encapsulation methods over the first hop. The rest of the network layer signaling data and the link layer signaling data remain unprotected. The link is vulnerable to all attacks identified in the previous section.

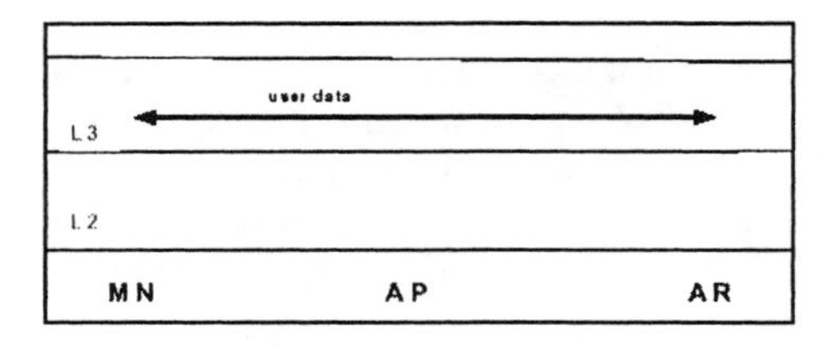

Figure 2.59 Protection only at network layer. (From: [47].)

Only the threats of hijacking can be prevented to some extent using the network layer protection.

While the wireless link can be hijacked by a malicious device or by a malicious network, the attempts to use the hijacked link for transferring user data will be detected at the network layer.

Option 2: Protection only at link layer (see Figure 2.60).

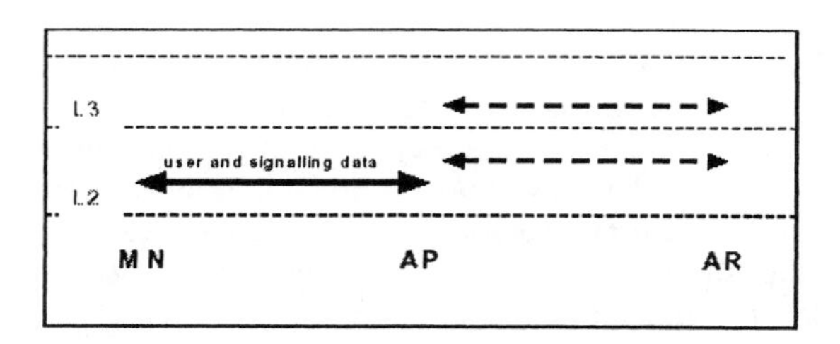

Figure 2.60 Protection only at link layer.

This option is the only possibility if the mobile node is not capable of performing security mechanisms at the network layer. In this case all transmitted data will be decrypted and integrity checked at the access point. If the access point is physically separated from the access router, then appropriate mechanisms need to be performed when transferring data between these two nodes. The protection can be at the network layer or at the link layer.

This option can offer some advantages if the mobile node has limited resources and if the access point and the access router are physically united.

Option 3: Protection at network and link layers (see Figure 2.61).

It was shown that option 1 does not protect tampering of the signaling messages, which may cause serious problems in QoS at the link layer. To prevent such problems from occurring it is advisable to protect the authenticity and integrity of the link layer signaling messages at the link layer.

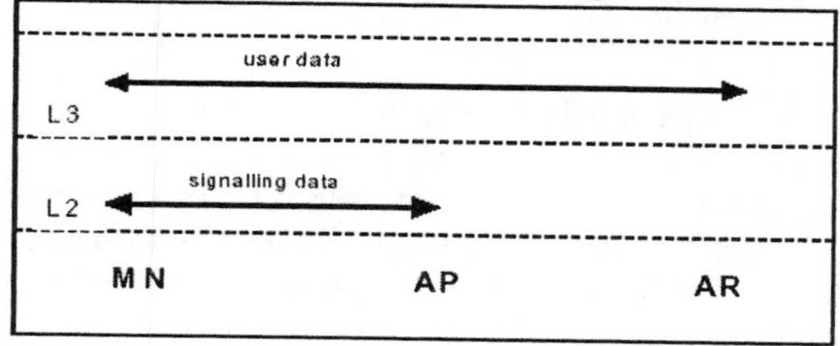

Figure 2.61 Protection at link and network layers.

The properties of the current wireless link technologies vary significantly with respect to providing authenticity and integrity of the signaling data. In UTRAN it is made possible to treat RRC signaling in a different way than user data. UTRAN provides specific procedures for RRC signaling, which forms the essential part of the link layer signaling. This is not possible in Bluetooth, where all different communication channels are mapped on the packet payload and encrypted. This means that in option 3, user data is subject to double encryption, which may significantly increase the processing overhead required by security mechanisms.

2.4.7 Summary

Optimal performance of wireless communication cannot be guaranteed without protecting the basic signaling messages at the wireless link. The type and purpose of the security mechanisms to be used for link layer protection remains for further study. In UTRAN the need for integrity protection (i.e., message authentication) of signaling messages at the RRC layer has been recognized and appropriate protection mechanisms have been implemented. Encryption is the appropriate method for the protection of confidentiality of the user data and signaling data. In modern applications user data is encrypted at the network layer or at a higher layer. To avoid multiple encryption, it may be a better strategy to suppress "bulk" encryption at lower link layers. Also, the requirements for efficient header compression must be taken into account when designing the architecture for confidentiality protection in different layers.

Peer entity authentication at the link layer, as implemented (e.g., for Bluetooth), where it also includes mechanisms for the derivation of the session keys, may be a very useful mechanism particularly in situations where the session keys, for link layer message authentication need to be refreshed during the session. Such a situation may occur in access point roaming when changing to a new access point. On the other hand, it may be argued that performing authentication at the middle of a session may cause unnecessary delay in session reestablishment, and should be avoided.

2.5 CONCLUSIONS

In future systems, access networks should be equipped to support high-speed IP mobility while maintaining negotiated QoS. The provision of QoS to a large number of users increases the need for an efficient utilization of radio resources both by the network and the users as well as the need for effective management of these networks in the context of a heterogeneous environment. For example, mobile network components should be able to monitor and evaluate the wireless channel conditions and adjust transmission parameters accordingly to avoid severe degradation of throughput [1].

Due to the nature of the radio channel as a shared, limited resource, wireless point-to-point medium transmission of high-bandwidth content is inherently inefficient, since it is effectively broadcast to all terminals within the region of coverage. However, assuming that within this coverage, several users are likely to demand the same content, this drawback can potentially be turned into an advantage: if the nature of the offered service lends itself to spatial and temporal bundling of the demands into one transmission, significant efficiency gains are possible by avoiding multiple transmissions of the same content. While, from a radio point of view, the content is broadcasted over the whole transmission area, the fact that the transmission is addressed to a group of users within this coverage area makes this scenario a radio multicast transmission.

Superficially, such multicasting is always desirable due to the obvious gain in spectral efficiency.

However, it creates significant challenges and trade-offs on all levels of the communication system.

Applications need to be designed such that the spatial and temporal bundling of transmissions is possible without significant impact on the user experience. Interactive applications are, generally, poor candidates for multicasting, since there is no temporal degree of freedom to "collect a group of users. Even inherently synchronized applications like video conferencing and shared whiteboard applications, which are prime candidates for multicasting in the wired Internet world, degenerate into point-to-point transmissions from the radio point of view once the partners are spatially distributed over several radio cells. However, in cases of a high spatial user density (stadiums, trade shows, etc.), they may still be feasible. Generally, noninteractive content distribution (video clip download, travel information) is a better candidate, because caching can be employed and delays are more acceptable.

Routing and multicast group management on the transport layer are more difficult in a mobile, wireless environment, since the route to a terminal constantly changes.

Mapping of transport layer multicast to radio multicast or radio unicast is highly dynamic at the RAN border, since each handover potentially represents a joining or leaving of a radio multicast group. Management of such handovers can significantly increase the signaling load in the system.

Power control (PC) loses efficiency if a radio transmission is addressed to several terminals with different propagation conditions. In the simplest case, the weakest link in the multicast group defines the required transmission power, thus leading to increased interference for all terminals in the cell, and therefore a capacity loss.

Radio resource allocation is more difficult, since more than one terminal is involved. QoS needs to be optimized for several terminals. At the same time, QoS for interactive, point-to-point communication must be maintained.

Given these challenges, it becomes increasingly clear that while multicasting can potentially increase spectral efficiency, there will be a trade-off between the drawbacks like lost PC efficiency and higher signaling load. Therefore, it is likely that an analysis of the suggested system will lead to a minimum number of members in a multicast group and other limiting factors (e.g., maximum user speed or minimum cell size) that have to be observed in order to obtain a resulting gain in spectral efficiency.

Determining this threshold is important both from the economic and technical points of view: the offered services must be suitable with respect to these limitations, and pricing must be according to the resulting efficiency of the service. The total obtainable revenue of the provider depends on the resulting efficiency of the transmission system. Operation of the system needs constant optimization of resource usage, which in turn needs prediction of efficiency figures.

Due to the complexity of the system, analytical treatment is not trivial except for simplified asymptotic cases. Therefore, a strong emphasis should also be placed on detailed simulations.

References

[1] ITU Draft Report, "Technology Trends," http://www.itu.org/, October 2003.

[2] IST project OVERDRIVE, http://www.overdrive-ist.org.

[3] IST project MONASIDRE, http://www.monasidre.com.

[4] IST projects BRAIN and MIND, http://www.ist-brain.org/.

[5] IST project SHAMAN, http:// www.ist-shaman.org.

[6] http://www.3gpp.org/.

[7] IST-2000-26144 Project MONASIDRE/ Deliverable TID/DS/02/1.0, "UMTS, MBS and DBS Network and Service Management System Architecture: Technical Requirements and Functionality Description," http://www.monasidre.com, December 2001.

[8] http://www.itu.org.

[9] Prasad, R., (ed.), *Towards the Wireless Information Society: Systems, Services, and Applications*, Norwood, MA: Artech House, 2006.

[10] Bernet, Y., et al., "A Framework for Integrated Services Operation over Diffserv Networks," RFC 2998, http://www.faqs.org/rfcs/rfc2998.html, November 2000.

[11] Holma and Toskala, *WCDMA for UMTS*, New York: John Wiley and Sons, 2000.

[12] Ojanpera, T., and R. Prasad, *Wideband CDMA for Third Generation Mobile Communications*, Norwood, MA: Artech House, 1998.

[13] Sipila, M., et al., "Estimation of Capacity and Required Transmission Power of WCDMA Downlink Based on a Downlink Pole Equation," *Proceedings of VTC Spring 2000*, Japan, May 2000, CD-ROM.

[14] Dahlman, et al., "WCDMA—The Radio Interface for Future Mobile Multimedia Communications," *IEEE Transactions on Vehicular Technology*, Vol. 47, No. 4, November 1998.

[15] Huang and Yates, "Call Admission Control in Power Controlled CDMA Systems," *Proceedings of VTC 96*, Atlanta, May 1996.

[16] Knuttson, et al., "Evaluation of Admission Control Algorithms for CDMA System in a Manhattan Environment," *Proceedings of the Second CDMA International Conference*, Seoul, South Korea, October 1997.

[17] Knuttson, et al., "Downlink Admission Control Startegies for CDMA System in a Manhattan Environment," *Proceedings of VTC98*, Ottawa, Canada, May 1998.

[18] Liu and Zarki, "SIR Based Call Admission Control for DS-CDMA Cellular System," *IEEE Journal on Selected Areas in Communications*, Vol. 12, 1994.

[19] Digital Video Broadcasting Web site, http://www.dvb.org, January. 2001.

[20] Digital Audio Video Council Web site, http://www.davic.org, January 2001.

[21] Donnelly, A., and C. Smythe, "A Tutorial on the Digital Audio-Video Council (DAVIC) Standardization Activity," *Electronics and Communications Engineering Journal*, Vol. 9, No. 1, February 1997, pp. 46–56.

[22] ETSI EN 300 744 v1.4.1 (2001-01), "Digital Video Broadcasting (DVB); Framing Structure, Channel Coding and Modulation for Digital Terrestrial Television."

[23] ETSI TR 101 190 v1.1.1 (1997-12), "Digital Video Broadcasting (DVB); Implementation Guidelines for DVB Terrestrial Services; Transmission Aspects".

[24] Burrow, R., P. Pogrzeba, and P. Christ, "Mobile Reception of DVB-T," *Deutsche Telekom Berkom*, http://www.dvb.org/resources/pdf/dvbtpaper.pdf.

[25] ETSI TR 101 202 v1.1.1 (1999-02), "Digital Video Broadcasting (DVB); Implementation Guidelines for Data Broadcasting."

[26] ETSI ETS 300 802 ed.1 (1997-11), "Digital Video Broadcasting (DVB); Network Independent Protocols for DVB Interactive Services."

[27] ETSI 101 194 v1.1.1 (1997-06), "Digital Video Broadcasting (DVB); Guidelines for Implementation and Usage of the Specification of Network Independent Protocols for DVB Interactive Services."

[28] IST Project MCP (Multimedia Car Platform), http://mcp.fantastic.ch/, January 2001.

[29] IST Project DRIVE (Dynamic Radio for IP-Services in Vehicular Environments), http://www.ist-drive.org, January 2001.

[30] Scalise, F., "TM2381: Overview of DVB-RCT: A Wireless Return Channel for Interactive DVB-T," September 2001.

[31] Tanton, N. E., et al., "The Design and Implementation of a System for the UK-Wide Distribution of BBC Digital Television Services," *Proceedings of the Interactive Broadcasting Conference*, 1998.

[32] ETSI ETS 300 813 ed. 1 (1997-12), Digital Video Broadcasting (DVB); DVB Interfaces to Plesiochronous Digital Hierarchy (PDH) Networks.

[33] ETSI ETS 300 814 ed. 1 (1998-03), "Digital Video Broadcasting (DVB): DVB Interfaces to Synchronous Digital Hierarchy (SDH) Networks."

[34] Sauvet-Goichon, D., "DVB-T Trends in France," *DVB 2001 Seminar*, Dublin, Republic of Ireland, March 2001

[35] "DAVIC 1.4.1 Specification Part 6: Management Architecture and Protocols," 1998.

[36] 3GPP TS 23.107 version 3.6.0 Release 1999, "Universal Mobile Telecommunications. System (UMTS); QoS Concept and Architecture."

[37] Third Generation Partnership Project (3GPP) Web site, http://www.3gpp.org, June 2001.

[38] ETSI TS 101 762 V1.1.1 (2000-10), "Broadband Radio Access Networks (BRAN)."

[39] "Management of TCP/IP-Based Internets: MIB-II," RFC 1213,

[40] http://snmp.cs.utwente.nl/ietf/rfcs/complete/rfc1213, June 2001.

[41] Pereira, J. M., "Re-Defining Software (Defined) Radio: Re-Configurable Radio Systems and Networks," Special Issue on Software Defined Radio and Its Technologies, *IEICE Transactions on Communications*, Vol. E83-B, No. 6, June 2000.

[42] IST Project OVERDRIVE, "Draft Proposal for Spectrum Efficient Multicast and Asymmetric Services in UMTS," OverDRiVE\WP1\D08, May 2003.

[43] 3GPP TS 22.146 v6.2.0, "Multimedia Broadcast/Multicast Services," Stage 1, Release 6, March 2003.

[44] 3GPP TS 25.346 v1.3.0, "Introduction of the Multimedia Broadcast/Multicast Services (MBMS) in the Radio Access Network," Stage2, Release 6, January 2003.

[45] 3GPP TS 25.992 v1.3.0, "Multimedia Broadcast/Multicast Services (MBMS): UTRAN/GERAN Requirements ," January 2003.

[46] 3GPP TSG R1-020930, Revised Study Item Description, "Analysis of OFDM for UTRAN Evolution," Nortel Networks, Wavecom.

[47] Caire, G., and D. Tuninetti, "The Throughput of Hybrid-ARQ Protocols for the Gaussian Collision Channel," *IEEE Transactions on Information Theory*, Vol. 47, pp. 1971–1988, July 2001.

[48] Findlay, D. A., et al., "3G Interworking with Wireless LANs," *3rd International Conference on 3G 2002 Mobile Communication Technologies (3G2002)*, London, England, May 2002.

[49] Kadelka, A., and A. Masella, "Serving IP Quality of Service with HIPERLAN/2," *Computer Networks*, Vol. 37, September 2001, pp. 17–24.

[50] http://www.ist-mind.org/.

[51] Resource management BRAIN deliverable 2.2, http:// www.ist-brain.org.

[52] IST-2001-35125 Project OverDRiVE Deliverable 08, "Spectrum Efficient Multicast and Asymmetric Services in UMTS," http://www.overdrive-ist.org, May 2003.

[53] 3GPP TSG RAN WG1 R1-021239, "MBMS Power Usage," Lucent Technologies.

[54] 3GPP TSG RAN WG1 R1-021240, "Power Usage for Mixed FACH and DCH for MBMS," Lucent Technologies.

[55] 3GPP TSG RAN WG1 R1-021325, "Comparison of DSCH and FACH for MBMS," Lucent Technologies.

[56] IST-2000-25350 Project SHAMAN, Deliverable 09, "Detailed Technical Specification of Security for Heterogeneous Access," SHA/DOC/SAG/WP1/050/1.0, http://www.ist-shaman.org, June 2002.

[57] IST-2000-25350 Project SHAMAN, Deliverable 08, "Intermediate Specification of Security Modules," http://www.ist-shaman.org.

[58] Haverinen, H., "EAP SIM Authentication," *Internet Draft*, draft-haverinenpppext-eap-sim-03.txt, February 2002.

[59] Arkko, J., and H. Haverinen, "EAP AKA Authentication," *Internet Draft*, draftarkko-pppext-eap-aka-03.txt, February 2002.

[60] Flykt, P., C. Perkins, and T. Eklund, "AAA for IPv6 Network Access," *IETF draft*, draft-perkins-aaav6-05.txt, March 2002.

[61] IEEE Draft 802.1X-2001, "Standards for Local and Metropolitan Area Networks: Port Based Access Control," June 2001.

[62] 3GPP TS 03.20 v8.1.0, 3rd Generation Partnership Project, Digital cellular telecommunications system (Phase 2+), Security related network functions, Release 1999.

[63] 3GPP TS 33.102 v4.3.0, 3rd Generation Partnership Project, Technical Specification Group Services and System Aspects, 3G Security, Security Architecture, Release 4.

[64] Bluetooth SIG, Specification of the Bluetooth System, Volume 1, Part B—Baseband Specification, v 1.1, 2001.

[65] IEEE Std 802.11–1997, "IEEE Standards for Local and Metropolitan Area Networks: Specific Requirements—Part 11: Wireless LAN Medium Access Control (MAC) and Physical Layer (PHY) Specifications."

[66] ETSI TR 101 683 v1.1.1, Access Networks (BRAN); HIPERLAN Type 2; "Broadband Radio System Overview."

[67] Mamakos, L., et al., "A Method for Transmitting PPP over Ethernet (PPPoE)," RFC 2516, February 1999.

[68] Aboba, B., and D. Simon, "PPP EAP TLS Authentication Protocol," RFC 2716, October 1999.

[69] Srisuresh, P., "Secure Remote Access with L2TP," RFC 2888, August 2000.

[70] Kent, S., and R. Atkinson, "Security Architecture for the Internet Protocol," RFC 2401, November 1998.

[71] Ferguson, N., and B. Schneier, "A Cryptographic Evaluation of IPSec," http://www.counterpane.com/ipsec.html.

[72] Mankin, A., et al., "Threat Models Introduced by Mobile IPv6 and Requirements for Security in Mobile IPv6," *IETF Draft,* draft-ietf-mobileip-mipv6-scrty-reqts-02.txt, November 2001.

[73] GSM 01.61 v8.0.0, Digital Cellular Telecommunications System (Phase 2+), General Packet Radio Service (GPRS), GPRS Ciphering Algorithm Requirements, Release 1999.

Chapter 3

Advanced Resource Management Techniques

3.1 INTRODUCTION

The world of wireless telecommunications is moving towards a new era, where information technology and communications are approaching the same goals. These goals can be summarized as communication anytime, anywhere, anyhow, at high data rates. On the other hand, the transition of existing cellular systems to wireless communication systems of third generation and beyond seems to be rather difficult. Operators have made huge investments in 3G licenses and cannot afford the costs of setting up next generation systems that will penetrate as much as GSM has in the population. Therefore, UMTS is considered as a hot spot network that will coexist with other wireless telecommunication systems.

One of the drivers for the development of telecommunications is for operators and manufacturers to find the means to achieve their goals and have revenues by offering valuable and useful services to their subscribers. The effort of a number of IST Fifth Framework projects was focused towards the provision of a highly innovative and scalable platform that will allow wireless systems of the same and different operators to be interconnected, and to achieve maximum utilization, increased network performance, and the ability to share operators' resources, rather than network's resources. In that way these projects have assisted the efficient rollout of new generation systems by guaranteeing low-cost and high quality services to the subscribers.

Next generation systems will be a conglomeration of different networking technologies and will support multiple broadband wireless access technologies. This will allow global roaming across systems constructed by individual access technologies. Such a scenario means that these systems will have a number of different characteristics to be taken into account when applying radio resource management (RRM) techniques to ensure the best possible resource allocation in a cost-effective manner. On the other hand, the effectiveness and efficiency of RRM are affected by the system characteristics at the physical, link, and network layers. Related to this topic, investigations were carried out in a number of projects within the IST Fifth Framework of Mobile and Satellite Communications; to name a few, the projects ROMANTIK [1], ARROWS [2], PACWOMAN [3], and CAUTION [4].

Short-range wireless systems have recently gained a lot of attention to provide seamless, multimedia communications around a user-centric concept in so-called wireless personal area networks. Ultra wideband technology presents itself as a good candidate for the physical layer of WPAN, both for high and low data rate applications. Although the PHY issues for UWB in WPAN environments have been extensively studied, the problems related to the medium access techniques remain largely unexplored, especially for targeted WPAN scenarios. This chapter discusses, among other issues, how to design medium access techniques in order to achieve efficient resource sharing in WPAN environments.

The role of medium access techniques in wireless networks is to coordinate transmission access to common radio resources so that the interference among different transmissions is avoided or decreased and capacity (number of communication links with satisfied QoS) is maximized in the network. In a WPAN scenario, there are multiple collocated WPANs, each of which is composed of user personal devices, such as PDAs and laptop PCs. Since WPAN technology uses license-free wireless links, the radio resources should be shared among the collocated and uncoordinated WPANs. Thus, devices belonging to a single WPAN should actually share the channel not only among themselves but also with the devices belonging to the neighboring, physically collocated networks. In the design of medium access techniques in such a WPAN context, two different control levels of interference must be considered. The first level of interference is intra-WPAN interference, which is the interference among links located in single WPAN. This level of interference can be controlled by employing traditional MAC protocols. For instance, in a Bluetooth-based WPAN, a polling scheme with a master/slave operation, which is a contention-free MAC technique, has been adopted for this coordination [3, 5]. Also, random access techniques such as carrier sense multiple access (CSMA), or a combination of contention-free and random access techniques can be employed. The second level of interference is inter-WPAN interference, which is the interference among links located in different WPANs. Since it is difficult to coordinate the transmissions and completely avoid interference among devices in these "uncoordinated" WPANs, the multiple collocated networks must share the resources in a sense to minimize the mutual interference, which can be achieved by adopting the transmission technologies that have the inherent immunity to the interference, for example, spread spectrum technologies.

This chapter discusses how to provide capacity in wireless networks, and uses the IST project CAUTION [4] and its follow-up, CAUTION ++ [4] as examples.

The IST project ARROWS [2] developed detailed resource management algorithms for the delivery of wireless services with sufficient QoS. These algorithms and the methods for their evaluation are discussed in detail in Section 3.3.

This chapter is organized as follows. Section 3.2 covers capacity research investigations, and Section 3.3 covers wireless devices and suitable radio resource algorithms for the delivery of services. Section 3.4 describes advances in the area of multihop networks optimization, and Section 3.5 focuses on optimization of PANs. Section 3.6 concludes the chapter.

3.2 CAPACITY UTILIZATION IN CELLULAR NETWORKS

The extension of mobile cellular communication to the wireless multimedia network as a seamless extension of the present Internet is seen by many as one of the most important driving forces for the success of a future commercial UMTS deployment. While voice services can be provided with sufficient quality and grade of service in the present second generation cellular networks, the greater flexibility of UMTS promises better suitability for optimization and adaptation for different QoS needs [1].

At present, apart from some densely populated metropolitan areas where capacity is already scarce, GSM can provide sufficient voice service for most purposes [6]. However, observing the growing demand for ubiquitous wireless data services as an extension of the Internet for the mobile user, it is clear that the GSM system is no longer adequate: in the face of scarce spectral resources, circuit-switched delivery of Internet traffic is too inefficient and therefore not economically feasible. GPRS as an extension of the GSM system can provide some improvement for the delivery of relatively low bandwidth Internet traffic. But due to the design restrictions that it had to be developed on top of the existing GSM systems, GPRS suffers from its inflexibility regarding the assignment of capacity to individual connections, as well as its poor QoS capabilities. Both factors make the delivery of the high bandwidth multimedia content of today's Internet applications difficult and uneconomical to the mobile customer.

At the same time, consumers will not buy mobile Internet connectivity at any price. Given the current pricing models, even GPRS is difficult to establish in the market; consumers are likely to question the necessity of large coverage wireless data connectivity if the price is too high, and will gravitate towards inexpensive, local solutions like WLAN hotspots.

Given both the observable trend towards high bandwidth data communication on the one hand, and the economic constraints of scarce, expensive transmission capacity and a limited consumer budget on the other, it is obvious that maximization of the efficiency of the transport system will be the key to economic success of next generation systems. Their flexibility in capacity assignment, together with good support for QoS mechanisms, represents a major improvement over existing radio systems, and these need to be exploited to succeed in the market. Therefore, a strong focus in research and development of 3G and Beyond 3G systems is on highly efficient, flexible data delivery.

3.2.1 Capacity and Network Management Platform for Next Generation Systems

Under the framework of the IST projects CAUTION and CAUTION ++ [4], a system platform was developed and extended to UMTS and systems beyond 3G for the smooth transition from existing wireless systems to next generation ones. The main goal was to design and develop a novel, low-cost, flexible, highly efficient and scalable system able to be used by mobile operators to increase the performance of all network segments. To provide enhanced capacity, quality, and reduced costs, radio access resources need to be monitored by means of distributed monitoring of a variety of access systems, including legacy (e.g., UMTS, WLAN, and so forth) and new ones.

This section describes a multielement system that monitors traffic, predicts and recognizes shortcomings, and reacts to cellular network congestion situations in overloaded sectors of cellular networks [7]. The presented system introduces new network units: the interface traffic monitoring unit (ITMU) and the resource management unit (RMU). ITMU monitors the available and used resources in all the system interfaces. This data is then forwarded to the RMU, which consists of a traffic load scenario tracker to find the correct scenario for the congested situation and a decision-making tool to apply the appropriate management technique. The overall system is described from the following viewpoints:

- Real-time monitoring;
- Traffic-load scenario identification;
- Decision-making for resource management;
- Application of resource management techniques.

The proposed system enhanced with the two new network units is shown in Figure 3.1. The ITMU is responsible for monitoring the resources in the system's interfaces. The measurements are then forwarded to the RMU, where they will be used for making decisions for the management of the system's capacity. The RMU consists of a traffic load scenario tracker that identifies the appropriate scenario for the congested situation and a decision-making tool that applies the appropriate management technique. The designed capacity management architecture is based on the general cellular network architecture, with additional components for traffic monitoring and decision execution.

In the architecture in Figure 3.1, the elements that deal with the packet-switched communications and their interfaces are not indicated. The ITMU monitors the whole network, having the functionality of a distributed monitoring unit. The measurements are being continuously evaluated and, when necessary, indications are forwarded to the RMU. This element collects all traffic information and performs a set of computations, which will lead to decisions for capacity allocation in the critical situations.

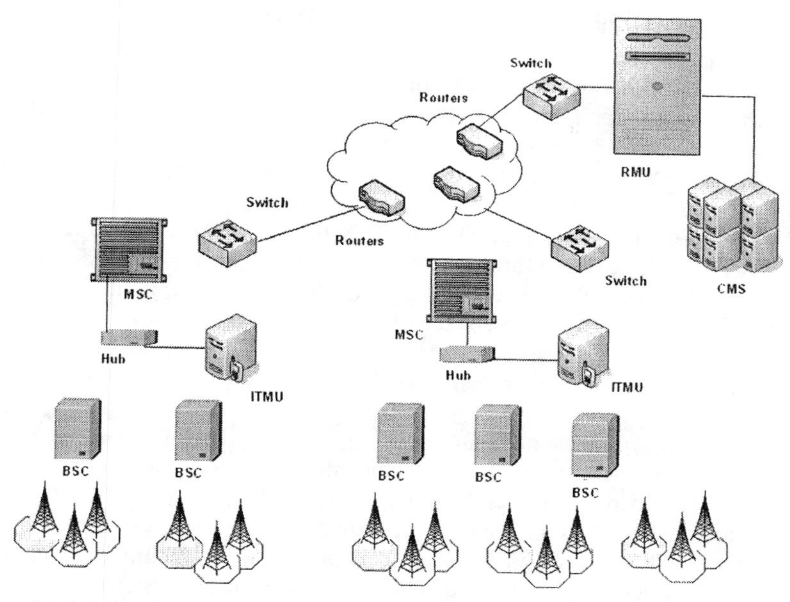

Figure 3.1 Cellular network architecture with capacity management components. (*From:* [7].)

The distributed traffic monitoring will assist the efficient bandwidth allocation if it can be performed dynamically. More specifically, the ITMU has to be connected with all network interfaces, and the management program must be able to grasp and display the traffic flows in real-time. The decision support offered by the RMU gives priority to solutions targeting the air interface in order to reduce the traffic congestion that usually prevents the full capacity utilization of the network.

3.2.1.1 Interface Traffic Monitoring Unit (IMTU)

Monitoring is one of the most important aspects for the efficient management of an operational network. In general, monitoring helps to obtain useful and in-time information about the state and behavior of a system. Related to capacity use, the main concern is the monitoring of the resources of a mobile cellular network, the use of the channels, the number of cells, and the number of base station controllers (BSCs) and mobile switching centers (MSCs) that the network has. It is also important to know, at any time, if a call request was established successfully or not, if a call was ended successfully or not, and if network resources are sufficient to accommodate the increased demand during emergency situations. If a connection was cleared unsuccessfully, we want to know the reason that caused the problem and to propose solutions. Accomplishment of the above tasks will be the duty of the ITMU, as described in Figure 3.1. The ITMU collects the various real-time traffic (RTT) reports, and registers the number and the kind of the occupied logical channels. This registration informs us about the use of each logical channel at any time. The ITMU fulfills its mission by processing two inputs. The first input is a file, which contains all the information about the clear codes. The proper completion of this file is very important and quite a difficult task. It contains all the clear codes with their identifiers, and the indicator for calling party/called party. It also contains the kind of channels [e.g., the traffic channels (TCH), the paging channel (PCH), the random access channel (RACH), the stand-alone dedicated control channel (SDCCH), the slow associated control channel (SACCH), and so forth], that a specific clear code seizes and the respective durations of channel occupancy. All the above information will result from the prestudy of the operation of the clear codes.

3.2.1.2 Resource Management Unit (RMU)

The RMU analyzes the data related to a congestion event and decides on the technique to be applied in order to alleviate the emerged problems. The main requirements that drive the RMU's design are that this element shall detect promptly any network shortcoming occurrences, and it shall execute the appropriate recovery action as soon as possible in order to limit the damage caused by the network failure. The RMU collects alarm messages from the ITMU and reacts to them by executing the most suitable resource management technique. The RMU can also be triggered when a scheduled event occurs, for instance a sporting match event. In this case the RMU is aware of a possible congestion situation and two different actions may be performed:

- The RMU applies the proper RM technique.
- The RMU waits for ITMU alarms and then it applies the RM technique that fits the particular event.

3.2.1.3 Radio Resource Management in Heterogeneous Environment

The system presented in Figure 3.1 can be enhanced to be able to support next generation system management. The ITMU, as described above, exploits management and billing reports. The evaluation of these reports results in the real-time key performance indicators (KPI) monitoring. ITMU should be modified in order to attach it to SGSN nodes and achieve a GPRS monitoring as well. The same should be done for UMTS monitoring as well as for WLAN monitoring.

Once a technique has been selected to deal with a congestion event, the parameters necessary to instantiate the technique can either be obtained by similar application cases stored in the database or be computed with a fine-tuning model-based approach that optimizes the resource management technique (RMT) to meet the network operator goals for the current scenario. The knowledge gained through experience is stored in the RMU database, and it will be reused to enhance future decision-making. Moreover, the database information will also be processed to obtain hints on the quality of the basic knowledge provided by the operator and make suggestions for possible modifications and updates.

Once the RMT is applied, the RMU shall monitor the overloaded cells in order to verify that the selected RMT was effective. Examples of RMT are dynamic signaling channel allocation, cell-breathing, modification of BCCH list, and so forth. All RMU decisions and the consequent results shall be recorded by the knowledge base manager in its internal database, which is used as a source of information for the tuning over time of the RMU operation.

The architecture of the system that monitors several network segments and makes decisions based on the user's location and the predefined business models is shown in Figure 3.2. This architecture manages resources of each system separately and enables system cooperation for efficient capacity management [8, 9].

The third component of the management hierarchy is the global management unit (GMU). The GMU, which is a new network element, has a coordinating role in the general architecture. It communicates with all the RMUs so that access and resource management is achieved considering all available network segments. The GMU controls the traffic rerouting as well as the seamless handover from one system to another. The various wireless access technologies have different mobility management techniques and protocols, each designed specifically for a certain network. An opportunity arises to explore the solution to a common mobility management coordination requirement that also exists in today's and tomorrow's disparate networks. The GMU retrieves location-related information, and the mobility management procedure may, for example, detect that a WLAN/cellular multimode terminal is approaching an area with WLAN coverage. In this case the traffic to and from the mobile terminal could be directed through the WLAN.

On the basis of these statements, the new module is responsible for choosing when and in which conditions a "system reselection" (in idle mode or in connected mode, better known as handover) has to be done by MS or not. The GMU coordinates different wireless networks by collecting changes and rationalizing interservice manipulations between systems. Data services of cellular networks will be reasonably provided by hotspot high data rate WLAN services in a dynamic and seamless manner. To avoid an unnecessary search for a WLAN beacon, the mobile terminal should be aware of the whereabouts of the overlay system to be visited.

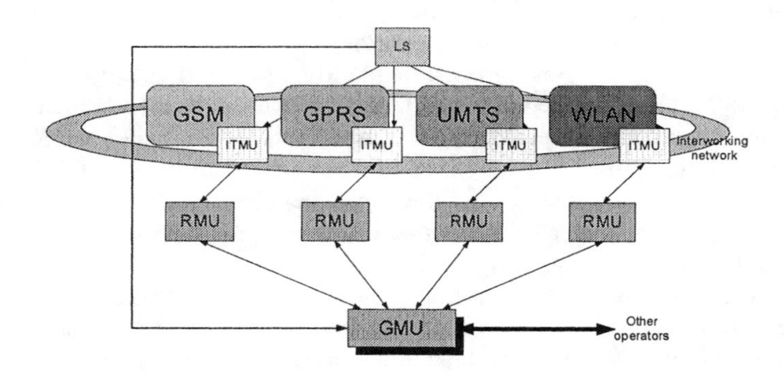

Figure 3.2 System architecture for dynamic reconfigurability in heterogeneous wireless systems. (*From*: [8].)

One of the most important aspects of the presented architecture is a global management system on top of the hierarchy that will enable the routing of traffic of foreign networks over the GMU. Therefore, the GMU, apart from the management role within the same operator, will act as a gateway to other operators and obviously to all kinds of networks they operate. As it is shown in Figure 3.3, there is a need for one GMU for each operator involved in the hierarchy.

In the context of all-IP networks, the radio access network can be enriched by introducing the concept of virtual wireless node-Bs [10]. The configuration can be based on traffic dynamics to allow for management of micro/picocell hotspot situations. An enhanced architecture may be designed such as to provide plug-and-play node-Bs. A cooperative resource management scheme can be designed to support a layered architecture and the virtual wireless node-Bs.

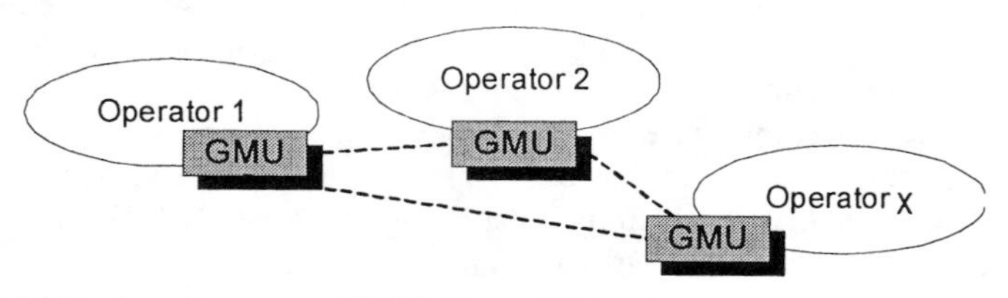

Figure 3.3 Handover by means of GMU. (*From*: [11].)

One of the most significant advantages of the presented architecture is flexibility, in the sense that it can be easily upgraded to function not only in 3G networks but also in composite radio environments where for each user there are several networks that can be defined according to his location and his demands. In the following section we present some results from an in-depth study of an operational system that shows the constraints and limitations. A set of KPIs is used to evaluate the network performance. Finally, the portability of these RRMs together with additional resource management techniques is proposed for 3G. These RRMs are validated with the use of a network simulator.

3.2.1.4 Network Performance Indicators

The CAUTION KPIs were chosen as a result of an analysis of network operational data, for the help that they can provide in understanding the performance of GSM logical channels, in particular when congestion arises.

The logical channels that are primarily used in today's voice traffic in cellular networks are the TCH and the SDCCH, often referred to as the signaling channel. A new call cannot be initiated if the SDCCH channels are not available and the same happens when the SDCCHs are available but all TCHs are already in use. The SDCCH appears to be the most important resource in a congestion situation, and it becomes a clear bottleneck when the traffic grows. For example, in the IST project CAUTION, an approach was adopted that allowed blocking of the SDCCH and the TCH among the KPIs for an operational GSM network. Other channels that were taken into account were the PCH, the RACH, and the access grant channel (AGCH). Use of each of these logical channels can be considered as one distinct KPI.

Other KPIs can include the success rate for a call setup (CSSR) and for handover (HOSR), and its counterpart, the handover failure rate (HOFR). The CSSR and the HOSR are chosen in order to appreciate the impact of congestion in the two most important procedures during a call attempt directly affecting the QoS offered to the subscribers. Finally, the SDCCH blocking rates and the TCH blocking rates (SDCCH BR and TCH BR) can be chosen in order to understand when and where congestion appears, because the respective channels are the most vulnerable.

The number of emergency calls in a cell or in a specific area of the network represents another important source of information, which can help identify some congestion situations in the network. A specific KPI can be defined for tracking the percentage of emergency calls with respect to the total traffic offered by each cell [9].

To understand how the behavior of these channels affects the performance of the network, real examples of congestion problems based on data provided by cellular network operators can be studied. The curves in Figure 3.4 map the variations of three KPIs (SDCCH BR, TCH BR, HOFR) as the workload increases in the network. The network throughput is also shown. The three vertical lines represent some thresholds of low, medium, and high congestion situations, based on size of the affected network area.

Let us first consider the throughput of the network. As seen in Figure 3.4, it increases beyond the threshold of the medium congestion situation, but reaches a limit before the threshold of the high congestion condition. This limit represents exactly how much traffic the network can handle (or the maximum throughput the network can achieve). After this limit, which is normally slightly different in absolute numbers from network to network, throughput slightly decreases, as the resources required for call establishment (SDCCH mostly) dramatically diminish.

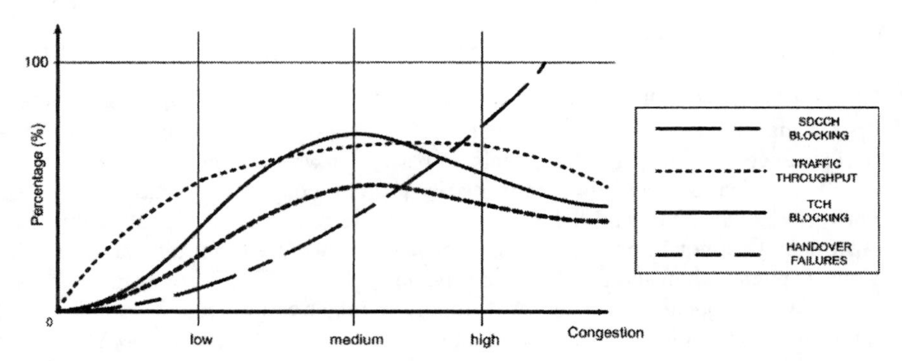

Figure 3.4 Behavior of performance indicators in congestion situations.

As far as the SDCCH blocking is concerned, it is clearly marked in Figure 3.4. The curve increases proportionally to the increase of the congestion level. This is normal, because the demand for establishment resources increases, while resources remain fixed and limited. The TCH blocking also increases as the traffic (or the demand) increases beyond the low congestion threshold, and it reaches a peak around the medium congestion threshold. Approaching high congestion situations, the TCH blocking clearly reduces its figures. This happens when the network runs out of call establishment resources (SDCCH mostly), and thus, the sustained traffic is reduced because the TCHs cannot be fully exploited. It must be noted that in cellular networks, the TCHs outnumber the SDCCHs in every single TRX. Finally, the HOFR (the same applies for the call setup failure rate, which is 1-CSSR) follows the curve of TCH blocking. It reaches its peak when the TCH blocking is at its highest point and diminishes at high congestion conditions, because the additional calls cannot be established.

From the above analysis it is clear that the elimination of the call establishment resources (SDCCH mostly) triggers a chain effect phenomenon that affects traffic resources and procedures, such as call setup or handover. Notice that, beyond a certain point, (approaching the high congestion situation and the maximum handling of traffic), service requests begin to be not fulfilled, thus decreasing the TCH blocking and handover failure rates. If conditions worsen, going even further beyond the high congestion threshold, then the user cannot even reach the network. That means that the RACH blocking starts to appear and the user cannot send a channel request message at all. The network then cannot even understand these attempts, and that is why the above indicators are not affected beyond that point.

Based on the above facts, what happens to the SDCCH and TCH use as congestion in the network increases can be examined. The results of the analysis are depicted in Figure 3.5. When monitoring a channel, one will first perceive a change in use, followed by a change in the blocking/success rates. This is very important and it shows that in order to assess the network's condition effectively, we have to look first at the use of the respective channels and then at the blocking rates.

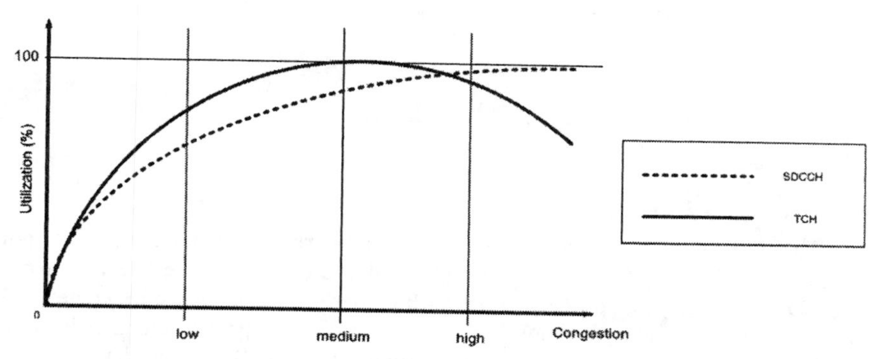

Figure 3.5 Utilization of SDCCH and TCH in congestion situations.

As shown in Figure 3.5, the use of TCH and SDCCH is similar in both low and medium congestion situations. When the SDCCH gets highly congested, the TCH use degrades because the TCH channels become available from terminated calls but cannot be reused because the system lacks call setup resources. If the traffic load becomes even bigger, the situation worsens, RACH and PCH overcome the highest level of use, and SDCCH use starts to degrade also, as the network does not perceive attempts for call setup.

In Figure 3.6, the use of RACH, PCH, and AGCH in relation to congestion is depicted. These channels follow a smoother increase of their use, as there are usually enough resources to handle the requests made. The AGCH rarely encounters problems in terms of blocking. The most common phenomenon, though, as far as the RACH/PCH are concerned, is that problems are encountered in the *A*bis interface (air interface's resources are usually adequate to handle most situations), leading to congestion situations. Normally, all three channels (PCH, RACH, AGCH) remain in medium or low utilization.

Utilization is the key indicator that will characterize a congestive scenario, as it is more representative of what is really happening and more independent from the situation itself than blocking. On the other side, blocking will provide additional information and, thus, extra help in the identification and characterization of a congestion scenario.

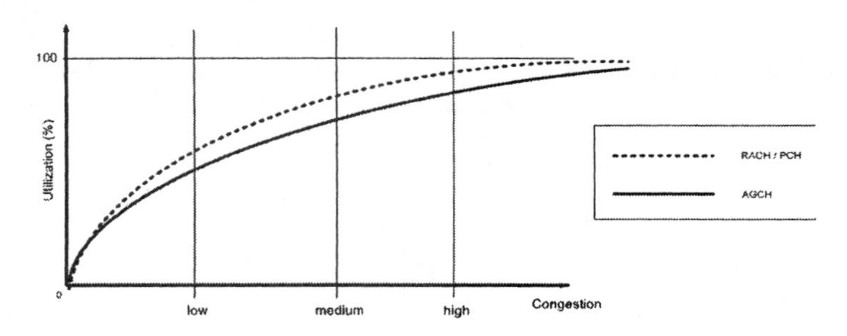

Figure 3.6 Use of RACH, PCH, and AGCH in congestion situations.

The blocking rate depends on certain network parameters, such as radio coverage of the selected area, overlapping of the cells, and the use of directed retry. Directed retry is a procedure used when there is a congestion in the network: in the call setup phase, the mobile station is assigned a traffic channel in a cell other than the serving one. This feature applies when a cell does not have any free TCHs for allocation but there are SDCCHs available. The use of blocking rate can offer extra help in characterizing a congestion situation. For instance, when utilization reaches high values, it is very possible for blocking to remain at a low level because of a good cell overlapping and proper use of the directed retry feature. Another possibility is that, while utilization decreases to very low values giving the impression that everything works fine, the blocking rate reaches very high values, raising in that way the alarm that channels are not used because of congestion. Thus, it is apparent that the use of utilization in combination with blocking for the proper definition and characterization of each congestion scenario is required.

3.2.1.5 Limits and Portability of Resource Management Techniques to All-IP Networks

As already explained in the previous section, even if immediate porting is not possible, most of the time because of different technology, the basic concept that subsumes the RMT definition may remain valid even in a completely different network.

For example, the RMTs implemented by the CAUTION system [4] are as follows:

- Half-rate and full-rate usage, which allows the RMU to determine the distribution between TCH/FRs and TCH/HRs according to the cell load;
- Forced handover, which permits the RMU to transfer an ongoing call from one excessively loaded cell to another one;

- Dynamic SDCCH allocation, by which the RMU can change the ratio of traffic/signaling channels to alleviate the bottleneck effect of the SDCCH channel;
- TRX prioritization in TCH allocation, which rationalizes the allocation of TCH channels over the available TRXs of the cell;
- Dynamic cell resizing with the use of C2 values, which allows the RMU to control the size of a cell, for decongestion under heavy telecommunication traffic situations;
- Bandwidth reservation, by which the RMU manages resource allocation to implement a fair sharing of traffic channels among a set of predefined classes of users;
- Enhanced multilevel precedence and preemption, which enables the RMU to define different priority levels for call setup and for call continuity in case of handover;
- Location update, which optimizes the location and paging functionalities depending on the current network workload;
- Queuing, which when enabled allows the network to increase the number of successfully completed calls in temporary congestion situations;
- Time limited calls, which permit more users to share the limited radio resources during congestion situations;
- FACCH call setup, which alleviates the SDCCH congestion by using the FACCH, the call setup signaling procedures;
- RX level control, which can be used by the RMU to reduce the offered traffic load to a congested cell;
- Cell broadcast messages, which are used to transmit useful information to the users in emergency situations;
- Directed retry, used during the call setup phase, to assign to a mobile station a traffic channel in a cell other than the serving cell;
- Access control by user groups, which allows the RMU to selectively block the admission of certain groups of users under congestion situations;
- Modification of BCCH frequency list, which allows the RMU to limit the number of mobile users willing to camp on a specific cell.

Some of the resource management techniques proposed here for GSM networks can be reused in all-IP networks more extensively. For example, the half-rate/full-rate traffic channel allocation resource management technique can be seen in a bigger QoS context in packet-switched networks. For instance, in UMTS, adaptive multirate codec (AMR) can be used to provide various levels of speech call quality and resource reservation. Thus, it can provide a trade-off between call quality and resource consumption in the same way as half-rate/full-rate in GSM. Moreover, since a lot of different service types are foreseen in all-IP networks, more extensive measures for the same purpose can be seen. The QoS levels of new calls could be negotiated in a call admission phase if the network load is high and the user is willing to settle for a lower QoS level. Also, the QoS levels of existing users might be downgraded by performing QoS renegotiation in order to reduce the network load if it has been driven to too high a level.

Handovers will be more complicated in networks such as UMTS with respect to GSM, but at the same time will provide more possibilities for resource management. The soft handover used in UMTS implies that the user can be communicating with multiple base stations at the same time, especially if he resides in the cell border where signals from multiple base stations are likely to be received approximately with the same signal level. Support for some handover operations like link addition, dropping, and replacement has been specified in 3GPP specifications, but the handover algorithm can be implemented freely based on those operations and the measurements provided by the network. Thus, algorithms relying on the common pilot channel powers or qualities can be envisaged and even more complicated algorithms that explicitly take into account the traffic load situation and distribution between cells might be used. The subsequent actions could encompass changing soft handover threshold values in such a way that the links to some congested base stations are not added so easily.

Functionality for some of the techniques proposed for GSM [9], such as user groups, is directly available in UMTS networks as well, according to the 3GPP specifications. Also, some techniques, such as time-limited calls, can be applied in a similar manner to UMTS networks.

Due to underlying differences in the access technologies of GSM (TDMA/FDMA) and UMTS (CDMA), some techniques are different in UMTS. Cell resizing is one technique that should partially take place automatically in UMTS: there is an inherent coupling between coverage and capacity in UMTS and thus if the load of a cell increases, the radius of the cell decreases since the interference level gets higher. However, it is also possible to provide additional control on the cell sizes in UMTS by modifying the common pilot channel (CPICH) transmission power and thus affecting the cell selection/reselection process as well as handovers.

Usage of priorities and subsequent preemption procedures are possible in UMTS as well. The eMLPP priority definitions are contained in the 3GPP specifications, and furthermore, the RAB priority levels are defined. However, a mapping between eMLPP and RAB priority levels can be seen.

- *QoS negotiation for a new call:* a new call is started with a lower QoS level than what the user originally requested.
- *Preventive QoS negotiation:* lower QoS level is automatically chosen for new calls if the network load is too high.
- *QoS renegotiation to reduce the network load:* if the network has gotten to a point where it is too heavily loaded, the QoS levels of existing calls are downgraded.

- *QoS renegotiation when a new call arrives:* the QoS level of lower priority (or non real-time) ongoing calls is downgraded in order to accommodate a new arriving higher priority (or real-time) call.
- *Time limited calls*: this is for constraining the duration of calls and thus being able to complete the calls of more users.
- *Access classes:* this is to constrain the proportion of call requests that are accepted.
- *Cell barring:* this is to deny completely call starts and handovers in a congested cell.
- *Soft handover active set size limit:* limit the number of soft handover branches that are allowed for calls if the diversity provided by soft handover seems to reserve too much overhead capacity.
- *Soft handover threshold value adjustment:* this is to adjust the thresholds values in soft handover.
- *Priorities and preemption:* enable the usage of priorities for calls and subsequent preemption.
- *Scheduling:* this is to control the downlink transmission of nonreal-time users.

3.2.2 Summary

The RRM techniques presented above are all based on a dynamic system reconfiguration and require a platform of the type developed by the project CAUTION [4, 11]. This platform detects network shortcomings (mainly traffic congestion), and by means of knowledge management mechanisms, it applies one or more of the techniques simultaneously for a set of cells, in order to tackle the network overload problem. Therefore, some of the techniques already exist, but the on-the-fly application can result in a more stable system.

3.3 ADVANCED RADIO RESOURCE MANAGEMENT FOR WIRELESS SERVICES

For the optimization of the radio interface utilization, RRM functions should consider the differences among the different services, not only in terms of QoS requirements but also in terms of the nature of the offered traffic, bit rates, and so forth. The RRM functions include the following [12, 13]:

- *Admission control:* it controls requests for setup and reconfiguration of radio bearers.
- *Congestion control:* faces situations in which the system has reached a congestion status and therefore the QoS guarantees are at risk due to the evolution of system dynamics (mobility aspects, increase in interference).

- *Mechanisms for the management of transmission parameters:* are devoted to decidingthe suitable radio transmission parameters for each connection (i.e., TF, target quality, power).
- *Code management:* for the downlink it is devoted to managing the orthogonal variable spreading factor (OVSF) code tree used to allocate physical channel orthogonality among different users.

3.3.1 Radio Resource Management for 3G Access Networks

Within the UMTS architecture, RRM algorithms are carried out in the RNC. Decisions taken by RRM algorithms are executed through radio bearer control procedures (a subset of radio resource control procedures) such as [14]:

- Radio bearer setup;
- Physical channel reconfiguration;
- Transport channel reconfiguration.

WCDMA access networks that are considered in the UTRA-FDD standard [1], provide an inherent flexibility to handle the provision of future and 3G mobile multimedia services. 3G offers an optimization of capacity in the air interface by means of efficient algorithms for radio resource and QoS management. The RRM entity is responsible for use of the air interface resources and covers power control, handover, admission control, congestion control, and packet scheduling [13]. These functionalities are very important in the framework of 3G systems because the system relies on them to guarantee a certain target QoS, to maintain the planned coverage area, and to offer a high capacity. RRM functions are crucial because in WCDMA-based systems there is not a constant value for the maximum available capacity, since it is tightly coupled to the amount of interference in the air interface. Moreover, RRM functions can be implemented in many different ways, thus having an impact on the overall system efficiency and on the operator infrastructure cost, so that RRM strategies will definitively play an important role in a mature UMTS scenario. Additionally, RRM strategies are not subject to standardization, so that they allow for differentiation among manufacturers and operators.

To cope with a certain QoS a bearer service with clearly defined characteristics and functionality must be set up from the source to the destination of the service, perhaps including not only the UMTS network but also external networks. Within the UMTS bearer service, the role of the radio bearer service is to cover all the aspects of the radio interface transport.

The radio interface of the UTRA is layered into three protocol layers: the physical layer (L1), the data link layer (L2), and the network layer (L3). Additionally, layer 2 is split into two sublayers, the RLC and the MAC. On the other hand, the RLC and layer 3 protocols are partitioned in two planes: the user plane and the control plane. In the control plane, layer 3 is partitioned into sublayers where only the lowest sublayer, denoted RRC, terminates in the UTRAN [14].

Connections between RRC and MAC as well as RRC and L1 provide local interlayer control services and allow the RRC to control the configuration of the lower layers. In the MAC layer, logical channels are mapped to transport channels. A transport channel defines the way in which traffic from logical channels is processed and sent to the physical layer. The smallest entity of traffic that can be transmitted through a transport channel is a transport block (TB). Once in a certain period of time, called a transmission time interval (TTI), a given number of TBs will be delivered to the physical layer in order to introduce some coding characteristics, interleaving, and rate matching to the radio frame. The set of specific attributes are referred to as the transport format (TF) of the considered transport channel. Note that the different number of TBs transmitted in a TTI indicates that different bit rates are associated to different TFs. As the UE may have more than one transport channel at the same time, the transport format combination (TFC) refers to the selected combination of the TF. The network assigns a list of allowed TFC to be used by the UE in what is referred to as the transport format combination set (TFCS).

3GPP has provided a high degree of flexibility to carry out the RRM functions, so that the parameters that can be managed are as follows:

- TFCS, which is network controlled and used for admission control and congestion control;
- TFC, which in the case of the uplink is controlled by the UE-MAC;
- Power, which is the fundamental physical parameter that must be set according to a certain quality target (defined in terms of an SIR target) and taking into consideration the spreading factor used and the impact of all other users in the system and their respective quality targets;
- OVSF code.

The next section describes in detail the different RRM algorithms.

3.3.2 RRM Algorithms Description

3.3.2.1 Admission Control

The admission control procedure is used to decide whether to accept or reject a new connection depending on the interference (or load) it adds to the existing connections. For example, it is responsible for deciding whether a new RAB can be set up and which are its allowed TFCs.

Uplink

Within a CDMA cell, all users share the common bandwidth and each new connection increases the interference level of other connections, affecting their quality expressed in terms of a certain (E_b/N_0) value. For n users transmitting simultaneously at a given cell, the following inequality must be satisfied [15]:

$$\frac{P_i \times SF_i}{P_N + \chi + [P_R - P_i]} \geq \left(\frac{E_b}{N_o}\right)_i r \quad i = 1 \ldots n \tag{3.1a}$$

$$P_R = \sum_{i=1}^{n} P_i \tag{3.1b}$$

where P_i is the kth user received power at the base station, SF_i is the ith user spreading factor, P_N is the thermal noise power, χ is the intercell interference, and $(E_b/N_0)_i$ stands for the ith user requirement, r is the channel code rate, and P_R is the total received own-cell power at the base station.

The theoretical spectral efficiency of a WCDMA cell is measured by the load factor, given by:

$$\eta_{UL} = \left(1 + \frac{\chi}{P_R}\right) \sum_{i=1}^{n} \frac{1}{\frac{SF_i}{\left(\frac{E_b}{N_o}\right)_i r} + 1} = \frac{P_R + \chi}{P_R + \chi + P_N} \tag{3.2}$$

By manipulating the above expressions [15], the view from the transmitter side (mobile terminal) gives

$$P_{T,i} = L_p(d_i) \frac{(P_N + \chi + P_R)}{\frac{SF_i}{\left(\frac{E_b}{N_o}\right)_i r} + 1} = L_p(d_i) \frac{P_N \frac{1}{1 - \eta_{UL}}}{\frac{SF_i}{\left(\frac{E_b}{N_o}\right)_i r} + 1}$$

$$i = 1 \,..\, n \tag{3.3}$$

with $P_{T,i}$ being the power transmitted by the ith user and $L_p(d_i)$ the path loss (including shadowing effects) at distance d_i.

Capacity and coverage are closely related in WCDMA networks, and therefore both must be considered simultaneously. The coverage problem is directly related to the power availability, so that the power demands deriving from the system load level should be in accordance with the planned coverage. So, it must be satisfied that the required transmitted power will be lower than $P_{T\max}$ allowed and high enough to be able to get the required (E_b/N_0) target even at the cell edge. Consequently, and due to the limited power available at mobile terminals and also for efficiency reasons, the cell load factor must be controlled. This is done by the admission control entity.

From the implementation point of view, admission control policies can be divided into modelling-based and measurement-based policies [17]. Also, hybrid solutions are possible. In the following, different possibilities are further detailed.

Measurement-Based Admission Control

In the case when the air interface load estimation is based on measurements, and assuming that K users are already admitted in the system, the $(K+1)$th request should verify

$$\eta_{UL} + \Delta\eta \le \eta_{max}$$

with

$$\eta_{UL} = \frac{P_R + \chi}{P_R + \chi + P_N}$$

and

$$\Delta\eta = \frac{1 + \dfrac{\chi}{P_R}}{\dfrac{SF_{K+1}}{\nu_{K+1} \cdot \left(\dfrac{E_b}{N_o}\right)_{K+1} r} + 1} \tag{3.4}$$

with v_{K+1} being the activity factor of the (K+1)th traffic source. In the case of the voice service this factor is typically set to 0.67. For interactive services, like Web surfing, this factor should be estimated on a service by service basis. In the measurement-based approach, η_{max} is obtained from the radio network planning so that the coverage can be maintained.

Statistical-Based Admission Control

In the case when the air interface load is estimated in statistical terms, it is the cell throughput that is maintained, and the cell breathing effects may arise due to the fact that intercell interference cannot be directly and precisely included.

For the statistical-based approach and assuming that K users are already admitted in the system, the (K+1)th request should verify

$$(1+f)\sum_{i=1}^{K}\frac{1}{\dfrac{SF_i}{v_i\cdot\left(\dfrac{E_b}{N_o}\right)_i r}+1}+(1+f)\frac{1}{\dfrac{SF_{K+1}}{v_{K+1}\cdot\left(\dfrac{E_b}{N_o}\right)_{K+1} r}+1}\le\eta_{max} \tag{3.5}$$

where other-cell interference power is modelled as a fraction of the own-cell received power ($\chi = f \times P_R$). According to (3.5), different admission strategies arise by balancing the following parameters:

- *The spreading factor:* by setting SF_i as an estimated average value the user will adopt during a connection time, the assumed load will be closer to the real situation at the expense of relying on the statistical traffic multiplexing. In its turn, considering the SF_i as the lowest SF in the defined RAB, one can cover the worst case at the expense of overestimating the impact of every individual user and, consequently, reducing the capacity.
- *The activity factor of the traffic source:* by setting $v_i < 1$ the admission procedure can be closer to the real situation of discontinuous activity (typical in interactive-like services) at the expense of relying on the statistical traffic multiplexing. In turn, $v_i = 1$ covers the worst case at the expense of overestimating the impact of every individual user and, consequently, reducing the capacity.
- *The overall load level:* by setting η_{max} the admission procedure allows for some protection against traffic multiplexing situations above the average (for example, having more active connections than the expected

average number, or having more users making use of low *SF* than the expected number).

Downlink

In the case of a downlink direction some differences compared to the uplink case arise. In particular, the intercell interference is user-specific because it depends on the user location, the base station transmitted power is shared by all users, and the power allocations depend on the user location as well. So, for the base station transmitting to n users simultaneously, the constraints can be expressed as

$$\frac{P_i \times SF_i}{P_N + \chi_i + \rho \times \left[\frac{P_T - P_{Ti}}{L_p(d_i)} \right]} \geq \left(\frac{E_b}{N_o} \right)_{i_k} r \qquad i = 1 \,..\, n \tag{3.6}$$

$$P_T = \sum_{i=1}^{n} P_{Ti} \tag{3.7}$$

$$P_i = \frac{P_{Ti}}{L_p(d_i)} \tag{3.8}$$

with P_T being the base station transmitted power, P_{Ti} being the power devoted to the ith user, and χ_i representing the intercell interference observed by the ith user.

The load factor is defined as [15]

$$\eta_{DL} = \sum_{i=1}^{n} \frac{\left(\rho + \frac{\chi_i \times L_P(d_i)}{P_T} \right)}{\frac{SF_i}{\left(\frac{E_b}{N_o} \right)_i r} + \rho} \tag{3.9}$$

It can also be found that

$$P_T = \frac{P_N}{(1-\eta_{DL})} \sum_{i=1}^{n} \frac{L_p(d_i)}{\frac{SF_i}{\left(\frac{E_b}{N_o}\right)_i r} + \rho} = \frac{P_N}{(1-\eta_{DL})} \cdot X \tag{3.10}$$

where it can be observed that as the load factor increases, the power demands also increase. That is, the cell load and the transmitted power are interrelated. For a maximum transmitted power, $P_{T\max}$, the maximum cell load shows quite a sharp variation depending on X (see Figure 3.7, where the background noise considered to be –103 dBm and the maximum node-B transmitted power is 43 dBm) as given by the following expression [note that X is defined as the summation term in (3.10)]:

$$\eta_{DL} = 1 - \frac{P_N}{P_{T,\max}} X \tag{3.11}$$

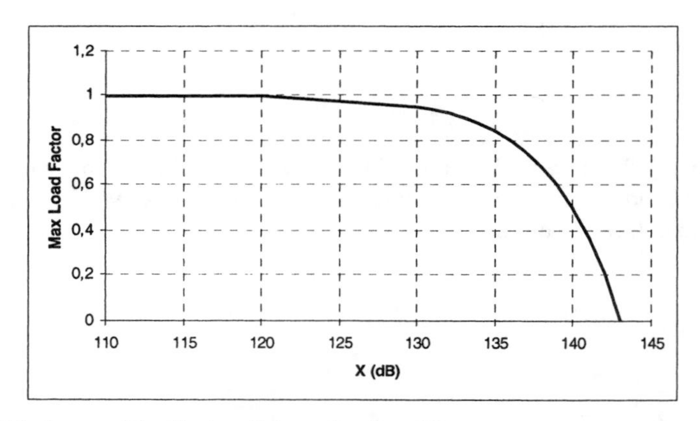

Figure 3.7 Maximum cell load in downlink as a function of *X*.

It can be concluded that the maximum cell load allowable in the cell for a sufficient transmitted power may change quite rapidly depending on how users are distributed in the cell, given that even for a low load the probability of X being higher than 125 dB is not negligible, as shown in [17]. Therefore, the maximum cell load factor and the transmitted power are coupled in the downlink and related by a time varying factor, X, that depends on the user locations. Consequently, it may seem more reasonable to control the downlink operation through the transmitted power rather than through the cell load factor, as is the case in the uplink. Furthermore, according to (3.9), a control based on the load factor would eventually require that mobile terminals report intercell measurements unless statistical average values were considered.

Within this context, the considered admission control algorithm checks the following condition to decide on the acceptance of a new connection request in the system, arriving at the *i*th frame:

$$P_{AV}(i) + \Delta P_T(i) \leq P_T^*(i) \tag{3.12}$$

$$P_{AV}(i) = \frac{\sum_{j=1}^{T} P_T(i-j)}{T} \tag{3.13}$$

where $P_{AV}(i)$ is the average transmitted power during the last *T* frames, $\Delta P_T(i)$ is the power increase estimation due to the new request (notice that it may vary along time), and $P_T^*(i)$ is the admission threshold that may also be adaptive.

3.3.2.2 Admission Control for Handoff Users

One of the most important connection-level issues in mobile communications is how to reduce handover drops due to the lack of available resources in the new cell and still retain a satisfactory blocking probability for the new requesting calls. In the past, a priori resource reservation in each cell to deal with handoff request was often proposed as a suitable admission strategy. Predicting the number of new handoff connections for a specified time interval ahead can help in the dimensioning of the number of guard channels, so that the number of guard channels can be fitted to the number of required handoff channels. A resource reservation strategy in UTRA-WCDMA is the introduction of a reserved set region per cell, surrounding this cell coverage area. Then, the system will take into account only the mobiles included in this region as susceptible of entering the active set, and consequently, object of resource reservation. The reservation set region could alleviate rigidity when dealing with the UTRA-CDMA system in the following way:

- By anticipating handoff connection needs, when these connections still reside in the reservation set region. Then, in the case when the continuity of these connections cannot be guaranteed, a congestion mechanism can be triggered if needed (for example, to release the resources occupied for lower priority connections in benefit of the higher priority ones, or to negotiate different QoS for the handoff connection). Notice that these resource exchange mechanisms were not enabled in 2G connections since all the connections had in practice the same priority and no bandwidth changes were allowed.

- By adjusting the resources reserved to those really required for a time ahead. Then the blocking of new calls does not increase too far, and, in turn, the system load can be somehow maximized.

To define how the reserved set region is determined is certainly a key point in the process. It relies on the utilization of real-time measurements. This set could be established in a similar way to the one that is used to build up the active set, by using a similar trigger event strategy as indicated in [13], with adequate thresholds and hysteresis parameters.

If a new user enters the reserved set region and it goes towards the active set region (obtained somehow, i.e., direction estimation by means of several thresholds crossed, successive positioning), then, for the uplink and with an intercell factor estimated as f, it should be verified at the reserved set level:

$$(1+f)\sum_{i=1}^{K}\frac{1}{\dfrac{SF_i}{v_i\left(\dfrac{E_b}{N_o}\right)_i r}+1}+\frac{(1+f)}{\dfrac{SF_{K+1}}{v_{K+1}\left(\dfrac{E_b}{N_o}\right)_{K+1} r}+1}\le\eta_H<1 \tag{3.14}$$

where a new handover-related load factor η_H is introduced. In this way all the necessary radio resources will be provided when the user joins the active set region.

In order to make the triggering of the handover activation and the managed resources more accurate, a second procedure is introduced. The reservation for a handover user can be implemented progressively as the user gets closer to the new BTS according to the following rule:

$$(1+f)\sum_{i=1}^{K}\frac{1}{\dfrac{SF_{,i}}{v_i\left(\dfrac{E_b}{N_o}\right)_i r}+1}+(1+f)\sum_{i=1}^{m}\mu_{H,i}\frac{1}{\dfrac{SF_{,i}}{v_{,i}\left(\dfrac{E_b}{N_o}\right)_i r}+1}\le\eta_H<1 \tag{3.15}$$

with $1\ge\mu_{H,k+1}\ge 0$ and m is the number of users in transit in the reserved set region. The different quantified values of the parameter $\mu_{H,k+1}$ can be related to the crossing of different thresholds in the value of the E_c/I_0 reflecting the degree of proximity to the active set region.

In the case when the above condition is not fulfilled, a congestion detection can be triggered as a mechanism to provide enough bandwidth to the handoff connection according to its priority and to the QoS negotiation capacity. Then the above-described approach can end up in a congestion solution to manage the initial rejection of a handover.

Notice that $\Delta\eta_{HO} = \eta_H - \eta_{\max}$ is the reservation made in terms of load for the mobiles involved in the handover process.

3.3.3 Congestion Control

Congestion control mechanisms should be devised to face situations in which the system has reached a congestion status and therefore the QoS guarantees are at risk due to the evolution of system dynamics (mobility aspects, increase in interference, etc.). Congestion occurs when the admitted users cannot be satisfied with the normal services agreed for a given percentage of time because an overload arises. The congestion state then has to invoke a congestion control procedure, including the following parts:

- *Congestion detection:* Some criterion must be introduced in order to decide whether the network is in congestion or not. A possible criteria to detect when the system has entered the congestion situation and trigger the congestion resolution algorithm is when the load factor increases over a certain threshold during a certain amount of time, ΔT_{CD} (i.e., if $\eta \geq \eta_{CD}$ in 90% of the frames within ΔT_{CD}). The load factor η measures the theoretical spectral efficiency of a WCDMA cell and must be $0 \leq \eta < 1$. Usually the network is planned to operate below a certain maximum load factor $\eta_{\max}$ and the congestion detection threshold, η_{CD}, should be set in accordance with the maximum planned value.
- *Congestion resolution:* When congestion is assumed in the network, some actions must be taken in order to maintain network stability. The congestion resolution algorithm executes a set of rules to lead the system out of congestion status. Many possibilities exist to carry out this procedure. In any case, four steps are identified:

    - *Prioritization:* Ordering the different users from lower to higher priority (i.e., from service to those with more stringent QoS requirements) in a prioritization table.
    - *Load reduction:* Two main actions can be taken: No new connections are accepted while in congestion or the maximum transmission rate for a certain number of users already accepted in the network is reduced, beginning from the top of the prioritization table.

- *Algorithm 1:* The user is not allowed to transmit any more while in congestion period. In the case of UTRA-FDD, this would be carried out through the layer 3 RRC protocol message "Transport Channel Reconfiguration."
- *Algorithm 2:* The user is allowed to transmit at most at half the normal maximum transmission rate, so that, for example, users are not allowed to transmit at more than 32 Kbps when under in normal conditions the maximum rate is 64 Kbps. Similarly, this is carried out through the layer 3 RRC protocol message "Transport Channel Reconfiguration."

- *Load check:* After the above actions are taken, one would check again the conditions that triggered the congestion status. If congestion persists, one would go back to the prioritization step for the ordering the group of users in the prioritization table. This step is carried out on a frame by frames basis. It could be considered that the overload situation has been overcome if, for a certain amount of time ΔT_{CR}, the load factor is below a given threshold: $\eta \leq \eta_{CR}$ (i.e., if $\eta \leq \eta_{CR}$ in 90% of the frames within ΔT_{CR}).
- *Congestion recovery:* A congestion recovery algorithm is needed in order to restore to the different mobiles the transmission capabilities before the congestion was triggered. It is worth mentioning that such an algorithm is crucial because depending on how the recovery is carried out the system could fall again into congestion. A "time scheduling" algorithm, restoring the former transmission capabilities on a user-by-user approach, is considered. That is, a specific user is again allowed to transmit at maximum rate (i.e., a "Transport Channel Reconfiguration" message indicating this decision has been sent). Once this user has emptied the buffer, another user is allowed to recover the maximum rate, and so on.

3.3.4 Short-Term RRM

Focusing on the uplink direction, centralized solutions (i.e., RRM algorithms located at the RNC) may provide better performance compared to a distributed solution (i.e., RRM algorithms located at the UE) because much more RRM relevant information related to all users involved in the process may be available at the RNC. In return, executing decisions taken by RRM algorithms would be more costly in terms of control signaling because in this case the UE must be informed about how to operate. Consequently, strategies are faced with the performance/complexity trade-off, which usually finds a good solution in an intermediate state where both centralized and decentralized components are present. The 3GPP approach for the uplink could be included in this category, as it can be divided in two parts:

- *Centralized component (located at the RNC).* Admission and congestion control are carried out.
- *Decentralized part (located at UE-MAC).* Short-term RRM is carried out. This algorithm autonomously decides a TF (or TFC if a combination of RABs exists) within the allowed TFCS for each TTI, and thus operates at a "short" term in order to take full advantage of the time varying conditions.

In the downlink case, the centralized operation arises in a natural way and consequently, short-term oriented strategies (packet scheduling) are naturally included in the overall RRM.

3.3.4.1 Uplink

The main responsibility of these algorithms relies on the selection of the most appropriate TFC among the TFCs for each TTI. It will be assumed that the different TFCs are ordered from 0 (lowest bit rate) to TFC_{max} (highest bit rate). There are several possibilities for designing these algorithms, mainly targeted to delay-tolerant services (e.g., interactive and background), depending on the QoS to be guaranteed:

- Delay-oriented strategy
- Taking into account that certain delay bounds should be guaranteed, a possibility relies on selecting the TFC that allows the transmission of the information in the buffer within the specified delay bound.

For example:

- Assume the maximum delay bound is TO ms.
- Assume a total of L_b bits to be transmitted within this delay bound
- Assume TB_{max} as the maximum number of transport blocks allowed in a TTI.

In order to transmit these bits in a maximum of TO ms, the number of bits to be transmitted per TTI would be

$$L = \frac{L_b}{TO} \cdot TTI \quad (3.16)$$

The number of transport blocks to be transmitted per TTI would be

$$numTB = \min\left(TB\max, \left\lceil \frac{L}{TBsize} \right\rceil\right) \tag{3.17}$$

TBsize is the number of bits in a transport block for the considered RAB. Consequently, the TFC selected would be the one allowing that *numTB* transport blocks are transmitted. $\lceil x \rceil$ denotes the lowest integer value higher than x. This algorithm seems in principle suitable for limited traffic volume and restrictive delay.

Rate-Oriented Strategy

When a certain mean bit rate should be guaranteed, a new possibility arises that makes use of the "service credit" (*SCr*) concept. The *SCr* of a connection accounts for the difference between the obtained bit rate and the expected bit rate for this connection. Essentially, if $SCr < 0$: the connection has obtained a higher bit rate than expected; if $SCr > 0$: the connection has obtained a lower bit rate than expected.

In each TTI, the *SCr* for a connection should be updated as follows:

$$SCr(n) = SCr(n-1) + \text{Guaranteed\_rate} / \text{TB\_size} - \text{Transmitted\_TB}(n-1) \tag{3.18}$$

where:

- $SCr(n)$: service credit for TTI=n;
- $SCr(n-1)$: service credit in the previous TTI;
- Guaranteed_rate: number of bits per TTI that would be transmitted at the guaranteed bit rate;
- TB_size: number of bits of the transport block for the considered RAB
- Transmitted_TB(n−1): number of successfully transmitted transport blocks in the previous TTI;
- At the beginning of the connection: $SCr(0) = 0$.

The ratio guaranteed_rate/TB_size reflects the mean number of transport blocks that should be transmitted per TTI in order to keep the guaranteed mean bit rate.

As a result, $SCr(n)$ is a measure of the number of transport blocks that the connection should transmit in the current TTI to keep the guaranteed bit rate. This algorithm seems in principle suitable for mass transmission.

Then, assume that in the buffer there are L_b bits, that *TBmax* is the maximum number of transport blocks that can be transmitted per TTI, and that *TBsize* is the number of bits per transport block. The number of transmitted Transport Blocks in the current TTI = n would be

$$numTB = \min\left(\left\lceil \frac{L_b}{TBsize} \right\rceil, SCr(n), TB\max\right) \quad (3.19)$$

Individual-Oriented Policies

Selecting the TFC that allows the highest transmission bit rate according to the amount of bits L_i to be transmitted. This approach will tend to reduce the delay for these users but it will also tend to increase the interference the rest of users (in any case, if the TFCS has been suitably selected by the RNC, this would not pose a problem).

Collective-Oriented Policies

Selecting the TFC according to:

- A set of thresholds, depending on the measured interference level I_o that is broadcast in the BCH.
- Each terminal's own experience, depending on the number of consecutive simultaneous successful or erroneous transmissions: after max_tr consecutive erroneous transmissions: $TFC = TFC-1$ (if $TFC > 0$) (i.e., it is inferred that the channel load is relatively high because the user transmission is experiencing errors and so the UE MAC decides to reduce the transmission rate in order to increase the CDMA processing gain); after *min_suc* consecutive successful transmissions: $TFC = TFC+1$ (if $TFC < TFC\max$) (i.e., it is inferred that the channel load is relatively low because the user transmission is not experiencing errors and so the UE MAC decides to increase the transmission rate as it is expected that the interference level is low and there is no need to transmit with a high spreading factor).

It can be assumed at this step that the different acknowledgments at the RLC layer can be made available at the MAC layer.

This strategy will tend to adapt the performance in terms of delay according to the overall system status.

In fact, some nonreal-time services could be provided through common channels (for example, short e-mails through RACH in UL). Also, for larger e-mails (with attachments) it could be suitable to take advantage of Transport Channel Type Switching (from RACH/FACH to DCH).

3.3.4.2 Downlink

At first glance, the short-term RRM mechanisms clearly differ from those of the uplink because UTRAN has information about all the involved users and thus it can schedule their transmissions appropriately.

Actually, the management of the transmission parameters and the congestion control are carried out jointly by the same type of algorithms with the following inputs and outputs.

The management of the transmission parameter algorithms in the downlink can be carried out in the following steps in each 10-ms frame.

(1) *Prioritization:* The different RABs with information to be transmitted should be ordered according to certain criteria in a "priority table." The prioritisation should be carried out at two levels:

- *Prioritization among services:* A possible approach would be to consider this order:

  1. Real-time services;
  2. Streaming services (basic layer);
  3. Interactive services;
  4. Streaming services (enhancement layer);
  5. Best Effort services.

- *Prioritization among RABs of the same service:* Some way to define a priority function should be devised. Different criteria also arise at this stage:

(a) Priority as a function of the number of information bits to be transmitted L_i and its timeout TO_i, for example

$$\phi_i = \begin{cases} \dfrac{L_i}{TO_i} & TO_i > 0 \\ L_i(2 - TO_i)^n & TO_i \le 0 \end{cases} \tag{3.20}$$

In general, this strategy will be appropriate for interactive and even best-effort users (although in the latter no QoS is guaranteed, a certain timeout could also be defined).

(b) Priority as a function of the service credits (i.e., $\phi_i = SCr,i$, among other possibilities). This strategy would be suitable for interactive and for streaming services.

(1) *Preferred TFC selection:* For each RAB in the priority table, an appropriate TFC should be selected according a suitable criterion.

(2) *Acceptance check:* It should be decided whether the transmission at this frame is allowed for each RAB depending on the selected preferred TFC. The checks should be performed for one RAB. If a certain RAB does not verify the previous conditions with any of the available TFCs, transmission is not allowed in the current frame and the procedure continues, considering the next RAB in the priority table.

3.3.5 Example of RRM Algorithms Evaluation

In this section results of an example evaluation of several RRM algorithms are presented. Due to the huge complexity of a complete WCDMA simulator—which requires for some signal level issues the chip rate resolution, while for traffic and mobility the need of simulation run in the order of minutes—the usual approach is to split the simulator in two: link and system level simulators. According to Figure 3.8, the system level simulator is fed by the link level simulator outputs as well as the scenario parameters.

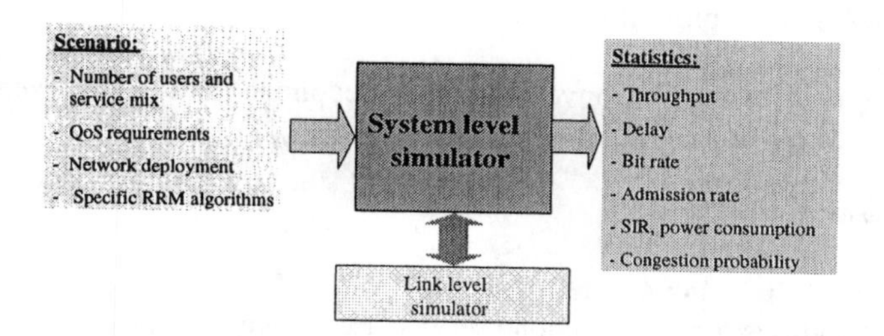

Figure 3.8 System level simulator.

The outputs are a range of performance statistics regarding the different RRM strategies evaluated. This was done with the help of an OPNET simulator as developed in the IST project ARROWS [16].

The developed OPNET simulator allows the support of a wide range of RABs, traffic models as well as deployment scenarios. Selected services and corresponding transport channels are summarized in Table 3.1. Detailed definitions of the RABs are in [18].

Table 3.1

Services and Corresponding Transport Channels

Service	WWW	Videophone	Streaming Basic	Streaming Enhanced
TrCH type (DL)	DSCH	DCH	DCH	DSCH
Rate range Kbps (DL)	0–256	64	32	0–128
TrCH type (UL)	DCH	DCH	—	—
Rate range Kbps (UL)	0–64	64	—	—

Physical layer performance, including the rate 1/3 turbo code effect and the 1,500-Hz closed loop power control, is taken from [19] to feed the system level simulator presented here with block error rate (BLER) statistics. The mobility model and propagation models are defined in [20], taking a mobile speed of 50 km/h and a standard deviation for shadowing fading of 10 dB. A detailed description of the developed simulators is in [21].

3.3.5.1 UTRA-FDD Uplink Results

In this section initial results for conversational (videophone) and interactive (Web browsing) services are presented.

Videophone Service

Statistical-Based Admission Control

Admission probabilities for a videophone service are shown in Table 3.2.

Table 3.2

Admission Probabilities for a Videophone Service

Number of Videophone Users	Admission Probability
15	1
20	0.98
25	0.79
30	0.59

Because videophone maintains a constant bit rate, the admission procedure is quite easy to handle. For 15 users in the cell, the system is still below the planned load factor; all users can be admitted and the performance is good, as shown in Figure 3.9 and Figure 3.10. Figure 3.9 shows the power limitation probability (i.e., the probability that the required transmitted power is above the maximum value), and Figure 3.10 shows the TB error rate also as a function of the distance to the cell site. If the offered load is above the planned load factor, as could be easily demonstrated that it is the case for 30 users, the admission procedure is able to assure the system stability, planned coverage, and planned quality by rejecting connection requests. Figures 3.9 and 3.10 show a little increase in the power limitation probability as well as the TB error rate due to the increase in the load factor. In turn, Figure 3.11 shows the load factor distribution in the system, with an average value around 0.5 and a certain deviation, due to the traffic generation process and the intercell-to-intracell interference factor variation linked to the user's mobility.

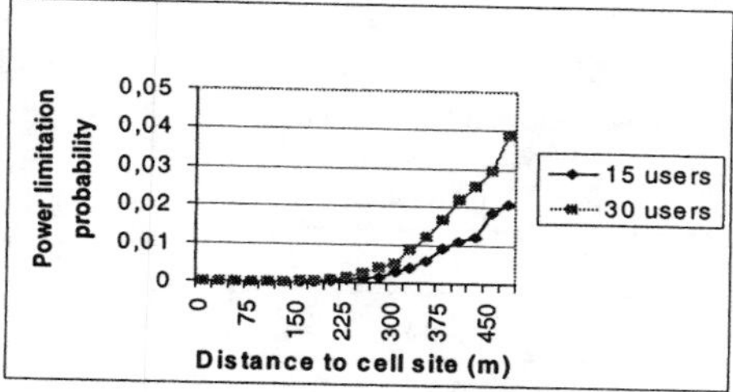

Figure 3.9 Dynamic power limitation probability.

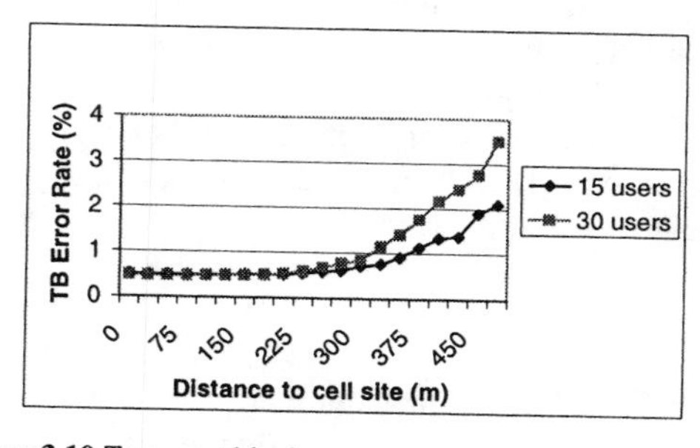

Figure 3.10 Transport block error rate.

Hybrid Admission Control

The admission strategy for new users and handover users performs better when the f factor is fixed to 0.6 than when it is measured and averaged in periods of 0.1s, as shown in Table 3.3.

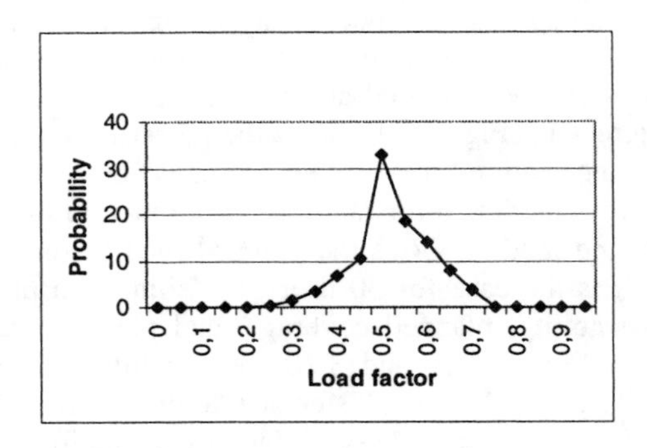

Figure 3.11 Load factor distribution for 20 videophone users.

Table 3.3

Admission Strategy for New Users and f Fixed to 0.6

$\eta_H = 0.75$	users/km^2
$F = 0.6$	7.94
$f_{measured}$ (0.1s)	2.08
$f_{measured}$ (100s)	7.39

The small averaging time produces at some times high values of f that put difficulties in the admission process. Additionally, users who try handover but who are not allowed contribute making a higher f value as they approach the neighbor cell. When the measurement period is much higher, there is a significant improvement with respect to the case of 0.1s. Then, for estimating the average value of the f-factor, long windows are required, which is basically equivalent to fixing f to a convenient value.

Handover Impact

It is also of interest to study the impact of different handover parameters on the overall system performance. To this end, the time for averaging the measurements for handover purposes (measurements are taken over the CPICH) can be varied: either 0.05 seconds or 0.5 seconds. The number of times an event has to occur for triggering a handover can be set to 1. The handover margin for exchanging cells in the Active Set (parameter Hyst_ev_1C) is 1 dB.

The system was studied up to the point where 1% of the calls are dropped and the performance figure retained is the average number of active users per square kilometer under the 1% dropping condition [18]. A call is dropped if the achieved

(E_b/N_0) is at least X dB below the E_b/N_0 target for a certain period of time T. It is worth noting that, due to the sharp behavior exhibited by the BLER versus the E_b/N_0 plot and caused by the turbo code effect, the influence of the specific value for X is not relevant and X = 1 dB was taken for the simulations. Several values for T (1s and 0.1s) were studied, although T = 1s is retained as a representative result.

According to Table 3.4, the time to decide a handover is very critical. Particularly, it should be as low as possible. Otherwise, a user can interfere a lot with the neighbor base station, which may impact the admission process (especially if f is measured) and the dropping probability; 50 ms may be a suitable window time for Rayleigh fading averaging at 50 km/h.

Table 3.4

Decision of Handover

Users/km^2	T_{HO}=0.05s	T_{HO}=0.5s
η_H=0.75, f=0.6, soft HO	8.21	6.68
η_H=0.75, $f_{measured}$ (0.1s), soft HO	2.41	< 1.53
η_H=0.75, $f_{measured}$ (100 s), soft HO	7.04	3.41

Handover procedures and soft handover in particular tend to reduce call droppings by handing a call over to another base station when the quality degrades. Reducing dropped calls can be further strengthened by assigning priority to handover requests over new call requests for admission control purposes. Thus, a different cell load threshold may be used. In this case, two possibilities are compared in Table 3.5: η_{max} = 0.75 (same threshold for handover and new requests) and η_H =0.9 (threshold for handover while new calls have η_{max} = 0.75).

Table 3.5

Possibilities for Reducing the Dropped Call

	η_H=0.75	η_H=0.9
Users/km^2	8.21	12.35

The fixed reservation strategy (i.e., reserving for handover users the margin $\eta_H - \eta_{max}$) results in an increase of capacity thanks to reducing the dropping probability of handovers, as it is appreciated in the cases η_{HO} = 0.75 and η_{HO} = 0.9. Nevertheless, when reserving resources for handover, the BLER at the cell edge increases slightly. The results in Table 3.5 are for the case T_{HO} = 0.05s, f = 0.6 and soft HO.

On the other hand, it is of interest to study the system up to the point where 10% of the calls are blocked and the performance figure retained is the average number of active users per square kilometer under the 10% blocking condition.

The fixed reservation strategy (i.e., reserving for handover users the margin $\eta_H - \eta_{max}$) does not increase capacity from the blocking point of view, as revealed in Table 3.6. This is because this strategy tends to avoid handover droppings but does not facilitate the access to the system for new users. The cases $\eta_H = 0.75$ and $\eta_{max} = 0.9$ give the same results.

Table 3.6

Fixed Reservation Strategy

Users/km^2	Soft HO
η_H=0.75	5.87
η_H=0.9	5.87

Finally, the system is studied for both conditions: the parameters assured a dropping probability below 1% and a blocking probability below 10%. Table 3.7 and Table 3.8 show the obtained results [it is indicated in parentheses if the system is limited by 10% blocking (b) or by 1% dropping, (d)].

Table 3.7

Dropping Probability Below 1%

Users/km^2	$f = 0.6$	$f_{measured}$ (0.1s)	$f_{measured}$ (100s)
η_H=0.75, soft HO	5.87 (b)	2.41 (d)	5.52 (b)

Table 3.8

Blocking Probability Below 10%

Users/km^2	Soft HO
η_H=0.75	5.87 (b)
η_H=0.9	5.87 (b)

The system is usually limited by blocking instead of dropping. This makes sense because the different parameters studied tend to reduce call dropping: $T_{HO} =$ 0.05 makes the handover procedure faster and so droppings can be reduced; handover itself (either hard and soft) is a procedure needed precisely to avoid dropping as the user gets farther from the actual cell site. On the other hand, call retries (i.e., responding to a call request with several admission control attempts before the call is rejected) would reduce the blocking probability on the presumption that the available retry time is long enough to face different system conditions.

Reserved Set Impact

Some preliminary results on the reservation strategy for advancing handover needs are presented. In particular, Figure 3.12 compares the dropping probability when the reserved set region is considered in the case where no advance reservation is done. The particular parameters are 10-dB threshold (i.e., the reservation is done when the quality of the new candidate cell differs from the current cell by less than 10 dB) and $\mu = 0.3$ from (3.15) as the reservation factor. As expected, the dropping probability can be improved with the anticipation procedure. On the other hand, Figure 3.13 shows the blocking probability in the same conditions and no significant degradations are observed. Consequently, the present case reveals that the reserved set results in a dropping gain at negligible blocking degradation.

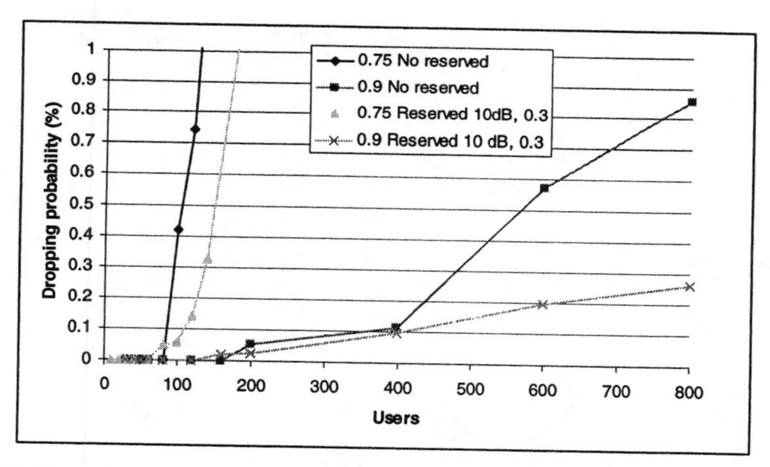

Figure 3.12 Dropping probability for no advance reservation.

3.3.5.2 Interactive Service

UE-MAC Algorithms

For a better understanding of the admission control phase, several figures regarding UE-MAC algorithms should be detailed. One important measurement to understand the behavior of the different UE-MAC strategies is the transport format distribution used. For the interactive RAB, the UE-MAC has the freedom to choose among TF 0 (when the buffer is empty or when $SCr < 0$), TF 1, TF 2, TF 3, or TF 4, ranging from 0 Kbps to 64 Kbps. For *SCr* 24 (see Figure 3.14, *SCr X* standing for a service credit strategy with a guaranteed rate *X* Kbps) it can be observed that most of the time TF 1 or TF 2 is used because the UE buffer queues several packets and so it tends to transmit the information at 24 Kbps. In turn, during the periods that the UE buffer is empty, the UE is gaining service credits, and when a new packet arrives, the transmission rate is increased over the guaranteed one (i.e., TF 3 or TF 4 is used). For the MR strategy, as it chooses the TF according to the buffer occupancy and tries to transmit the information as fast as possible, most of the time TF 4 is used (see Figure 3.14).

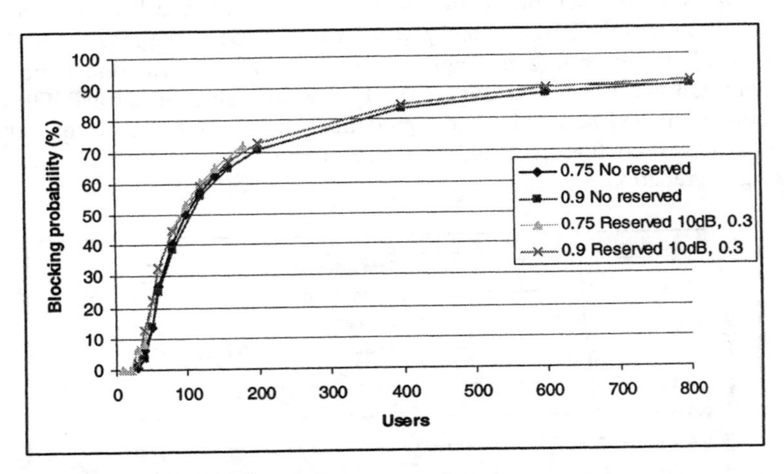

Figure 3.13 Blocking probability for no advance reservation.

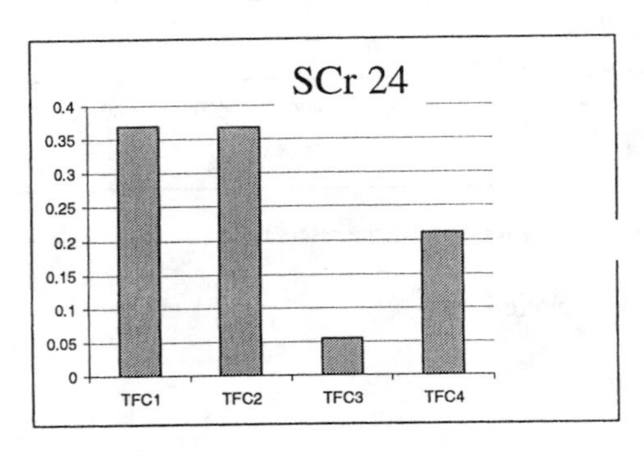

Figure 3.14 TF distribution for *SCr* strategy.

It can be observed from Table 3.9 that the MR strategy provides the highest rate per page. It is worth mentioning that since no admission procedure is considered in this simulation, the system is observed under high (but not heavy) load conditions (i.e., number of users), so that it is possible to observe the behavior of the different strategies avoiding mixing effects with admission and/or congestion control decisions. Since the *SCr* strategy does not take into account the buffer occupancy, it provides a better control of the transmission rate reflected in a low rate per page jitter.

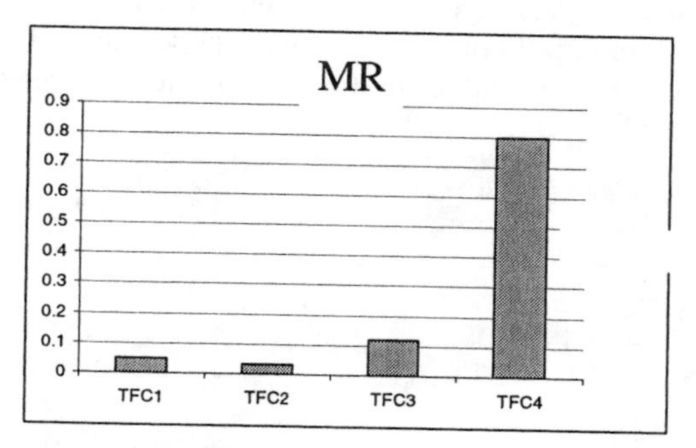

Figure 3.15 TF distribution for MR strategy.

Table 3.9
Delay and Rate for Different Strategies

	Average Packet Delay (s)	Packet Delay Jitter (s)	Rate per Page (Kb/s)	Rate per Page Jitter (Kb/s)
SCr16	1.8	2.28	14.2	2.1
SCr24	0.54	0.95	19	5.0
MR	0.12	0.18	23.6	11.3

Admission Control

For the interactive service the situation is not so easy to handle because of two dynamic and difficult to predict issues affecting the system behavior: the statistical traffic multiplexing (the interactive service is of a discontinuous nature, and consequently, the number of simultaneous users in a given frame in principle is not known in advance) and the TF used in uplink transmissions (it is decided in a decentralized way by UE-MAC, and consequently, the set of SF used by simultaneous users in a given frame in principle is not known in advance). Table 3.10 shows the admission probabilities for different values of the admission TF (equivalent to SF) and η_{max} for both MR and *SCr* strategies. The activity factor is assumed to be the average value coming from the traffic model. Since a step-by-step approach is followed here to avoid mixing effects, congestion control will be dealt with later and in this section simplified assumptions are considered. In particular, the criterion for considering the system under a congested situation is when the condition expressed below holds for more than 90 out of 100 consecutive frames, revealing that the CDMA capacity has been overcome. Note that depending on the specific congestion detection and congestion resolution algorithms, the system could continue operating under normal conditions or not, and the interest of the present criterion is only for establishing a basis for comparison purposes.

$$(1+f)\sum_{i=1}^{n}\frac{1}{\dfrac{SF_i}{\left(\dfrac{E_b}{N_o}\right)_i r}+1}>\eta_{th} \tag{3.21}$$

It can be observed from Table 3.10 that for a proper admission procedure the characteristics of the decentralized algorithm being applied at UE-MAC layer should be taken into account, or in other words, we cannot decouple short-term transmissions from admission itself. By observing the three different columns it can be said that:

- If TF2 and η_{max} = 0.75 are considered in the admission phase for MR strategy, and since the dynamic behavior of this algorithm tends to use TF4 in most cases, the system enters congestion with less than 500 users because the admission is too soft. In turns, for *SCr* the TF considered for admission purposes is much better adjusted to the real dynamic value, so that admission allows for more than 550 users to enter in the system while maintaining a controlled performance.
- On the other hand, if TF4 is considered for admission purposes, congestion is avoided because from the transmission rate point of view the worst case is considered. From the traffic multiplexing point of view (i.e., the impact of considering an average activity factor v_i) η_{max} = 0.75 is low enough to absorb traffic fluctuations above the average without causing congestion. Nevertheless, for the *SCr* strategy this is not so suitable because the admission is too strict.
- The value for η_{max} eventually allows for a softer or stricter admission as shown in the example in Table 3.10, where increasing the value up to

$\eta_{max} = 0.9$ improves the percentage of admitted users for *SCr* strategy compared to the TF4 and η_{max} =0.75 case.

Congestion Control

Comparisons for the two load reduction algorithms presented in Section 3.3.3 are summarized in Table 3.11 and Table 3.12.

Table 3.10

Admission Probabilities for Different Cases

Number of Web users	Admission probability TF2 $\eta_{max} = 0.75$		Admission probability TF4 $\eta_{max} = 0.75$		Admission probability TF4 $\eta_{max} = 0.9$	
	MR	SCr	MR	SCr	MR	SCr
450	1	1	0.98	0.98	1	1
500	Cong.	1	0.93	0.91	1	1
550	Cong.	1	0.84	0.82	Cong.	0.98
600	Cong.	Cong.	0.76	0.74	Cong.	0.93

Table 3.11

Results for η_{CD} =0.8, η_{CR} =0.7, ΔT_{CD}=10, ΔT_{CR} =10

	Algorithm 1 (TF0)		Algorithm 2 (TF2)	
Number of Web users	Admission Probability	Time in Congestion (%)	Admission Probability	Time in Congestion (%)
600	1	≈ 0	1	≈ 0
650	1	≈ 0	1	≈ 0
700	1	0.13	0.99	0.57
750	1	0.33	0.97	2.34

The performance figures are the admission probability (i.e., the probability that a user request is accepted into the system), the time in congestion (i.e., the percentage of time while the network is congested), and the delay distribution of the transmitted packets during the congestion period (measured as the packet delay percentiles during congestion periods). It can be observed that the "softer" load reduction action for algorithm 2 leads to more time in congestion and, consequently, a reduction in the admission probability (notice that the first action in congestion is to reject all connection requests). It seems that the firmer actions taken by algorithm 1 result in shorter congestion periods. Additionally, one of the expected impacts of the congestion status is delay degradation due to the transmission rate capabilities limitation. It can be observed in Table 3.12 that algorithm 1 provides a better delay distribution compared to algorithm 2, especially for the 95% percentile.

Table 3.12

Results for η_{CD} =0.8, η_{CR} =0.7, ΔT_{CD}=10, ΔT_{CR} =10, 700 users

Packet delay percentiles During Congestion Periods	Algorithm 1	Algorithm 2
50%	< 0.12s	< 0.16s
75%	< 0.84s	< 1.12s
95%	< 2.94s	< 6.62s

It is also of interest to study the sensitivity to the detection and resolution thresholds. Thus, Table 3.13, Table 3.14, and Table 3.15 show performance results for two different options: thresholds for an "early" congestion detection and a conservative resolution (represented by η_{CD}= 0.8 η_{CR} = 0.7) and a representative case for a "late" detection and a less conservative resolution (represented by η_{CD} = 0.9, η_{CR} = 0.8).

Table 3.13

Results for ΔT_{CD} = 10, ΔT_{CR} = 10

Packet Delay Percentiles During Congestion Periods	η_{CD} = 0.8 η_{CR} = 0.7 ("early")	η_{CD} = 0.9 η_{CR} = 0.8 ("late")
50%	<0.12 s	<0.2 s
75%	<0.84 s	<0.98 s
95%	<2.94 s	<3.14 s

The delay distribution appears somehow nicer for the early detection, especially for the 50% percentile case. However, the lower thresholds of the early detection lead to closer congestion situations, as shown in the cumulative distribution of the time between congestions and more time in congestion. This means that a higher signaling load would be necessary for the early detection case.

Table 3.14

Results for $\Delta T_{CD} = 10$, $\Delta T_{CR} = 10$

Cumulative Probability of the Time Between Congestions	$\eta_{CD} = 0.8$ $\eta_{CR} = 0.7$ ("early")	$\eta_{CD} = 0.9$ $\eta_{CR} = 0.8$ ("late")
< 1s	20%	3%
< 100s	60%	12%
< 1000s	99%	48%

Table 3.15

Results for $\Delta T_{CD} = 10$, $\Delta T_{CR} = 10$

Number of Web users	$\eta_{CD} = 0.8$ $\eta_{CR} = 0.7$ Time in congestion (%)	$\eta_{CD} = 0.9$ $\eta_{CR} = 0.8$ Time in congestion (%)
600	≈ 0	≈ 0
650	≈ 0	≈ 0
700	0.13	0.02
750	0.33	0.07

Finally, it is also of interest to study the impact of the observation time for triggering congestion actions (see Table 3.16).

Table 3.16

Results for $\eta_{CD} = 0.8$, $\eta_{CR} = 0.7$

	$\Delta T_{CD} = 10$, $\Delta T_{CR} = 10$		$\Delta T_{CD} = 10$, $\Delta T_{CR} = 100$	
Number of Web users	Admission Probability	Time in Congestion (%)	Admission Probability	Time in Congestion (%)
600	1	≈ 0	1	0.1
650	1	≈ 0	1	0.32
700	1	0.13	0.99	0.95
750	1	0.33	0.98	2.34

Table 3.17 compares the case of two different congestion resolution windows. According to Table 3.16, the case $\Delta T_{CR} = 100$, which represents a safer congestion overcome decision, leads to longer congestion periods and,

consequently, lower admission probabilities because the system is blocked during congestions.

In terms of time between congestions, the safe margin only achieves to reduce very close congestion situations while the probability for longer periods is almost the same.

Table 3.17

Results for $\eta_{CD} = 0.8$ $\eta_{CR} = 0.7$

Cumulative Probability of the Time Between Congestions	ΔT_{CD}= 10, ΔT_{CR} = 10	ΔT_{CD}= 10, ΔT_{CR} = 100
< 1s	20%	4%
< 100s	60%	58%
< 1000s	99%	99%

3.3.5.3 UTRA-FDD Downlink Results

Conversational Service

The considered admission control algorithm checks the following condition to decide the acceptance of a new connection request in the system, arriving at the ith frame:

$$P_{AV}(i) + \Delta P_T(i) \leq P_T^*(i) \tag{3.22}$$

$$P_{AV}(i) = \frac{\sum_{j=1}^{T} P_T(i-j)}{T} \tag{3.23}$$

where $P_{AV}(i)$ is the average transmitted power during the last T frames, $\Delta P_T(i)$ is the power increase estimation due to the new request (notice that it may vary along time), and $P_T^*(i)$ is the admission threshold that may also be adaptive.

Despite the simplicity of the algorithm, it can offer an efficient performance provided that their design parameters are suitably set. Particularly, in this report we focus on the following aspects to be analyzed:

- How the increase in power demand ΔP_T is estimated.
- The impact of the power admission threshold P_T^*. By modifying this threshold the admission can become softer or stricter at the expense of the performance perceived by the admitted users.
- The impact of the measurement period *T*. In order to overcome the high variability of the mobile radio channel as well as interference patterns, transmitted power measurements should be time averaged and again a trade-off will appear.

As mentioned before, in the case of downlink direction some differences compared to the uplink case arise. In particular, the intercell interference is user-specific since it depends on the user location, the base station transmitted power is shared by all users, and the power allocations depend on the user location as well. In order to observe the differences that this behavior originates, Figure 3.16 plots the probability density function of the required transmitted powers to each user in the cell, P_{Ti}. High deviations from the average value are observed.

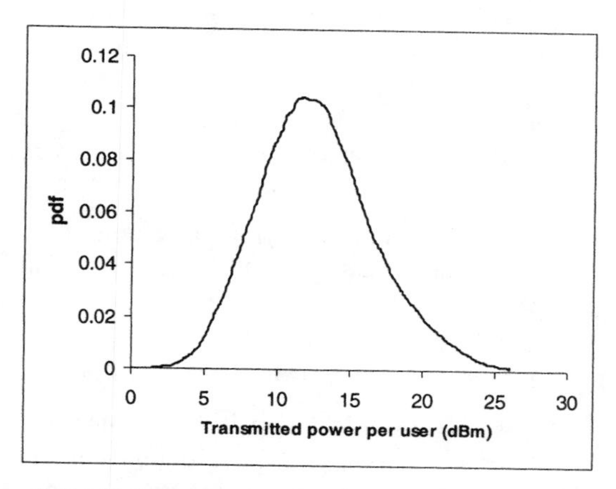

Figure 3.16 Pdf of the transmitted power devoted to every single user.

With respect to the admission control algorithm parameters, the impact of the estimated power increase due to the requesting user, ΔP_T, is first studied. A reasonable criterion could be to estimate the power demand as a time average along the last *T* frames of the required transmitted power to every user, so that it leads to an adaptive estimation:

$$\Delta P_T(i) = \frac{\sum_{j=1}^{T}\left(\frac{\sum_{k=1}^{n_{i-j}} P_{Tk}(i-j)}{n_{i-j}}\right)}{T} \tag{3.24}$$

$n_{i\text{-}j}$ being the number of users transmitting in the $(i\text{–}j)$th frame, $P_{Tk}(i\text{–}j)$ being the transmitted power devoted to the kth user in the $(i\text{–}j)$th frame. For comparison purposes, and in order to assess the importance of the term ΔP_T in the overall admission procedure, a pessimistic estimation is considered:

$$\Delta P_T = P_{Ti}(90\%CDF) \tag{3.25}$$

In this case the estimation is fixed and assumed to be the 90% percentile of the required transmitted power per user. Deriving the cumulative distribution function (CDF) from Figure 3.16, it is found that ΔP_T=17.48 dBm for an offered load of 160 Erlangs in the overall scenario. Also, from further simulations it is obtained that ΔP_T= 9.63 dBm for 120 Erlangs and ΔP_T= 22.1 dBm for 200 Erlangs.

Table 3.18 presents the admission probability (i.e., the probability that a connection request is accepted) for the two different criteria and in the case of $P_{AV}(i)$ averaged over T = 100 frames and a fixed threshold P_T^*= 35 dBm. No significant differences are found. From the achieved performance point of view, Table 3.18 also presents the packet error rate (PER). In both cases, a quite similar performance is achieved, and it is concluded that the ΔP_T estimation has a limited impact on the overall admission procedure.

The impact of the maximum power, P_T^*, is shown in Figure 3.17 and Figure 3.18. At this stage, fixed thresholds of P_T^* = 25, 30, and 35 dBm are considered. In particular, Figure 3.17 plots the admission probability for different numbers of users in the system and it can be observed that for a restrictive value of 25 dBm, many requests are rejected for 160 Erlangs in the scenario. On the other hand, softer admission policies (35 dBm) provide a much higher power limitation probability (i.e., the probability that the base station does not have enough power at a given frame to serve all users), and as a consequence the PER increases (see Figure 3.18).

Table 3.18

Admission Probability and PER for Two Different ΔP_T Estimations

	Admission Probability		Packet Error Rate	
Offered Load (Erlangs)	Time Average	90% CDF	Time Average	90% CDF
120	100%	100%	2.00%	2.00%
160	94.33%	93.97%	2.87%	2.83%
200	56.31%	55.34%	8.20%	7.83%

The previous results indicate the existence of an optimum threshold for each load level, suggesting an adaptive admission control procedure based on a load estimation. Table 3.19 summarizes the approximated optimum threshold for a twofold objective: (1) to obtain a controlled performance (i.e., PER<2.5%), and (2) to obtain as high as possible admission probability.

Another important issue in the admission phase is the estimation of the transmitted power $P_{AV}(i)$ used in (3.22) because as mentioned earlier, the instantaneous transmitted power may exhibit significant fluctuations. Figure 3.19 to Figure 3.22 analyze the impact of different averaging periods, focusing on the $P_T^* = 25$ dBm case. In all cases a representative 5-minute segment is shown. Different cases are analyzed:

(1) Medium load (i.e., 120 Erlangs).

(1.a) If T is low ($T = 100$ in Figure 3.19), there are short periods where, due to the high variability on the required node-B transmitted power, it happens that $P_{AV}(i) > P_T^*$ and, consequently, some calls are rejected (in this case the admission probability is around 98% and PER=2%).

(1.b) If T is high (T = 5,000 in Figure 3.20), the smoothing due to the longer averaging periods avoids unnecessary call rejections (in this case the admission probability is 100% and PER=2%).

Consequently, a relatively high averaging period T is desirable to avoid effects from instantaneous and seldom high transmitted power situations (if the load is low, the required power will also be low).

(2) High load (i.e., 200 Erlangs).

(2.a) If T is low (T = 100 in Figure 3.21), it is possible to take advantage of the periods where the required node-B transmitted power is lower than usual (in this case the admission probability is around 41% and PER=2.25%).

(2.b) If T is high (T = 5000 in Figure 3.22), a wave-like effect arises in the $P_{AV}(i)$ form and the periods where calls can be accepted are reduced (in this case the admission probability is 35% and PER= 2.23%).

Consequently, a relatively short averaging period T is desirable to avoid long memory effects, which would produce a high rejection rate because the required transmitted power will usually be high.

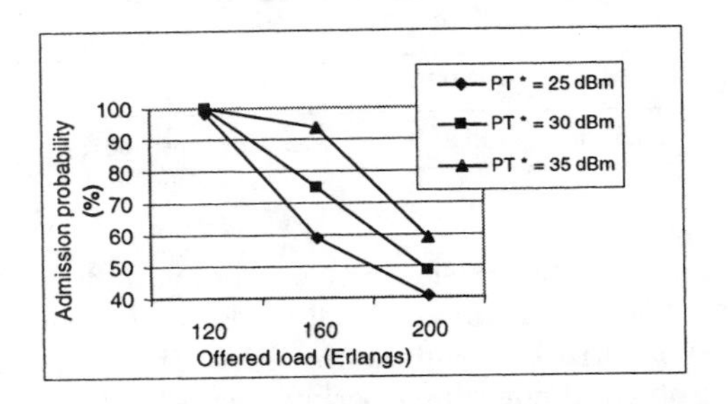

Figure 3.17 Admission acceptance ratio for different power thresholds.

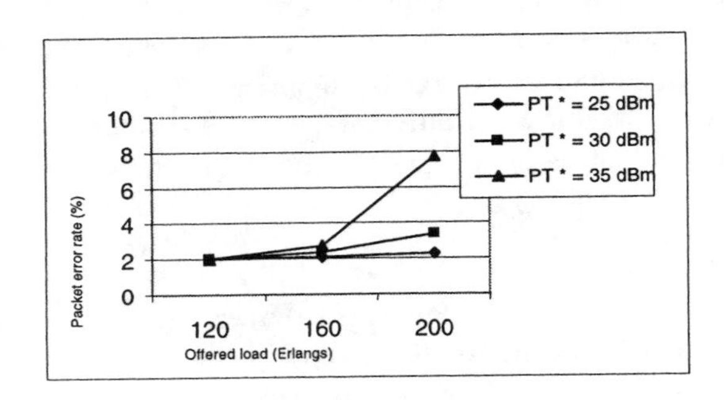

Figure 3.18 Packet error rate for different power thresholds.

Table 3.19
Optimum Threshold for Different Loads

Offered Load (Erlangs)	Optimum Threshold
120	> 35 dBm
160	30 dBm
200	25 dBm

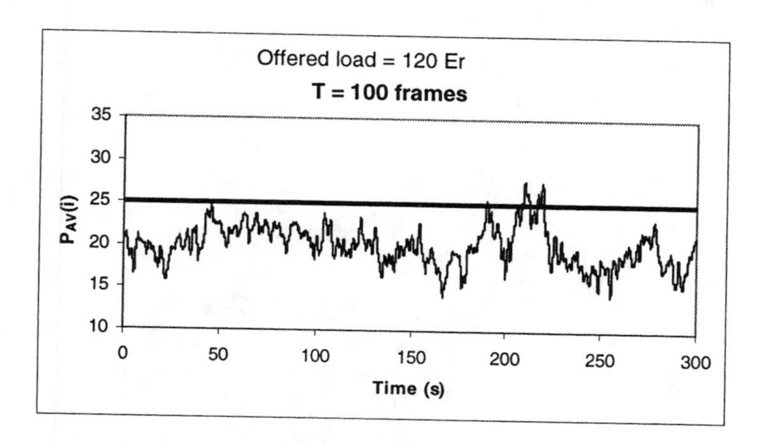

Figure 3.19 Plot of the time averaged transmitted power.

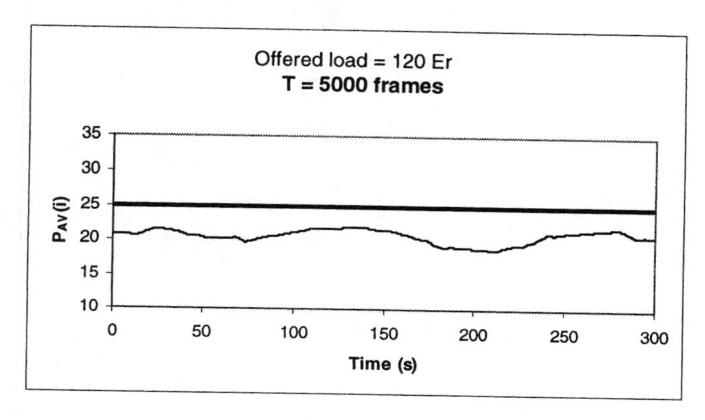

Figure 3.20 Plot of the time averaged transmitted power.

The proposed strategy allocates resources to the different flows that make use of the DSCH channel. It operates on a frame-by-frame basis (i.e., a frame is 10 ms) after the current transmissions for users in DCH channels are known. So the input parameters for this algorithm are:

- The number of users sharing the DSCH;
- The number of transport blocks x waiting for transmission in the buffer of each user;

- The required E_b/N_0 target for each user, which depends on the BLER target to be achieved;
- A measurement of each user's path loss $L_p(d_i)$.

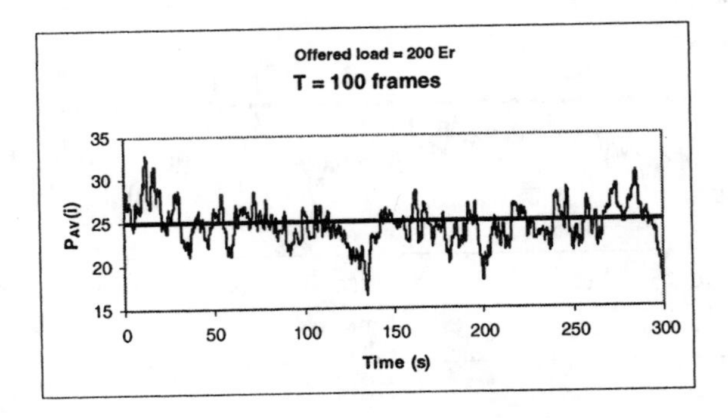

Figure 3.21 Plot of the time averaged transmitted power.

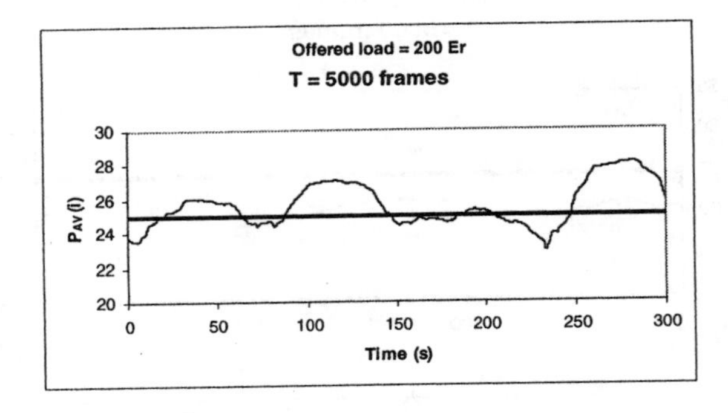

Figure 3.22 Plot of the time averaged transmitted power.

Streaming Service

A measurement of the other-to-own cell interference factor for each user can be given by

$$f_{DL,i} = \frac{\chi_i L_p(d_i)}{P_T}$$

The number of current transmissions in DCH channels, together with their corresponding TF and E_b/N_0 targets.

Taking into account all these parameters, the algorithm performs the following steps in each frame.

Prioritization

The first step consists in ordering the requests of different users in the DSCH depending on some priority criterion that takes into account the required QoS of each user. In particular, the priority table is derived from higher to lower priority according to the following:

- The higher the number of basic layer TBs to be retransmitted the higher the priority will be.
- For the same number of basic layer TBs, the priority is established according to the service credit concept, explained below. The higher the service credit of the enhancement layer, the higher the priority.

The service credit concept consists of monitoring the QoS that each flow has received in terms of bit rate and measuring the difference between the expected bit rate and the offered bit rate. The higher the difference, the higher the resources to be allocated (or equivalently the number of transport blocks that should be transmitted). So the credit that the system owes to the flow should be computed. This leads to the definition of the service credit, that accounts for this difference and can be computed as follows in each TTI:

$$SCr(k) = SCr(k-1) + \frac{R_G}{TB} - NumTx(k-1) \tag{3.26}$$

where $SCr(k)$ is the service credit for TTI = k, $SCr(k-1)$ is the service credit in the previous TTI, R_G is the guaranteed bit rate measured in bits/TTI, TB is the number of bits of a Transport Block for the considered RAB, and $NumTx(k\text{-}1)$ is the number of successfully transmitted Transport Blocks in the previous TTI. It is worth noting that the quotient R_G/TB reflects the mean number of transport blocks that should be transmitted per TTI in order to keep the guaranteed mean bit rate. As a result, $SCr(k)$ is a measure of the number of Transport Blocks that the connection should transmit in the current TTI to keep the guaranteed bit rate. For example, if TB = 320 bits, R_G = 32 Kbps, and TTI=40 ms, four service credits are added in each TTI.

After computing the service credit, and assuming a total of x transport blocks in the buffer, the number of transport blocks to be transmitted in the current TTI would be

$$numTB = \min(x, SCr(k), TB\max) \tag{3.27}$$

where *TBmax* is the maximum number of transport blocks that can be transmitted per TTI depending on how the RAB is defined. Finally, the selected TF would be the one that allows to send *numTB* blocks.

The output of this phase is an ordered list of requests for the different users, each containing a *TF* value.

Resource Allocation

Once requests are ordered, the next step consists deciding whether or not they are accepted for transmission in the DSCH channel and which is the accepted TF. The limitations explained previously dealing with interference and code availability are taken into account in this phase. To this end, it is required to estimate the expected load factor and transmitted power level once all the requests are accepted. Then, the expected load factor whenever there are n transmissions in the system in frame t (including both DCH and DSCH transmissions) is:

$$\tilde{\eta}(n,t)=\sum_{i=1}^{n}\frac{\left(\rho+f_{DL,i}(t-1)\right)}{\dfrac{SF_i}{\left(\dfrac{E_b}{N_o}\right)_i r}+\rho} \tag{3.28}$$

Similarly, the expected power is given by

$$\tilde{P}_T(n,t)=\frac{P_N}{\left(1-\tilde{\eta}(n,t)\right)}\sum_{i=1}^{n}\frac{L_p(d_i)}{\dfrac{SF_i}{\left(\dfrac{E_b}{N_o}\right)_i r}+\rho} \tag{3.29}$$

The differences between the expected load factor and the real value can be caused by the inaccuracies of the measurement of the other-to-own-cell interference factor $f_{DL,i}$ and the path loss.

With these restrictions in mind, the algorithm executes for each request the rules in Figure 3.23, assuming a total of n already granted transmissions. At the beginning, for the initially selected TF, the Kraft's inequality is evaluated (in order to check the availability of OVSF codes); afterwards, the expected load factor is compared with a threshold ϕ, and finally the expected transmission power level should be below a fraction δ of the maximum transmitted power. If all three conditions hold, transmission is granted for this request during one TTI; otherwise, the transport format is reduced by one, or equivalently, the transmission bit rate is reduced. If this is not possible, the request should wait for the next frame.

The control parameters ϕ and δ (both < 1) should be appropriately set in order to take into account the possible fluctuations between the expected values and the real measurements.

One of the most relevant parameters in the design of the packet scheduling algorithm relies on the threshold ϕ of the estimated load factor $\tilde{\eta}(n+1,t)$ when deciding the granted transmissions. Particularly, if ϕ is too high, the difference between the estimated load and the real load values can lead the system to a situation where no available power exists that satisfies all the user's requirements at the same time.

On the other hand, if ϕ is too low, fewer problems will exist for basic transmissions and a lot of enhancement requests will be postponed. This trade-off can be observed in Figure 3.24 and Figure 3.25. Figure 3.24 presents the average bit rate obtained during a streaming session for the enhancement layer depending on threshold ϕ, $\delta = 1$ has been assumed. The basic layer is not presented since its achieved bit rate is almost always 32 Kbps. Figure 3.25 presents the percentage of lost packets for both basic and enhancement layers.

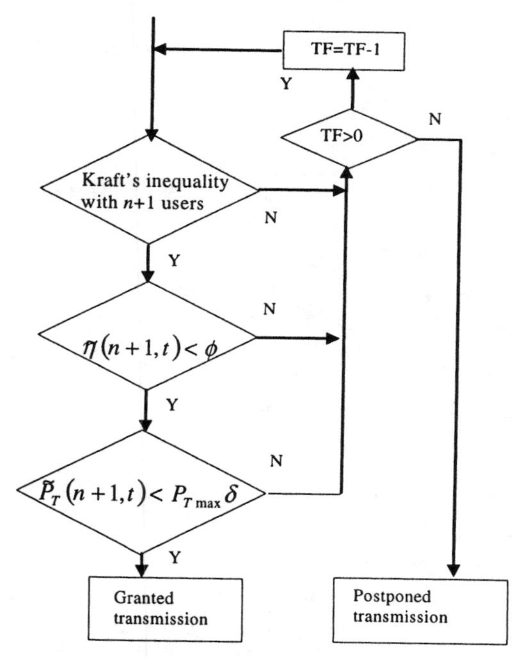

Figure 3.23 Resource allocation process.

From both figures it can be concluded that the selection $\phi = 0.95$ provides the best behavior since it achieves the maximum bit rate for the enhancement without degrading the quality of the basic layer. Notice also that thanks to the retransmissions, the packet loss ratio is zero for the basic flow whenever $\phi \leq 0.95$ even for high loads.

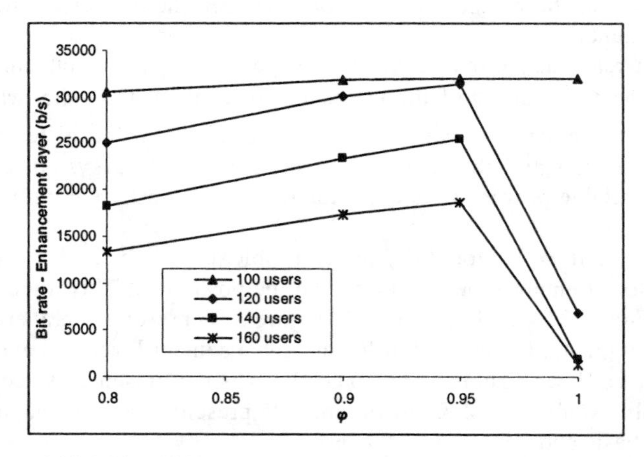

Figure 3.24 Achieved bit rate. (*From:* [21].)

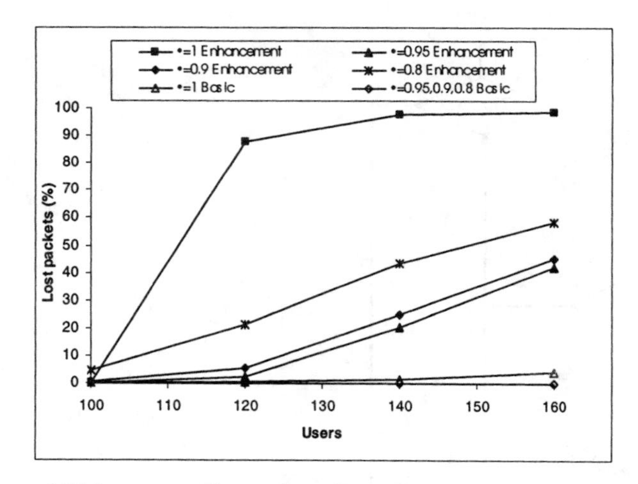

Figure 3.25 Percentage of lost packets. (*From:* [21].)

Furthermore, Figure 3.26 shows the jitter of the packet delay, which is one of the main QoS requirements for a streaming service. For the basic flow, it can be observed that, apart from the case $\phi = 1$, the jitter is below 1 TTI. For the enhancement case, the jitter is somewhat higher than the basic due to the packet scheduling operation.

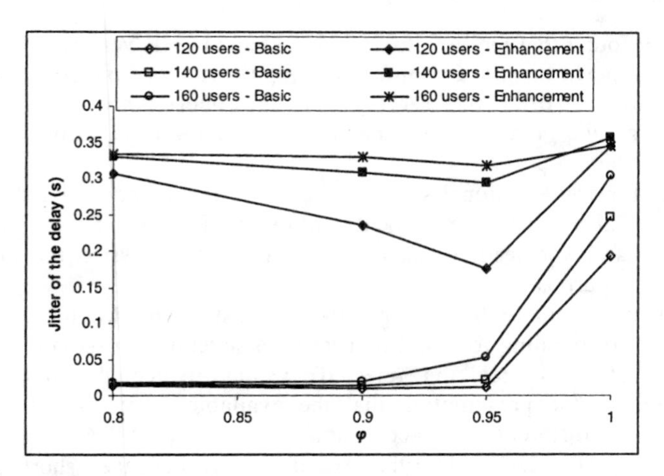

Figure 3.26 Jitter of the packet delay. (*From:* [21].)

The value of the jitter impacts the buffer dimensioning at the receiver side to guarantee that each flow is delivered to the user in a continuous way. Consequently, the maximum allowable jitter will depend on the specific buffer capabilities.

3.3.6 Summary

The experience gained in the framework of the IST project ARROWS has confirmed that RRM is a complex problem affected by many different issues, with crossed effects among the different parameters of a given RRM strategy as well as among the different RRM strategies themselves. Changes in the scenario or the system condition may lead to different RRM suitable solutions. Consequently, simple RRM algorithms with a reduced set of key parameters instead of complex algorithms with large sets of parameters to be adjusted are preferable. In the latter case, suitable values for the different parameters would vary significantly from one scenario condition to another, and parameters disadjustment could have a strong impact on the network performance.

Thus, taking simplicity as the starting point for RRM algorithms definition, the focus is placed on identifying the key elements influencing the performance as well as providing some design guidelines in the parameters optimization process. In this respect, we offer some relevant observations:

- Uplink admission control based on statistical cell load control is able to assure system stability, planned coverage and planned quality, with the aid of congestion control mechanisms if needed. In the admission algorithm, the intercell to intracell interference factor is better to be considered as an average value rather than considering real-time measurements of the f-factor.

- It has been found that the time to decide the handover is very critical for the network performance, as it influences the interference pattern (as the user moves far from the serving cell the interference caused to neighboring cells increases). The handover averaging time should be as low as possible.
- Handover for conversational services should be facilitated to avoid call droppings. This can be done by the definition of a Reserved Set region to anticipate handover needs or also by adjusting the admission threshold for handover purposes.
- The system is usually limited by blocking instead of dropping. Call retries (i.e., responding to a call request with several admission control attempts before the call is rejected) would reduce the blocking probability on the presumption that the available retry time is long enough to face different system conditions.
- The uplink UE-MAC algorithm used to decide the short-term transmission rate cannot be decoupled from the admission control procedure because different algorithms may lead to very different interference pattern behaviors, which need to be monitored from the admission control phase.
- Regarding congestion control, results reveal that in general firmer actions (i.e., high reductions of the TFS) and early congestion detection algorithms lead to shorter congestion periods with smoother delay degradations. Also, the recovery algorithm after a congestion period is quite critical and should follow a time scheduling approach.
- Downlink admission control should be based on the monitoring of node-B transmitted power instead of controlling the cell load factor.
- The required downlink transmitted power averaging period depends on the load level relative to the admission threshold. For low loads long averaging periods are more suitable (in the order of a minute), while for high load situations shorter averaging periods (in the order of a second) lead to a better performance.

3.4 RESOURCE MANAGEMENT AND ADVANCED TRANSCEIVER ALGORITHMS FOR MULTIHOP NETWORKS

3G communication systems theoretically make it possible for users to have universal access to a range of high data rate multimedia services. To achieve these goals, systems beyond conventional 3G cellular networks are required to provide the desired uniform coverage and high data rate [22].

Deployment of 3G networks that incorporate the concept of intelligently deployed relay units was a topic investigated within the framework of the IST project ROMANTIK.

The high price paid in many countries for 3G spectrum has added to the pressure to deploy the next generation of cellular networks [23, 24]. For this reason, the need to achieve higher levels of coverage and capacity at a minimum cost is key. The concept of intelligent relays has been proposed as an attempt to achieve higher capacities and/or improved area coverage areas compared to traditional microcellular networks. The method can also be used in an attempt to minimize the overall cost of the network. Further benefits, such as minimizing the necessary transmit power, are also believed to be possible. Also, the use of fixed relay nodes enables the use of direction antennas that can minimize the emission of unwanted interference. Fixed nodes enable the network to planned, rather than having to relying on the presence of mobile relay terminals.

Intelligently deployed relays can be used to reduce the number of traditional base stations required and thus to reduce the cost of the required infrastructure. The idea behind the relaying concept is to forward to communications locally using a network of wireless relays. In the most general case, these relays may be other mobile terminals, or dedicated fixed nodes deployed by the operator to transfer data from the user terminals to the dedicated base stations. In this section, the analysis concentrates on the use of fixed relay nodes (rather than other mobile terminals) and in particular on the potential cost/performance benefits of intelligently determining the location of these nodes.

The relaying architecture can be used in an attempt to enhance coverage and/or capacity while minimizing the cost of the network. The use of fixed relays is particularly well suited to complex dense urban environments (where heavy shadowing occurs and the benefits of a relay link are strong). The key to the success of relaying depends on the strategic positioning of the relay units and their associated base stations. However, it is well known that the optimum location of base stations is not an easy issue and the positioning of relays can prove to be just as difficult. To ease the process of base station and relay positioning, a new site-specific deployment algorithm was developed within the IST project ROMANTIK [22, 25].

The optimization algorithm is based on the idea of combinatorial theory, although other techniques are also used to improve the performance of the algorithm. The developed algorithm estimates the optimum location of base stations and relay units in a specific microcellular environment. The algorithm interacts with two external modules and these are used to provide information on capacity and coverage for a given topology. Using this information, the algorithm is able to optimize the location of the base stations and relay units according to a set of supplied user requirements. The combinatorial algorithm is referred to here as the combinatorial algorithm for total optimization (CAT) algorithm. A newly developed cost model has also been added to the final stages of the optimization process. The model allows a cost/performance analysis of conventional and mixed networks to be performed.

3.4.1 The Site-Specific Deployment Algorithm-CAT

This section examines the substitution of base stations in favor of relay units, and examines the potential cost/performance benefits of this approach. In order to quantify the potential coverage, capacity and cost benefits offered to a microcellular network by relaying technology, conventional networks, and mixed networks have been deployed to allow comparisons to be drawn. The network deployments have all been performed using the CAT algorithm [22].

The CAT optimization algorithm [24–26] was developed to meet a need to minimize the infrastructure cost of cellular networks while guaranteeing capacity and coverage levels in complex topographical environments. The algorithm also reduces the complexity and removes the guesswork involved in the planning of a modern cellular network.

The optimum placement algorithm is based on a combinatorial approach; however, other techniques such as the Greedy algorithm and the "split and merge" process are also used to enhance performance [26–28]. To provide accurate planning, the algorithm requires as much information about the operational environment as possible. For this reason, the algorithm makes use of two external modules (coverage and capacity) to provide the reliable information that is necessary to drive its operation.

The coverage module uses the University of Bristol's RaySim propagation tool, which determines the propagation characteristics between a potential base station or relay site to each of the specified control nodes. The quality of the propagation information plays a key role in the optimization process and the subsequent value of the final solution. For microcells, traditional statistical propagation models are inaccurate and more advanced ray-based propagation tools are required. However, despite their improved accuracy, ray models require longer prediction times (often in the order of seconds per prediction point). The use of a ray model can result in unacceptable run times for optimization techniques requiring iterative propagation updates between many deployment sites and control nodes. The optimization algorithm used in this project has been designed to interact noniteratively with the external propagation module (although the resulting propagation tables are used iteratively within the processing of the CAT).

The site-specific deployment algorithm also makes use of an external capacity module. The capacity simulator model is used in a similar way to the propagation model. The optimization algorithm selects the final locations of base stations and relay units according to the information acquired in the capacity simulator. Firstly, the algorithm optimizes the location of base and relay stations according to the coverage requirement. Once a number of possible solutions have been obtained, and depending on the user requirements, the capacity of the system is computed. There are two variants of the capacity module: originally a capacity module based on the use of Erlang tables was developed in [29, 30]; more recently, a capacity module based on a system level Monte-Carlo simulation is used to determine the interference, transmit power, capacity, and spectrum efficiency benefits of the mixed cellular–relaying network over the conventional cellular network. The choice of the capacity module depends on the environment and technology used in the optimization process.

Before any optimization takes place, a number of constraints must be defined. The choice of the parameters used in the optimization process defines the user problem and configures the placement tool. These parameters are described in the following paragraphs.

The planning area is described by a universal set of discrete points, U, which contain all locations where coverage/capacity is required. The points can take the form of control nodes (CNs), relay units (RUs), or base stations (BSs) and are defined by (3.1).

$$U = \{CNc, n, BSb, s, RUr, u\} \tag{3.30}$$

A number of discrete user supplied points, or CNs, are used to represent the capacity and coverage requirements in the area. This set is mathematically denoted by $CN_{c,n}$ and is defined by (3.32), where c represents the location of each control node (x,y) and n the total number of such nodes.

The number and location of all possible RU locations in the deployment area is also user supplied and represented by the $RU_{r,u}$ set. $RU_{r,u}$ represents an over specified set of RU locations and is defined by (3.32), where r represents their locations (x, y) and u the total number in the study area.

$$CN_{c,n} = \{ci : i = 1 : n, ci \in U\} \tag{3.31}$$

$$RU_{r,u} = \{rj : j = 1 : u, rj \in U\} \tag{3.32}$$

The number and location of all possible BSs is user supplied and represented by the $BS_{b,s}$ set. $BS_{b,s}$ represents an over specified set of BS locations and is defined by (3.33), where b represents their locations (x, y) and s the total number in the study area.

$$BS_{b,s} = \{bj : j = 1 : s, bj \in S\} \tag{3.33}$$

The specification for user-supplied base stations and relay unit locations is used as a matter of practicality, due to the fact that network operators cannot deploy base stations and/or relay units in arbitrary locations, such as protected buildings, inappropriate geographical locations, and so on. The user supplied locations avoid this problem and assure that a sensible solution based on locations where deployment is viable (e.g., where planning permission has been or can be obtained, locations with adequate power supply, established sites, etc.) is produced. The use of discrete locations for base stations, relays and coverage points also enables the iterative use of detailed ray tracing propagation tools (since coverage tables can now be prebuilt for all possible connections). Since the quality of a planned deployment is only as good as the underlying propagation tool, the use of highly accurate tools is a strong step forward.

The three elements described above (BSs, RUs, and CNs) provide the basis for the algorithm's definition. Given a set of CNs, RUs, and BSs, the optimization algorithm must deploy a minimum number of BSs and RUs to satisfy the operator' requirements (which are defined in U in the form of CNs). Every CN defined in the $CN_{c,n}$ set has to meet the planning requirements, R_n as defined by the user in the form

$$R_n = \{P_{\min}, \tau_{\max}, C_{\min}\} \tag{3.34}$$

where P_{min} represents the minimum power (dBm) target for each CN; τ_{max} indicates the maximum rms delay spread (ns) that the deployed system can tolerate; and C_{min} represents the minimum capacity requirements in the area. The user can define one, two, or all three of the parameters described in (3.34), and consequently the final solution will vary based on the number of constraints applied.

The CAT algorithm searches for a network solution that satisfies the planning criteria of all CNs according to the requirements introduced by (3.34). For example, for a scenario that only requires acceptable coverage (i.e., for a low number of high value subscribers), (3.34) would configure the algorithm to find a solution that covers all CNs with a minimum number of BS and RUs. Depending on the relative cost of a relay node to a BS, the ratio of BSs to relay numbers can also be varied. This type of solution might suit a rural area with a low number of subscribers.

However, this type of solution may be inadequate for a busy city center area since the topology may offer insufficient capacity. To deal with this type of scenario, capacity requirements can be added to the network requirements. Once a number of potential solutions are obtained, the target capacity values can be evaluated and the best solution can be selected. This allows a proper dimensioning of the network by trying to achieve a balance between coverage and capacity.

The *site-specific deployment algorithm* can be used to optimize the planning of a network at different stages in its growth. That is, the algorithm can optimally plan the network for a new area (green field site) or alternatively it can be used to improve the performance of a mature network.

The algorithm can also be used to optimize the location of relay units that can be used in a cellular environment to improve performance. Relay units are generally used to reduce the number of necessary base stations and can also be introduced at different stages in the network's growth.

The techniques and approaches used for the development of the optimization algorithms used here were presented in [22].

3.4.1.1 Propagation Module

A state-of-the-art propagation model can be used to provide channel data for evaluating cellular outdoor networks. The deterministic model uses geographic data (terrain, building, foliage, and ground cover) to predict power, as well as time, frequency, and spatial dispersion in the radio channel [31, 32]. It is optimized for intracellular coverage as well as intercellular predictions (interference) between different cells in a mixed-cell network. Propagation data is supplied for each base site in a list of potential base site locations as well as for each potential relay. This data is then passed on to the optimization module and is used to optimize the number and locations of cellular base stations and fixed relay nodes.

3.4.1.2 Capacity Module

A system level MonteCarlo simulation can be used to determine the interference, transmit power, capacity, and spectrum efficiency benefits of the mixed cellular-relaying network over the conventional cellular network. The simulation is static (i.e., it considers snapshots of the system at particular time instants, and does not consider signaling or other network overheads). For the experiments described here, the following parameters are assumed. The 3G simulation assumes a 3.84-Mcps UTRA TDD type system [33], with a fixed and ordered time slot allocation, but with asymmetric sharing of slots between the uplink and the downlink. Three classes of user are considered: class A users support voice at 15 Kbps, class B users support data at 144 Kbps, and class C users support data at 384 Kbps. The traffic mix is specified as: 60% class A: 30% class B: 10% class C. Users are uniformly deployed in the area of Bristol, and the propagation model is used to determine path losses to the various base stations (BS-UE) and relay units (RU-UE) as well as losses between base stations and relay units (BS-RU) and between different relay units (RU-RU) for interference calculations. The optimal transmission path from user to BS is then determined as the path with the minimum total path loss. The admission control, resource allocation, power control, and joint detection processes followed for the development of this model are described in [26].

3.4.1.3 The Optimization Algorithm for Base Stations and Relay Units Optimization

The CAT algorithm allows the optimization of base stations and relay units in the cellular environment. The algorithm can be used to optimize the minimum number of base stations or the minimum number of both base stations and relay units needed in the area. It performs its computations by using the information provided by the propagation and capacity modules.

For the optimization of a scenario that includes base stations only, the process followed has been previously described in [24, 25]. For mixed scenarios (base stations and relays) the process is similar but a number of different steps are taken. The basics of the algorithm remain the same; a predefined number of control nodes, relay units, and base stations must still be defined by the user before the algorithm can be executed. Once the basic parameters have been entered, the optimization process begins. The approach taken for the optimization of relay units and base stations is described below.

The initial process of optimizing the minimum number of base stations needed to cover all the control points is performed. The combinatorial algorithm provides a ranked list of solutions that contain different combinations (some solutions could contain a different number of base stations).

The ranked list will contain solutions that fulfill the requirements and solutions that partially meet the requirements. A breakdown (or ranked list of solutions) based on the percentage of coverage provided by each base station within a solution can be obtained. With this information, it is therefore possible to select a combination that provides a good trade-off between the number of base stations and the coverage and/or capacity achieved. At this point, a user choice must be made on the number of base stations to deploy. Choosing a combination with fewer base stations than the number recommended by the algorithm will leave a number of control points uncovered and thus result in coverage black spots. It is under these conditions that the optimized deployment of cheaper relay units can take place.

According to the locations of the uncovered control points, the user can make an informative guess as to the location of potential relays to use. An oversized set of potential relay units must be located in the area. The CAT algorithm must take into account three things before offering a solution:

1. The chosen relay units must have a strong communication link with at least one base station, to make sure that they can connect to the core network. Each potential relay will be assessed and the ones that do not fulfill this requirement will be dropped from the optimization process. This initial step removes unsuitable relay units.
2. After the previous step has been executed, an optimization process to select the minimum number of relay units is performed. This is achieved by using the site specific deployment algorithm.
3. A final sanity check is performed to make sure that the solution(s) offered are useful and fulfill the user's requirements. This check completes the optimization process.

To summarize, the method consists of reducing the number of required base stations in favor of potentially cheaper relaying technology. To assess the potential cost benefits of mixed (base stations and relay units) scenario solutions, a cost model has been developed. The next section describes how the model operates.

3.4.2 Cost Model Used in the Optimization of Base Sites and Relay Units

A cost model was developed as part of the IST project ROMANTIK [1] to aid the planning of traditional and mixed cellular networks. This model allows the user to quantify the cost of solutions offered by the CAT algorithm. Of particular interest is the ability of the model to compare the cost of a traditional network solution with that of a mixed solution. This section investigates and identifies the costs of installation and maintenance of base stations and relay units in a cellular environment. There are a number of costs associated with the deployment of a network; however, in this research they have been reduced to four main costs: the equipment/hardware costs (i.e., cost of the transceiver units/base stations or relay units), the deployment costs (cost of deploying the units), the site-acquisition costs, and finally the maintenance costs of the units. For the success of mixed networks the cost of relay units is a key aspect for the successful adoption of relaying technology in a cellular environment.

Network equipment suppliers are seldom willing to discuss the cost of building a cell site or network; hence, there is some difficulty in finding the data required. Although this information is not critical for this research task, it helps to put things into perspective to have an indication of the actual cost of deploying a network. The cost of developing a cellular network might vary for different equipment suppliers [23].

To obtain an insight into the long-term pricing of infrastructure from leading 3G suppliers, the WCDMA radio network contract was awarded in equal shares to main suppliers. This information is very useful when trying to price the deployment of base stations and relay units. The network operator's total budget dedicated to equipment for the construction of a network comprising 6,000 base stations over the next 5 years was equal to $900 million. This readily infers that the unit cost per base station amounts to $150,000. The author of [23] assumes that over this period of time, radio networks will represent some 80% of the total equipment spent, therefore the actual cost for the base station plus the radio controller portion falls in the region of €120,000. The above provides the means to estimate the price of a base station unit.

However, this does not include the acquisition, deployment, and maintenance costs. We assume (based on deductions from [23]) without loss of generality, that the acquisition, installation, and maintenance costs are 10%, 5%, and 10% of the base station cost, respectively. These ratios will obviously differ for different operators and different environments, but the values do not affect the results and conclusions in this analysis.

Table 3.20 shows the total cost per BS unit assumed for this study. These figures are based on the following assumptions: the cost of the BS transceiver as identified above, the cost of acquiring the base station site assumed to be 10% of the base station transceiver cost, the cost of deployment believed to be 5% of the transceiver cost, and the maintenance cost assumed to be 10% of the transceiver cost (which would cover the maintenance costs for 10 years).

Table 3.20

Assumed Base Site Costs (in US$)

BS Transceiver Cost	BS Site Acquisition Cost	BS Deployment Cost	BS Maintenance Cost	Total BS Cost
120,000	10% (12,000)	5% (6,000)	10% (12,000)	150,000

It is also necessary to consider the cost of the relay units. The price of the units will depend on the characteristics of the equipment and it can go from a few hundred EUR to tens of thousands of EUR. For this reason, here, the cost of the relay units was calculated as a fraction of the cost of a base station unit. The cost of the relay units was considered to be a variable. This allows one to consider if the deployment of relay units is a cheaper option, or if the costs compromise the success of structured relaying.

The cost of the relay units were calculated as a percentage of the cost of the base station units.

To investigate the acceptable cost of the relay units, it was assumed that the cost of a relay unit can vary from 10% to 80% of the cost of the base station. In addition, the total cost of using a relay unit was similarly a composite of costs that included the acquisition, maintenance, and deployment cost plus the cost of the relay unit itself.

Table 3.21 shows the potential costs of the relay units as a percentage of the cost of the base station units.

The acquisition site cost for the relay units was assumed to be fixed for any relay unit cost. The cost was fixed at 10% of the acquisition cost calculated for a base station (10% of US$12,000). It was assumed that the physical size and power consumption requirements of the relay units are lower than those of the base stations and therefore it would be expected to accrue a lower cost for the site acquisition. The deployment cost of the relay units was assumed to be 5% of the relay unit cost (as assumed for the base station case). The cost of maintaining the relay units was assumed to be 10% of the relay unit cost (this would cover the maintenance costs for 10 years).

Table 3.21

Cost Summary of Relays and Base Stations (in US$)

	Transceiver unit cost	Acquisition cost (10%)	Deployment cost (5%)	Maintenance Cost (10%)	BSs/RUs Total cost
Base station unit	120,000	12,000	6,000	12.000	150,000
Relay unit at 10% BS cost	12,000	1,200	600	1,200	15,000
Relay unit at 15% BS cost	18,000	1,200	600	1,800	21,600
Relay unit at 20% BS cost	24,000	1,200	600	2,400	28,200
…….	………	……..	………	………	………
Relay unit at 80% BS cost	96,000	1,200	600	9,600	107,400

Figure 3.27 shows the total costs for different numbers of base station and relay units.

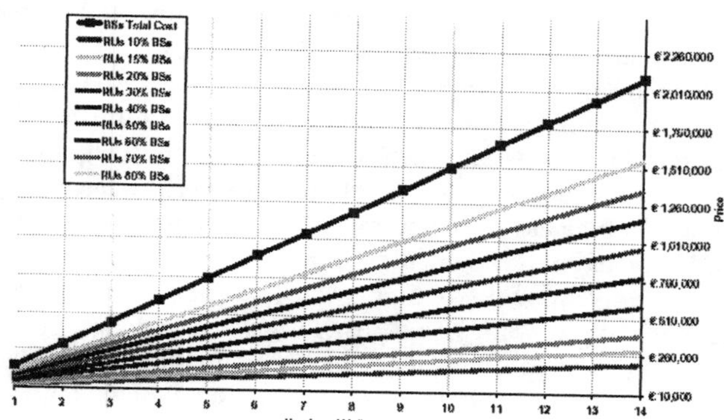

Figure 3.27 Number of units (BSs and RUs) versus cost.

3.4.3 Routing Strategies for Multihop Networks

In mobile wireless networks, where communication terminals are mobile and transmission media is wireless, routing is one of the main problems. Mobiles in a wireless ad hoc network share the same frequency channel, like IEEE 802.11 [34, 35] and ETSI HIPERLAN Type 1 [36]. The limitations on power consumption imposed by the portable wireless radios coupled with the fact that the communication infrastructure does not rely on the assistance of centralized stations, implies that terminals must communicate with each other either directly or indirectly using multihop routing techniques. As nodes move about, this results in a distributed multihop wireless network with a time-varying topology.

Before delving into the details of the properties underlying dynamic routing in wireless networks, our primary issue is to find out whether a conventional routing protocol, like link-state or distance-vector, could apply in a wireless multihop environment. To respond, we first need to list several outstanding structural differences that exist between wired and wireless mobile networks that make routing very different in the two environments [37–40].

In a mobile wireless network, the rate of topological changes is relatively high compared to that of wired networks. As is the case in wireline networks, the procedures for route selection and traffic forwarding in wireless mobile networks require accurate information about the current state of the network (e.g., node interconnectivity and link quality) and the session (e.g., traffic rate, end-point locations) in order to direct traffic along paths that are consistent with the service requirements of the session and the service restrictions of the network.

However, changes in network or traffic sessions are likely to occur more frequently in mobile wireless networks than in stationary wireline networks. The degree of dynamism in route selection depends on several factors, including the type and frequency of changes in network and session state; the limitations on response delay imposed in assembling, propagating, and acting upon this state information; the amount of network resources available for these functions; and the expected performance degradation resulting from a mismatch between selected routes and the actual network and session state.

The routing mechanism must be able to quickly detect and respond to such state changes in order to minimize service degradation of existing traffic sessions whereas, at the same time, the algorithm must do so using a minimal amount of network resources, in order to maximize the overall network performance [41].

On one hand, the effectiveness of a routing protocol increases as network topology information becomes more detailed and up to date. To maintain up-to-date routing tables, a conventional routing protocol should be forced to continuously send and receive topology updates. In wireless ad hoc networks, however, the topology may change quite often, requiring frequent exchanges of control information (e.g., routes, route updates, or routing tables) among the network nodes. In an event-triggered link-state protocol any topological change would trigger a flooding, resulting in a flooding rate equal to the topological change rate. In this scenario, a blind route update mechanism could unnecessarily waste network resources since updates are sent even when no data transmission at all occurs in the network. In addition, as the number of network nodes can be large, the potential number of destinations is also large, thus requiring a high volume of control information exchanged among the network nodes. As a consequence, the amount of update traffic can be even higher, the distribution of which can eventually saturate the network.

Notably, radio spectrum is a scarce resource, which means that packet-radio networks typically have limited bandwidth available. Because the wireless devices must share access to the radio channel, the bandwidth available to any node is even more limited. Relatively low bandwidth combined with the potential for routing algorithms to generate large numbers of packets means that efficiency is paramount in designing packet-radio routing algorithms. This observation, however, is in contradiction with the fact that all updates in the wireless communication environment travel over the air and are then costly in resources. An even more disappointing fact is that as the network size increases and as the nodal mobility increases, a smaller and smaller fraction of this total amount of control traffic will even be used.

3.5 OPTIMIZED PERSONAL AREA NETWORKS

Being able to communicate anywhere, anytime, and with any device is a global trend in today's development of communication and computing systems. Ubiquitous computing is becoming possible with the advance of wireless communication technology and the availability of many light-weight, compact, portable computing devices. The IST project PACWOMAN [3] performed the necessary research, development, and optimization on all OSI layers, enabling the design of a low-cost, low-power, and flexible WPAN-based system. This chapter presents the results of the studies undertaken in the context of the project PACWOMAN towards MAC layer development, parameter extraction, and optimization. MAC mechanisms for WPANs, for both high and low data rate environments are discussed. The main part of the discussions focuses on the development of MAC mechanisms for ad hoc networking, and new solutions for medium access control in the ad hoc networking domain are proposed. A simulation model and approach to investigate the MAC protocols is also presented.

3.5.1 Research Issues for Future WPAN Technology

The new emerging communication technologies will be centered on the user, adapting to user's preferences and improving his living and working environment. The future WPAN is seen as a communication paradigm with a high level of personalization and ubiquitous access to information on demand and in an ad hoc manner (opportunity driven) through personal and public networking resources.

The future WPAN should provide the user with the following communication capabilities [42]:

- Wireless connectivity within the personal operating space (POS); in-bubble networking;
- Access to sensors and actuators (mobile, wearable, or fixed);

- Access to the other wireless/wired networks (i.e., Internet via LAN access points, 3G network, organizations intranet) and/or other POS.

Several application scenarios are foreseen for the WPAN environment. They should be seamlessly available wherever the user is, supported by a variety of future services. In order to realize the future WPAN dynamic and adaptive networking concept, research activities will continue to concentrate on the topics that will provide optimized architecture and protocol solutions. The requirements of the WPAN connections will develop in synergy with the pervasiveness of computing in general. In general, the ubiquity achieved by the future short-range wireless technologies will depend on their degree of adaptation to the computing environment: fast service/resource discovery, fast autoconfiguration, and seamless interworking with communication entities from infrastructure wireless/wired networks. Hence, the design of the WPAN should essentially depend on the legacy wireless systems of wider coverage with respect to coexistence and compatibility. Most important, the WPAN will also have the role of a complementary and integrating technology to the other technologies, being the "closest meter" to the user.

Some examples of relevant future research areas include the following:

- *QoS/multimedia support.* The support of multimedia services will most likely be required within and throughout the ad hoc WPAN. However, the QoS will gain different flavor in the future short-range wireless networks, due to the emergence of unique applications. For instance, there may be applications for which the timeliness and quickness of the ad hoc connection establishment is critically important, while the actual rate that the application will use over the already established connection may be quite modest.
- *Trade-off QoS/power efficiency.* It has repeatedly been stressed that low power consumption is of paramount importance for the technology intended for use by handheld battery-powered devices. Therefore, the WPAN systems should be flexible in a way that will enable them to trade QoS for power saving. To illustrate this point, recall that the inherent stochastic communications quality in a wireless ad hoc network makes the provision of service guarantees difficult. Thus, a power-aware protocol should not perform transmission when the packet is highly likely to be received in error (due to a certain reason, which can be predicted to some extent) despite the fact that the QoS guarantee requires the transmission of that packet. Eventually, as soon as the communication conditions are stable, the WPAN should adapt and provide the user with better QoS, for example, as one defined for the access network. The illustration above is another example of cross-layer optimization, something that should be a fundamental approach in designing power-aware ad hoc communication systems. In this particular

case, the increased cross-layer interaction consists of control of the activity of an upper layer (QoS provision entity) with information from a lower layer (bad channel conditions). Many challenging research issues are produced by the need for building richer interfaces among the layers of the protocol stack in the wireless ad hoc systems. An optimal interplay of QoS and power efficiency can only be achieved if the optimization regards the protocol layers jointly.

- *Integration and cooperation with other networks.* The future WPAN should provide seamless integration and mobility management for heterogeneous infrastructure and other ad hoc networks. It will be IP oriented, and should support the coexistence of IPv4 and IPv6. Mobility of terminal devices, within or with their POS, addresses issues such as vertical/horizontal handover, location awareness, and roaming. The network infrastructures will strongly influence the addressing, routing, and security solutions.
- *Interference/coexistence in unlicensed band.* At the physical layer and for baseband conditioning, the research efforts and the solutions will largely depend on what is "low cost and simple" on the DSP market, thus determining how complex and expensive solutions for PAN transceivers will be acceptable. Research on new radio interfaces has already started in Japan and in Europe to increase the data rates of third generation mobile radio systems by more than one order of magnitude. Japan has proposed to reach at least 10 to 20 Mbps in a cellular environment and 2 Mbps for moving vehicles, having the multicarrier CDMA (MC-CDMA) as one of the candidate multiple access techniques. Generalized multicarrier (GMC) CDMA systems are capable of multiuser interference (MUI) elimination and intersymbol interference (ISI) suppression, irrespective of the encountered wireless frequency selective channel. Other license-free bands should be considered, such as UNII 5 GHz or the unlicensed portion around 60 GHz. Finally, UWB is a highly promising research area on its own, because its transmission characteristics may be uniquely exploited by the upper layers (again cross-layer optimization).
- *Advanced link adaptation and MAC techniques.* The great volume of research work in ad hoc routing protocols should be put in the context of concrete MAC layer realization. The future work on the short-range wireless network must apply the routing protocols in a way that is adapted both to the channel access/transmission conditions and the application requirements. The research in MAC layer development for PAN should focus on cross-layer optimization, ensuring that the MAC layer mechanisms include functionality to ensure guaranteed service levels (QoS) and power efficiency. This requires MAC layer awareness of the requirement put on the data streams originating from the network layer, and also mechanisms for adjusting the parameters of the physical

link. This means that the MAC layer should be able to exchange information with both the network layer and the physical layer. The requirement to offer QoS also implies the implementation of a MAC layer that is able to support different service classes each with their own characteristics. Typically some kind of scheduling is required for that; however, this is in contrast to the random nature of most current MAC protocols. Actually the scheduled users could override the best-effort users; a delicate trade-off is required in this case. There are still many open issues when it comes to the development of random access protocols that differentiate according to the user class. Especially in the case of multiple hop connections, the assurance of end-to-end QoS levels becomes a delicate issue. The mobility of the users makes a reservation scheme difficult to maintain; however, in the case of a WPAN of limit scale some specialized solution might be developed.

- *Context discovery protocols and initialization techniques.* The area of context discovery and initialization concentrates on developing different initialization algorithms and discovery protocols such as: location discovery, single device discovery, discovery of infrastructure and noninfrastructure networks, service discovery, and environment discovery (e.g., security aspects).
- *Protocols and techniques for self-organization.* The area of self-organization and reconfigurability for the support of wireless and mobile communication is recently receiving growing attention. The protocols and algorithms for self-organization should be combined with energy-aware routing and cooperative information processing techniques.
- *Robust and optimized protocol stack.* The overall protocol stack should be optimized towards power consumption, rate adaptation capability, end-to-end QoS, improving TCP/IP performances, and security requirements. It includes development of cross-layer optimization techniques, distributed TCP-aware PEPs, new intelligent PEPs implementation, development of robust and adaptive protocols that can cope with packet losses, and bandwidth adaptation.
- *Security.* With the introduction of many wireless systems and users of those different systems, security of those connections and the network as such, as well as authentication, becomes extremely important. The exact implementation of the security mechanism influences the design of the network architecture and could also be taken into account when developing link layer and network layer protocols. Complete stand-alone operation is difficult to achieve since all security protocols rely on a shared secret; this secret is usually exchanged via a different manner. For example, many security mechanisms known today rely on common access to a third party; however in WPAN networks such a third party is not necessarily available.

The overall concept of WPAN in the future will develop and offer novel solutions for new applications, services, and higher data bit rates.

3.5.2 Overview of MAC Techniques

The introduction of multihop links introduces new research challenges to the design of short-range wireless networks. The first design problem is the inherent presence of the hidden terminal effect. Since nodes make multihop networks by definition, not all nodes are within radio range of each other and thus carrier sensing mechanisms alone are not sufficient.

Also, in higher layers problems can occur: multihop connections introduce the need for routing protocols to determine to which neighboring node the packets have to be forwarded [42]. These protocols should be able to deal with the characteristics of ad hoc networks as discussed in the previous section. Several routing protocols have been proposed in recent days under which the most promising for practical application are dynamic source routing (DSR) and ad hoc on-demand vector (AODV) routing (for more details see [43, 44]).

TCP or UDP are typically applied as a transport layer protocol over these ad hoc routing protocols. The performance of TCP/IP over erroneous links can be unstable due to lost acknowledgments and time-outs. The time-outs occur due to route discoveries that are invoked by nodes that are not able to contact their adjacent nodes (e.g., due to set network allocation vectors). This effect is enhanced by the slow-start mechanism of TCP.

Broadcasting is expected to be a common operation in ad hoc networks, for example for paging, route discovery, or other kinds of signaling information. A straightforward manner to implement broadcasting is *flooding*, however, if the flooding is done blindly it will result in many collisions since nodes will receive and try to forward the data almost simultaneously, resulting in many colliding channel accessing attempts. This effect is called the *broadcast storm problem* and was first addressed in [45].

A particularly important issue to be investigated with respect to multihop ad hoc networks is power awareness throughout the design. Here we will focus on the power awareness of the MAC protocols. The first method to reduce the power consumption of the device is the implementation of a spectrum efficient protocol, limiting the amount of unnecessary retransmissions and collisions to the absolute minimum. A second method is control of the transmission power; by reducing the transmission power, the interference is reduced and the throughput increased. However a minimum power is required to maintain connectivity, and between these two boundaries an optimal topology can be derived. A third method is power-aware routing, where the paths are optimized according to a certain cost function (e.g., battery lifetime, or minimum power usage in the path). Low-power modes can also be applied, where, for example, the devices only wake up periodically [46].

The multihop connections might cause serious *unfairness* problems; differences in round-trip time may cause TCP connections with a shorter round-trip time to starve out TCP connections with a larger one [47] But there are also unfairness issues arising from the MAC layer, in particular because the sensing range is typically larger than the reception range. This means that a data flow that is being transmitted starves out connections that are in its sensing range, even though the flow is not received there and would not interfere with an ongoing connection. Basically this is due to the exposed terminal problem. Another example of MAC layer unfairness is the fact that the exponential back-off mechanism favors the latest successful node, since the back-off window of all unsuccessful nodes is increased while the window of the successful one remains the same [47]. QoS (i.e., bounded service levels) is particularly difficult. In this chapter MAC protocol extensions will be investigated and developed to offer bounded service levels in a multihop environment, while reducing the number of collisions and thus limiting the power consumption of the devices.

The MAC layer is responsible for the establishment and control of the physical channels provided by the PHY layer. The efficiency of the MAC layer mechanisms is especially important for packet transmission since only a small overhead can be tolerated here. In the context of WPANs, power awareness is a crucial requirement.

The MAC layer has two main functions: to coordinate the access of a link by the involved terminals (multiplexing of the different users on the link), and to reduce the interference between links within the system and between different systems. Recently several groups are investigating MAC layer performance in multihop networks [48, 49].

3.5.3 Architecture Design

The reference architecture for short-range wireless systems consists of three main connection concepts and was derived within the IST project PACWOMAN: personal area networks (PAN), community area networks (CAN), and wide area networks (WAN). Where a PAN provides connectivity in the POS of a user, the CAN provides direct connectivity between the PANs, and the WAN is used if no direct communication is possible. In this case an alternative system such as GSM/GPRS or UMTS is used for connectivity. Three different types of terminals are distinguished: basic terminals, advanced terminals, and master terminals. The basic terminals have a limited protocol stack and communicate in a master-slave manner; the master terminals can communicate with basic terminals, as well as with the advanced terminals, in a master-slave and ad hoc manner, respectively. The advanced terminals can communicate with each other and with the master terminals in an ad hoc manner.

Within the PACWOMAN architecture two domains were defined: the master-slave domain and the ad hoc domain. The master-slave domain deals with the communication within a virtual device consisting of a master and several slaves;

the ad hoc domain deals with the communication between master devices and advanced terminals. These domains are also referred to as low data rate (LDR) and high data rate (HDR). Figure 3.28 presents a high level overview of the main concepts in the PACWOMAN architecture as defined in [50, 51].

Besides the HDR domain, a medium data rate (MDR) domain has also been defined; basically here bit rate has been exchanged for transmission range as compared to the HDR domain. Table 3.22 summarizes the main system requirements of the different domains as defined for the architecture described here.

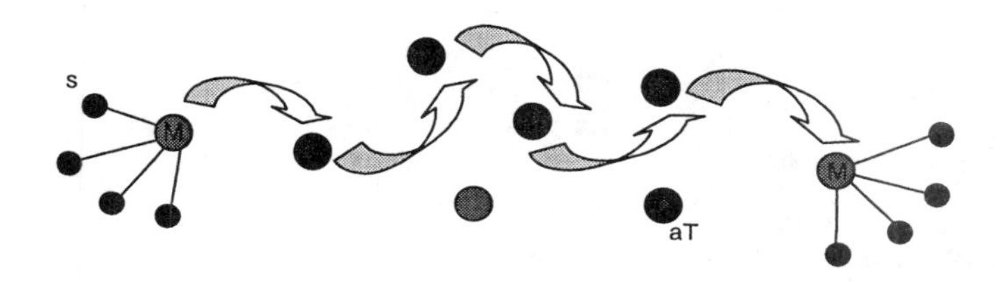

Figure 3.28 Overview of the high-level PACWOMAN architecture. (*From:* [3].)

Additionally, the maximum number of hops for the MDR and HDR domain is set to 10. The LDR environment applies a star topology, whereas the MDR and HDR use a meshed topology.

Table 3.22

Characteristics for the Different Domains: Low, Medium, and High Data Rate.

	LDR	MDR	HDR
Data rate	10 bps–10 Kbps	10 kbs–1 Mbps	> 1Mbps
Required BER	10^{-3}–10^{-7}	10^{-3}–10^{-7}	10^{-3}–10^{-7}
Maximum distance	1m–5m	< 20m	< 10m
Autonomy (power consumption)	months / years	days / months	hours / days (main supply)

Within the PACWOMAN architecture different technologies are applied for these different links. This is one of the characteristics of WPAN systems—heterogeneity of the access techniques. Within work package 4 two new access methods for the physical layer are being investigated, Hybrid-OFDM (H-OFDM) and UWB for the ad hoc and master-slave domain, respectively. The project developed a demonstrator where the concept was prototyped using Bluetooth and IEEE 802.11 WLAN.

Figure 3.29 depicts the protocol stacks for both the basic and the advanced terminal, with the different possible physical layers. The basic terminal has possibly an even more limited protocol stack since there is no need for a simple sensor to run a full TCP stack.

In this section the requirement and design issues are addressed for the MAC layer of both the basic and advanced terminal. One of the main requirements taken into account is the need for power-saving mechanisms.

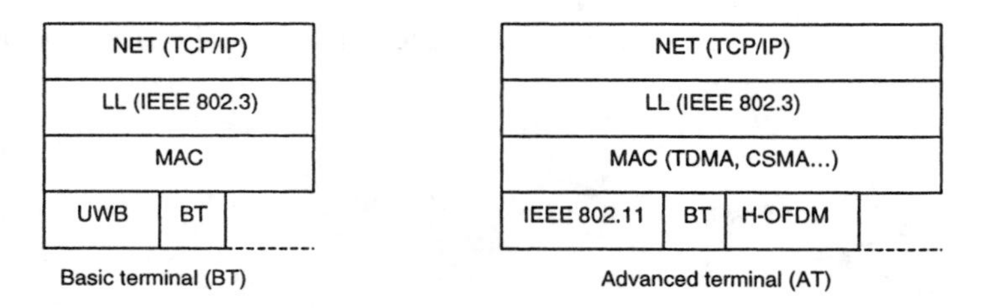

Figure 3.29 The different PHY layers within the PACWOMAN project.

3.5.3.1 Power-Saving Mechanisms

A critical resource for all kinds of portable devices when operated on batteries is power. Without battery power those devices cannot function, and therefore power-saving mechanisms are very important. In addition, it is expected that the efficiency increase of batteries will not be as fast as the increase of computational complexity of computing devices. Generally, power-saving mechanisms for short-range wireless systems can be classified as follows [46].

Transmission Power Control

In wireless communication, transmission power has a strong impact on bit error rate, transmission rate, and inter-radio interference. These are typically contradicting factors. In [52], power control is adopted to reduce interference and improve throughput on the MAC layer. How to determine the transmission power of each mobile host so as to determine the best network topology, known as topology control, is addressed in [53]. How to increase network throughput by power adjustment for packet radio networks is addressed in [53].

Power-Aware Routing

Power-aware routing protocols have been proposed based on various power cost functions. In [56], when a mobile host's battery level is below a certain threshold, it will not forward packets for other hosts. In [58], five different metrics based on battery power consumption are proposed. A hybrid environment consisting of battery-powered and outlet-plugged hosts is considered in [57]. Two distributed heuristic clustering approaches for multicasting are proposed in [58] to minimize the transmission power.

Low-Power Mode

More and more wireless devices can support low-power sleep modes. IEEE 802.11 [35] has a power-saving mode in which a radio only needs to be awake periodically. HIPERLAN allows a mobile host in power-saving mode to define its own active period. An active host may save power by turning off its equalizer according to the transmission bit rate. Comparisons are presented in [55] to study the power-saving mechanisms of IEEE 802.11 and HIPERLAN in ad hoc networks. Bluetooth [60] provides three different low-power modes: sniff, hold, and park. Another method for saving power (for nontime-critical applications) in a mobile system is to hold transmission until the channel is favorable (i.e., strong).

3.5.3.2 Relevant Standards

In this section we briefly summarize the main features of already existing standards for short-range wireless communication systems. We introduce IEEE 802.11 wireless LAN, Bluetooth, and IEEE 802.15.x wireless PAN [42].

Wireless LAN: IEEE 802.11

In 1997 the IEEE adopted IEEE Std. 802.11-1997, the first wireless LAN standard. This standard defines the MAC and PHY layers for a LAN with wireless connectivity. It addresses local area networking where the connected devices communicate over the air to other devices that are within close proximity to each other. This section provides an overview of the 802.11 architecture and the different topologies incorporated. The standard is similar in many respects to the IEEE 802.3 ethernet standard. Specifically, the 802.11 standard addresses:

- Functions required for an 802.11 compliant device to operate either in a peer-to-peer fashion or integrated with an existing wired LAN;
- Operation of the 802.11 device within possibly overlapping 802.11 wireless LANs and the mobility of this device between multiple wireless LANs;
- MAC level access control and data delivery services to allow upper layers of 802.11 networks to function transparently;
- Several physical layer signaling techniques and interfaces;
- Privacy and security of user data being transferred over the wireless media.

The IEEE 802.11 Wireless LAN Architecture

The IEEE 802.11 architecture is comprised of several components and services that interact to provide station mobility transparent to the higher layers of the network stack. The basic architecture is depicted in Figure 3.30. Several entities can be distinguished, namely the wireless station or mobile host, the basic service set, the distribution system and the access point.

- *Wireless LAN Station:* The *station* [or mobile host (MH)] is the most basic component of the wireless network. A station is any device that contains the functionality of the 802.11 protocol, that being MAC, PHY, and a connection to the wireless media. Typically the 802.11 functions are implemented in the hardware and software of a network interface card (NIC). A station could be a laptop PC, a handheld device, or an AP. Stations may be mobile, portable, or stationary, and all stations support the 802.11 station services of authentication, de-authentication, privacy, and data delivery.
- *Basic service set (BSS):* This is the basic building block of an 802.11 wireless LAN. The BSS consists of a group of any number of stations. The BSS is not a very interesting topic until we take the topology of the WLAN into consideration.
- *Distribution system (DS):* This is the means by which an access point communicates with another access point to exchange frames for stations in their respective BSSs, forward frames to follow mobile stations as they move from one BSS to another, and exchange frames with a wired network. As IEEE 802.11 describes it, the distribution system is not necessarily a network nor does the standard place any restrictions on how the distribution system is implemented, only on the services it must provide. Thus the distribution system may be a wired network like 803.2 or a special purpose box that interconnects the access points and provides the required distribution services.
- *Access point:* This is a fixed entity in the network that connects the mobile hosts with the distribution system.

The coverage area of the wireless LAN can be extended via an extended service set (ESS). An extended service set is a set of infrastructure BSSs, where the access points communicate among themselves to forward traffic from one BSS to another to facilitate movement of stations between BSSs.

IEEE 802.11 Topologies

The 802.11 standard describes two modes of deployment for the WLAN: the infrastructure mode and the ad hoc mode (Figure 3.31).

The most basic wireless LAN topology is a set of stations, which have recognized each other and are connected via the wireless media in a peer-to-peer fashion. This form of network topology is referred to as an independent basic service set (IBSS) or an ad hoc network.

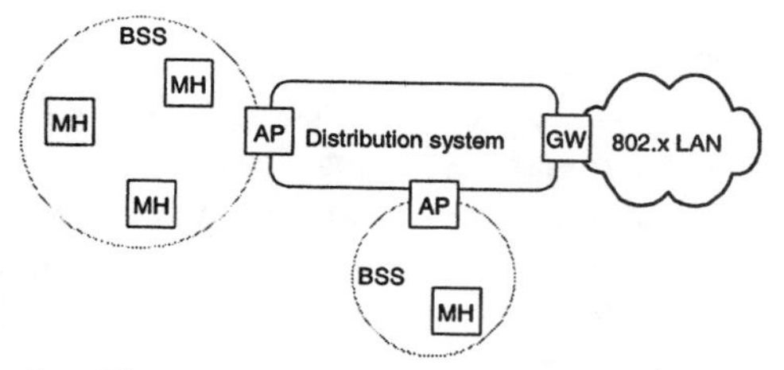

Figure 3.30 The basic architecture of an IEEE 802.11 wireless LAN.

In an IBSS, the mobile stations communicate directly with each other. Every mobile station may not be able to communicate with every other station due to the range limitations. There are no relay functions in an IBSS; therefore, all stations need to be within range of each other and communicate directly.

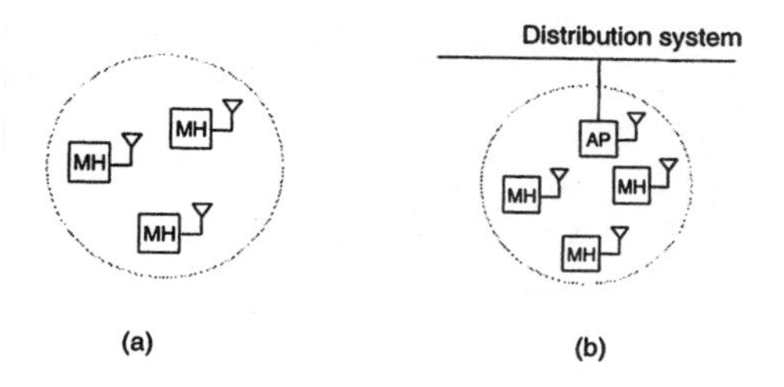

Figure 3.31 (a) The ad hoc mode and (b) the infrastructure mode of IEEE 802.11.

An Infrastructure Basic Service Set is a BSS with a component called an access point (AP). The access point provides a local relay function for the BSS. All stations in the BSS communicate with the access point and no longer communicate directly. All frames are relayed between stations by the access point. This local relay function effectively doubles the range of the IBSS.

IEEE 802.11 Media Access Control

The MAC layer has to fulfill several tasks. First of all, it controls, of course, the access to the channel, but it can also offer support for roaming, power conservation, and authentication. In basic mode the system offers asynchronous data services, and an optional mode exists for offering time-bounded services.

Which modes can be used depends on the applied topology. In the infrastructure mode both options are available, whereas in the ad hoc mode only the asynchronous data services are available.

The following three basic access schemes have been defined for IEEE 802.11:

- The mandatory basic access method based on CSMA/CA;
- An optional method avoiding the hidden terminal problem;
- A contention-free polling method for time-bounded service.

The first two methods are implemented as the distributed coordination function, the third method is referred to as point coordination function. These MAC mechanisms are referred to as distributed foundation wireless medium access control (DFWMAC).

Several parameters have been defined for implementing the access methods (see Figure 3.32).

DCF interframe spacing: this is the longest waiting time available and thus grants the lowest priority for accessing the medium; after this time-out CSMA/CA is used for accessing the channel.

PCF interframe spacing: this is the medium waiting time and it is used for the time-bounded services, which have priority over the asynchronous CSMA/CA services. Thus, after the IFS expires the system can move into polling mode.

Short interframe spacing: this is the shortest waiting time and thus it has the highest priority. This waiting time is used for control messages (e.g., acknowledgments, polling responses, and RTS/CTS messages).

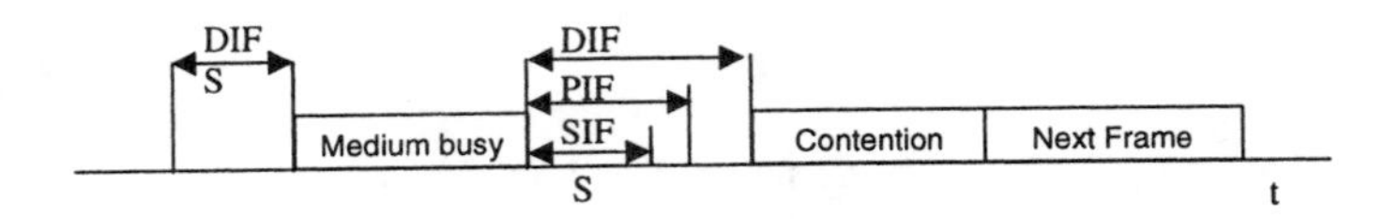

Figure 3.32 Interframe spacing for the IEEE 802.11 medium access.

Bluetooth

The Bluetooth specification [1] has been produced by the Bluetooth Special Interest Group (SIG). The Bluetooth SIG is an industry group consisting of leading manufacturers in telecommunications and computing. Bluetooth technology was initiated as a means to connect cell phones to laptop computers, but it is expected that it will gradually replace cables and infrared as the primary means of wirelessly exchanging information between devices. Bluetooth is also set to become a link between a multitude of future devices. Examples include using Bluetooth as the link between personal e-payment devices and point-of-sale terminals. There are even proposals for Bluetooth motion-sensing rings that would allow fingers to act as electronic pointing devices. These examples affirm the fact that Bluetooth is the first technology that embodies the PAN concept. However, Bluetooth will only deliver best-case real-world data speeds of 720 Kbps, which would be insufficient for a number of applications, such as real-time video. This is one more argument for further extensive research in personal area networking solutions.

For the scenarios envisioned by Bluetooth, it is highly likely that a large number of ad hoc connections will coexist in the same area without any mutual coordination. This is different from ad hoc scenarios considered in the past, where ad hoc connectivity focused on providing a single (or very few) network(s) between the units in range, as it is the case with the wireless LANs. For Bluetooth applications, typically many independent networks overlap in the same area. This is referred as a *scatter ad hoc environment* [60]. The difference between a conventional cellular environment, a conventional ad hoc environment, and a scatter ad hoc environment is illustrated in Figure 3.33.

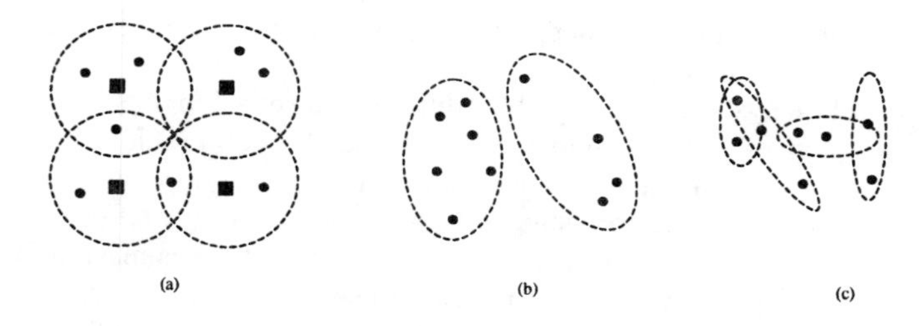

Figure 3.33 Example topologies for: (a) cellular radio systems with squares representing base stations; (b) conventional ad hoc systems; and (c) scatter ad hoc environment.

Bluetooth is a de facto standard for present WPAN technology. In addition, the functional requirements of the Bluetooth specification are very similar to the ones proposed by IEEE 802.15 (more precisely 802.15.1). A large portion of Bluetooth will be a key ingredient of the potential 802.15.1 standard. The use of Bluetooth is expected to be so widespread that each future PAN technology must consider Bluetooth compatibility.

Bluetooth Specifications

The Bluetooth specification is divided into two parts: the *Core* and the *Profiles*. The Core portion specifies the architectural details of the Bluetooth system, hardware components, protocols, and interoperability with different legacy communication protocols. The Profiles portion specifies the protocols and procedures required for different Bluetooth applications. The most important aspects of the Bluetooth specification are examined in [1, 60, 61].

The Bluetooth system operates in the unlicensed ISM frequency band at 2.4 GHz. Regarding the choice of the multiple access technique, CDMA offers the best properties for ad hoc radio systems intended to operate in an unlicensed spectrum. However, direct sequence (DS)-CDMA is less attractive for the short-range wireless connectivity due to the excessive usage of battery power for DSP and power control.

Bluetooth is based on frequency hopping (FH)-CDMA. In the ISM band, a set of 79 hop carriers has been defined at a 1-MHz spacing.[1] The hopping rate is 1,600 hops per second, which provides good immunity against the sources of interference in the 2.4-GHz band. The applied Gaussian frequency shift keying (GFSK) modulation achieves link speed of 1 Mbps.

A *channel* is defined by a particular pseudorandom FH sequence. Two or more (up to eight) active Bluetooth units that share the same channel form a *piconet*. Among all units participating in a piconet, only one is distinguished as a master, and the others are slaves. The master determines the particular hopping sequence associated with the piconet, and the phase of the FH sequence is defined by the master's native clock.

The Bluetooth system provides full duplex transmission based on slotted TDD. The duration of a single hop is 625 μs and the communication within a single slot is performed on one frequency. Since the nominal hopping rate is 1,600 hops per second, the working frequency changes in every slot (exceptions are the multislot packets, explained further). There is no direct transmission between slaves in a Bluetooth piconet, only from master to slave and vice versa.

Communication in a piconet is organized so that the master polls each slave according to a polling scheme. A slave is only allowed to transmit after having been polled by the master. The slave will start its transmission in the slave-to-master time slot immediately after it has received a packet from the master. The master may or may not include data in the packet used to poll a slave. However, it is possible to send packets that cover multiple slots. The multislot packets may be either three or five slots long, and during the multislot packet the frequency is not changed. The frequency applied after a multislot packet is the same as the frequency that would be applied in the observed slot if there was not a multislot packet. In general, the slots in which the master starts the transmission do not change their ordering parity (the same is valid for the slaves). Since the master schedules the traffic in both the uplink and downlink, intelligent scheduling algorithms have to be used that take into account the slave characteristics.

The number of units that can participate in a piconet is deliberately limited to eight (seven active slaves) in order to keep high-capacity links between all the units. If the master needs to communicate with more than seven devices, it can do

[1] At the time the specification was produced for France, Spain, and Japan, a reduced set of 23 hop carriers had been defined at a 1-MHz carrier spacing.

so by first instructing active slave devices to switch to low-power park mode and then inviting other parked slaves to become active in the piconet. However, there may be a need for simultaneous connection of more than eight devices. Thus, two or more interconnected piconets form a *scatternet*. A Bluetooth unit can simultaneously be a slave member of multiple piconets, but only a master in one. In principle, it is best for the bridge node to be a slave in both piconets, since otherwise the throughput within the piconet where the bridge node is a master is degraded. The bridge node schedules its presence in the piconets in which it participates in a time division manner. Therefore, the global connectivity and throughput in the scatternet depend on the individual schedules of the bridge nodes, as well as the schedules determined by the piconet masters.

The scatternet functionality is important to allow a flexible forming of ad hoc PANs. Scatternets are an instance of the ad hoc network concept. When the interpiconet nodes forward packets between piconets, Bluetooth ad hoc PANs belong to the class of multihop ad hoc networks. These scatternet-based PANs may be used when information needs to be spread widely among the PANs residing within reach through a "reasonable" number of radio hops. Here, the reasonable number of hops depends on the type of information in terms of required data rate and end-to-end delay.

Bluetooth uses a procedure known as *inquiry* for discovering other devices; it uses *paging* to subsequently establish connections with them. Inquiry and paging are conceptually simple operations, but the frequency-hopping nature of the physical layer makes the low-level details quite complex. Two nodes cannot exchange messages until one of them becomes a master, thus determining the frequency hopping sequence that will be used in the piconet. During inquiry, both nodes (one is the listener and the other is the sender) hop using the same sequence; but the sender hops faster than the listener, transmitting a signal on each channel and listening between transmissions for an answer. When more than one listener is present, their replies may collide. To avoid the collision, listeners defer their replies until expiration of a random back-off timer. Eventually the sender device collects some basic information from the listeners, such as the device address and the clock offsets. This information is subsequently used to page the selected listener device.

Low-Power Modes

Bluetooth offers different low-power modes for improving battery life. Piconets are formed on demand when communication among devices is ready to take place. At all other times, devices can be either turned off or programmed to wake up periodically to send or receive inquiry messages. It is possible to switch a slave into a low-power mode whereby it sleeps most of the time and wakes up only periodically. Three types of low-power modes have been defined:

- *Hold mode* is used when a device should be put to sleep for a specified length of time. As described earlier, the master can put all its slaves in the hold mode to suspend activity in the current piconet while it searches for new members and invites them to join.
- *Sniff mode* is used to put a slave in a low-duty cycle mode, whereby it wakes up periodically to communicate with the master.
- *Park mode* is similar to the sniff mode, but it is used to stay synchronized with the master without being an active member of the piconet. The park mode enables the master to admit up to 255 slaves in its piconet.

IEEE 802.15.3

The IEEE P802.15.3 high rate task group (TG3) for wireless personal area networks is chartered to draft and publish a new standard for high rate (20 Mbps or greater) WPANs. The task group defined the PHY and MAC specifications for high data rate wireless connectivity with fixed, portable, and moving devices within or entering a POS. A goal of the high rate WPAN task group was to achieve a level of interoperability or coexistence with other 802.15 task groups. It is also the intent of this project to work toward a level of coexistence with other wireless devices in conjunction with coexistence task groups such as 802.15.2. Currently the draft standard entitled *Wireless Medium Access Control (MAC) and Physical Layer (PHY) Specs for High Rate Wireless Personal Area Networks (WPAN)* has been completed. The main features and characteristics of the draft standard are:

- Data rates: 11, 22, 33, 44, and 55 Mbps;
- Quality of service isochronous protocol;
- Ad hoc peer-to-peer networking;
- Security;
- Low power consumption;
- Low cost;
- Designed to meet the demanding requirements of portable consumer imaging and multimedia applications.

More details and updates on the balloting results can be found on the task group Web site located at http://www.ieee802.org/15/pub/TG3.html.

IEEE 802.15.3a

The IEEE 802.15 high rate alternative PHY task group (TG3a) for wireless personal area networks is working to define a project to provide a higher speed PHY enhancement amendment to 802.15.3 for applications that involve imaging and multimedia. Out of initially 32 proposals for the alternative PHY layer for the

IEEE 802.15.3 high rate WPAN that were presented in March 2003, 10 proposals have been withdrawn and seventeen proposals were merged into only six new ones. All these proposals are based on PHY layers that mostly operate with a channel bandwidth of 500 MHz, such that the IEEE 802.15.3a can be qualified as an UWB-based system. The PHY layer option presented in the different proposals include pulse-based modulation schemes such as pulse position modulation (proposal no. 4) and soft spectrum adaptation (proposal no. 5), as well as continuous modulation schemes such as DS-CDMA (proposal no. 1), direct sequence spread spectrum (proposal no. 2), differential QPSK (proposal no. 3), and OFDM (proposal no. 6). The latter, the so-called multiband OFDM proposal, is the merged proposal, which is likely to play an important role due to its large support from industry. Note that these different proposals also provide a description of the MAC layer. Details can be found in [62].

3.5.3.3 IEEE 802.15.4

The IEEE standard 802.15.4 [63] specifies the medium access control and the physical layer for low rate wireless personal area networks (see also [64]). A low-rate WPAN is a simple, low-cost communication network that allows wireless connectivity in applications with limited power and relaxed throughput requirements. The main objectives of a low-rate WPAN are ease of installation, reliable data transfer, short-range operation, extremely low cost, and a reasonable battery life, while maintaining a simple and flexible protocol. Some of the most important characteristics of a low-rate WPAN specified in the IEEE 802.15.4 standard are:

- Over the air data rates of 250 kbps, 40 Kbps, and 20 Kbps;
- Star or peer-to-peer operation;
- Allocation of guaranteed time slots;
- CSMA/CA channel access;
- Fully acknowledged protocol for transfer reliability;
- Low power consumption;
- Energy detection;
- Link quality indication;
- Sixteen channels in the 2,450-MHz band, 10 channels in the 915-MHz band, and 1 channel in the 868-MHz band.

There are two different device types that can participate in a low-rate WPAN network: a full function device (FFD) and a reduced function device (RFD). The FFD can operate in three modes serving either as a PAN coordinator, a coordinator, or a device. An FFD can talk to RFDs or other FFDs, while an RFD can only talk to an FFD. An RFD is intended for applications that are extremely simple, such as a light switch or a passive infrared sensor; they do not have the need to send large amounts of data and may only associate with a single FFD at a time. Consequently, the RFD can be implemented using minimal resources and memory capacity.

A system conforming to IEEE 802.15.4 consists of several components, the most basic being the device. A device can be an RFD or an FFD. Two or more devices within a POS communicating on the same physical channel constitute a WPAN. However, a network shall include at least one FFD, operating as the PAN coordinator. An IEEE 802.15.4 type network is part of the WPAN family of standards though the coverage of a low-rate WPAN may extend beyond the POS, which typically defines the WPAN. A well-defined coverage area does not exist for wireless media since propagation characteristics are dynamic and uncertain.

The MAC sublayer provides two services: these are the MAC data service and the MAC management service interfacing to the MAC sublayer management entity service access point. The MAC data service enables the transmission and reception of MAC protocol data units across the PHY data service. The features of the MAC sublayer are beacon management, channel access, guaranteed time slot management, frame validation, acknowledged frame delivery, association, and disassociation. In addition, the MAC sublayer provides hooks for implementing application appropriate security mechanisms.

The low-rate WPAN standard allows the optional use of a *superframe* structure. The coordinator defines the format of the superframe. The superframe is bounded by network beacons, it is sent by the coordinator, and is divided into 16 equally sized slots. The beacon frame is transmitted in the first slot of each superframe. If a coordinator does not wish to use a superframe structure it may turn off the beacon transmissions. The beacons are used to synchronize the attached devices, to identify the PAN, and to describe the structure of the superframes. Any device wishing to communicate during the contention access period (CAP) between two beacons shall compete with other devices using a slotted CSMA/CA mechanism. All transactions shall be completed by the time of the next network beacon.

The superframe can have an active and an inactive portion. During the inactive portion the coordinator shall not interact with its PAN and may enter a low power mode. For low latency applications or applications requiring specific data bandwidth, the PAN coordinator may dedicate portions of the active superframe to that application. These portions are called guaranteed time slots (GTSs). The guaranteed time slots comprise the contention-free period (CFP), which always appears at the end of the active superframe starting at a slot boundary immediately following the CAP, as shown in Figure 3.34. The PAN coordinator may allocate up to seven of these GTSs, and a GTS may occupy more than one slot period. However, a sufficient portion of the CAP shall remain for contention-based access of other networked devices or new devices wishing to join the network. All contention-based transactions shall be complete before the CFP begins. Also, each device transmitting in a GTS shall ensure that its transaction is complete before the time of the next GTS or the end of the CFP.

There are three types of data transfer transactions. The first one is the data transfer to a coordinator in which a device transmits the data. The second transaction is the data transfer from a coordinator in which the device receives the data. The third transaction is the data transfer between two peer devices. In star topology only two of these transactions are used, since data may be exchanged only between the coordinator and a device. In a peer-to-peer topology data may be exchanged between any two devices on the network, and so all three transactions may be used in this topology.

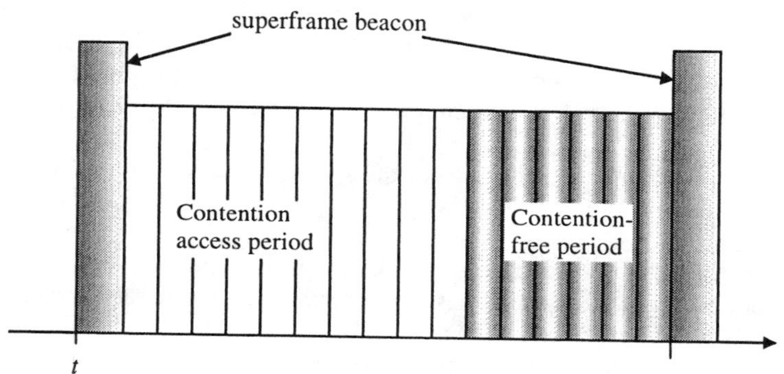

Figure 3.34 The IEEE 802.15.4 WPAN superframe structure.

The mechanisms for each transfer type depend on whether the network supports the transmission of beacons. A beacon-enabled network is used for supporting low latency devices such as personal computer peripherals. If the network does not need to support such devices, it can elect not to use the beacon for normal transfers. However, the beacon is still required for network association purposes.

3.5.4 Medium Access Control for the Ad Hoc Link

This section discusses the medium access control layer for the HDR or ad hoc domain of the PACWOMAN architecture. Within the PACWOMAN architecture for this domain the H-OFDM physical layer is being designed and analyzed for its suitability. Also the use of wireless LAN (IEEE 802.11) is investigated by means of demonstration. Therefore, here we will focus on these two types of systems. In particular, optimizations and extensions to the IEEE 802.11 MAC layer will be considered, as well as design issues for the H-OFDM MAC layer.

First, the requirements for the HDR domain are reviewed and from this design constraints are derived. Next, design considerations are discussed, as well as suitability of techniques for the ad hoc domain, and required modifications and adaptations. For the ad hoc domain from several proposed extensions to the IEEE 802.11 MAC one idea is designed in more detail: a receiver-initiated scheme based on channel jamming with energy pulses.

3.5.5 Requirements and Constraints

The HDR link operates between different PANs and possibly between several high data rate devices within the PAN. The maximum distance between the individual devices is 10m, however a larger range is possible by using multiple hops to transmit packets, which means that intermediate nodes are used to forward the packets to the final destination.

Typical applications are telemonitoring, business applications with moderate bit rate requirements, and interactive video services or time-critical computer data transfers with high bit rate requirements. These services require a bandwidth of more than 1 Mbps with bit error ratios of 10^{-3}–10^{-7}. The delay requirements are set to up to 0.5s for the MDR domain and in the range of 10 to 50 ms for the HDR domain.

General requirements that apply to the medium access control in the ad hoc mode are as follows:

- Energy constrained operation is of the utmost importance in the WPAN context. The MAC schemes should meet the power requirements of the advanced terminals.
- Collisions should be limited to prevent unnecessary transmissions. This way transmission power is saved, especially under high load conditions. Here it is assumed that the transmit power associated with packet transmission dominates the power consumption (over other sources of power consumption such as the signal processing and channel sensing power).
- Increased efficiency, overhead reduction. By preventing unnecessary control messages from being exchanged, both the protocols are more efficient and transmission power is saved.
- Flexibility, fast changing topologies, caused by new nodes arriving and others leaving the network. The developed MAC mechanisms should be able to cope with these fast changing topologies and be able to quickly adapt to the new situation.
- Scalability, in the ad hoc network a growing number of nodes can be incorporated (at the moment ranging up to 10 hops).
- Coexistence of links (interlink interference), which limit the interference between links so that the spectrum can be used efficiently

For the ad hoc domain two types of techniques are being considered, carrier sensing and the H-OFDM. The following sections will review their basic properties, which are needed to derive the design criteria.

3.5.6 Extensions to CSMA/CA

Typically carrier sense networks are designed such that there is a difference between the receiving and the sensing range. The threshold for sensing the channel is typically set (much) lower than the power required for receiving a frame correctly. Due to these different ranges, a wireless multihop network can be described by defining the following three types of links:

- *Communication links*, defined as links between two nodes in a transmission range that have packets to be exchanged between them.
- *Interfering links*, defined as links between two nodes in a transmission range that have packets to send to different destinations, but in overlapping time instances. These packets are said to have collided.
- *Sensing links*, defined as links between two nodes such that if one of the nodes starts a transmission, the wireless link is sensed busy by the other node.

These three types of links can be described by connectivity matrices G: $G_C = (N, L_C)$ for the matrix with communication links, $G_I = (N, L_I)$ for the interfering ones, and $G_S = (N, L_S)$ for the graph with the sensing links, where N is the nodes in the system and L_C, L_I, and L_S are the links (edges) that communicate, interfere, and allow sensing. The communication graph is a directed graph; the interference and sensing graphs are not; the effect is reciprocal. G_I is a spanning subtree of G_S, $L_I \subset L_S$.

Each link between nodes i and j has a propagation or path delay τ_{ij} associated with it. Because of these delays, the channel is not immediately sensed busy the moment the transmissions start.

The basic black burst scheme is proposed and investigated in [65]. The black burst mechanism is an extension that can be applied to the CSMA system. Applying this mechanism, users compete for the common radio channel by means of pulses of energy, referred to as black bursts (BB). The length of these bursts depends on the time that users have been waiting for transmission as of their scheduled moment of transmission. The algorithm is distributed and completely based on channel sensing. It distinguishes two classes of users, one priority class and one best-effort class.

The protocol distinguished two different events: (1) obtaining the channel either as a best-effort user or a nonscheduled priority user, and (2) obtaining the channel as a scheduled priority user. In the first case normal CSMA/CA is applied. In case of a priority class user the following transmission is scheduled, in case of best-effort nothing further is done and the next time the user has to transmit it is again done in the same manner. In the second case, the priority users compete among each other for channel access; they do this by jamming the channel using the BBs, proportional to the waiting time. This effectively means that the user that has been waiting the longest obtains the channel. The two events also have a priority between them; if there are scheduled priority users they will transmit before the users that are of type (1). This is ensured by making the time before the black bursts are transmitted less than the time the users of type (1) sense if the channel is empty before going into contention by means of the contention window. This means that the BB mechanism can be applied on top of a normal CSMA/CA algorithm. The black burst mechanism is depicted in Figure 3.35.

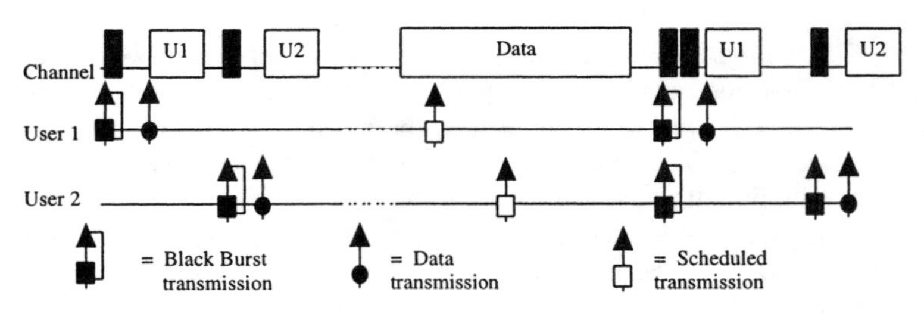

Figure 3.35 Basic operation of the black burst scheme.

Although there is some scheduling of the priority class users, the mechanism is still random. The disadvantage is that the priority users have hard priority over the best-effort users. The work is done, however, with some relatively strong assumptions, the most important one being that the G_I and G_S are identical during the analysis, meaning that all interfering nodes can be sensed, thus effectively assuming that the analysis is done in the absence of the hidden terminal effect. If the BB system is put in a situation where hidden nodes do exist, the performance is severely degraded by their presence.

One important issue for investigation is the interaction between the MAC layer and the higher layer. In order to investigate this, two different operation modes are defined:

- *Operation with feedback*: An RT source can generate blocks of data at regular intervals. The MAC layer can convey all information blocks created up to a certain moment in a packet and transmit the packet when the channel access instance comes. This way the packet size varies with the encountered delay. In practice, however, it is difficult to create packets in this manner.
- *Operation without feedback*: This is a simpler communication architecture, where already assembled packets are passed onto the MAC layer for transmission one by one.

This is also the case when a node acts as a relay station. Here the operation without feedback is assumed, since this is the most general one (both originating and relaying sources can apply it), and the most feasible one from an implementation point of view.

Considering the case where a RT-node has an interarrival time for the packets to be transmitted of t_{rdy}. To keep the delay bounded, the time between two scheduled transmissions of packets (t_{sch}) should be less than the interarrival time, especially since the channel may be occupied by other RT sources or best-effort traffic at the scheduled instances. Now t_{sch} can be found as

$$t_{sch} = t_{rdy} - t_{bslot} - t_{obs} - \delta \tag{3.35}$$

with $\delta > 0$. δ is being called the slack time and should be larger than the time it takes to transmit a maximum length data packet since on the scheduled instance it is possible that a BE transmission prevents the other users from transmitting immediately.

3.5.6.1 MAC Layer for H-OFDM

The H-OFDM system as it is being considered in the project PACWOMAN uses a hybrid of two multiple-access schemes: OFDMA and TDMA. This means that the OFDM signal is frequency and time shared in order to achieve a larger granularity of bit rates than is achievable with pure OFDMA or TDMA. This higher granularity allows for a large flexibility and scalability on the high data rate links.

The parameters that characterize the H-OFDM system are thus of two natures. There are parameters that describe the OFDM signal and parameters that describe the TDMA frames. In OFDM the total available bandwidth is divided into N subcarriers, which are orthogonal. The symbols that are to be transmitted are multiplexed in frequency by modulation on the subcarriers. To prevent adjacent channel interference, some of the subcarriers can remain unused as frequency guards and some can be used for channel estimation.

The OFDM parameters may be adapted depending on the channel state information (CSI); the main parameters that could then be modified are:

- The number of subcarriers N;
- The length of the OFDM symbol T;
- The number of frequency guards N_{fg};
- The number of pilot subcarriers N_p;
- Length of the cyclic prefix.

Note that the number of available or useful carriers (N_u) is influenced by these choices: $N_u = N - N_{fg} - N_p$. Note, however, that the parameters N and T cannot be changed independently if the bandwidth and cyclic prefix are given.

The multiple-access features for OFDM are introduced by assigning a subset of the N subcarriers to a certain user. For example, when one subcarrier is assigned to each user the system can accommodate N users. In the H-OFDM proposal the number of subcarriers that can be assigned to a user i is flexible and referred to as s_i. The concept is depicted in Figure 3.36.

The bit rate r of a subcarrier depends on the modulation alphabet; for now this is assumed to be constant. However, a flexible modulation alphabet would increase the degrees of freedom in the system and thus the adaptability. The H-OFDM proposal will apply convolutional coding to exploit channel diversity.

There is also a TDMA structure; the OFDM symbols are grouped into time slots, which have a header/preamble. Frames consist of between 0 and l of these slots, with a maximum of L. This structure is explained in Chapter 4 of this book.

Depending on how the time slots are grouped, different bit rates can be achieved. Also, the polling frequency influences how often a frame is assigned to a certain user.

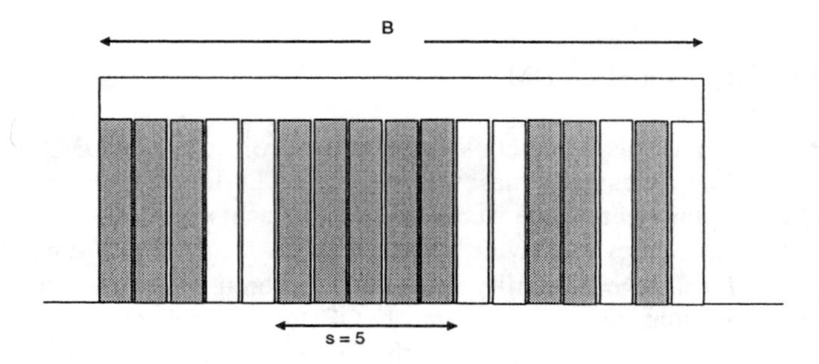

Figure 3.36 Division of the N subcarriers over multiple users with different data rates.

With these two sets of parameters (i.e., the OFDMA and TDMA ones), three different approaches for link adaptation are proposed:

- Assigning all subcarriers to a single user during some slot ($s_i = N_u$), the data rate can be varied by l.
- Assigning some subcarriers to a user for the whole duration of a frame ($l = L$), the data rate can be varied by changing s.
- Assigning some subcarriers to some slots, the bit rate can be determined by varying both s and l.

The first option is basically a combination of OFDM and TDMA. The second option is OFDMA with TDMA, and the third option is a full hybrid with complete flexibility.

3.5.7 Design Considerations

The two considered technologies have their own additional peculiarities.

3.5.8 Extensions to CSMA

The carrier sensing based protocols can be executed in a completely distributed manner. There is no need for a centralized node that schedules the transmissions. This allows for a large flexibility in the system; nodes can leave and enter at their convenience without the need to register and deregister. The exponential backoff mechanism automatically adjusts the size of the contention window depending on the size of the population of communicating terminals (more communicating terminals correspond to more collisions).

However, this flexibility comes at a price, for a larger number of nodes the MAC scheme operates in a collision limited domain, and the performance is degraded severely. Additionally the basic carrier sensing MAC introduces unfairness: the last successful node has an advantage over the other nodes, since its back-off window is smaller

Another performance degrading factor is hidden terminals, which are not sensed by nodes that want to communicate, causing collisions. Similarly, exposed terminals prevent nodes from transmitting when at the receiver no collision would occur [66].

Finally, at present in IEEE 802.11 there is no possibility for association with two BSSs with one interface; two network cards are required for this. This means that it is quite inefficient to use this interface for true ad hoc networking.

3.5.8.1 H-OFDM

The H-OFDM scheme is explained in detail in Chapter 4 of this book. Here, H-OFDM is mentioned in connection to the design of the MAC layer in WPANs.

In general, a centralized entity is required and, complete distributed implementation of the MAC scheme is impossible. However, the mechanism is more efficient for higher data load.

Synchronization is required at the MAC level; users need to know which blocks of subcarriers are assigned to which user. If they want to communicate, the length of the time frames should be determined, both for the transmission and for the other communicating users to be able to determine when the channel becomes available again.

To apply carrier sensing mechanisms, all the subcarriers in the block assigned to the user should be monitored. If such a block was not assigned, then the whole spectrum can be monitored to see if there are sufficient subcarriers available for the transmission. To prevent these sensing operations, some logical link (subset of time slots and subcarriers) could be dedicated as the control channel to exchange the parameters required for the transmission.

In [67] a proposal was made for a leader election procedure based on the H-OFDM concept. In this procedure a terminal is elected as a leader and this terminal is used to synchronize and coordinate the transmissions. For this purpose a leader channel (LCH) has been defined, which is a logical channel mapped on the first slot of every frame.

One advanced feature of the MAC is the implementation of load-dependent degradation. If a high rate user wants to enter the system and the system is already quite loaded, the bandwidth assigned to the newly entering user and possibly some of the other users is reduced so that everybody has at least some level of service. Applicability of this feature depends on the user requirements to the connection.

3.5.9 MAC Alternatives

In this section several options for extending the CSMA protocol to make it better suited for a multihop environment are considered. The extensions are based on channel jamming using energy pulses as is done, for example, in the Black Burst extension of CSMA [65]. A receiver-initiated version of the BlackBurst protocol for multihop scenarios is proposed.

Another system based on energy pulses has been proposed in [68]. Collisions are avoided by transmitting short sequences of energy bursts that are transmitted as busy tones. Each node produces a unique pattern of tones. Each node i counts the number of successful transmissions a_i, since its last successful transmission, modulo the total number of nodes N. Since no two nodes can transmit successfully at the same time, each node counts a unique number of successful packets a_i. This number is used as the unique binary identifier. The identifier is transmitted by on-off keying the black bursts (for each "1" an energy pulse is transmitted, for a "0" the station listens to the channel), starting with the most significant bit. The user with the highest identifier (waiting the longest) captures the channel. The disadvantage of this method is that the exact number of nodes needs to be known to determine the optimal length of the contention period; furthermore all nodes need the following:

- *Black burst combined with RTS/CTS:* The most straightforward modification is the integration of BB with the RTS/CTS mechanism. However the additional protection offered by the BB mechanism is limited in this case.
- *Black burst combined with FAMA*: A black burst energy pulse can be used to protect the RTS within the FAMA scheme. The CTS is already protected by an additional length. This way the collisions of RTS messages are reduced.
- *Coded black burst*: On-off keying with BB to transmit additional information (e.g., to replace RTS).
- *Multicast black burst*: Each node that receives a one-hop BB broadcasts it once. Geographically this scheme is very inefficient.
- *Setting the NAV based on black burst*: this does not provide complete protection and requires modulation of the black burst pulses.

- *Making the mechanism receiver initiated*: after the RT flow starts, it can be assumed that packets arrive at regular intervals. The system can be made receiver initiated. A CTS is sent to reserve the channel at the moment that a packet is expected. The advantage is that we can use "exposed" transmitters, so geographically these will be more optimal solutions, and it can be still an efficient mechanism [69].

Here we evaluated the last method and compared it with the basic operation in a single and multihop environment [70]. The basic idea here is to move the black burst scheme on the receiver side and combine it with a *floor acquisition* technique. The suggested solution solves the hidden-node problem, enforces a round-robin discipline among real-time nodes, and gives priority to real-time traffic, thus achieving QoS requirements. The proposed access scheme will be referred to in the following as Black Burst++, or simply BB++.

Using bursts of energy to resolve access contentions among priority nodes has proven to be an excellent technique, with two main advantages:

- Every contention has a unique winner.
- It is completely distributed.

The fact that it is completely distributed is of the utmost importance in ad hoc networks, where it is not desirable to have a central coordinator to control traffic flows, so that the network can self-organize and operate in every condition. The RTS/CTS technique is also the building block of many well-studied solutions to the hidden-node problem. The BB++ scheme gets rid of the RTS control frame and uses a CTS modified frame as an invitation for real-time nodes to send their data. This reduces the signaling load and thus the overhead of the protocol.

In combining these two techniques, we take advantage of the good properties of both of them. The BlackBurst++ scheme makes an effort in this direction. There are still many design issues to address, but preliminary simulations and an analysis of the proposed scheme suggest that this direction is worth following.

3.5.9.1 Basic Protocol Operation

Real-time flows have an important characteristic: they are predictable. Assuming that real-time packets are generated every t_{rdy} units of time, the receiver of a real-time flow can take advantage of that. After receiving the first packet of a real-time flow, the receiver can assume that new packets will arrive (approximately) every t_{rdy} units of time. On the other hand, the same receiver does not know if at the time when a RT-packet is expected, it will be busy in a communication with another neighbor sending.

What the common receiver can do is to protect the arrival of a real-time packet with a Clear to Send (CTS) control frame, ensuring that none of the neighboring nodes transmit simultaneously. The CTS frame is an invitation to the real-time node to send its packet. At the same time, the CTS acquires the floor for the incoming RT-packet any other node in the range of the receiver, but (eventually) not in the range of the real-time user, is prohibited to transmit for a period of time specified in the CTS frame. This time is equal to the maximum duration of a real-time packet transmission. Neighbor nodes that receive a CTS frame, set their network allocation vector according to that duration. Hence, the real-time packet can be sent free of collisions.

The basic signaling of the BB++ scheme is depicted in Figure 3.37. Node *A* initiates a communication session with node *B* by sending a RTS_b, with a flag set that the BB++ protocol will be applied. If this message is received correctly by node *B*, then this node replies with a CTS_a. Now node *A* can send a data packet to node *B*. From this moment onwards node *A* and *B* have established a BB++ context and node *B* will schedule regular Invite-to-Send messages (ITS), that are transmitted to node *A* indicating that it can send the next data packet. An Invite-to-Send is composed of a CTS packet protected by a Black Burst transmission. Nodes to which the ITS is not addressed refrain from communicating (here Node *E* and *C*).

A hidden node *D* is out of range of the ITS message; however, it will sense the channel is busy when node *A* transmits its data and refrain from transmitting at those moments. If the ITS collides with an ongoing transmission in range of node *A* (e.g., a data transmission or a RTS), then the invite is resent later, since the receiver can see from an expiring time-out indicating that no data is arriving as a response to the ITS that the message was lost. Node *A* has priority over all NRT nodes since a shorter time-out is applied, and with the RT nodes the BB++ scheme divides the bandwidth fairly. So, as in the original BB, an additional delay of one BE packet has to be taken into account.

Referring to the typical hidden-node scenario shown in Figure 3.37(D), how the BB++ access scheme works in this case is explained in detail by the time diagram in Figure 3.38.

Both the transmitted (Tx) and received (Rx) packets at a given node are shown on the same axis ($y > 0$ for Tx packets, $y < 0$ for Rx packets). Node 1 and node 2 are hidden from each other and they both need to communicate with node 0 in the middle, so they both receive node 0's transmissions. Node 2 has bursty data traffic to send, while node 1 has real-time, delay-constrained packets to send every t_{rdy} seconds. Node 1 sends its first real-time packet using the usual DATA/ACK exchange with the destination. After receiving the first packet of a real-time flow, the destination assumes that a new packet will arrive in the near future (approximately in t_{rdy} seconds) and schedules an invitation for node 1 in t_{sched} seconds from the instant a real-time packet has been correctly received.

The invitation (ITS, composed of a BB+CTS) should arrive as soon as the sender has a new packet available for transmission. On the other hand, the sender will wait until it is invited to transmit by node 0, and defers without trying to access the channel until the ITS arrives. On the scheduled invitation attempt, if the channel has been idle for t_{med} consecutive seconds, node 0 can successfully send its ITS frame, just in time for node 1 to transmit the newly arrived real-time packet. Node 2 is hidden to node 1, but it receives the BB+CTS frame sent by node 0 and can allocate its NAV accordingly,[2] thus avoiding collision with the incoming real-time packet at node 0. Moreover, the interframe space between a BB and a CTS is always t_{obs}, which guarantees that the CTS is not going to collide with any other data packet from a neighbor in the radio range of node 0. Hence, despite that it is not its main use, the BB also works as a protection for the CTS. The real-time packet is sent free of collisions, and the hidden-node problem is alleviated.

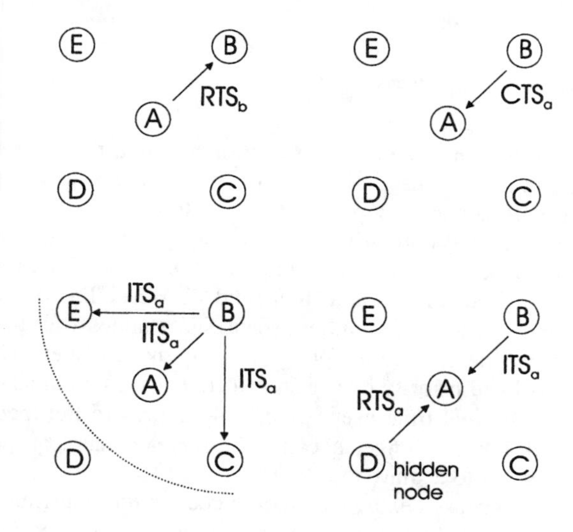

Figure 3.37 Basic protocol operation.

[2] For now, it is assumed that the channel does not introduce any error, so that every packet is correctly received if it does not collide with another packet. With this assumption, node 2 always correctly receives a CTS.

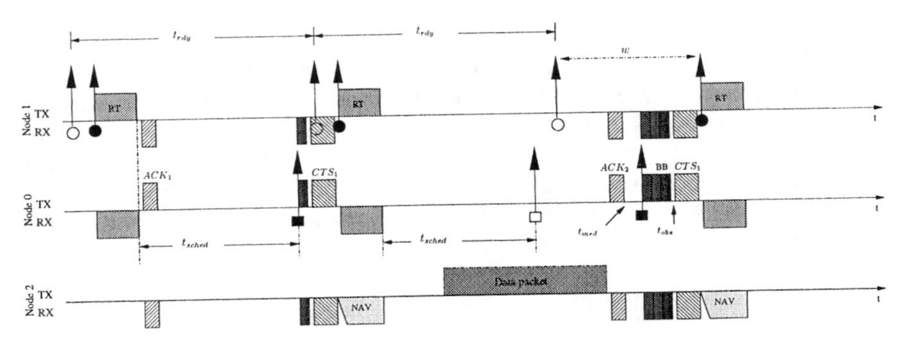

Figure 3.38 The proposed scheme: Black burst++.

It can happen that on a scheduled invitation attempt, the air around node 0 is busy with a data packet transmission from a node hidden to the real-time user. Since the latter will not start a packet transmission unless invited, there is no chance that its packet will collide with the currently received data packet at node 0. When the data packet reception is completed (DATA+ACK), node 0 will wait until the channel is idle for t_{med} and then send its BB+CTS. A new data packet transmission cannot start in the middle, because data nodes wait for a longer inter frame space (DIFS) before reattempting to grab the channel. This is how the BB++ RT stations have priority over best-effort. Thus, as soon as the data packet frees the channel, the real-time user has priority access to the medium. Now the length of the black burst is longer because the *receiver* has experienced a longer delay while attempting to acquire the floor.

In this case the access delay (w) experienced by the real-time user(s) cannot necessarily be neglected; however, it can be shown that it is bounded. In fact, in the worst case, the real-time packet is available for transmission by the sender at the same instant a data packet transmission starts. So, at maximum, it has to wait until a maximum length data packet transmission time, plus some fixed timings:

$$w \leq t_{data} + \text{SIFS} + t_{ACK} + t_{med} + b_{max} + t_{obs} + t_{CTS} + \text{SIFS} \tag{3.36}$$

where b_{max} is the maximum burst length associated to a contention delay

$$d_{max} = t_{data} + \text{SIFS} + t_{ACK} + t_{med} \tag{3.37}$$

Obviously, the choice of t_{sched} is a critical design issue for proper behavior of the scheme. Besides, it is still an open issue. Some design evaluations on this point will be discussed later.

3.5.9.2 Contention Resolution

Consider now the situation shown in Figure 3.39, where two real-time *receivers* may need to synchronize themselves in such a way that they will not disturb each others reception of a real-time packet. In Figure 3.39, Rx_1 and Rx_2 are both destinations for two separate real-time flows. The two flows start from Tx_1 and Tx_2, which are hidden to each other. By using the black burst contention mechanism, Rx_1 and Rx_2 can organize themselves in a TDMA structure, with the one experiencing the longest contention delay being sure to win an eventual contention with other receivers.

In Figure 3.40 it is shown how the contention resolution mechanism is employed. On the left side of the diagram, the receiver-oriented scheme described above is enforced either by Rx_1 and Rx_2, until (on the right side) they are not disturbed by a data packet transmission, which may arrive from a fifth node not shown in the figure. Both receivers have an invitation attempt scheduled for that time, but since the channel is busy, they defer until the channel can be detected idle for t_{med} consecutive seconds. At this later time, they both start a contention phase by jamming the channel with a number of black slots proportional to the experienced contention delay. It is exactly as the original black burst scheme, but receiver oriented.

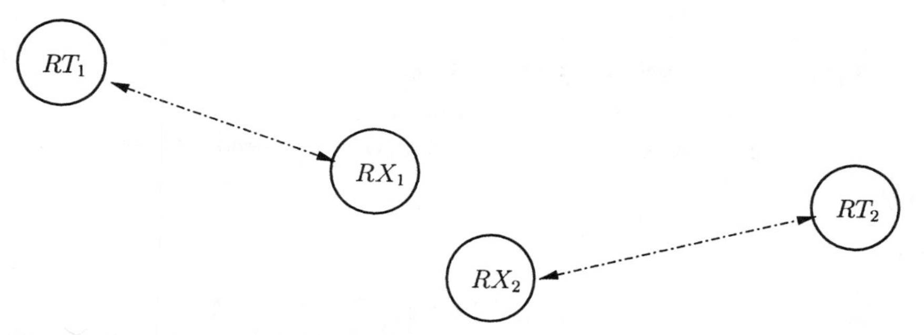

Figure 3.39 Contention resolving: a sample scenario.

The length of the burst is still given by the following equation:

$$l_{bb} = \left(1 + \left\lfloor \frac{d_{cont}}{t_{unit}} \right\rfloor\right) t_{bslot} \tag{3.38}$$

After sending their bursts, they both observe the channel for t_{obs} seconds, to see if any other contender is sending a longer burst: Rx_1 will win the contention because it has experienced a longer delay ($d_{cont1} > d_{cont2}$); it thereby transmits a longer burst and can transmit its invitation.

The contention phase will lead to a unique winner. In fact, as in the original version of the protocol, in this BB++ scheme two scheduled invitation attempts are staggered in time by at least t_{pkt}, the transmission time of a maximum length real-time packet transmission, and so are their contention delays. Therefore, counting the burst length in terms of t_{pkt} (i.e., choosing $t_{unit} = t_{pkt}$) ensures that two different receivers always contend by means of a black burst that differs for at least one black slot.

The BB++ scheme described above proposes a solution to the hidden node problem that is also able to give priority access to real-time traffic. The main advantages of this scheme are that real-time packets are sent free of collisions and that they are no longer affected by the presence of hidden nodes. Furthermore, multiple real-time flows are organized in a TDMA structure without requiring a central coordination (totally distributed).

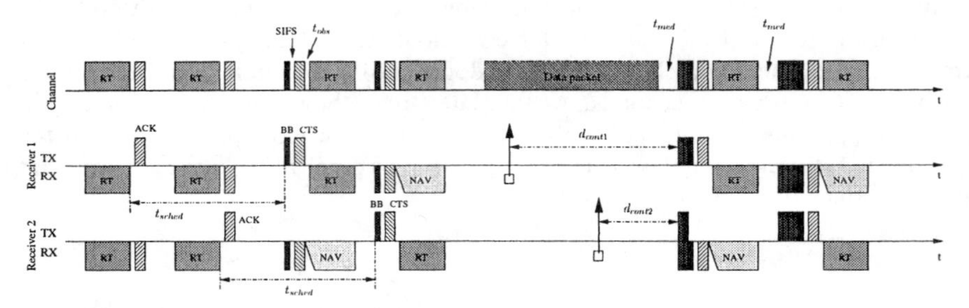

Figure 3.40 The proposed scheme: contention resolution.

- Access delay for real-time flow is bounded.
- High real-time throughput because the scheme is ideally collision-free.

3.5.9.3 The Choice of t_{sched}

The main critical design parameter in the proposed scheme is finding an appropriate value for t_{sched}. It is a critical choice, because this parameter depends on the correct behavior of the protocol and its stability. Clearly, t_{sched} should be chosen such that the access delay of real-time packets always stays bounded. Referring to Figure 3.41, an analysis of the scheduling time is made. As can be seen from Figure 3.41, the scheduling time (t_{sch}) is defined as the time between the end of a real-time transmission and beginning of the contention for the next transmission.

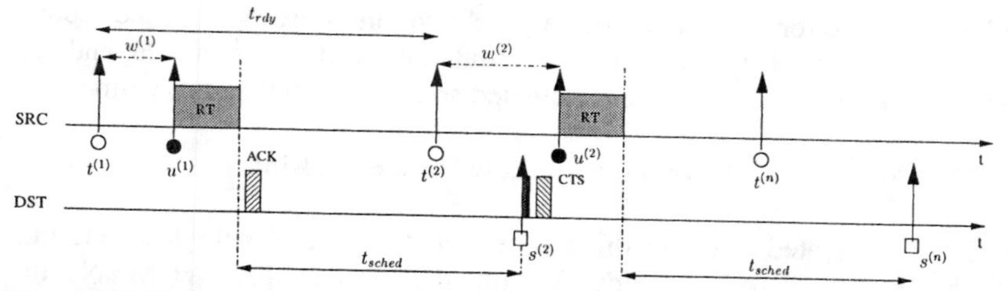

Figure 3.41 The proposed scheme: the choice of t_{sched}.

The source node (SRC) generates new packets every t_{rdy} units of time, with nth packet being generated at time:

$$t^{(n)} = t^{(1)} + (n-1)\, t_{rdy} \tag{3.39}$$

The access delay $w^{(1)}$ of the first packet, which cannot be predicted, is given by

$$w^{(1)} = u^{(1)} - t^{(1)} \tag{3.40}$$

A first goal in the design of t_{sched} would be choosing it in order to recover from the initial delay, reducing the access delay of the second packet, and leading to the smallest possible access delay for subsequent packets. The condition is:

$$w^{(2)} < w^{(1)} \tag{3.41}$$

where $w^{(2)} = u^{(2)} - t^{(2)}$, and the access instant of the second packet is given by

$$u^{(2)} = s^{(2)} + (t_{bslot} + t_{obs} + t_{CTS} + SIFS) = u^{(1)} + t_{pkt} + t_{sched} + \xi \tag{3.42}$$

where $\xi = (\, t_{bslot} + t_{obs} + t_{CTS} + \text{SIFS})$ is a fixed quantity of approximately 500 to 600 μs. Substituting (3.39), (3.40), and (3.42) into (3.41), an upper bound can be obtained for t_{sched}:

$$t_{sched} < t_{rdy} - t_{pkt} - \xi \tag{3.43}$$

So a possible choice would be introducing a parameter to satisfy this bound. For coherence with the standard black burst scheme, this parameter will be called the *slack time* $\delta > 0$ and then t_{sched} is given by:

$$t_{sched} = t_{rdy} - t_{pkt} - \xi - \delta \tag{3.44}$$

With this choice for t_{sched}, it can be easily shown that starting from the second packet the access delay decreases from the initial value with a rate that depends on δ. The slack time should also be upper-bounded so that $t_{sched} > 0$ at every time.

3.5.10 Medium Access Control for the Low Data Rate Link

This section is dedicated to the medium access control layer for the low data rate link of the PACWOMAN system. Within the project PACWOMAN, the suitability of UWB technology for this low data rate link is evaluated. UWB-based systems are most likely to operate in the frequency band between 3.1 and 10.6 GHz (see also [71]) with a nominal bandwidth of at least 500 MHz. With respect to the MAC layer it is important to note that it is very likely that UWB systems are going to operate in TDD mode. Furthermore, given the relatively large bandwidth and the low power spectral density (the upper limit is set –41.25 dBm/MHz; see [71]), UWB systems are vulnerable to interference from narrowband systems operating in the same frequency range as the UWB system. Therefore, interference detection and avoidance mechanisms, which are not an issues for conventional wireless systems, are important aspects in the context of the MAC layer design for UWB-based systems.

In this section we will briefly review the general requirements for the low data rate link and based thereon the more specific constraints for the MAC layer are derived. Furthermore, we are going to detail the different design aspects and discuss the suitability of MAC layers of existing and emerging wireless communication systems. The actual design and performance analysis of the MAC layer for the UWB-based LDR component is described in the PACWOMAN deliverable D4.3.1.

3.5.10.1 Requirements and Constraints

The LDR system is envisioned to operate in the so-called body-to-body as well as in the body-to-access point environment (see also D2.1 for a more detailed description of the system requirements). The body-to-body environment shows a maximum link distance in the range of 1 to 2m, whereas the maximum link distance of the body-to-access point environment is in the range of 3 to 5m. The system is operating in the regulated UWB frequency bands ranging from 3.1 to 10.6 GHz. Accordingly, the system is expected to experience interference from narrowband systems also operating in these frequency regions, like, for instance, the WLAN systems operating at 5 GHz.

Typical applications for the LDR system include the transmission of data from body-worn sensors to a central and also body-worn terminal (body-to-body environment). Depending on the sensors, the amount of data to be transmitted varies from several tens of bits (e.g., for a temperature sensor) to several kilobits (e.g., for electrocardiogram data). Two basic types of services are distinguished, namely the very low rate service and the low-rate service. The associated data and tolerated bit error rates range from 100 to 10 Kbps and 10^{-3} to 10^{-7}, respectively. The tolerated delay is in the order of 1 sec for both services. Note that the support of voice traffic is optional. Furthermore, it is worth noting that data exchange takes place in a rather sporadic and infrequent fashion, thus resulting in relatively low overall traffic.

The basic mode of operation is in master-slave mode, where the central body-worn so-called smart terminal coordinates the data transmissions. An optional ad hoc communication mode is also foreseen for communication among multiple smart terminals. The number of maximum users that can be supported simultaneously is set to 10.

A low overall power consumption is of foremost importance, and it is anticipated that terminals operating in body-to-body environments run on a single alkaline battery for at least 1 to 2 years. This constraint especially applies to the sensor terminals. The smart terminals, like the central master terminal, are expected to run on more sophisticated and expensive batteries that can be recharged or exchanged. Therefore, terminals capable of operating in the body-to-access point environment are granted a power autonomy of 1 to 2 months.

The system requirements relevant for the MAC layer design of the LDR component of the PACWOMAN system are summarized in Table 3.23.

Table 3.23

MAC Related System Requirements for Envisioned LDR Services.

	Very Low Rate Service	Low Rate Service
Amount of data	100 bits	10 kilobits
Data rate	100 bps	10 Kbps
Required BER	10^{-3}	10^{-7}
Tolerated delay	1s	
Number of users	Maximum of 10 simultaneous users	
Autonomy	Body-to-body environment: 1–2 years Body-to-access point environment: 1–2 months	

3.5.10.2 Design Considerations

In general, medium access protocols for wireless communication systems are designed with respect to network performance metrics such as throughput, efficiency, quality-of-service support, and packet delay. Given the above-presented constraints, energy consumption at the MAC level is an additional major performance metric that has to be taken into consideration during the design process for the development of the LDR component of the PACWOMAN system.

The premise is that dumb terminals have a very limited power, whereas smart terminals are provided with larger power reserves.

The chief sources of energy consumption due to MAC related activities are the CPU, the transmitter, and the receiver. CPU usage in dumb terminals may be reduced by relegating most of the high-complexity computation related to medium access to the smart terminals. Therefore, the focus of the medium access protocol design and optimization is on transceiver usage. The radio can operate in three modes: standby, receive, and transmit. In general, the radio consumes more power in the transmit mode than in the receive mode, and consumes the least power in the standby mode. In general, four directions to reduce the energy consumption have been identified [72]:

- Collisions should be eliminated as far as possible since they result in retransmissions leading to unnecessary energy consumption and possible higher delay. Note, however, that retransmissions cannot be completely avoided in cases of high link error rates. This aspect is also of importance in collision-free medium access schemes, where new terminals registering with the master terminal may have to use some form of random access protocol. Techniques such as reservation and polling can help meet the requirement that collisions be minimized.
- Protocol overhead should be kept as small as possible. The size of the control packets has to be adapted to the size of the data packet in order to yield a sufficient efficiency. The time associated with protocol overhead depends on the synchronization and channel estimation requirements of the PHY layer and the synchronization requirements of the MAC layer, as well as on the amount of necessary control information for the protocol itself. On the other hand, the time spent to transmit the actual data depends on the spectral efficiency of the PHY layer and the amount of data to be transmitted, which in turn is application-dependent. Especially for low data rate applications, where small amounts of data are exchanged, the reduction of the protocol overhead is of foremost importance and subject to meticulous design.
- The time spent in receive mode should be reduced as much as possible. In common wireless communication systems, the receiver has to be powered on at all times [e.g., to analyze the data received control information (to determine whether the information is for other terminals or not) or to keep track of channel status through constant monitoring]. This results in significant energy consumption. Two different categories can be identified:

    - Listening to a data transmission that is destined to someone else, also referred to as overhearing. Techniques to prevent overhearing have been presented in [72, 73].

- Listening to the transmission channel during time intervals without data transmission, which is also referred to as idle listening. To reach terminal lifetimes of several months or even years with a single alkaline battery, the transceiver will have to remain in standby mode most of the time.

In the context of the envisioned LDR component of the PACWOMAN system, all the above-mentioned points will have to be considered in order to reach the requirements. However, since efficient techniques reducing the number of collisions (e.g., collision avoidance) and preventing overhearing are known, special emphasis will have to be given to the second and the last point. The small amounts of data as well as the rather low traffic load will have a crucial impact on the performance of the medium access control. Concerning point, it is anticipated that the use of TDMA-based protocols provides an efficient means to avoid idle listening [74].

Furthermore, it is anticipated that the aspect of interference detection and appropriate interference avoidance or mitigation strategies play a major role for the design of an adequate MAC layer for the UWB-based LDR component.

3.5.10.3 MAC Alternatives

The MAC layers of WLAN systems like the IEEE 802.11 standard are not suitable for the LDR component of the PACWOMAN system since the primary optimization criterion of these MAC layers was to maximize the throughput and to minimize the delay. Low power consumption was not considered as a primary design goal. This is reflected by the fact that WLAN systems need to listen most of the time in order to determine whether the channel is free or not, which in turn accounts for a huge portion of the overall power consumption.

As outlined in the previous section, reduction of the protocol overhead and idle listening are of foremost importance for the MAC layer design of the LDR system component. It is anticipated that the IEEE 802.15.4 could effectively limit idle listening due to its TDMA component. However, due to the differences in the PHY layers the protocol is not directly applicable. The PHY specific synchronization procedures and the associated protocol issues (preamble design), in particular, need to be evaluated carefully. Furthermore, the use of the IEEE 802.15.4 protocol needs to be investigated in the context of the envisioned LDR scenarios in order to derive an adequate parameter set that optimizes the overall power consumption. These and other issues, the use of simple ALOHA protocols for the ad hoc option, were studied and evaluated in [67].

An initial study was performed to consider more flexible MAC layers that could be used in an ad hoc manner. The MAC layers apply *link adaptation* to adjust themselves to the interference conditions of that moment and can adjust the setting to a varying number of nodes. In a master-slave setting with a limited number of nodes, these advanced features might not be required, but in future scenarios with larger numbers of nodes they are mandatory. A time-hopping spread spectrum (TH-SS) UWB communication system can be considered, but it is important to have a picture of how to design medium access techniques in order to achieve efficient resource sharing in WPAN environments. The throughput performance is evaluated by means of computer simulation in a multinetwork scenario, and the impact of TH parameters (such as the number of bins per frame and the processing gain) on the throughput performance is discussed. Furthermore, a link adaptation mechanism is proposed which can adapt TH parameters according to interference conditions among collocated WPANs. This mechanism aims to maximize the throughput achieved by WPANs while keeping the robustness with respect to the basic timing mechanisms used in the TH-SS communications. The proposed mechanism is based on *chip puncturing* (i.e., transmitting less chips than determined by the hopping code, leaving some of the chips unused and thus reducing the interference in the system). The work done so far is an initial exploration of the topic, identifying the most important parameters and developing the link adaptation mechanism.

3.6 CONCLUSIONS

A highly complex communications environment will require advanced RRM algorithms. It could be beneficial to have a central intelligent unit that can maximize resource utilization and ensure sufficient capacity. This capability can also be provided by the so-called bunched systems [76].

Multilink/multinetwork radio resource provisioning and control is a way to ensure optimal next generation mobile communications. This involves definition, evaluation, and optimization of appropriate (existing or new) mechanisms that have an influence on the achieved systems performance in terms of delays (BER, PER) and capacity. Therefore, basic requirements for ensuring successful operation of future communication systems are that RRM mechanisms are activated from radio access level. In the case of multiple networks, a far-reaching RR manager can be introduced, which encapsulates the advances in the physical layer within the overall framework.

References

[1] IST 2001-32549 Project ROMANTIK, Resource Management and Advanced Transceiver Algorithms for Multihop Networks, at http://www.ist-romantik.org.

[2] IST Project ARROWS, Advanced Radio Resource Management for Wireless Services, at http://www.arrows-ist.upc.es/.

[3] IST 2001-34157 Project PACWOMAN, Power Aware Communications for Wireless Optimized Personal Area Networks, at http:// www.pacwoman-ist.org.

[4] IST Project CAUTION and CAUTION ++, at http://www.telecom.ntua.gr/CautionPlus/.

[5] Bluetooth SIG, *Bluetooth Core Specification*, Version 1.1, February 2001, http://www.bluetooth.com

[6] IST OVERDRIVE http/overdrive-ist.org.

[7] Kyriazakos, S., A. Mihovska, and J. M. Pereira, "Dynamic Network Reconfiguration of Heterogeneous Wireless Systems," *ANWIRE Workshop,* Mykonos, Greece, September 2003.

[8] CAUTION++ EU Project, IST-2001-38229, Section D-2.1, "Resource and Mobility Management in Multisystem Environment," May 2003.

[9] CAUTION++ EU Project, IST-2001-38229, Section D-2.2, "System Requirements Specifications," May 2003.

[10] Ponnekanti, S., et al., "Flexible Cross-Layer Radio Design for Systems Beyond 3G," *IST Mobile and Wireless Telecommunications,* Aveiro, Portugal, June 2003.

[11] Kaldanis, V., et al., "Hierarchical Resource Management in Heterogeneous Radio Environments—The CAUTION++ Approach," *Proceedings of the IST Mobile Summit,* Aveiro, Portugal, June 2003, CD-ROM.

[12] 3GPP TS 25.211, "Physical Channels and Mapping of Transport Channels onto Physical Channels (FDD)."

[13] 3GPP TR 25.922 v4.0.0, "Radio Resource Management Strategies."

[14] 3GPP TS 25.331 v4.0.0, "RRC Protocol Specification."

[15] Holma, H., and A. Toskala (eds.), *WCDMA for UMTS,* New York: John Wiley and Sons, 2000.

[16] ARROWS D06 section at http://www.arrows-ist.upc.es/publicationssections.htm.

[17] Phan-Van, V., and S., Glisic, "Radio Resource Management in CDMA Cellular Segments of Multimedia Wireless IP Networks," *Proceedings of WPMC'01,* Aalborg, Denmark, September 2001, CD-ROM.

[18] ARROWS D04 section at http://www.arrows-ist.upc.es/publicationssections.htm.

[19] Olmos, J., and S. Ruiz, "UTRA-FDD Link Level Simulator for the ARROWS Project," *Proceedings of the IST'01 Conference,* pp. 782–787.

[20] 3GPP TR 25.942 v.2.1.3, "RF System Scenarios."

[21] ARROWS D08 section, http://www.arrows-ist.upc.es/publicationssections.htm.

[22] IST-2001-32549 ROMANTIK Section 333, "Technical and Financial Benefits for a Site Specific 4G Microcellular Urban Network," July 2004.

[23] Dresdner Kleinwort Wasserstein Research, "Mobile Infrastructure: Restoring Investment Confidence," White Paper 2002, at http:/www.3gnewsroom.com/html/whitepapers/index.shtml, 2002.

[24] Molina, A., G. E. Athanasiadou, and A. R. Nix, "Optimized Base Station Location Algorithm for Next Generation Microcellular Networks," *IEE Electronic Letters,* Vol. 36, No. 7, March 2000, pp. 668–669.

[25] Molina, A., "A New Automatic Base-Station Optimization Algorithm for Next Generation Cellular Networks," Thesis, University of Bristol, December 2000.

[26] Tameh, E., A. Molina, and A. Nix, "The Use of Intelligently Deployed Fixed Relays to Improve the Performance of a UTRA-TDD System," *Proceedings of IEEE VTC Fall 2003*, CD-ROM.

[27] Cormen, T. H., C. E. Leiserson, and R. L. Rivest, *Introduction to Algorithms,* New York: McGraw Hill Book Company, 1990.

[28] Papadimitrion, C. H., and K. Steglitz, *Combinatorial Optimization: Algorithms and Complexity,* Upper Saddle River, NJ: Prentice Hall, 1982.

[29] Baier, A., and K. Bamdelow, "Traffic Engineering and Realistic Network Capacity in Cellular Radio Networks with Inhomogeneous Traffic Distributions," *Proceedings of IEEE 47th Vehicular Technology Conference,* Phoenix, Arizona, Vol. 2, May 1997, pp. 780–84.

[30] Sarker, J. H., "An Analysis of Multidimensional Erlang Loss Formula for Channel Sharing in Cellular Environment," *Proceedings of the 14th Nordic Teletraffic Seminar,* Copenhagen, Denmark, August 1998, pp. 75–89.

[31] Tameh, E. K., "The Development and Evaluation of a deterministic Mixed Cell Propagation Model Based on Radar Cross-Section Theory," A Thesis at the University of Bristol, December 1998.

[32] Tameh, E. K., A. R. Nix, and M. A. Beach, "A 3-D Integrated Macro and Microcellular Propagation Model, Based on the use of a Photogrammetric Terrain and Building Data," *Proceedings of IEEE 47th Vehicular Technology Conference,* Phoenix, Arizona, Vol. 2, May 1997.

[33] ETSI, "UMTS UTRA (UE) TDD: radio Transmission and Reception," 3GPP TS 25.102 V4.3.

[34] IST-2001-32549 ROMANTIK Section 5.2.1, "Routing Strategies for Multihop Networks," July 2004.

[35] IEEE 802 LAN/MAN Standards Committee, "Wireless LAN Medium Access Control (MAC) and Physical Layer (PHY) Specifications," IEEE Standard 802.11, 1999 edition, 1999.

[36] ETSI TC-RES, Radio Equipment and Systems (RES); HIgh PErformance Radio Local Area Network (HIPERLAN); Functional Specification. ETSI, Draft prETS 300 65206921 Sophia Antipolis France, July 1995.

[37] Johnson, D.B., "Routing in Ad Hoc Networks of Mobile Hosts," *Proceedings of the IEEE Workshop on Mobile Computing Systems and Applications (WMCSA),* Santa Cruz, California, December 1994, pp. 158–163.

[38] Perkins, C. E., (ed.), *Ad Hoc Networking*, Reading, MA: Addison-Wesley, 2000.

[39] Prakash, R., "Unidirectional Links Prove Costly in Wireless Ad hoc Networks," *Proceedings of the 3rd International Workshop on Discrete Algorithms and Methods for Mobile Computing and Communications*, Seattle, Oregon, August 1998, pp. 15–22.

[40] Toh, C.-K., *Ad Hoc Mobile Wireless Networks, Protocols and Systems*, Englewood Cliffs, NJ: Prentice Hall, 2002.

[41] Ramanathan, S., and M. Steenstrup, "A Survey of Routing Techniques for Mobile Communications Networks," *ACM/Baltzer Mobile Networks and Applications*, Vol. 1, No. 2, 1996, pp. 89–103.

[42] PACWOMAN Consortium, "State-of-the-Art of the WPAN Networking Paradigm," *Section 5.1, PACWOMAN IST-2001-34157,* 2002.

[43] Wilting, C. S., and R. Prasad, "Evaluation of Mobile Ad Hoc Network Techniques in Cellular Systems," *IEEE Conference on Vehicular Technology Conference Fall 2000,* Boston, Massachusetts, September 2000, pp. 1025–1029.

[44] Broch, J., et al., "A Performance Comparison of MultiHop Wireless Ad Hoc Network Routing Protocols," *Proceedings of the ACM/IEEE MobiCom*, October 1998.

[45] Ni, S., et al., "The Broadcast Storm Problem in a Mobile Ad Hoc Network," Proceedings *of ACM MOBICOM 99,* August 1999.

[46] Tseng, Y.-C., C.-S. Hsu, and T.-Y. Hsieh, "Power-Saving Protocols for IEEE 802.11-Based MultiHop Ad Hoc Networks," *Proceedings of IEEE INFOCOM*, 2002, pp. 200–209.

[47] Xu, S., and T. Saadawi, "Does the IEEE 802.11 MAC Protocol Work Well in Multihop Wireless Ad Hoc Networks?" *IEEE Communications Magazine*, Vol. 39, No. 6, June 2001, pp. 130–137.

[48] Royer, E. M., S.-J. Lee, and C. E. Perkins, "The Effects of MAC Protocols on Ad Hoc Network Communication," *Proceedings of IEEE Wireless Communications and Networking Conference*, Vol. 2, September 2000, pp. 543 –548.

[49] Gerla, M., et. al., "TCP Over Wireless MultiHop Protocols: Simulation and Experiments," *Proceedings of IEEE International Conference on Communications,* June 1999, pp. 1089–1094.

[50] PACWOMAN Consortium, "System Requirements and Analysis," *Section 2.1, PACWOMAN IST-2001-34157,* 2002.

[51] PACWOMAN Consortium, "Functional and Technical Specifications," *Section 2.2, PACWOMAN IST-2001-34157,* 2002.

[52] Wu, S. L., Y. C. Tseng, and J. P. Sheu, "Intelligent Medium Access for Mobile Ad Hoc Networks with Busy Tones and Power Control," *IEEE Journal on Selected Areas in Communications*, Vol. 18, September 2000, pp. 1647–1657.

[53] Ramanathan, R., and R. Rosales-Hain, "Topology Control of Multihop Wireless Networks Using Transmit Power Adjustment," *Proceedings of IEEE INFORCOM*, 2000, pp. 404–413.

[54] Huang, C. F., et al., "Increasing the Throughput of Multihop Packet Radio Networks with Power Adjustment," *Proceedings of International Conference on Computer, Communication, and Networks*, 2001.

[55] Woesner, H., et al., "Power-Saving Mechanisms in Emerging Standards for Wireless LANs: The MAC Level Perspective," *IEEE Journal of Personal Communications,* June 1998, pp. 40–48.

[56] Gomez, J., et al., "A Distributed Contention Control Mechanism for Power Saving in Random Access Ad Hoc Wireless Local Area Networks," *IEEE International Workshop on Mobile Multimedia Communications,* 1999, pp. 114–123.

[57] Ryu, J. H., and D. H. Cho, "A New Routing Scheme Concerning Power Saving in Mobile Ad Hoc Networks," *Proceedings of IEEE International Conference on Communications*, Vol. 3, 2000, pp. 1719–1722.

[58] Ryu, J. H., S. Song, and D. H. Cho, "A Power Saving Multicast Routing Scheme in 2-Tier Hierarchical Mobile Ad Hoc Networks," *IEEE Vehicular Technology Conference,* Vol. 4, 2000, pp. 1974–1978.

[59] Singh, S., M. Woo, and C. S. Raghavendra, "Power Aware Routing of Mobile Ad Hoc Networks," *International Conference on Mobile Computing and Networking,* 1999, pp. 181–190.

[60] Haartsen, J. C., "The Bluetooth Radio System," *Journal of IEEE Personal Communications,* Vol. 7, No. 1, February 2000, pp. 28–36.

[61] IEEE Std. 802.15.1-2002, "Wireless MAC and PHY Specifications for Wireless Personal Area Networks (WPANs™)."

[62] http://grouper.ieee.org/groups/802/15/pub/TG3a_CFP.html.

[63] IEEE Std. 802.15.4-2003, Wireless Medium Access Control (MAC) and Physical Layer (PHY) Specifications for Low Rate Wireless Personal Area Networks (LR-WPANs).

[64] Callaway, E., et al., "Home Networking with IEEE 802.15.4: A Developing Standard for Low-Rate Wireless Personal Area Networks," *IEEE Communications Magazine,* August 2002.

[65] Sobrinho, J. L., and A. S. Krishnakumar, "Quality-of-Service in Ad Hoc Carrier Sense Multiple Access Wireless Networks," *IEEE Journal on Selected Areas in Communications,* Vol. 17, No. 8, August 1999.

[66] Tobagi, F. A., and L. Kleinrock, "Packet Switching in Radio Channels: Part II: The Hidden Terminal Problem in Carrier Sense Multiple Access Modes and the Busy-Tone Solution," *IEEE Transactions on Communications*, Vol. 23, No. 12, 1975.

[67] PACWOMAN Consortium, "Analysis and Simulation of Physical Layer Requirements. Synchronization, PAR Reduction and Resource Assignment Algorithms. RF Specifications," *Section 4.2.3., PACWOMAN IST-2001-34157,* 2002.

[68] Papachristou, C., and F. N. Pavlidou, "Collision-Free Operation in Ad Hoc Carrier Sense Multiple Access Wireless Networks," *IEEE Communications Letters*, Vol. 6, No. 8, August 2002, pp. 352–354.

[69] Garcia-Luna-Aceves J. J., and A. Tzamaloukas, "Receiver-Initiated Collision-Avoidance in Wireless Networks," *Journal of ACM Wireless Networks, Special Issue on Best Papers from ACM/IEEE Mobicom 99,* Vol. 8, 2002, pp. 249–263.

[70] Zhenyu, T., and J. J. Garcia-Luna-Aceves, "Collision-Avoidance Transmission Scheduling for Ad Hoc Networks," *IEEE International Conference on Communications*, June 2000, pp. 1788–1794.

[71] Porcino, D., and W. Hirt, "Ultra-Wideband Radio Technology: Potential and Challenges Ahead," *IEEE Communications Magazine*, July 2003.

[72] Ye, W., J. Heidermann, and D. Estrin, "An Energy-Efficient MAC Protocol for Wireless Sensor Networks," *USC/ISI Technical Report ISI-TR-543*, September 2001.

[73] Singh, S., and C. S. Raghavendra, "PAMAS: Power Aware MultiAccess Protocol with Signaling for Ad Hoc Networks," *ACM CCR,* Vol. 28, No. 3, July 1998, pp.20–26.

[74] El-Hoiydi, A., "Spatial TDMA and CSMA with Preamble Sampling Aloha with Preamble for Low Power Ad Hoc Wireless Sensor Networks," *Proeedings of IEEE International Conference on Computers and Communications (ISCC 2002)*, Taormina, Italy, July 2002, pp. 685–692.

[75] Yomo, H., et al., "Medium Access Techniques in Ultra-Wideband Ad Hoc Networks," *Proceedings of ETAI 2003*, Ohrid, Macedonia, September 2003.

[76] ITU Draft Report, "Technology Trends," www.itu.org/, October 2003.

[77] Findlay D. A, et al., "3G Interworking with Wireless LANs," *IEE 3rd International Conference on 3G 2002 Mobile Communication Technologies (3G2002),* London, England, May 2002.

[78] Kadelka, A., and A. Masella, "Serving IP Quality of Service with Hiperlan/2," *Computer Networks*, Vol. 37, September 2001, pp. 17–24.

[79] http://www.ist-brain.org/.

[80] http://www.ist-mind.org/.

Chapter 4

Advanced Radio Technologies

4.1 INTRODUCTION

Future generation systems will require various supporting multinetwork/multiaccess supporting mechanisms. The multiaccess functionality impacts the whole mobile system, so important research issues are the terminal architecture, radio access and radio network, authentication and security, subscription and policies, mobility support, services, transport, and signaling, so that such a system will be easy to use and economically viable to offer.

Throughput, defined as the data rate successfully received, is a key measure for the performance of wireless data transmission systems (e.g., in terms of QoS). Throughput is affected by the channel environment such as the distance between the transmitter and the receiver, the fading state of the channel, and the noise and interference power characteristics. It is also influenced by the choice of design parameters, for example, symbol rate, modulation and coding, constellation, power level, multiple-access scheme, and many others.

Adaptive modulation has been proposed as an efficient technique to improve the data rate of mobile wireless systems by adapting some of those design parameters to the time-varying channel environment to maintain an acceptable bit error rate (BER) [1–3]. Most works in adaptive modulation, however, have aimed only to improve the link layer performance. As application requirements become more complex, there is a need to take into account higher layer metrics. For instance, several works in the literature have developed adaptive techniques that support different types of traffic with different QoS requirements [4].

Technologies for improving bandwidth efficiency include OFDM [5, 6], UWB, and flexible frequency sharing. In pedestrian and indoor environments, there will be severe fluctuations in traffic demands, high user mobility, and different traffic types.

Recently proposed technologies (e.g., enhanced TDD structure, adaptive antenna, and MIMO techniques) will allow the mobile uplink to access a reconfigured channel to transmit large files to the network. This reconfiguration can assign a large portion of the time slots to the uplink transmission, but whenever the transmission is completed the time slots can be instantly assigned to other users, probably downlink users. Since the uplink is not expected to be used as frequently as the downlink, it is spectrally efficient to have the flexibility to reassign the same spectrum between uplink and downlink.

The motivation of a number of IST projects was to investigate the potentialities of the MIMO technology from a physical, propagation point of view. MIMO technology involves a state-of-the-art combination of multielement array (MEAs) and digital signal processing. The MIMO concept is defined as a radio link with M elements at one end and N elements at the other end. The benefit of the MIMO technology comes from the creation of $min(M, N)$ orthogonal information subchannels, the combination of Tx and Rx diversity, and the increased antenna gain.

The performance of the MIMO technology depends directly on the propagation properties of the MIMO channel, through the correlation coefficients, the branch power ratios (BPRs) at the MEAs, the polarization, and so forth. It is therefore of importance to gain a good insight into these properties. A thorough analysis of measured data was collected by the ACTS projects Smart Universal Beam Forming, Overview Contributions to Standards Results (SUNBEAM) [7, 8] and the IST project METRA [9, 10] projects. The results of this analysis performed within the scope of the IST project I-METRA [11, 12] give a better understanding of the MIMO propagation channel and on the performances to be achieved through it. They also deliver sets of significant parameters that characterize MIMO channels and enable their simulation.

The gains that MIMO systems can deliver strongly depend on the characteristics of the radio environment. During the ACTS SUNBEAM [7] and the IST METRA [13] projects, extensive measurement campaigns were performed in a variety of environments: outdoor-to-outdoor (macrocell), outdoor-to-indoor (microcell), and indoor-to-indoor (picocell). A subset of key parameters (correlations matrices, Doppler spectrum) was extracted from these measurements to be fed in the stochastic MIMO channel model proposed by METRA. This stochastic model was experimentally validated by comparing measured data to the outcome of simulations. Section 4.2 is based on input from the IST project I-METRA and discusses research results and challenges for the development of adaptive antenna arrays.

UWB radio communications are an extreme form of spread spectrum communication systems, generally defined as operating with a fractional bandwidth greater than 0.25 (i.e., a −3-dB bandwidth, which is at least 25% of the center frequency used). Alternatively, the Federal Communications Commission (FCC) in the United States defines a UWB device as any radio with a −10 dB fractional bandwidth greater than 0.20 or occupying at least 500 MHz of the spectrum [14]. In this chapter, and particularly in Section 4.3, we consider UWB systems based on impulse radio (IR-UWB), which use very short pulses and have been known for a long time [15]. IR-UWB meets the requirements for short-range wireless systems and additionally possesses the following desirable features: mitigated multipath fading effects, the possibility of high bit rates, and unique location ability. Indeed, the IEEE 802.15 Task Group 2, which deals with high rate PHY for WPAN, has formed a study group (SG3a) to consider UWB as a candidate for this PHY [16]. The contributions to Section 4.3 are based on input from the IST project PACWOMAN [17].

Support of broadband fixed wireless services suggest that a wireless system will have to address two key issues: (1) an optimized use of available spectral resources through enhanced technologies such as multiple transmit multiple receive (MTMR) techniques using preequalization and joint detection allowing simultaneous communications from and towards multiple subscribers, and (2) a seamless interworking and cooperation of existing standards (e.g., WMAN and WLAN), ensuring both an end-to-end quality of service and the in-house and out-house coverage. These points were studied and demonstrated in the IST project STRIKE [18] for the fixed wireless access relying on interworking between HIPERMAN and HIPERLAN. In fact, since HIPERLAN (see Chapter 2) is foreseen to be proposed in the major high-density areas, a standard switching from FWA to HIPERLAN is attractive and will lead to low-cost information delivery in buildings. These issues are discussed in Sections 4.4 and 4.5 of this chapter.

4.2 ADAPTIVE ANTENNA TECHNIQUES

In conventional communication systems, one antenna at the transmitter (Tx) and one at the receiver (Rx) [also called single input single output (SISO) antenna system] creates a bottleneck in terms of capacity. Whatever the modulation scheme used, the coding strategy employed, or other system characteristics, the radio channel will always set the limit for the telecom engineers. This situation is rather critical in the current wireless communication market since the user demand for higher data rates is increasing. For instance, the wireless Internet connection is the next step for users to be connected to the Web while being on the move. To increase the capacity of wireless systems, three options are usually considered: the deployment of additional base stations, the exploitation of more bandwidth, and/or transmissions with higher spectral efficiency [10].

More BSs involve tighter network planning and the deployment of narrower cells, which is a very onerous way of increasing the capacity. On the other hand, some believe that the answer to the demand for high data rates lies in the millimeter wave band because more bandwidth is available there. However, this technology is still expensive. Furthermore, both UMTS [13] and WLAN [19], regarded today as suitable wireless technologies depending on the scenario, work at microwave band (e.g., 2 GHz for UMTS and 2 and 5 GHz for WLAN systems). As a result, the use of higher frequencies does not seem to be the short-term answer to today's wireless needs.

The motivation for deploying MIMO technology in wireless systems is the possibility of achieving orthogonal subchannels between the two ends of the path in a rich scattering environment and subsequently to increase the offered capacity.

in a rich scattering environment and subsequently to increase the offered capacity. The concept of orthogonality emphasizes the strength of the MIMO system because it indicates that these multiple channels are independent from each other. The MIMO technology is also beneficial in increasing the antenna gain and in providing combined Tx/Rx diversity.

Claude E. Shannon (1916–2001) derived a formula to evaluate the information theory capacity C of a transmission channel, normalized with respect to the bandwidth and expressed as a function of the signal-to-noise ratio (SNR). The obtained relation meant that for SNRs >> 1, a 3-dB increase in SNR results in an enhancement of the capacity by 1 bps/Hz. In a single user scenario, the SNR is basically related to the thermal noise. A method to increase it, and subsequently the capacity, is to increase the Tx power. However, this solution is not recommended, not only for health reasons [20] due to the exposure to electromagnetic radiation, but also due to the fact that transmitting higher powers requires that power amplifiers operate linearly in higher regime. This is a difficult task for the hardware designer, not to mention the heat dissipation due to higher Tx powers. Moreover, in interference-limited cellular scenarios, the noise floor, usually higher than the thermal noise, is due to the interfering users. Increasing the Tx power would therefore not help to increase the capacity.

Another method to increase the capacity would be to improve the received SNR by using diversity techniques along with some optimum combining technique. For some years, the use of MEAs at the Rx, namely the fixed terminal, considering one single element at the Tx, has been the subject of a lot of research activity ([21, 22] among others), and was designated as the single input multiple output (SIMO) system. Similarly at the Tx, the UMTS WCDMA standard supports transmit diversity using a single element at the Rx, which is equivalent to a multiple input single output (MISO) system [23].

The natural evolution of the SIMO and MISO technologies is to provide MEAs at both ends of the radio link, hence creating a MIMO system. Telatar [24] showed that a huge capacity gain compared to SISO can be achieved using the generalized form of the Shannon equation when considering MIMO technology. The benefit of using the MIMO technology results from a combination of multiplexing, combining, and antenna gains.

This section presents the technical as well as the theoretical background on the efficiency of the MIMO system compared to the SISO system. The eigenvalues are presented as a quantitative approach to the parallel subchannelling concept along with the capacity formula derived for the MIMO technology. Definitions of correlation coefficient and BPR are presented.

Additionally, this section addresses the characterization of the radio channel. It presents the analysis of the measured data with respect to SISO and SIMO antenna topologies in order to understand the propagation properties of the radio channel. It includes the study of the statistical distribution in terms of the Ricean K-factor and a characterization of the direction of arrival (DoA). A detailed analysis of the power correlation coefficient is performed so that the performance of the MIMO technology developed later in the deliverable can be explained. Finally, several measured paths are selected and their properties are listed so that they can be used as reference scenarios in the rest of the document.

This section describes the update performed on the IST METRA MIMO radio channel model [10, 11] with emphasis on its empirical validation of the polarization domain. Also, results of the validation of the model are presented in the spatial domain. The validation is based on the comparison between measured and simulated eigenvalues; the simulated results are based on the characterization of the measured environment.

I-METRA [11] introduced the MATLAB implementation of the MIMO channel model initially proposed by METRA [9], and updated during I-METRA according to the findings of the characterization on one hand, and the evolution of 3GPP standardization on the other hand. The package enables its user to derive correlation coefficients for linear arrays in a variety of scenarios. These coefficients can later be used in the actual channel model to simulate the corresponding MIMO channel. The limitations of the current version of the implementation are then discussed. Since the package is distributed as freeware, the distribution terms and the status of the diffusion so far are also addressed in there.

4.2.1 Intelligent Multielement Transmit and Receive Antennas

The MIMO concept is defined, throughout this section, as a radio link with M elements at the BS and N elements at the MS as pictured in Figure 4.1. For uplink the Tx is at the MS and the Rx at the BS, while for downlink, the roles are reversed. In the current context for this section, the actual role (Tx/Rx) of the MS and the BS is not important with regard to the conclusion made on the MIMO technology.

The term "element" is used instead of antenna to avoid potential confusion when dealing with polarization diversity since sometimes an antenna can be dual polarized with two elements.

The received signal vector $y(t)$ at the BS antenna array is denoted by

$$y(t) = [y_1(t), y_2(t), \ldots, y_M(t)]^T \tag{4.1}$$

where $y_M(t)$ is the signal at the Mth antenna element and $[\ldots]^T$ denotes the transpose operation. Similarly, the transmitted signals at the MS, $s_N(t)$, define the vector $s(t)$:

$$s(t) = [s_1(t), s_2(t), \ldots, s_N(t)]^T \tag{4.2}$$

The vectors $y(t)$ and $s(t)$ are related by the following expression:

$$y(t) = \mathrm{H}(t)s(t) + n(t) \tag{4.3}$$

where $n(t)$ is additive white Gaussian noise and $\mathrm{H}(t) \in C^{MxN}$ is the instantaneous narrowband (NB) MIMO radio channel matrix. $\mathrm{H}(t)$ describes the connections between the MS and the BS and can be expressed as

$$\mathrm{H}(t) = \begin{bmatrix} \alpha_{11}(t) & \alpha_{12}(t) & \Lambda & \alpha_{1N}(t) \\ \alpha_{21}(t) & \alpha_{22}(t) & \Lambda & \alpha_{2N}(t) \\ \mathrm{M} & \mathrm{M} & \mathrm{O} & \mathrm{M} \\ \alpha_{M1}(t) & \alpha_{M2}(t) & \Lambda & \alpha_{MN}(t) \end{bmatrix} \tag{4.4}$$

where $\alpha_{MN}(t)$ is the complex NB transmission coefficient from element n at the MS to element m at the BS.

A graphical representation of two-antenna arrays in a scattering environment is given in Figure 4.1.

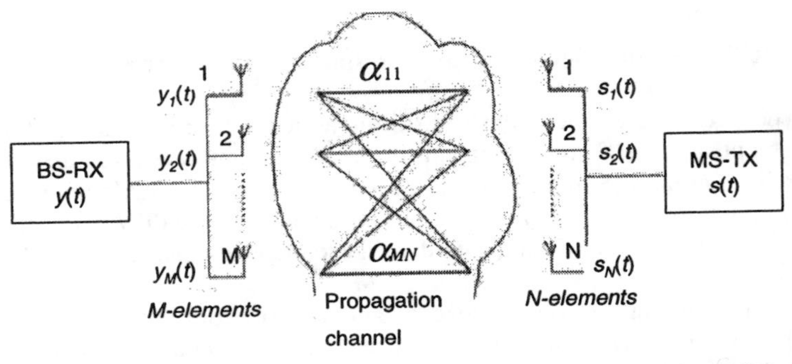

Figure 4.1 Two-antenna arrays in a scattering environment. Representation of an uplink situation.

The term "instantaneous" refers to a snap-shot of the radio propagation channel. In the context of mobile measurement, one snap-shot is equivalent to one sample of the recorded multipath in time (or distance). In the rest of this section, in order to keep the reading simple, the explicit time dependency (t) of $\mathrm{H}(t)$ and $\alpha_{MN}(t)$ is dropped.

Mathematically, the number of independent subchannels between two terminals can be estimated by using the singular value decomposition (SVD) of the matrix **H** or the eigenvalue decomposition (EVD) [26] of the instantaneous correlation matrix **R,** defined as

$$\mathbf{R} = \mathbf{H}\mathbf{H}^{\mathrm{H}} \text{ or } \mathbf{R} = \mathbf{H}^{\mathrm{H}}\mathbf{H} \tag{4.5}$$

where $[\mathbf{H}]^H$ represents the Hermitian transposition, (i.e., the transpose conjugate).

The derivation of the parallel independent channels is summarized below where U and V are unitary matrices, Σ and Γ are diagonal matrices, and u and v are the left and right singular vectors, respectively. On the other hand, K is defined in (4.8). There is an important relationship between the SVD of **H** and the EVD of **R**, such that where $\sigma_K k \gamma \sigma = 2k$ is the kth singular value and γ_k is the kth eigenvalue. Γ_{ij} denotes the elements of the matrix Γ.

SVD

$$H = U \Sigma V^H$$

EVD

$$HH^H = U \Gamma U^H \text{ or } H^H H = V \Gamma V^H$$

$$\Gamma_{ij} = \Sigma_{ij}^2$$

where

$$\Sigma = diag(\sigma_1, \ldots, \sigma_k) \qquad \Gamma = diag(y_1, \ldots, y_k)$$

$$\sigma_1 \geq \sigma_2 \geq \ldots \geq \sigma_k \geq 0 \qquad \gamma_1 \geq \gamma_2 \geq \ldots \geq \gamma_K \geq 0$$

with

$$\begin{aligned} \mathrm{U} &= [\mathrm{u}_1, \ldots, \mathrm{u}_M] \in C^{\mathbf{MxM}} \\ \mathrm{V} &= [\mathrm{v}_1, \ldots, \mathrm{v}_N] \in C^{\mathbf{NxN}} \end{aligned} \tag{4.6}$$

In the rest of the chapter, the normalized kth eigenvalue λ_k is used instead of γ_k. Unless otherwise mentioned, the normalization is made with respect to the mean power $|\alpha_{MN}|2$ between all possible pairs of MS-BS elements so that λ_k is defined as

$$\lambda_k = \frac{\gamma_k}{\mathrm{E}\left[\frac{1}{MN}\sum_{m=1}^{M}\sum_{n=1}^{N}|\alpha_{mn}|^2\right]} \tag{4.7}$$

where E[.....] denotes the expectation over time (or distance). In order to keep the reading simple, the explicit term normalized is dropped.

Irrespective of the numerical method used to perform the analysis, a channel matrix **H** may offer **K** parallel subchannels with different power gains, λ_k, where

$$K = Rank(\mathbf{R}) \leq \min(M, N) \tag{4.8}$$

and the functions $Rank(\mathbf{R})$ and $\min(M, N)$ return the rank of a matrix and the minimum value of the arguments, respectively [26].

In engineering terms, **u** and **v** are also referred to as the weight vectors while the *k*th eigenvalue can be interpreted as the power gain of the *k*th orthogonal subchannel [26]. This is illustrated in Figure 4.2 for a 4×4 MIMO antenna topology.

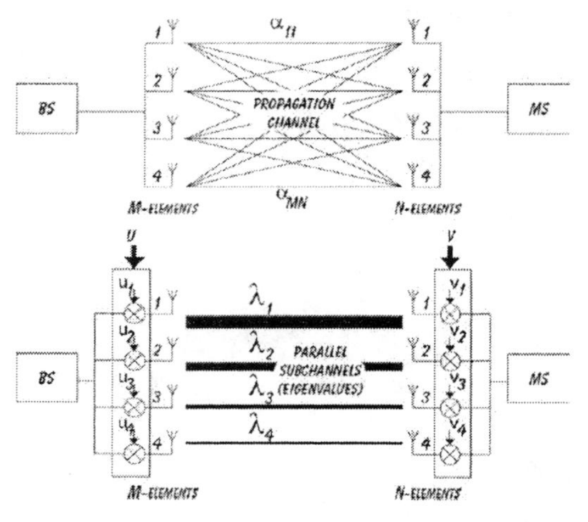

Figure 4.2 Illustration of parallel subchannels for a 4×4 MIMO MEA topology.

In such a configuration, MN=16 radio links, α_{MN}, are created, but only four orthogonal subchannels with power gain λ_1 to λ_4 are available. The difference in the thickness of the lines emphasizes the difference in gain of the parallel subchannels so that

$$\lambda_1 \geq \lambda_2 \geq \ldots \geq \lambda_k \geq 0 \tag{4.9}$$

To get the weight vectors, it is numerically more convenient to use the SVD of **H**, while to obtain the eigenvalue it is easier to use the EVD. The EVD technique is the optimal way to extract the power gain of the MIMO subchannels. However, if this technique is to be optimal in a practical system implementation, the proper unitary matrix **U** and **V** must be applied at the respective ends of the link. Consequently, the EVD method is only useful when the channel is known (i.e., when the information of the radio channel is available at both the Tx and the Rx), a situation which can be assumed reasonable for TDD systems, while for FDD systems significant feedback information is required.

4.2.1.1 MIMO Propagation Scenarios

This section presents the different MIMO propagation scenarios, which have been extensively studied to date in the MIMO literature. Before enumerating them, the notion of angular dispersive channels needs to be explained. The angular power density distribution of the environment (i.e., the power at each angle of arrival, fully describes the angular spreading of the signal). This is also often referred to as the power azimuth spectrum (PAS). It is obtained as the Fourier transform of the spatial correlation function [27, 28]. The azimuth spread (AS) is defined as the root second central moment of the PAS [21]. The PAS is closely related to the spatial correlation at the MEA (i.e., depending on the spatial separation between the elements of the MEA), a low AS is equivalent to a high spatial correlation coefficient and vice versa. Expressions of the spatial correlation function have been derived in the literature assuming that the PAS follows a cosine raised to an even integer [29], a Gaussian function [30], a uniform function [31], and a Laplacean function [32].

Three MIMO scenarios, each exhibiting a different propagation characteristic, are listed below:

- *Uncorrelated scenario.* In a rich scattering environment, the elements of H are fully decorrelated. This corresponds to a full rank scenario where the maximum number eigenvalues is achieved (see Figure 4.3).
- *Correlated scenario.* In a situation with line-of-sight (LOS) or low AS at one or both ends of the radio link, the elements of H exhibit a certain correlation that results in a low rank scenario, hence providing a lower number of eigenvalues (see Figure 4.4).
- *Pin-hole [33] or keyhole [34] scenario.* This is where the signals at the MEAs at both the Tx and Rx are decorrelated but the rank of H is one. This is analog to a propagation scenario where the radio link is interrupted by an infinite metallic plate with a hole in the middle, which is the only way for the signal to pass through, subsequently creating only one subchannel (see Figure 4.5).

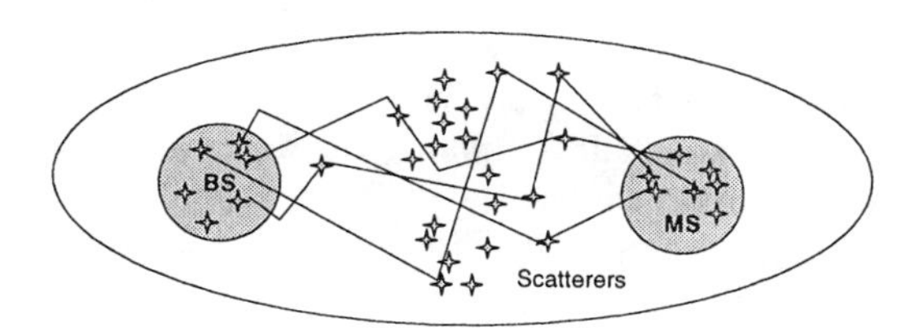

Figure 4.3 Uncorrelated scenario. (*From:* [12].)

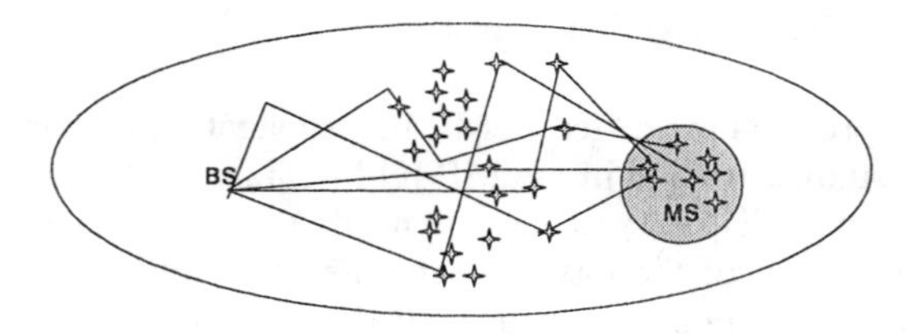

Figure 4.4 Correlated scenario. (*From:* [12].)

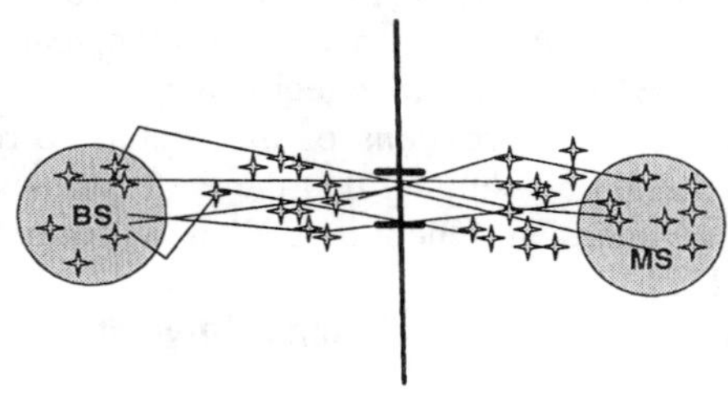

Figure 4.5 Pin-hole scenario.

Practically, this would be the case when Tx and Rx are on opposite sides of a mountain and the signal can only pass through a tunnel. Another practical interpretation was reported in [35] where a diffraction-induced keyhole could appear. However, as conceived by the same authors, such a scenario may be very rare and difficult to encounter. Moreover, the pin-hole effect has also been considered for situations with large Tx/Rx separation, a situation where the low SNR would degenerate MIMO capacity. In [36, 37], the concept of keyhole embraces the three scenarios presented above where the first and third scenarios are extreme cases.

4.2.1.2 The Antenna and Diversity Gain Aspect

The enhancement in the radio link from the MIMO technology can be expressed in terms of antenna gain, diversity order, and throughput performance. As already presented in [38], the antenna gain and the diversity order obtained from a MIMO system depend on the scattering environment and especially on the AS at each end of the MIMO radio link. These findings are summarized in Table 4.1 since they outline MIMO performance very well for different scenarios. In Table 4.1, the antenna radiation pattern of each element of the MEA is assumed isotropic and the mean gain is the strongest eigenvalue λ_1.

Temporal Illustration of the Eigenvalue

A quantitative description of the MIMO gain performance is the temporal representation of the eigenvalues. Figure 4.6 (a) presents the behavior of the eigenvalues derived from a measured 4×4 MIMO setup where the AS is high at both the BS and MS. Figure 4.6 (b) presents a situation where the AS is low at the BS and high at the MS.

Table 4.1

Antenna Gain and Diversity Order in a MIMO Context for Different Scenarios [38]

AS at the MS	AS at the BS	Scenario	Mean Gain	Diversity Order
Low	Low	Correlated	MN	1
Low	High	Correlated	MN	M
High	Low	Correlated	MN	N
High	High	Decorrelated	$\approx \left(\sqrt{M}+\sqrt{N}\right)^2$	MN

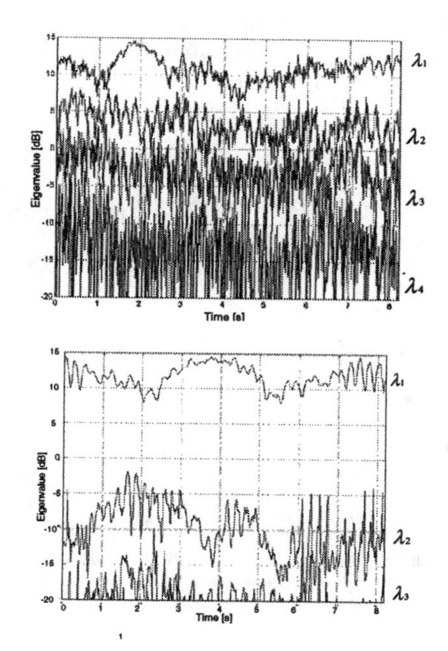

Figure 4.6 Variation of the eigenvalues over time. (a) Decorrelated scenario: AS-BS high, MS high, and (b) correlated scenario: AS-BS low, MS high.

The behavior of the eigenvalues can be summarized as follows: they fluctuate with time and the strength of each λ_k strongly varies depending on the propagation scenario (i.e., correlated or uncorrelated).

Antenna Gain

Another method to present the eigenvalues is by using their cumulative distribution function (cdf). Throughout the rest of the deliverable, unless otherwise mentioned, the cdf is computed over time. The logarithm of the cdf is shown to be coherent with the original eigenvalue representation reported in [26].

Figure 4.7 presents the empirical cdf of the normalized eigenvalues compared to a Rayleigh SISO channel. The graphs should read as follows. At the 10% cdf level a power gain of 19 dB is achieved using MIMO compared to a SISO setup. The 10% cdf level is used here because it is a typical measure for the system-level performance. The mean value of the strongest eigenvalue of Figure 4.7 is 11 dB, which is in line with the value of 10 dB for a fully decorrelated 4×4 case.

The cdf representation is also very useful when interpreting the diversity performance in a MIMO system. The steepness of the slope of the cdf reflects the diversity order. The steeper the slope, the less amplitude of the fluctuation in the signal (i.e., the fades due to the multipath are less deep), and therefore a higher degree of diversity is obtained, as shown in Figure 4.7.

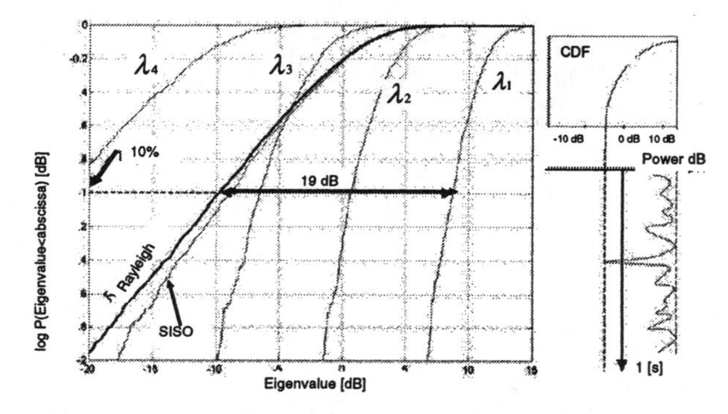

Figure 4.7 The cdf of the eigenvalues. The slope is an indication of the amplitude of the eigenvalue fluctuations. The same decorrelated scenario presented in Figure 4.6 (a).

Diversity Gain

Figure 4.8 illustrates the improvement in antenna gain and diversity order from a measured MIMO system for a correlated scenario (i.e., low AS at the BS and a high AS at the MS). Three different M×N antenna setups are compared: 1×1, 1×4, and 4×1. Recall the following notation: BS $\rightarrow$ M×N $\leftarrow$ MS.

A simulated Rayleigh curve is plotted and compared to the measured 1×1 radio link suggesting that the 1×1 measured signal is Ricean distributed since its curve is steeper than the Rayleigh curve. Relative to this measured SISO case, a gain of 6 dB at 10% level is achieved for a 4×1 setup. This is the classical gain of 10log10(M) with $M = 4$ at the BS (i.e., no Rx diversity gain is obtained, the slope of the 4×1 is the same as for the 1×1 situation).

However, the slope is steeper for a 1×4 topology, the diversity gain at the MS is added to the array gain, and 9 dB gain is achieved at 10% level. In the 4×1 case, the four antenna elements at the BS act as a single high gain antenna while in the 1×4 case the four antenna elements at the MS are uncorrelated and therefore they achieve both antenna gain and diversity.

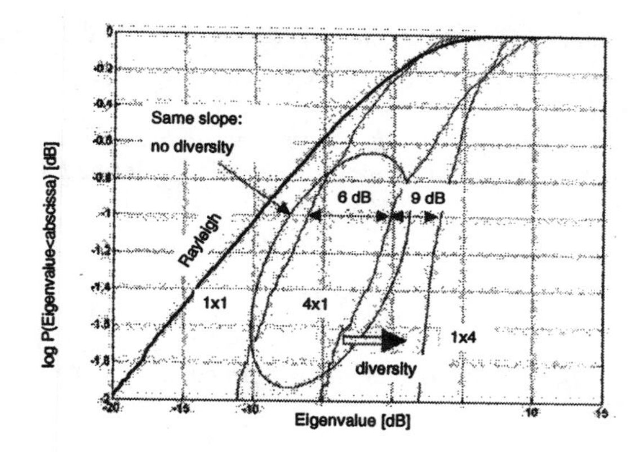

Figure 4.8 Empirical illustration of the antenna gain and the diversity gain for different spatial antenna array configuration set-ups for the correlated scenario. AS-BS: low, MS: high.

Capacity Aspect

The concept of orthogonal subchannels is now understood. This permits us to introduce the capacity concept that MIMO systems can offer compared to a conventional SISO system. This section defines the spectral efficiency of MIMO channels based on the channel capacity as defined in information theory [39, 40]. This is the theoretical maximum amount of information that can be transmitted over a bandwidth limited channel for which error-free transmission is possible in the context of Gaussian channel [41].

In real system implementation, the achievable capacity is limited due to coding, detection, and constellation size, among others [42]; therefore, the capacity results presented here are to be seen as upper bound values.

This study is limited to a single-user scenario where two power allocation strategies are compared: the water-filling and the uniform power allocation strategies. The Shannon's formula for a SISO radio channel is expressed as:

$$C = \log_2(1+\zeta) \tag{4.10}$$

where ζ is the SNR.

For the MIMO technology, [24] presented the total Shannon's capacity, for uniform power allocation strategy, as

$$C = \log_2\left[\det\left(\boldsymbol{I} + \boldsymbol{Q}\boldsymbol{H}^H\boldsymbol{H}\right)\right] \tag{4.11}$$

where $\boldsymbol{I}$ is the identity matrix, $\boldsymbol{Q}$ is the signal covariance matrix , $\det\left(\boldsymbol{I} + \boldsymbol{Q}\boldsymbol{H}^H\boldsymbol{H}\right)$ is the determinant and $\log_2$ is the logarithm base 2. This formulation is often used by the MIMO community.

This equation can be rewritten so as to emphasize the influence of the $\boldsymbol{K}$ parallel subchannels [26] and the total Shannon's capacity is defined as

$$C = \sum_{k=1}^{K} \log_2(1+\zeta_k) \tag{4.12}$$

where ζ_k is the SNR for the kth subchannel and is defined as

$$\zeta_k = \lambda_k \frac{P_k}{\sigma_n^2} \tag{4.13}$$

where P_k is the power assigned to the kth subchannel and is the noise power. Hence

$$C = \sum_{k=1}^{K} \log_2\left(1+\lambda_k \frac{P_k}{\sigma_n^2}\right) \tag{4.14}$$

Depending on the power allocation scheme employed, the total transmitted power is distributed in a different manner between subchannels.

When considering the total capacity offered by the MIMO setup, the total mean SNR per Rx antenna is defined as

$$SNR = \frac{\mathrm{E}[P_{Rx}]}{\sigma_n^2} = \frac{\mathrm{E}[PT_x]}{\sigma_n^2} \tag{4.15}$$

where the time-averaged channel power gain is assumed to be 0 dB so that there is no loss in average between P_{Tx} and P_{Rx}.

4.2.2 MIMO Summary

The purpose of the characterization work in I-METRA was to go deeper into the exploitation of the measurement data collected during the ACTS projects SUNBEAM and the IST project METRA, in order to get a better insight in the properties of MIMO radio channels. This insight has been gained through large-scale post-processing of the measurement data. It has lead to the study of the influence of a large set of parameters and contexts, such as correlation, polarization, directional information, power imbalance, and presence of a LOS component.

Additionally, in parallel to this characterization, the MIMO channel model proposed by METRA evolved in two ways within the project I-METRA. On the one hand, it was modified so as to embed the findings of the characterization activities, mainly the impact of power imbalance and polarization diversity. On the other hand, it was maintained inline with the evolution of standardization in 3GPP.

The advantage of exploiting MIMO techniques is to increase the system throughput data rate for the same total radiated power and channel bandwidth.

In highly scattering propagation environments, the theoretical maximum data rate for MIMO algorithms increases directly in proportion to the number of antennas, rather than only being proportional to the logarithm of the number of antennas when conventional phased array beam forming methods are used.

4.3 ULTRAWIDEBAND CONCEPTS FOR WIRELESS NETWORKS

Salient requirements of WPAN systems include low-power operation, use of license-free wireless links, low cost, scalability, and the use of ad hoc networking techniques. Many gigahertz of bandwidth have been authorized for license-free WPANs using UWB [43]. This technology has the potential to provide unprecedented high-connectivity consumer products in the home, such as video conferencing, wireless video and audio distribution systems, new home entertainment appliances, diskless computers, and position location and navigation applications. The concept of UWB originated with Marconi, in the 1900s, when spark gap transmitters induced pulsed signals having very wide bandwidths. Spark transmissions created broadband interference and did not allow for coordinated spectrum sharing, and so the communications world abandoned wideband communication in favor of narrowband, or tuned, radio transmitters that were easy to regulate and coordinate [44].

In the mid-1980s, the FCC encouraged an entirely new type of wideband communications when it allocated the Industrial Scientific and Medical (ISM) bands for unlicensed spread spectrum and wideband communications use. This revolutionary spectrum allocation is most likely responsible for the tremendous growth in WLAN and Wi-Fi today, as it led the communications industry to study the merits and implications of wider bandwidth communications than had previously been used for consumer applications. The Shannon-Hartley theorem states that the channel capacity grows linearly with the bandwidth and decreases logarithmically as the SNR decreases. This relationship suggests that radio capacity can be increased more rapidly by increasing the occupied bandwidth than the SNR. Thus, for WPANs that only transmit over small distances, where signal propagation loss is small and less variable, greater capacity can be achieved through greater bandwidth occupancy. UWB technology presents itself as a good candidate for the physical layer of such systems, both for high and low data rate applications. Possible applications are: medical monitoring, office automation, sensor networks, information services, and banking/financial applications.

UWB radio communications are an extreme form of spread spectrum communication systems, generally defined as operating with a fractional bandwidth greater than 0.25 (i.e., a −3-dB bandwidth which is at least 25% of the center frequency used). Alternatively, the FCC [45] defines a UWB device as any radio device with a −10-dB fractional bandwidth greater than 0.20 or occupying at least 500 MHz of the spectrum [14]. This chapter considers UWB systems based on IR-UWB, which use very short pulses and have been known for a long time. IR-UWB meets the requirements for short-range wireless systems and additionally possesses the following desirable features: mitigated multipath fading effects, possibility of high bit rates, and unique location ability. Indeed, the IEEE 802.15 Task Group 2, which deals with the high rate PHY layer for WPAN, has formed a study group (SG3a) to consider UWB as a candidate for this PHY [16].

Recently, the PHY issues for UWB in WPAN environments have been extensively studied [46–48]. However, the problems related to the medium access techniques remain largely unexplored, especially regarding the targeted ad hoc networking scenarios. In ad hoc networking, the devices are interconnected via spontaneously created, disposable connections, without relying on a pre-existing infrastructure. These scenarios pose seriously challenging research tasks, since the same medium should be used by many mutually interfering WPANs under the stringent synchronization conditions imposed by the IR-UWB. The two basic modulation schemes used in IR-UWB are pulse position modulation (PPM) and bi-phase modulation (BPM), while the two common channelization techniques are time-hopping spread spectrum (TH-SS) and direct sequence spread spectrum (DS-SS). This work analyzes only IR-UWB communication systems using TH-SS and PPM. Since WPAN technology uses license-free, wireless links, the radio resources should be shared among the collocated and uncoordinated WPANs. Then, devices belonging to a single IR-UWB WPAN could actually share the channel not only among themselves but also with the devices belonging to the neighboring, physically collocated networks. In such a situation, there are two different levels of multiplexing we need to achieve. First, the MAC protocols should coordinate the transmissions of all devices within a network by using contention-free techniques (e.g., polling), random-access techniques (e.g., carrier-sensing), or a combination of both. At the second level, the multiple collocated networks must share the resources in a sense to minimize the mutual interference, which can be achieved through adapting the parameters of TH used within the networks. In this section, a TH-SS UWB communication system is considered, and it is shown how to design TH parameters in order to achieve efficient resource sharing at the second level of multiplexing or resource sharing among multiple networks. The throughput performance is evaluated by computer simulation in a multinetwork scenario, to show the impact of TH parameters on the throughput performance. Furthermore, by applying a link adaptation mechanism [49], TH parameters can be adapted according to interference conditions. This mechanism can maximize the throughput achieved by WPANs while keeping the robustness with respect to the basic timing mechanisms used in the TH-SS communications.

4.3.1 ULTRAWIDEBAND PHYSICAL LAYER (UWB PHY)

4.3.1.1 UWB Radios

The fractional bandwidth for UWB, F_{BW}, as defined by the FCC is given by the following expression [14]:

$$F_{BW} = 2\frac{f_H - f_L}{f_H + f_L} \geq 0.20 \tag{4.16}$$

where the upper, f_H, and lower, f_L, frequencies correspond to the −10-dB bandwidth.

The FCC also regulated the spectral shape and maximum power spectral density ($\approx$−41-dBm/MHz) of the UWB radiation in order to limit the interference with other communication systems. The ETSI regulations in EU are expected to follow the FCC but with a more restrictive spectral shape, motivated by a different management of the available spectrum [14].

The UWB signal generation methods can be grouped in two major categories:

- Single-band (SB) based: employing one single transmission frequency band;
- Multiband (MB) based: employing two or more frequency bands, each with at least 500-MHz bandwidth.

In the SB solution, the UWB signal is generated using very short, low duty-cycle, baseband electrical pulses with an appropriate shape and duration. Because of the carrier-less characteristics (no sinusoidal carrier to raise the signal to a certain frequency band), these UWB systems are also referred to as carrier-free or IR-UWB communication systems [47].

The MB UWB systems can be implemented as carrier-less (different pulse shapes/lengths are used according to the frequency band) or carrier-based (multicarrier like) [46].

In the IR-UWB solutions, typically the radiated pulse signals are generated without the use of local oscillators or mixers; thus, a potentially simpler and cheaper construction of the Tx and Rx is possible, as compared to the conventional narrowband systems. The characteristics of the pulse used (shape, duration) determine the bandwidth and spectral shape of the UWB signals. The most common pulse shapes used in IR-UWB are: Gaussian monocycle (and its derivatives) and Hermitian pulses. The low transmit power levels together with the ultra-fine time resolution of the system can considerably increase the synchronization acquisition time and the complexity of the receiver.

The large transmission bandwidth in UWB, in the order of $n \times 100$ MHz (with $n = 5K\ 70$), has as a result a higher immunity to interference effects and improved multipath fading robustness. The multipath can be resolved down to differential delays of a nanosecond or less (i.e., 30 cm or less spatial resolution). In order to convey the information symbols in UWB communications, several approaches for the modulation techniques exist, mostly based on the classical base-band modulation types. Modulating the UWB pulse characteristics, such as amplitude (PAM), time position (PPM), phase (PM), shape (PSM), or any combination of these, can be used. Another direct consequence of the large communication bandwidth is the possibility to accommodate many users, even in multipath environments.

The two most common channelization and multiple access (MA) techniques in IR-UWB are:

- Direct sequence spread spectrum, similar to CDMA communication systems but using an UWB radio pulse as a chip pulse shape;
- Time hopping spread spectrum, which uses UWB pulses pseudo-randomly shifted in time domain.

For simplicity, here we refer to only TH-SS (SB) IR-UWB communication systems using binary PPM (BPPM). Furthermore, an abstraction of the pulse shape and of the corresponding optimal detection techniques that can be used is made.

4.3.1.2 TH-SS UWB Description

For a multiuser (device) scenario, the format of the transmitted TH-SS IR-UWB signal, $s_{tx}^{(k)}$, corresponding to the kth user (device) is given by:

$$s_{tx}^{(k)}(t^{(k)}) = \sum_{j=-\infty}^{\infty} w_p(t^{(k)} - jT_f - c_j^{(k)}T_c - d_j^{(k)}\delta) \tag{4.17}$$

where $t^{(k)}$ is the k th clock time of the transmitter; $w_p(t)$ is the used UWB pulse; T_f is the time frame allocated for each UWB pulse; T_c is the time shift step used in channelization / MA together with the $c_j^{(k)}$ TH code (sequence) allocated to each communication channel/user (device); and δ is the additional time shift used for M-PPM of the information symbols, $d_j^{(k)}$. To increase the reliability of the communication link, the same symbol can be repeated a certain number of times, N_p, increasing the processing gain of the system (i.e., $d_j^{(k)}$ =const. for $j = [p\mathrm{K}\ p + N_p]$ and any integer p). A TH sequence $c_j^{(k)}$ is typically a pseudo-random sequence with period N_p, where each element of the sequence is an integer in the range $[0\mathrm{K}\ N_h - 1]$. The time frame T_f is divided into equally spaced TH shifts, T_c, so that $T_f \geq N_h T_c$. The TH codes can be chosen based on their auto- and cross-correlation properties, number of estimated users, and so forth, similarly to the CDMA systems. The M-PPM is achieved by using the additional time shift δ and $T_c \geq M\delta$. In the rest of this section we consider only the BPPM case, thus $M = 2$ and $d_j^{(k)} = 0$ or $d_j^{(k)} = 1$.

In a multiuser scenario (ad hoc networks, and so on), several transmitted signals, as given by (4.17), propagate through the radio channel, interfere with each other, and add up at the input of the receivers of all the users. Additionally, due to the multipath fading, delay, attenuation, and other noise sources in the radio channel, the received signals are further distorted. The choice for the system parameters T_f, T_c, and δ is dependent on the desired system performances (throughput, reliability, and so on) and the radio propagation channel characteristics (multipath fading, dispersion, and so on).

In [15] the radio channel propagation mechanisms are abstracted in terms of signal fading, spread, and additional noise. The absolute signal attenuation and delays, introduced by the different relative geographical locations of the users/devices, and the multiple access pulse collisions are the only sources of link level errors that are considered and analyzed.

4.3.2 UWB Summary

The main concern regarding UWB emissions is the potential interference that UWB systems could cause to the incumbents in the frequency domain as well as to specific critical wireless systems that provide an important public service (e.g., GPS). There are many factors that affect how UWB impacts other narrowband systems, including the separation between the devices, the channel propagation losses, the modulation technique, or the pulse repetition frequency employed by pulse-based UWB system. In general, because of the wideband nature, a UWB transmitter can be viewed as a wideband interferer, which has the effect of raising the noise floor of narrowband receivers.

In the United States, the FCC determined and is still in the process of evaluating appropriate rules to regulate the use of UWB radio technology.

According to the FCC, emission limits are specified for different UWB applications. The two application classes of interest in the context of low data rate WPAN systems are the so-called indoor and hand-held UWB systems. For an LDR WPAN terminal to classify as an indoor UWB system, it must be demonstrated that the system units will fail to operate if they are removed from the indoor environment. The second category embraces miscellaneous UWB devices that are primarily handheld and intended to operate without restriction on location.

In Europe, UWB spectrum regulation is still ongoing, which is—at least partly—due to the different regulatory authorities and interest groups (CEPT, national bodies, ETSI, EC) involved. Currently the European regulatory authorities are, similar to the situation in the United States, discussing and evaluating the possibility to simply apply the electromagnetic compatibility (EMC) limits. For example, ETSI has proposed a spectral mask for Europe that is similar to the FCC spectral mask but with a smooth roll-off at the edges of the central band from 3.1 to 10.6 GHz. It is anticipated that the regulatory framework for Europe would have been finalized in the year 2005.

At the same time, Japanese regulators have issued the first UQWB experimental license allowing the operation of a UWB transmitter in Japan, and in Singapore, IDA UWB-FZ has allowed UWB systems to operate locally [44].

To summarize, the FCC ruling allows UWB devices to operate at low power (an EIRP of –41.3 dBm/MHz) in an unlicensed spectrum from 3.1 to 10.6 GHz, with out-of-band emission masks that are at substantially lower power levels.

The low in-band and out-of-band emission limits are meant to ensure that UWB devices do not cause harmful interference, or in other words, are able to coexist with "licensed services and other important radio operations," which include cellular, PCS, GPS, 802.11a, satellite radio, and terrestrial radio. The fact that the FCC specified that UWB is at a minimum bandwidth of 500 MHz is important—while UWB can occupy several GHz of bandwidth using small pulses, the 500-MHz bandwidth rule has provided the impetus for chip makers to consider channelization, or a multiband approach, in the UWB standardization activity [44].

4.4 INTERWORKING BETWEEN DIFFERENT RADIO TECHNOLOGIES

A number of developments in the market contribute to the understanding of how multistandard access might develop, or is being adopted now [50]. Not only have devices with multiple radios been demonstrated, but Wi-Fi 3G services are on the market and convergence of fixed and mobile telecommunications is getting serious attention.

Many of the expectations of future wireless technology needs were based on the roll-out and uptake of services based on 3G technologies. However, this was delayed for a number of reasons. Still uptake is slow, but operator's initial strategies for selling are becoming clearer. In the United Kingdom and other European countries, data access for business professionals from a laptop has been the leading market. In the meantime the 3 service from Hutchinson has attempted to break into the consumer market, but has had to focus on competition of price of minutes.

The public WLAN hotspot market has also grown, with a number of trends. Primarily, operators have moved into building or reselling public wireless hotspots and user subscriptions. From being a technology competitive with 3G networks, it is now seen as a key part of the bundle of services that an operator must be able to offer, especially to the traveling businesspeople. This has clearly opened a convergence space based on laptops having two radios. With 3G networks not delivering some of the promised benefits, in particular broadband Internet access that many are now accustomed to in the office, a multistandard service would appear to be necessary. However, operators are not providing dual radio data cards, just the 3G card, and a software "dashboard" to switch between 3G, public, and private WLAN services. They are riding on, or may be suffering from, the independent success of the wireless laptop, where many laptops already have a 802.11b/g card. However, the 3G cards are data-only, and one could say that they are crippled as a fully converged service. VoIP over public WLAN using a laptop is a viable and growing technology, but with a 3G data-card, handover of that voice call to the cellular network is not possible.

Three key drivers behind the success of WLAN are (1) the marketing of Wi-Fi-enabled laptop computers, (2) the growth in fixed broadband Internet access, and (3) investment in WLAN facilities by organizations, from offices to universities. Many laptops have WLAN facilities built in, thus encouraging their use. Cheap broadband access is complemented very well by cheap WLAN devices in the home and small office environment, where multiple computers are increasingly the norm.

Enterprises and organizations are investing to save costs on wiring and also to allow for flexible access by employees and students.

The time frame 2003–2004 saw the demonstration of a range of devices with Wi-Fi, 2G- and 3G-radios, from handset vendors and silicon vendors. However, achieving seamless handover of voice calls between radio systems would appear to be causing difficulties. VoIP on 802.11 is working and so is handover from access point to access point. Currently, the situation is that voice roaming between bearer standards does not work, but data sessions where seamless handover is not important can be done, opening up the possibility of simultaneous use. One of the delays fundamental to implementing 3G—providing handover between 3G systems and 2G standards—appears to have been largely solved.

The research community has focused considerable attention on new air interfaces, from IEEE standards committees, and extensions and evolutions of GSM, CDMA, and 3G systems. This has been driven by the technology firms behind the standards and pulled by the growth in Internet usage, especially the broadband access market, and the realization that existing 3G mobile systems will probably not be able offer the sort of network access for data that people have grown to expect from fixed broadband access. Broadband wireless Internet access is behind the IEEE 802.16 and 802.20 initiatives and the deployment of systems based on 3GPP TDD standards. The WiMax initiative to develop fixed wireless broadband has the very important backing of firms that are interested in the market for devices that will use the system, rather than supplying or operating the system. There are also projects developing completely new 4G air interfaces. There are thus multiple paths to the deployment of a broadband wireless access (BWA) system, fixed or mobile, and it can be deployed in competition or as a complement to existing mobile systems. The existence of complementary radio access systems and their ownership was a basic topic for investigation in the IST project FLOWS [50, 51], (see Figure 4.9).

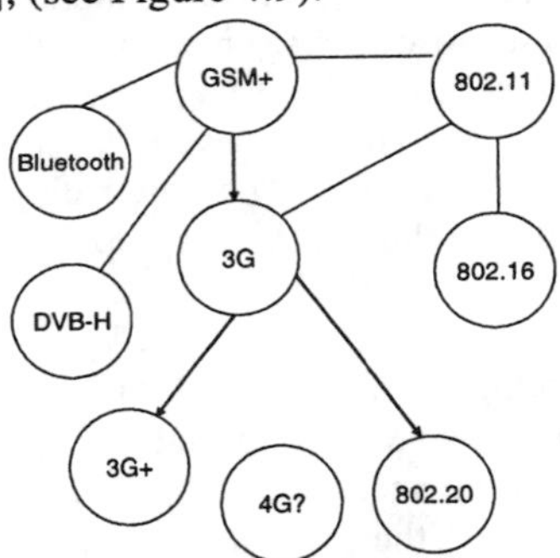

Figure 4.9 Path to the convergence of standards. (*From:* [50].)

One of the key arguments against convergence around access system interworking is that a single standard or family of standards is the easiest and most efficient way of providing radio access, compared to multiple standard systems, especially when it comes to implementing more advanced features, such as seamless handover of voice services. In the medium term these types of systems, are likely to remain separate, with new BWA systems built by firms that do not have advanced cellular networks.

Simultaneous use is an important issue for communication systems in the future as high demands on data rates and convergence of standards are anticipated. Simultaneous use enables the pool of available bandwidth to be shared among the users and their varying requirements. The concept of a convergence manager (CM) was introduced in [52]. This document reports on simulations that were carried out to evaluate various CM strategies that were proposed in the IST project FLOWS [51].

A multiservice, multisystem, and multioperator terminal is assumed to exist, enabling a large variety of simultaneous use concepts. The operator (Op) is the entity enabling services over systems. Multiple services (*nSe*) corresponds to a simultaneity of usage from a user perspective, where several services run simultaneously in the same terminal. Multiple standards (*nSt*) corresponds to the simultaneous use of several standards in a single terminal, which can correspond to multiple instances of the same standard.

Without having any objective of mathematically dealing with the subject, but rather helping in the understanding of the several concepts at stake, the three fundamental components (*Se*, *St*, *Op*) constitute an "orthogonal space" where each simultaneous use concept can be represented, as illustrated in Table 4.2 and Figure 4.10.

Table 4.2

Characterization of Single and Simultaneous Use Concepts

Space coordinate	Characterization
Single-use concept (1*Se*, 1*St*, 0*Op*)	Classical ad hoc
Single-use concept (1*Se*, 1*St*, 1*Op*)	Classical use of a telecommunication system
Simultaneous use concepts [(1*Se*, n*St*, 0*Op*)]	Generalized ad hoc network where several systems are involved

4.4.1 Examples of Simultaneous Use Concepts

4.4.1.1 Ad Hoc Environment

It is possible to have simultaneous use in an ad hoc network environment where, in the absence of an SP, several terminals establish a connection through different wireless standards (1*Se*, *nSt*, 0*Op*), as illustrated in Figure 4.11. In this example, three terminals can route communications through different standards, which is only possible with multistandard terminals.

4.4.1.2 Different Services Via a Single Standard

From a user's perspective, the simultaneous use of two or more services over the same standard (n*Se*, 1*St*, 1*Op*) may be seen as a concept of simultaneous use. When mapping services to standards, different criteria need to be evaluated to ensure that the mapping will benefit all involved actors (i.e., the end users, the network operators, and the service providers). The mapping of services to standards could be called feasible if the overall satisfaction is good—where the overall satisfaction is a composition of the satisfaction according to some measures of each involved party. In order to manage the simultaneous use of standards, a functionality is needed to manage the mapping between different services and multiple standards. The decision will be made based on different mapping parameters such as QoS requirements, network capacity, channel status, and the mapping policy. The functionality will be the CM that can have different locations and functions. The complexity of the functionality will differ in relation to the granularity of the convergence.

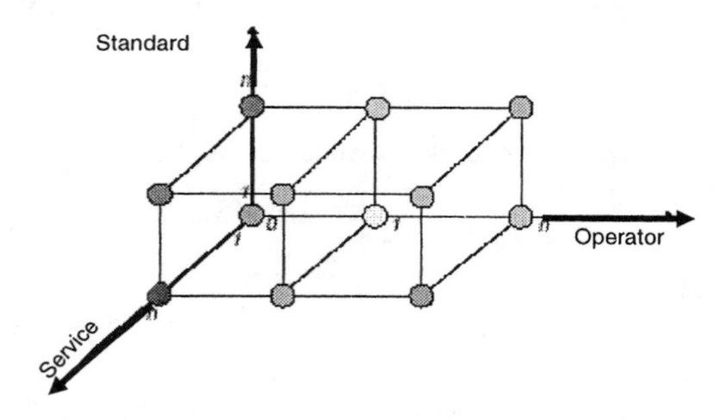

Figure 4.10 Space of simultaneous use.

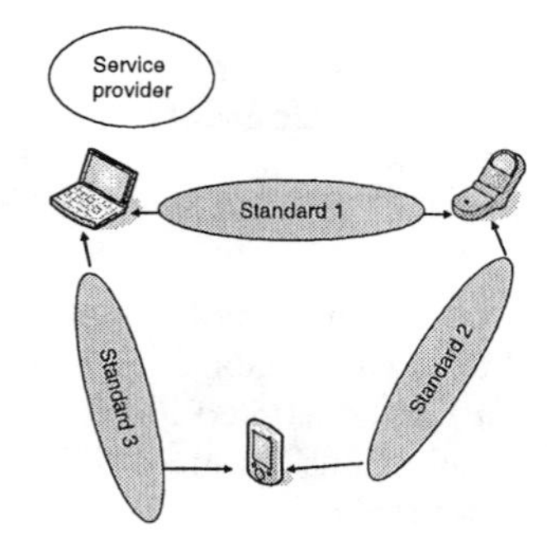

Figure 4.11 Generalized ad hoc network.

The benefits from the mapping services to different standards will depend on the location of the CM.

Several proposals on the possible locations of CM are given in [52], as follows:

- *The CM is located only in the MTs.* The CM can operate either in a generic way, as SIP is used to direct media streams or data channels via different access technologies, or in a specialized solution, enabling the CM functionality for a dedicated class of traffic, the file transfer traffic. Only the charging system needs to be changed, while the overall QoS would be increased. There is no additional standardization effort required for an enhancement or an introduction of protocols for the proposed solution. However, the session layer requires the CM entity and the policy base in the terminal to enable behavior according to the wishes of the user. The interactions required towards the lower layers, towards the session layer, and towards the policy base must be considered. Also, the way, such a policy base is created and managed is of importance.
- *The CM is located in the MTs and the RANs.* There are two different options to place the CM entity in the RAN: a BS/AP or a RNC/ BSC. This placement of the CM to this entity provides the optimization for radio access networks and MT transmission modes. But it brings difficulties as well: no loose coupling scenario can be supported. Another important drawback is the enormous standardization effort for this kind of solution.
- *The CM is located in the MTs and the core/backbone networks.*
- *The network sided CM is placed on the combined GGSN-PDG element.* To enable that a CM located at such a combined element spreads packets over or collects packets from the different access networks, a mapping between the different IP addresses has to be enabled. However, the interaction between the CM in the network and the CM on the UE would also have to be standardized. Therefore, an additional control plane protocol would have to be defined and to be included in the 3GPP procedures.
- *The CM as a border gateway to a single operator's public land mobile network (PLMN) domain.* The CM is placed as an entity interfacing all edge nodes (gateway nodes) of the various access networks. The service providers send applications addressed to the CM. The CM will investigate the network status, possibly the subscription with the user's preferences, the availability of access networks to the terminal of the user, and other information necessary to decide how to split the session/media streams onto the potential access networks available. In most cases, it will enhance the QoS, but there will be a need for a correlation of charges between the radio standards. There are also several open issues to be investigated in the future work.

- *The CM is at a border gateway to the PLMN domains of different operators.* In this case a user would be able to simultaneously receive session streams from two or more network operators to the same terminal device. Only minor enhancements need to be done to the network nodes to support this type of CM handling of session streams. The billing system among several operators will be quite complex and the QoS can cause problem in some cases.
- *The CM is located in the MTs and the corresponding server.* This approach suffers from its inflexibility. The CM on the remote server can provide CM functionality for the services hosted on this server. Therefore, it is not a generic approach in terms of service independence.

With the help of a CM, a mapping of different services can be performed onto the different service definitions found in the different radio network standards. The goal is to obtain common service mapping definitions. For reaching this goal, the concept of CM is raised up to manage the mapping from different services onto different standards. Abstract means in this context that the CM functionality is able to make use of all available lower layer technology to provide simultaneous use of standards according to either operator- or user-defined policies (see Figure 4.12).

It can be defined that simultaneous use of standards means parallel or alternate use of equipment and radio networks adhering to two or more physical layer transmission standards for the transmission of information related to the same communication [50]. The CM then will manage the mapping between different services, applications, and multiple standards. The decision will be made based on different mapping parameters such as QoS requirements, network capacity, channel status, and the mapping policy. The CM can have different locations and functions. The complexity of the functionality will differ in relation to the granularity of the convergence.

4.4.1.3 Location of the Convergence Manager

The location of the CM is a crucial issue for the architectural design of an environment, providing simultaneous usage of standards. It impacts the user's experience of service rendering and the service provider's ability to combine services into combined services increasing the value for the customer. The operator may also take advantage of the functionality of a CM (e.g., by using more efficiently the implemented networks). Therefore, the implementation of the CM functionality affects many players in the value chain (e.g., network operators, terminal manufacturers, terminal software vendors, and network equipment manufacturers). Based on the granularity of the introduced convergence, the end user may also be involved in the decision process of choice of network.

The basic assumption of the IST project FLOWS [50] was that the mobile terminal, enabled for a simultaneous usage of standards, has to contain a CM

instance in every case. Depending on the concept, this terminal CM can either work autonomously or cooperate with another CM entity anywhere in the network. It shall be explicitly emphasized at this point that the CM concept therefore can be used to fulfill both user-defined policies and dynamic operator policies. Solutions that make use only of a CM in the terminal are, without specific enhancements, only able to support user-defined policies. For solutions that make use of a pair of CMs, the interaction between these CM instances, which is implicitly required, enables both kinds of policies to be considered.
The following sections describe the various CM placement concepts and point out specific advantages and constraints.

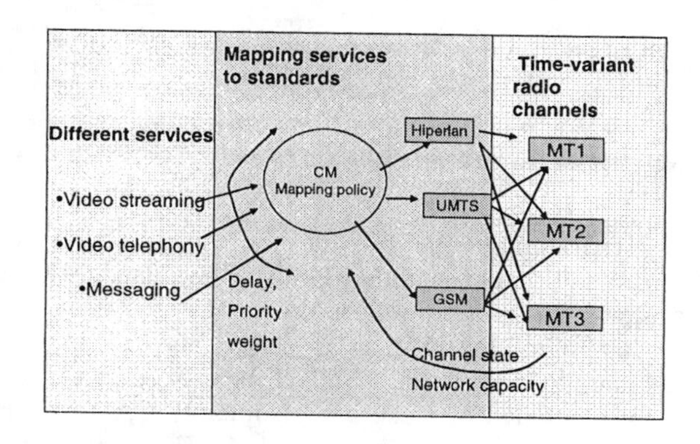

Figure 4.12 Mapping services to standards. (*From:* [52].)

CM Only in the Mobile Terminal

To satisfy user-based policies it is sufficient to make use of only one CM instance in the terminal (see Figure 4.13 for GSM and UMTS). An example for such a user-based policy might be to switch the video channel of an ongoing videoconference from UTRAN to WLAN, whenever WLAN is available.

Because of the higher available bandwidth in WLAN, a better video codec can be used that will increase the service quality for the user.

There are several possibilities for how the CM entity can operate on the terminal. One is to use SIP [53] to direct media streams or data channels via different access technologies. Because SIP runs at the session layer, this approach is called *the session layer concept*. Another approach is a specialized solution, enabling the CM functionality for a dedicated class of traffic, the file transfer traffic.

The first question that has to be answered when applying the session layer concept is, how the user will be reachable in the network, if more than one physical access point is available. The IP multimedia system (IMS) provides a proper solution here. The UE has to send registrations to a so-called serving-control session control function (S-CSCF) in his home network, which interacts with the HSS, the general network database in terms of subscription. The registration messages are passed via a so-called IP-CAN network to a so-called proxy (P)-CSCF, forwarding them to the home network of the user.

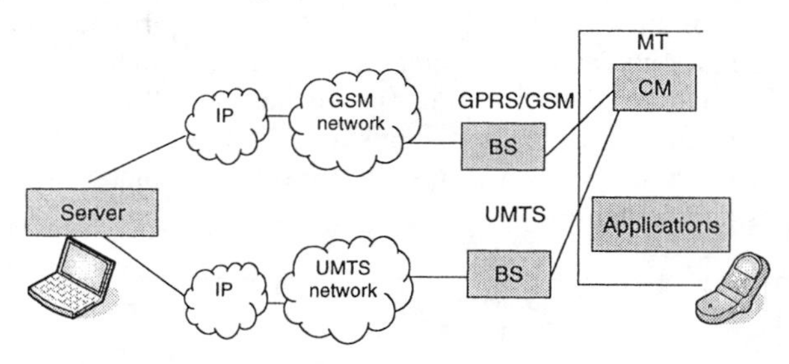

Figure 4.13 CM is located only in the mobile terminal.

During this registration, the UE provides the home S-CSCF and therefore also the home HSS with its IP address in the IP-CAN. The standard document for IMS given in [54] abstracts fully from the IP-CAN.

In [54], the use of GPRS [55] as an access network is explained and the required interactions to obtain an IP-CAN bearer service are demonstrated. However, it is not precluded to make use of other IP-CAN options, for instance the UMTS-WLAN interworking architecture [56]. Exact definitions of how this interaction of obtaining an IP-CAN bearer on top of the diverse access technologies takes place are not given in the IMS documents.

It is possible to adopt an architecture [52] where the user can choose with which access network the signaling takes place. Because the user might not have knowledge about signaling issues, this will most probably be a static policy on the UE. After the IP addresses are obtained via the IMS, the next step is to establish connections using available standards simultaneously. Including here the end-to-end scope of the session layer, it is no longer necessary to operate a CM element in the network. In comparison with former approaches, the explicit CM entity on the UE also has a reduced functionality. The reason is that the session layer now takes over several CM-related tasks. Let us consider again that there is a rule base or policy base related to the simultaneous use of standards available on the UE. How this rule base or policy base is created or managed is not explained here. The most important task of the CM is now to monitor the changing conditions on the lower layers and to cross-check changes with the policy base. If there are consequences for an ongoing session or for sessions newly initiated, the CM has to indicate this to the session layer. The principle architecture of the UE in regard to the CM is shown in Figure 4.14.

As it can be seen, the CM has to implement three different interfaces, one to the session layer, one to the lower layers, and one to the policy base.

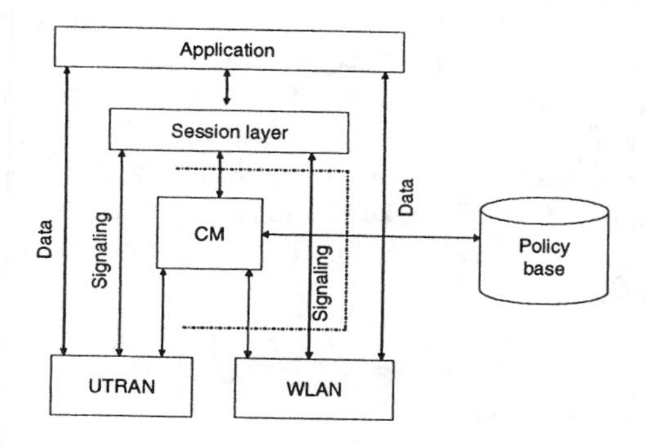

Figure 4.14 Terminal architecture including CM.

The next question to answer is how does the session layer establish connections, making use of the lower layer possibilities according to the rules defined in the policy base. The SIP INVITE request is used to initiate the session establishment process. It is delivered from the originating UE via several IMS instances to the called party UE. All the IMS entities on the signaling path have certain tasks including routing of signaling messages, checking of the subscription parameters, and charging for value-added services. If the INVITE request is passed to the called party, the SIP UA receives the message and inspects the payload. SDP can be used to define all the required parameters of the session.

It is important to understand that the SDP in the INVITE contains the originator sided endpoints of the media streams, which will be established. The IP addresses, the port numbers, and the codecs are required here. The called entity can try to renegotiate the parameters with a SIP response message. Several iterations can take place here and resources have to be reserved in the network for the final agreed media streams. In terms of QoS, a complex interaction between the two UA, as well as some other entities in the network can now take place. It is assumed here that after the completion of the signaling process by the ACK message, both entities know about the IP address and the port to which the packets of a certain media stream shall be sent. The codec, which shall be applied on that media, has also been agreed upon by both entities. That means that it is now possible to transfer media data. In our scenario, at first the UE has no WLAN coverage and therefore establishes both of the streams (audio and video), using the IP address obtained from the GPRS IP-CAN. For the example of Figure 4.15, the signaling goes via UTRAN.

During the session, the CM is now monitoring the lower layer conditions. If it recognizes that the WLAN can also be accessed and that the tunnel towards the PDG is established, it cross-checks the rules in the policy base. Let us assume again that the video channel shall be switched over to WLAN and the video codec shall be changed. That means that the SIP UA on the terminal is triggered by the

CM to change the media connection endpoint for the video channel from the GPRS IP-CAN IP address to the WLAN IP-CAN IP address. Additionally, the preference for the new codec may be obtained from the policy base. For such cases, the IMS, or more exactly SIP, provides the so-called re-INVITE procedure, which is used to change session parameters in an end-to-end manner.

Hereby, all intermediate signaling entities are informed about the changes and also all required QoS procedures can take place. Afterwards, the remote entity is informed about the changed IP address for the video connection and has also agreed on the new codec.

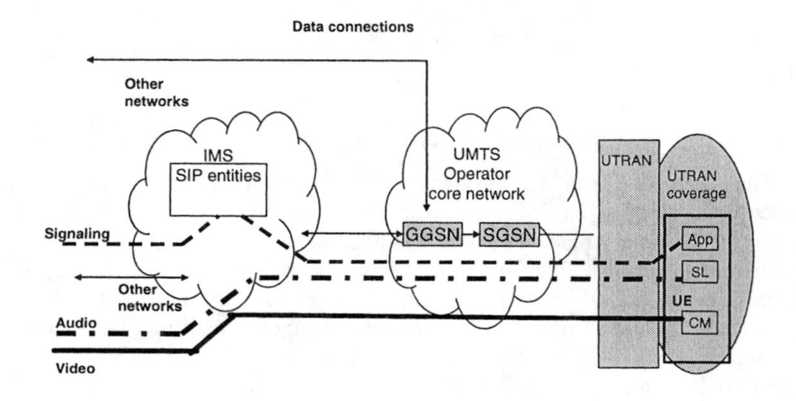

Figure 4.15 Example of a videoconference call without WLAN coverage.

A sequence diagram of the ongoing interaction is shown in Figure 4.16.

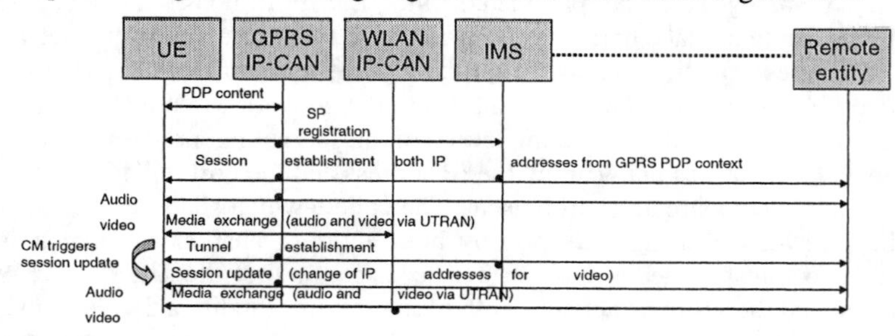

Figure 4.16 Interactions for session establishment and update.

Each of the interactions is a procedure consisting of several messages. This complexity is omitted in Figure 4.16. After the recognition of the CM, the WLAN becomes available too, and an end-to-end signaling interaction has to take place, including, additionally, several QoS-related message exchanges. This delay may not be favorable for several applications. This is, however, the procedure that the

3GPP has foreseen for such conditions of changing session parameters. The flexibility of the IMS and especially of SIP allows reusing the already defined session layer interactions and therefore the proposed solution provides the requested CM behavior. Figure 4.17 shows the resulting traffic flows.

Depending on the SLAs between the 3G operator and the WLAN operator, several slightly different scenarios are imaginable. Although it is not our focus to evaluate business relations or SLAs, these scenarios shall be outlined briefly.

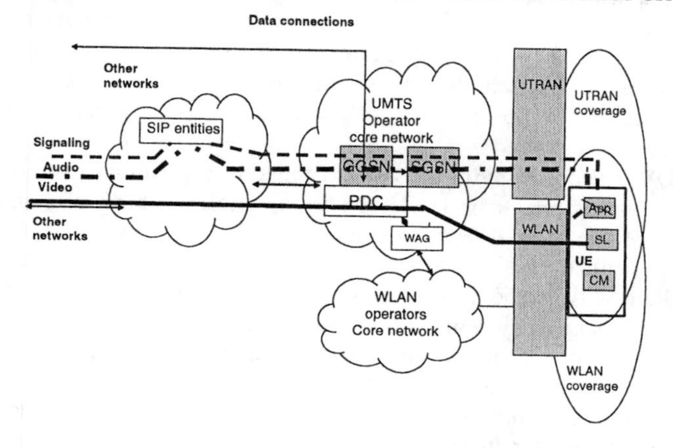

Figure 4.17 Videoconference session using WLAN and UMTS (video via 3G network).

If, for instance, the 3G operator does not mandate signal streams, which are not transferred via its network, the WLAN operator may also choose other connection points to other PDNs. Figure 4.18 shows such a scenario.

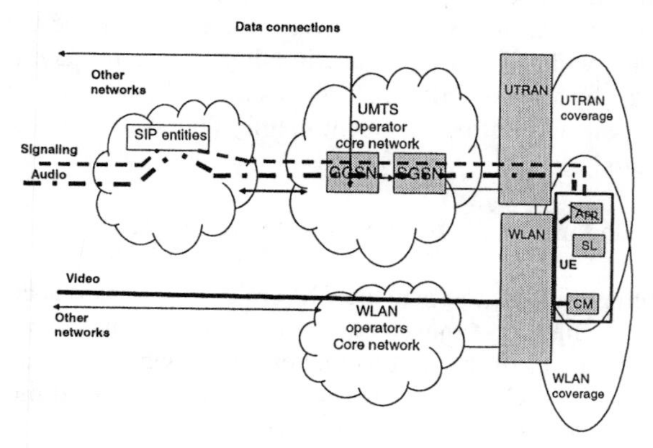

Figure 4.18 Videoconference session using WLAN and UMTS (video not via 3G network).

In this case, the signaling no longer may be able to take care of QoS issues for all the streams. In principle, such an interworking is therefore a contradiction of the 3G philosophy to establish sessions only in cases when QoS can be assured.

Another interesting question is whether or not the 3G network operators might agree on providing signaling effort in cases when all the streams are transferred via other access and core networks. In this case, SLAs have to be agreed upon, which define how the signaling effort can be charged. Exactly the same scenario, however, could also be seen from another perspective. If, for instance, the UTRAN of a 3G operator does not provide enough free capacity to satisfy users' service demands, it is imaginable that the 3G operator would ask the WLAN operator for additional bandwidth. In this case the direction of the business case has changed.

The DVD accompanying this book contains algorithms to allow switching between standards and their implementation. The switching of standards works well in the context of an FTP file download.

If the terminal is able to split the media streams onto different networks directly, there will not be a need for any additional functionality in the transport network. The charging systems will need some additional information if the media streams are to be combined into its original session and presented to the customer as one bill for the application rendered.

If the terminal is able to route the traffic directly onto different networks, only the charging system could be affected. The customer would like one bill for the application rendered.

If each of the media streams could be transported on a standard suited to support the characteristics of the media stream, the overall QoS would increase. Different networks would be used according to the transport requirements set by each media stream. The important part of the network is the access network and the choice of radio standard. In the core network and the backbone network there is normally high bandwidth. With QoS functionality to differentiate traffic types in the network (e.g., by use of DiffServ), the QoS would be supported. On the radio access part this would not be sufficient mainly due to the low bandwidth on certain radio standards.

4.4.1.4 CM Located in Mobile Terminal and the Radio Access Network

There are two different options to place the CM entity in the radio access network. Each investigated technology makes use of a base station or something comparable. The placement of the CM to this entity provides several difficulties, but is possible in principle. Cellular access networks are organized hierarchically. Therefore, the CM can also be placed on a RNC/BSC entity.

Figure 4.19 shows the architecture for placing the network sided CM entity to the BS.

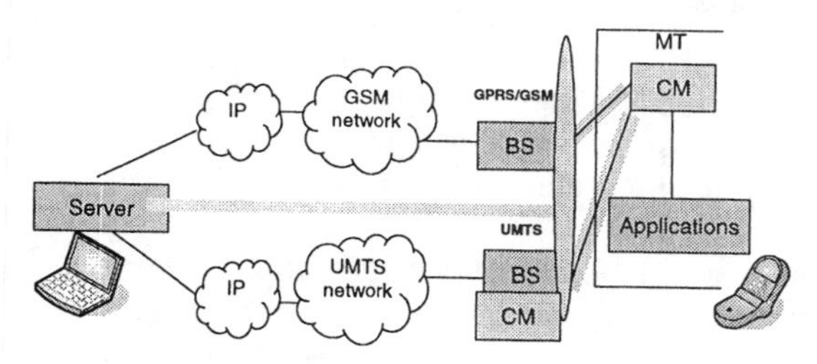

Figure 4.19 CM located in the BS and MT.

If the CM is located in the BS or AP, through the coordination between the two convergence managers, the MTs can access different standards simultaneously with the following optimization:

- Can optimize the radio access technology choice;
- Can optimize the network route (BS used);
- Can optimize the PHY modes (for all users).

The placement of the CM is a crucial issue for the architectural design of a simultaneous standard environment. With the decision to locate it on a multimode BS (a BS supporting several radio standards), there are several consequences and difficulties. To avoid these, the following methods can be applied.

The first option is to disallow the existence of all BSs that are not of the multimode type. Applied to an UMTS-WLAN interworking scenario, which seems to be one of the most important scenarios in the near future, we can identify a limitation in the way that only one operator can be involved in the scenario. Therefore, a cooperation of different business entities, as it will probably happen between UMTS and WLAN operators, is implicitly impossible in this approach. A second issue is the different coverage of the applied radio standards. Therefore, in the case of a simultaneous usage of UMTS and WLAN, either the UMTS cell areas have to overlap dramatically to provide full WLAN coverage or the WLAN areas cannot be designed continuously. A third important drawback is that in cases of handovers from one BS to another, the state of the CM has to be transferred too.

Therefore, a standardization effort for the interworking protocol has to be considered.

The second option is to allow the existence of single-mode BSs. The advantage compared with the previous option is that now the interaction of different operators can be supported too. Also, the coverage problem is solved because now the BSs of the different radio technologies can be located independently. However, to provide the required functionality, the CMs on the different BSs (for UMTS—the node B, and for WLAN—the access point) will probably have to interact intensively. In the cases of a multioperator scenario, this interaction has to be performed across the provider boundaries, which has two consequences. First, a lot of additional signaling traffic is generated between the distributed active CM entities, and, second, a lot of additional standardization effort has to be performed. Furthermore, another disadvantage can be identified as described below.

Let us consider a video streaming application, streaming video and audio from a server located somewhere in the network towards the terminal. The simultaneous use of standards enables the transport of the audio via the UMTS RAN and the more bandwidth consuming video via the WLAN.

The questions are now: Which entity in the network shall perform the split of the different streams? and Who will be responsible for that and who will charge for this effort?

Without an additional component this scenario cannot have a solution The only exception is the redirection of one media stream from one BS to the other, an action that will (to mention only the most obvious disadvantages) cause huge amounts of additional traffic as well as a different network delay for both streams. In the other direction the problem exists as well. If, for instance, the CM on the terminal decides to separate a stream over different standards, it is not clear which entity in the network will reassemble the different substreams.

Together with the standardization effort, these circumstances again promise a difficult migration scenario.

Convergence Functionality in the RNC/BSC

In the case when the session streams are split at the access network level, each session stream may also be split on a lower granularity level (e.g., at IP packet level), where the packets used for one media stream are split.

This functionality may take place in the BSC or RNC for GSM and UMTS, respectively. The common situation will be to split the media streams onto different RABs, which have different characteristics in terms of QoS, bandwidth, and so forth. UMTS already does this RAB selection related to different traffic classes. The CM would be an integral part of the BSC/RNC. The CM on a media stream level would function similarly to the RAB selection functionality in the UMTS RNC.

The described solution has several advantages and drawbacks. An advantage is the possibility for a combined WLAN-UTRAN access network in a very tight coupled scenario.

Spreading and reassembling of traffic can be kept on a low layer. Therefore, packet-based splitting or even byte or bit-based splitting seems to be possible. Furthermore, the gathering of channel condition values is simplified and therefore dynamic operator policies can be satisfied. Important drawbacks might be that, similar to the placement of the CM at the BS/node-B, no loose coupling scenario can be supported. The cellular network operator will probably also be the operator of the WLAN access points.

Another important drawback is the enormous standardization effort for this kind of solution.

4.4.1.5 CM Located in the Mobile Terminal and the Core/Backbone Network

In the core/backbone network there are several possible locations for the convergence manager. These are shown in Figures 4.20 to 4.23.

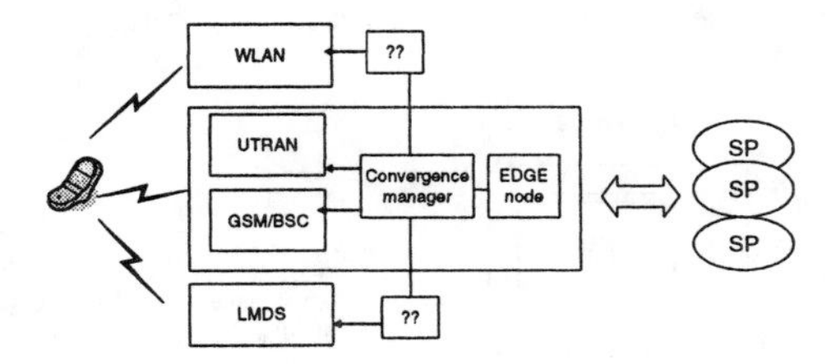

Figure 4.20 Convergence manager inside the core network a of GSM/UMTS network.

The CM in Figure 4.23 can therefore handle sessions and split them to the best network regardless of the operator owning the network. In this situation the user and terminal must be enabled to receive (and send) data flows from/to several operators simultaneously. The user must be a subscriber of several operators and be able to use their networks as best suited for all involved parties (i.e., the customer, the operator, the service providers). The service provider sends multimedia content (bundled media streams) to the CM and the CM directs the media streams or individual sessions according to predefined rules.

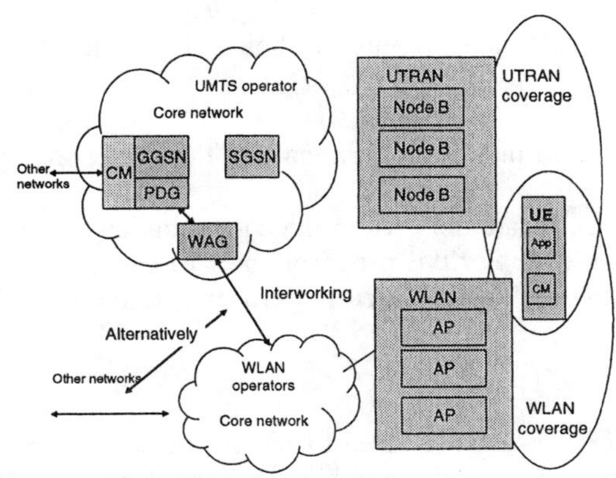

Figure 4.21 CM placement for UMTS-WLAN interworking.

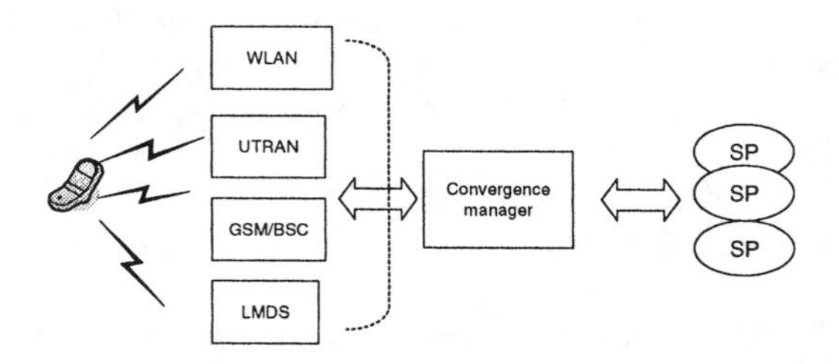

Figure 4.22 CM placed outside the edge nodes of various radio networks.

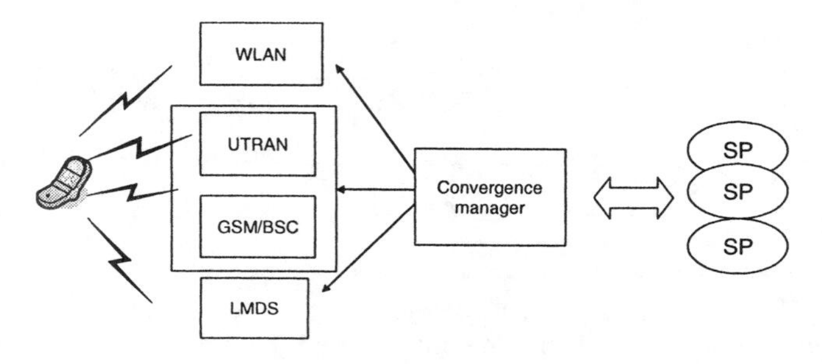

Figure 4.23 CM placed outside the border of different operators domain.

The rules may be linked to the subscription and take into account the personal preferences of the customers, the preferences of the operators regarding network efficiency and optimal usage, load balancing, and the interoperator agreements for usage in selected areas. Strict service level agreements need to be worked out to support each operator's interests and the choice of operator that a customer might make.

4.4.1.6 CM Located in Mobile Terminal and the Corresponding Server

It is now possible to locate the CM on the corresponding server. The architecture is shown in Figure 4.24. For this proposal, the CM can optimize the radio access technology choice (for all users) but the approach is rather inflexible.

4.4.2 Summary

The CM functionality was introduced to realize the simultaneous use of multiple standards, so that it is possible to achieve potential benefits for both users and network operators (see Figure 4.25). Two proposals are considered for further study here.

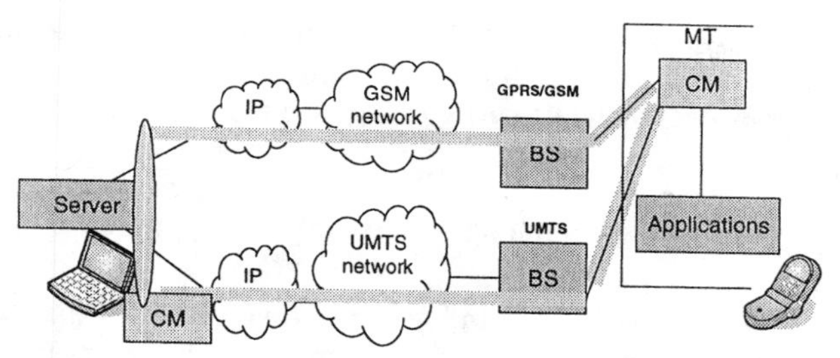

Figure 4.24 CM is located at server and MT.

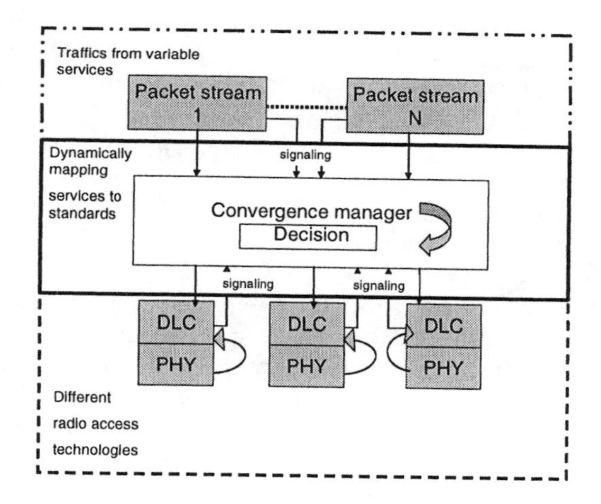

Figure 4.25 Function of the CM in the network layer.

In the first one, the CM is assumed to be a functionality located in the base stations of different standards, with access to information not only from different users and services but also from the network and radio air interface side. It works with some specified aims, applying different scheduling policies, and tries to find an optimized solution meeting the different end user QoS requirements using multiple mobile communication standards. In other words, the function of CM is to dynamically optimize the mapping from services to standards for the instantaneous system situation.

Alternatively, the CM may be assumed to be located deeper in the network, for example in the RNC, so that the mapping of services to standards can only be changed slowly relative to changes in the radio channel. In this case, as a simplification, the mapping of services to standards is considered to be static for a given user (or at least static for a substantial period). In either case the CM may be considered to be located in the network layer—referring to the ISO/OSI model by the International Organization for Standardization. It receives the information from both the upper layer, given as traffic signaling, and possibly the lower layers in the form of radio access technology signaling (see Figure 4.25).

4.5 FLEXIBLE WIRELESS INDOOR COMMUNICATIONS

802.11 and Wi-Fi standards have developed to solve many of the problems of security that worried corporate customers, opening the market for greater expansion [50]. Integration of WLAN standards in 3GPP is also progressing, creating a framework for integrating the technologies from the fixed and mobile domains. However, WLAN radios still have considerable problems with high power consumption, and there is concern over the limited number of channels available on 2.4-GHz bands. The uptake of 5 GHz, 802.11a, or the HIPERLAN technology have been very low, despite the U.S. government's opening up of more spectrum for this use. 802.11b/g products still provide the functionality that almost all users need, with greater range. Economies of scale in products have also made this equipment very cheap.

Another technology is also showing potential in the fixed-mobile voice device market, and that is Bluetooth.

One of the key economic costs in the wireless deployment model has traditionally been the installation of the base station and the provision of fixed backhaul. The development of cable, xDSL, and the wireless broadband systems, and their rapid uptake in developed countries, has made a turn upside down. Now it is very cheap to install a broadband wireless hotspot in most urban and suburban areas, and many people are doing so, with reports of up to 30% of broadband customers in the United States adding local Wi-Fi networking. This has changed the picture for home wireless, as alternative wireless media distribution technologies (such as those based on UWB) start to compete in or complement the network home market.

Bringing broadband connections to the street and house level opens up new possibilities for extensive wireless coverage in populated areas, which can offer competition for BWA. An example is a firm looking at exploiting this for providing public broadband connections, but using the broadband fixed links of subscribers to attach publicly accessible WLAN hotspots that cover the public space outside the location. MESH systems using standardized WLAN technologies are also attracting attention as a way of providing public wireless access, even over whole cities, as is the plan in Philadelphia. Broadband into the home has also opened the doors for VoIP as a way of delivering telephony services, and the now-famous "triple play" of television, Internet, and telephony via one cable, a potential saviour for local loop owners facing consumer desertions to the mobile phone.

All recent wireless technology developments, although pulled by different market requirements and applications (e.g., WLANs for portable connectivity in office IT enterprises, WPANs for cable replacement and personal sphere interconnectivity), have been characterized by a common quest for higher capacity density and link speed on the one hand, as well as greater reliability and scalability in the presence of varying channel conditions and/or traffic loading and QoS system requirements on the other. Guided by strong evidence of these common persistent trends, the IST project WIND-FLEX [57] searched for solutions to fulfill the above requirements.

Due to the combined adoption of a high frequency (17 GHz) for the radio carrier (see Chapter 1 of this book), flexible radio interface schemes, plus maximization of modem functions/controls allocation in the digital domain, the specified WIND-FLEX radio interface provides dynamically adaptive reliable and efficient communications at a very high link speed, namely, up to four bundle-able channels, each providing a maximum speed of 215-Mbps gross data rate in the air, which translates to slightly more than 100-Mbps payload at the IP layer. The maximization of radio interface functions in the digital domain allows for the possibility to automatically and dynamically control whole radio interface and react at all layers to guarantee proper communication conditions, despite varying propagation, traffic loading, and networking topology conditions. To implement such features, modulation, coding and access schemes, and protocols have been designed in order to allow highly granular flexible variations of their parameters, instantiated by adaptive algorithms and reconfigurable subsystems.

The project WIND-FLEX provided a very attractive and innovative solution for the next generation of WLAN/WPAN developments. The general proposed architecture envisions the presence of multiple electronic devices in an indoor environment (i.e., office, house) which, when provided with a WIND-FLEX radio interface and colocated in the same short-range coverage area, can communicate at very high speed and, in addition, has the attractive possibility of automatically forming a meshed network. Such a self-configuring networking characteristic, which considers also the run-time dynamic assignment of the roles of network elements among devices, represents an additional value when requiring interconnectivity in an infrastructureless environment (e.g., WPAN at home) while still remaining compatible with infrastructure-based environments such as those typical of WLAN deployment in enterprises. A shorter-term opportunity for WIND-FLEX technology can be foreseen in synergy with existing WLAN technologies, by adopting WIND-FLEX as a wireless backbone for standard wireless LANs. Such a combined WIND-FLEX/WLAN proposition will enable full wireless network coverage of indoor environments, thus providing a much easier, faster-to-deploy, and flexible (distributing capacity according to needs) design and setup solution, compared to the current hybrid ethernet/WLAN solution.

So far, no specific actions have taken place versus regulatory organizations to discuss the use of the 17-GHz band and/or closer ranges (say +/−3 GHz) in which

WIND-FLEX technology could be applied without any change in its design. At any rate, the choice to address the 17.1–17.3-GHz frequency band was in line with existing ETSI recommendations (CEPT/T/R 22-06) [58] assigning them for very high-speed wireless LAN use on a nonprotected and noninterference basis. In addition, CEPT/ERC/REC 70-03 [59] recommended, among other options, the use of 17.1–17.3 GHz for short-range wireless connectivity [60].

The WIND-FLEX technology outperforms both current state-of-the-art WPAN and WLAN technologies working in the 2.4-GHz (i.e., Bluetooth and 802.11b) and the 5-GHz range (i.e., 802.11a), in terms of maximum bit rate (e.g., 802.11a can achieve in its higher profile a maximum air interface bit rate of 72 Mbps) capacity offering due to the easier channel reuse and the smaller cell sizes, spectral efficiency (about 20% better than 802.11a at PHY layer). Moreover, the proposed solution allows for joint and dynamic optimization of its RF, baseband, and MAC/DLC parameters according to time-varying channel conditions, traffic loading, and QoS requirements, in order to guarantee required performances at minimum power consumption, while also remaining backward compatible with other OFDM-based technologies (i.e., 802.11a), thus leveraging the huge OFDM know-how developed during recent years in the whole wireless industry.

OFDM [6] is a kind of multicarrier modulation (MCM) that has been standardized recently for several communication systems in which multipath propagation or interference constitutes a problem, due to its good behavior under this type of impairment.

Because of these properties, OFDM techniques have created an interest in the research community, and many papers can be found in the literature regarding implementation of this kind of modulation during recent years. Some of the most important issues have been found to be:

- Channel estimation and equalization;
- Time and frequency synchronization;
- Phase noise effects;
- Nonlinear effects.

It can be inferred that OFDM and UWB offer similar performances in terms of maximum capacity and spatial efficiency, although, because of their intrinsic nature, they provide very different figures in terms of power efficiency (much higher for UWB) and spectral efficiency (much higher for WIND-FLEX). Given the comparable global system performances, the WIND-FLEX technology offers the added values of keeping backward compatibility with all OFDM systems, of not requiring significant design challenges on the RF design (as in UWB), and of not producing any harmful interferences on other existing technologies, due to the limited portion of unoccupied spectrum used (see Table 4.3).

Table 4.3
WIND-FLEX Benchmarking with UWB

System	Units	WIND-FLEX	UWB
Modulation scheme		OFDM	TM-PPM
Frequency carrier		17 GHz	No carrier
Maximum capacity (PMD-SAP)	Mbps	102.00	110.00
Transmit output power (EIRP)	mW	1.0^{exp+01}	3.2^{exp-02}
Transmit output power (MAX)	mW	2.0^{exp+02}	1.6^{exp-01}
Bandwidth	MHz	50.00	7,500.00
Allocable band	MHz	200.00	7,500.00
Range	m	5.00	10.00
Area/user (office scenario)	m^2	12.00	12.00
Power efficiency (capacity/mW)	Mbps/mW	10.20	3,478.51
Bandwidth efficiency (capacity/Hz)	bps/Hz	2.040	0.015
Overlapping piconets	#	4.00	4.00
Aggregate BW efficiency (capacity/Hz)	Bps/Hz	2.040	0.059
Spatial efficiency (capacity/m^2)	Mbps/m^2	5.1975	1.4013

4.5.1 WIND-FLEX System Characteristics Overview

The most important WIND-FLEX parameters are given in Table 4.4.

4.5.1.1 Baseband Architecture

The WIND-FLEX baseband architecture (see Figure 4.26) is based on turbo channel coding, an OFDM modulation scheme, and a supervisor (SPV) unit for real-time system optimization. The available subcarrier modulation schemes are BPSK, QPSK, 16-QAM, and 64-QAM. The constellation is adaptively chosen among the various schemes and a variable number of subcarriers can also be adaptively switched off. Moreover, as a form of reconfigurability in order to save processing power, a scalable (I)FFT structure, likewise a change from turbo coding to convolutional coding or no coding at all (during favorable QoS and channel condition), can be adopted.

Table 4.4

Most Important WIND-FLEX System Parameters

System Parameters	RF Parameters	Baseband Parameters	Networking Parameters
Coverage range: LOS:≤100m (QPSK);≤50m (16-QAM); ≤30m (64-QAM) NLOS:≤10m (QPSK); ≤6m (16-QAM); ≤4m (64-QAM)	Frequency 17.1–17.3 GHz	Modulation scheme: OFDM	Access scheme: FDM/TDMA
Radio interface optimization strategy:	Channel bandwidth: 50 MHz	Modulation adaptivity: per frame, per user	Duplexing scheme: TDD
	Number of channels: 4	Subcarrier modulation schemes: BPSK, QPSK, 16-QAM, 64-QAM	Symbol slots/frame: 178
	Channels center frequencies: 17.125; 17.175; 17.225; 17.272	Nominal OFDM carriers number: 128	Frame length: 491.28 μs
	Upper guard frequency band: 5,469 MHz (14 suppressed carriers)	Unused carriers: 28 (27 + DC)	Preamble structure: 3 symbols
	Subcarrier spacing: 390.625 KHz	Active OFDM carriers numbers: 100	Network entities: master, slaves, bridge, gateway
	Max peak EIRP: 27 dBm Max average EIRP: 27 dBm Receiver sensitivity: −85 dBm	Pilot carriers: 0 Useful OFDM carriers numbers: 100–N OFDM useful symbol length: 2.56 μs Guard interval: 200 ns	Network entity allocation: dynamic
	Dynamic range: 60 dB Noise figure: 6 dB	Coding scheme: Turbo code	
		Turbo code scheme: parallel convolutional punctured	
		Coding rates: 1/2, 2/3, 3/4	

The coding scheme of the WIND-FLEX modem is a parallel convolutional turbo code. The available code rates are 1/2, 2/3, and 3/4 and the block length is adaptive and dependent on the triplet: code rate, constellation size, and number of ON subcarriers.

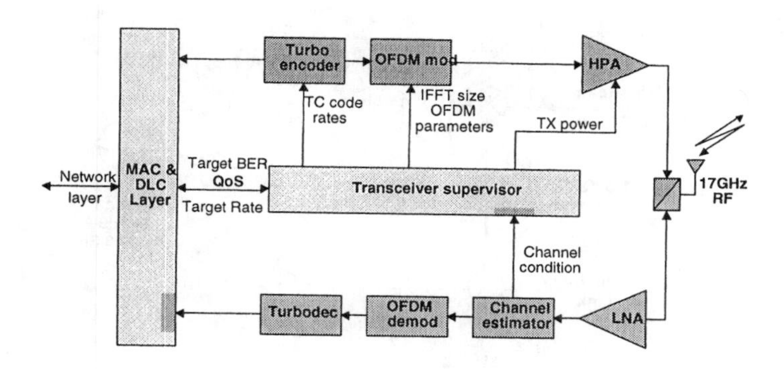

Figure 4.26 The WIND-FLEX baseband architecture.

The WIND-FLEX SPV's task is to solve the problem of fitting the QoS requirements from the MAC layer, given the current channel conditions, with the minimum transmitted/processing power. To get this result, the SPV finds the best configuration of the transmission system and sets the best values for the transmission parameters. The optimization procedure relies on the channel reciprocity hypothesis, satisfied because of the TDMA/TDD nature of the frame and of the adoption of single antenna transceivers. Channel reciprocity allows open loop adaptivity.

4.5.1.2 RF/IF Front-End Architecture

OFDM modulation is a demanding technique in terms of phase noise and frequency stability particularly for higher order constellations (i.e., 64-QAM). The RF/IF architectures (Figures 4.27 and 4.28) can be designed in order to keep a low cost design while achieving low phase noise and high frequency stability according to specifications.

As far as the filters are concerned, given the fact that broad channel bandwidth (i.e., 50 MHz) does not require the use of expensive components such as SAW or ceramic filters, low-cost filters based on microstrip or discrete devices can be adopted for the design. The use of a free running dielectric oscillator (13.950 LO in Figure 4.28) allows for low-phase noise and low costs.

The required frequency stability in the working temperatures range can be achieved by a coarse tuning obtained by an IF oscillator controlled by the baseband. Fine frequency tuning can be obtained within the baseband. The RF unit can be manufactured at very low costs by relying on advanced SiGe technology processes.

Because of the high peak to average power ratio nature of OFDM signals (about 13 dB), particular care must be given to the characteristics of the power amplifier

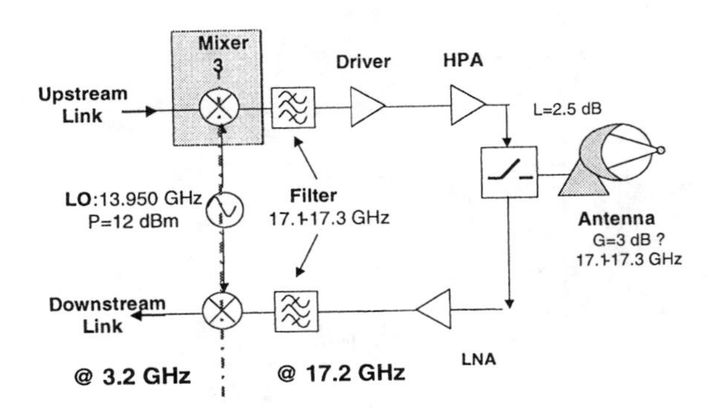

Figure 4.27 WIND-FLEX RF architecture.

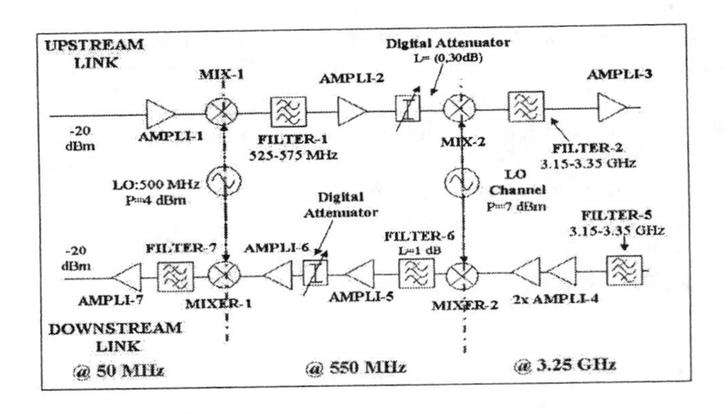

Figure 4.28 WIND-FLEX IF architecture.

In order to guarantee the presence of a reciprocal channel, allowing the transmitter to perform adaptive controls on the basis of channel information estimated by the colocated receiver, the design requires the presence of a single antenna, switched between transmitting and receiving phases. Given the mobile and ad hoc network nature of the architecture, an omnidirectional antenna is compulsory.

For this particular design, printed technology antennas provide a number of advantages: they are lightweight, low cost, and relatively easy to manufacture. Among the different possibilities the line-fed-slot-coupled microstrip patches are considered suitable, as this technique makes relatively easier the impedance matching and facilitates fed network design for array configurations. Restrictive specifications for bandwidth in terms of return losses have to be given in order to provide the required bandwidths.

For high volume production, the use of a multilayer PCB structure with some filters and the MMICs mounted in the top layer and other filters in the intermediate layers is envisaged. This structure would minimize the size of the PCB and avoid the radiation of the filters. Four MMICs are envisaged for the chip-set with the size and technology indicated in Table 4.5.

Table 4.5

Chip-Set for High Volume Production of the WIND-FLEX Modem

Unit	Ranges	Estimated Size	Technology
IF transmitter	50 MHz to 3.25 GHz	6 mm × 5 mm	Si-Ge
IF receiver	3.25 GHz to 50 MHz	6 mm × 5 mm	Si-Ge
Transmitter RF	3.25GHz to 17 GHz	2 mm × 2 mm	GaAs/ Si-Ge
Receiver RF	17 GHz to 3.25 GHz	2 mm × 2 mm	GaAs/ Si-Ge

A dielectric substrate of low losses should be used for the PCB. In order to reduce the size of the PCB, a substrate of high dielectric constant, such as a ceramic of dielectric constant around 6 should be used. With this kind of substrate the size of the PCB for both units RF and IF would be around 5 cm × 5 cm. Four layers are envisaged for the PCB with this substrate. The packages must be compatible with a high volume manufacturing process such as SMD. There are no problems for the IF chips for which plastic packages could be used, but these kind of packages are not available for 17 GHz. Wire bonding must be avoided because it requires very sophisticated facilities for volume production. Flip-chip technology may be the most adequate. The entire RF unit must have a metallic enclosure manufactured by an injection molding process.

4.5.1.3 Network Architecture

The WIND-FLEX network is composed of a set of clusters, which are a set of devices such that every pair can establish a direct connection; a cluster is then a full-meshed subnetwork. Every cluster consists of a master device and one or more (up to 63) slave devices, which must perform an appropriate association procedure to the cluster. The master carries out the synchronization and coordination of the cluster, and it is in charge of allocating the shared radio resources. There is no hardware or software difference among the master and slave devices, and each device can become a master. The master of a cluster is chosen among all the devices associated to the cluster: a master election procedure ensures that the master is the device having the best position within the cluster, meaning that its physical links with all the other devices associated to the cluster are characterized by the best quality. A proper algorithm guarantees that the master election is repeated, mainly due to device movement, the network topology significantly changes. In case a new master is elected, the master role is handed over in a seamless fashion from the old master to the new one [61, 62]. The system is able to reconfigure itself to react to changes in the network topology. The WIND-FLEX protocol stack is fully based on IP as depicted in Figure 4.29.

MAC Layer

At MAC layer the bit transmission is organized in a TDMA way. The time axis is divided into frames that, in turn, are divided into 178 time slots, characterized by:

- A fixed time duration (about 2.8 ms), which entails a frame duration equal to about 0.5 ms;
- A variable number of transported bits; this number depends on the adopted radio transmission mode (combination of modulation and channel coding schemes) and is communicated from the PHY to the MAC layer via a proper interface.

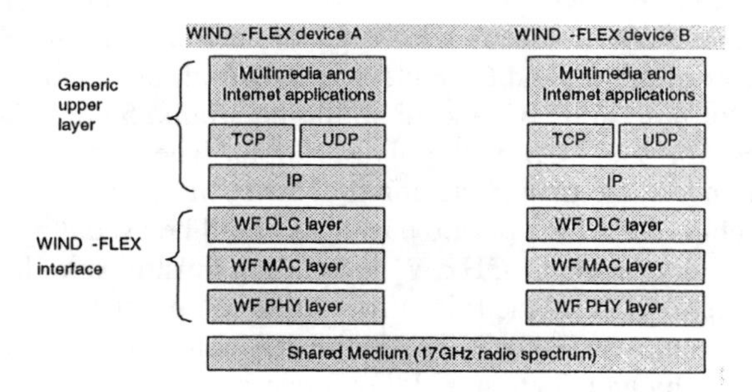

Figure 4.29 WIND-FLEX protocol stack.

In order to reduce the battery consumption, the data transmission speed is reduced whenever the traffic situation allows the delivery of the packets with an acceptable end-to-end delay and jitter delay using a lower transmission bit rate. On the contrary, when traffic is heavy and the buffer length is large, the system attempts to use a bigger constellation scheme (up to 64-QAM), provided that the channel status is so good that the required BER can be accomplished

Resource assignment is performed by the master using a novel greedy-type scheduler algorithm, where the useful slots are assigned frame-by-frame to the different connections of all the devices in the cluster. The frame information as described in the previous bullets is also the basis for the resource assignment.

4.6 ADVANCED MODULATION AND SIGNAL PROCESSING SCHEMES

4.6.1 OFDM and OFDMA

In this section, OFDM is examined and the state of the art regarding the physical layer implementation of OFDM as attempted in the IST project WIND-FLEX is summarized.

A new application of OFDM, [orthogonal frequency division multiple access (OFDMA)] is explored and the maintenance of the requirements and techniques available for OFDM are discussed [63].

4.6.1.1 OFDM and OFDMA Principles

Multicarrier modulations have been used for years under the names orthogonally multiplexed QAM [64] or multitone QAM [65]. Now they are being extensively used [66] for several wireless and wireline applications, generally under the names of OFDM and discrete multitone (DMT).

OFDM has been standardized for digital audio broadcasting (DAB) in terrestrial and satellite channels [67], terrestrial broadcasting of DVB-T [68], and high-speed WLANs (HIPERLAN) [69]. DMT is the choice for asymmetric digital subscriber lines [ADSL, very high bit rate DSL (VDSL)] [70].

The reasons for this new interest in MCM are their ease of implementation using *fast Fourier transform algorithms* [71] and their ability to cope with transmission impairments such as multipath. In OFDM (and DMT) the total available bandwidth is divided to N subchannels. Each of them is modulated with one symbol from a digital constellation (generally M-PSK or M-QAM), so that a set of N symbols are sent modulating N subcarriers and they are all multiplexed in frequency. The signals of the different subchannels are allowed to overlap, and the fact that they are kept orthogonal facilitates their separation in the receiver.

Let T be the duration of the OFDM symbol and B the total bandwidth. The separation between subchannels is chosen so that $B = N/T$. The complex envelope of the OFDM signal sampled with sampling frequency equal to B can be obtained by an inverse discrete Fourier transform (IDFT) of the complex input symbol sequence s_{ik} (where i is the time index and k is the subcarrier index) as [71]:

$$x(n) = \sum_{i=-\infty}^{\infty} \mathrm{DFT}^{-1}\{s_{ik}\}\Pi_N(n - iN) \quad (4.18)$$

Here $\Pi_N(n)$ represents a rectangular pulse with N samples of duration. The frequencies of the N subchannels are:

$$f_k = f_0 + k / T, \quad k = 0, 1, ..., N-1 \tag{4.19}$$

In the same way, the demultiplexing process can be accomplished by means of a discrete Fourier transform in the receiver [71].

Additionally, a cyclic prefix is introduced in the OFDM signal prior to transmission as a means of preserving orthogonality in multipath channels and avoiding intersymbol interference (ISI) and intercarrier interference (ICI). It consists of a repetition of a certain number of the last samples of the useful symbol that are added at its beginning. These goals are achieved as long as the length of the cyclic prefix is equal to or greater than the channel impulse response duration [67]. Besides, some of the subcarriers (frequency guards) may be kept unmodulated so as to avoid causing and receiving adjacent channel interference (ACI) and some others may be used as reference symbols for channel estimation and synchronization (pilots). Figure 4.30 shows the situation of frequency guards and pilots in the OFDM spectrum. The most important parameters are summarized as follows:

- The modulation used in each of the subchannels;
- The symbol length;
- Frequency guards and length of the cyclic prefix.

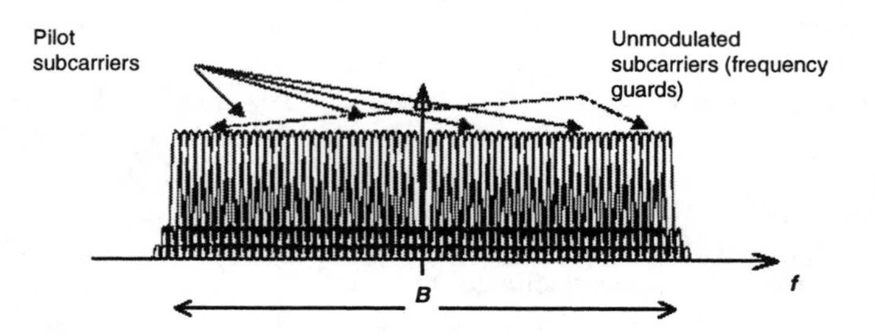

Figure 4.30 OFDM spectrum.

The modulation used for the different subchannels constitutes the main difference between OFDM and DMT. While in OFDM all subcarriers are modulated with the same modulation type and order (i.e., they carry the same number of bits), in DMT the number of bits in each subchannel is chosen so as to adapt the bit rate to the channel characteristics. The proper number of bits is determined by the so-called bit loading algorithms [72]. These algorithms need an estimation of the signal-to-noise ratio, and for practical reasons, they have traditionally been reserved for wire-line communications in which the channel is more stationary and easier to estimate than in wireless systems. However, bit loading has been recently proposed for OFDM in fixed wireless [73], WLAN [74], and even for the downlink of cellular systems [75], so that differences between DMT and OFDM are becoming just historical.

Several possibilities arise when an OFDM signal must be shared between several users in a multiuser environment. Given the inherent frequency-division nature of the signal, an FDMA approach is an immediate choice. When an OFDM signal is shared in an FDMA fashion—that is, different users transmit in different subcarriers—the OFDM-based multiple access scheme is commonly named OFDMA. This concept is illustrated in Figure 4.31.

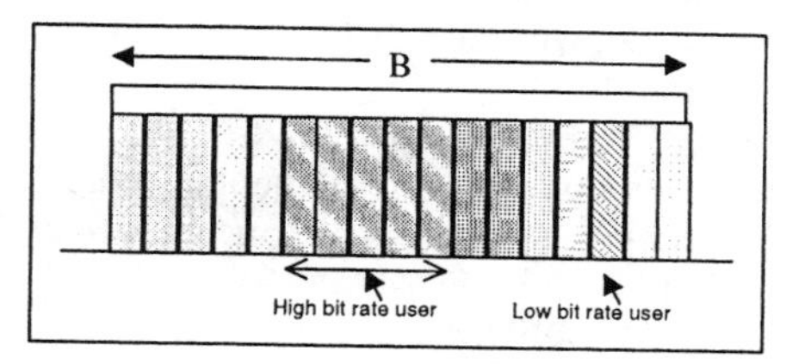

Figure 4.31 Concept of OFDMA.

In its simplest form, each user sharing an OFDMA channel will transmit a single-carrier signal, and the signals from different users will be separated with a carrier spacing of $1/T$, T being the OFDM symbol time. In this way, the composite signal with the contributions of N users will be an OFDM signal that may be demodulated with a discrete Fourier transform, which is much simpler than a set of N traditional FDMA demodulators.

This idea is easy to generalize to the case in which users may transmit a different number of subcarriers depending on their needs. Since in OFDM all subcarriers are loaded with the same number of bits, the first OFDMA proposals that appeared in the literature maintained the same idea and exploited the good protection against multipath and narrowband interference of this kind of signals[76, 77]. This kind of OFDMA also offers the advantage of flexibility in bit rate allocation for the different users. Indeed, since any number of subchannels can be allocated between 1 and N, different users may be granted different bit rate requirements with high granularity. For example, if the available bandwidth is 20 MHz and modulation is 16-QAM, a terminal using one subchannel out of 256 will transmit 312 Kbps whereas using all 256 subchannels will lead to 80 Mbps (frequency guards and cyclic prefix are not considered for the sake of simplicity in this example).

Obviously, allowing the terminals to choose the modulation index will increase the flexibility considerably. Moreover, because bit loading is a way to achieve capacity in a single-user spectrally shaped Gaussian channel (water-filling distribution) [78], bit loading algorithms designed for multiuser environments may achieve capacity. With this goal in mind, some bit loading algorithms using OFDMA have appeared recently in the literature for multiuser environments [73–75].

OFDM may be combined with other multiple access schemes such as TDMA [79], SDMA [80], or CDMA [81].

In OFDM-TDMA, the whole OFDM symbol is time-shared between several users. This suggests another strategy to obtain more flexible bit rates: the use of a hybrid multiple access scheme OFDMA-TDMA. In this H-OFDMA multiple access method, the OFDM signal is frequency-shared and time-shared so that the bit rate granularity is higher than with a pure OFDMA. This increased granularity may be necessary, for instance, in high-speed high-flexibility WLAN environments in which the number of subcarriers needed to obtain a high ratio between the maximum and minimum bit rates would be too high for low-cost terminal implementation [82].

When combining OFDMA with a TDMA frame of L time slots, a given terminal may transmit as little as one subchannel out of N in one time slot out of L, so that the ratio between the maximum and minimum bit rates increases by a factor of L with respect to pure OFDMA.

OFDMA or multiuser OFDM has been proposed for several applications in the literature. It constitutes one of the most feasible approaches to maximize the number of terminals and the aggregate data rate, hence filling the gap of medium to high data rate communications in WPAN environments. However, it has not yet been implemented, mainly due to difficulties that arise in such systems, some of them common to single user OFDM implementations and some others due to the multiuser nature of OFDMA.

Complete details about the PHY layer requirements of OFDM and OFDMA can be found in [6, 63, 83].

4.6.2 Analysis and Simulation of Physical Layer Requirements

The goal of the present section is to describe the signal and physical layer design carried out for an H-OFDM system considering channel estimation, phase noise sensitivity, and scalability issues. The performance of the different techniques that have been selected and alternatives for resource assignment and RF design are also presented. A hybrid OFDMA-TDMA scheme, here called a *hybrid OFDM* system, in which users are not only assigned different subchannels (OFDMA) but also different time slots (TDMA), was devised in order to comply with the requirements of the PACWOMAN MDR/HDR devices within the scope of the project [49].

OFDMA alone is not flexible enough to provide the bit rate scalability required in this project.

The goal of the present section is to describe the signal and physical layer design carried out for an H-OFDM system in order to comply with the requirements of PACWOMAN MDR/HDR devices. This issue was discussed a little in Chapter 3. However, here, the performance of the different techniques that have been selected and RF design are also presented.

4.6.2.1 H-OFDM Parameters

Since H-OFDM is a combination of OFDMA and TDMA, there are two types of parameters that must be configured in the system: those related to the OFDM signal and those related to the TDMA frame.

The main OFDM-related parameters are:

- The number of subcarriers: N (frequency bins);
- The OFDM symbol length: T (seconds);
- Number of frequency guards: N_{fg} (frequency bins);
- Number of pilot subcarriers: N_p (frequency bins);
- Number of useful frequencies: $N_u = N - (N_{fg} + N_p)$ (frequency bins);
- Length of the cyclic prefix: N_{cp} (samples) or T_{cp} (seconds).

The number of bits per symbol that are conveyed in every subcarrier is generally constant in OFDM. We will denote this number for a given terminal as r. The total number of bits per symbol for that terminal using s subcarriers will be

$$R = \sum_{k=1}^{s} r_k = s \cdot r \tag{4.20}$$

Allowing the terminals to choose the modulation index will increase the flexibility considerably. Moreover, bit loading is a way to achieve capacity in a single-user spectrally shaped Gaussian channel (the well-known water-filling distribution) and it is interesting to extend it to multiple user scenarios.

Channel coding (with interleaving) is necessary for OFDM in order to take advantage of the diversity inherent to frequency-selective channels. In the H-OFDM system, convolutional coding will be used. Different coding rates (R_c) will be selected depending on the target error probability.

The TDMA frame has been configured so that it provides a division in slots (constituted by preambles and a number N_{OFDM} of OFDM symbols) and a higher level organization (i.e., frame, super-frame). Figure 4.32 provides an illustration of this structure.

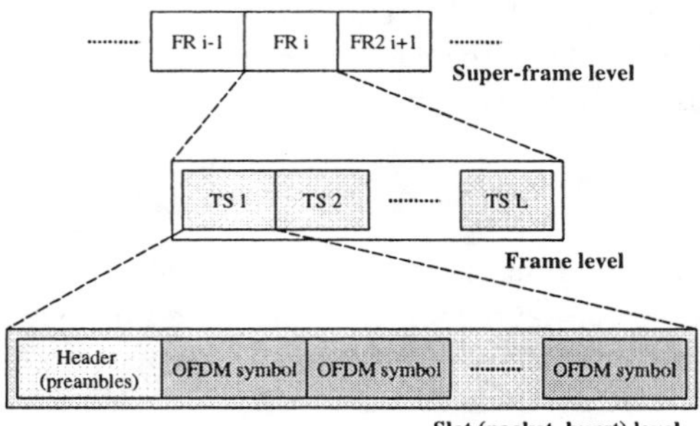

Figure 4.32 Illustration of TDMA frame.

We will denote as L the number of slots that constitute a frame. A given terminal will use a number l of slots (between 0 and L). Since H-OFDM is a combination of OFDMA and TDMA, three approaches are possible for the frame design, giving different granularity levels:

- All subcarriers are assigned to a given communication during a certain slot: OFDM-TDMA. This means that $s = N_u$ and l is assigned depending on bit rate.
- Some subcarriers are assigned to a given communication for a whole frame: OFDMA-TDM. This means that $l = L$ and s is assigned depending on bit rate.
- Some subcarriers are assigned to a given communication during a certain slot: OFDMA-TDMA. This means that s and l are assigned depending on bit rate.

In the following sections we define the performance requirements that the H-OFDM system is targeting and select the values of the parameters that are able to provide this performance.

4.6.2.2 Performance Requirements

H-OFDM is targeting medium and high data rate WPAN communications that are characterized by short-range ad hoc connectivity, generally limited to a 10 to 20m radius. Also, low power consumption and small size and cost are required.

Table 4.6 summarizes the main requirements that are envisioned for low, medium, and high data rate devices in the present project. More details can be found in [49].

Table 4.6

Target Characteristics for Physical Layer Design

	LDR	MDR	HDR
Data rate	10 bps – 10 Kbps	10 Kbps – 1 Mbps	> 1Mbps
Required BER	10^{-3}–10^{-7}	10^{-3}–10^{-7}	10^{-3}–10^{-7}
Maximum distance	1m – 5m	< 20m	< 10m

4.6.2.3 Parameter Selection

In order to make a selection of the optimum values of the H-OFDM signal parameters, we must consider channel estimation, phase noise requirements and scalability. Some of these effects have not been analyzed in the literature taking into account the multiuser nature of OFDMA.

4.6.2.4 Channel Estimation Considerations

Although a hexagonal pilot pattern provides better performance than rectangular and triangular pilot patterns [84], taking into account the packet-based nature of H-OFDM-WPAN communications, a full-pilot preamble providing channel estimation for the whole packet seems a most appropriate solution (this is also the choice for IEEE 802.11a [85]). The length of this packet must be short enough so that the estimation is valid for the OFDM symbols contained in it (i.e., the packet must be shorter than the channel coherence time).

The time-varying nature of the channel is usually characterized by the coherence time (T_C), which is the time duration over which the time-domain signal may be considered correlated (i.e., the time over which the channel does not experiment a significant variation). It can be expressed as [86]

$$T_C \approx \frac{1}{f_m} \text{ [secs]} \tag{4.21}$$

$$f_m = \frac{v}{\lambda} \text{ [Hz]} \tag{4.22}$$

where f_m is the maximum Doppler frequency deviation, v is the mobile terminal speed, and λ is the carrier frequency wavelength. Coherence time is actually a statistical measure, so there is a statistical characterization as the time over which the correlation function is above 0.5. Based on the zero-order Bessel function of the first kind $J_0(x)$, coherence time can be approximated by [86]

$$T_c \approx \frac{9}{16 \cdot \pi \cdot f_m} \text{ [secs]} \tag{4.23}$$

In practice, the signal may fluctuate severely and (4.21) is not very restrictive; on the other hand the (4.23) is often too severe. Therefore, it is usual to define the coherence time as the geometric mean of the two expressions for T_c [86]

$$T_c \approx \sqrt{\frac{1}{f_m} \frac{9}{16 \cdot \pi \cdot f_m}} = \sqrt{\frac{9}{16 \cdot \pi \cdot f_m^2}} = \frac{0.423}{f_m} \text{ [secs]} \tag{4.24}$$

Since f_m in (4.22) depends on the carrier frequency (through the wavelength), it is advisable to use a value that corresponds to a carrier frequency of 2.4 GHz (i.e., the ISM band) that is available for WPAN systems and has lower loss values than other possibly available higher frequencies. Figure 4.33 shows a representation of coherence time versus terminal mobile speed for a conservative design (using the most restrictive expression and the practical rule of the expression above).

Looking at Figure 4.33, it can be concluded that for a pedestrian speed of 5 km/h, the coherence time will be shorter than 15 ms.

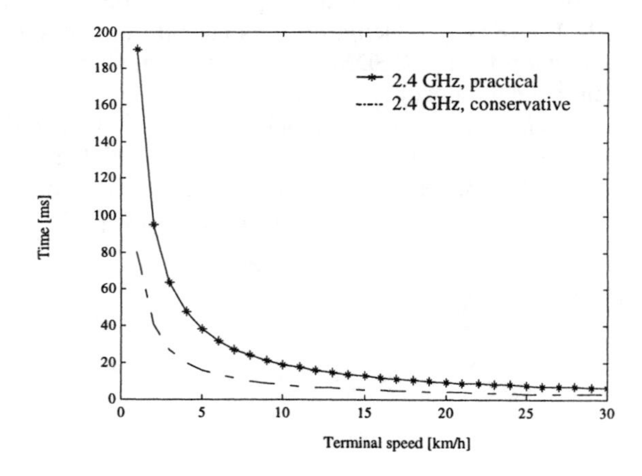

Figure 4.33 Coherence time versus terminal speed.

Since the packet must be shorter than the coherence time in any situation, it can be considered to be shorter than 5 ms. In order to improve the estimation when packet length approaches this limit, a small number of pilots will be included in each OFDM symbol to be used for this purpose.

The propagation channel must also be taken into account to design the length of the cyclic prefix. In order for the orthogonality of subcarriers to be maintained, this length must be longer than the duration of the channel response. Since the maximum range of the described system is 20m, the duration of the channel response can be expected to be less than 66 ns. Since the duration of the channel response and the propagation delay must be covered by the cyclic prefix, a value of 200 ns or greater is a good choice for the cyclic prefix length.

In order to maintain an adequate throughput, the useful part of the OFDM symbol can be chosen to be much longer than the cyclic prefix. Typically, it can be decided between:

$T > 16Tc$: $T > 3{,}200$ ns
$T > 8Tc$: $T > 1{,}600$ ns

4.6.2.5 Phase Noise Considerations

When designing an oscillator for OFDM, special care must be taken to minimize the phase noise power density in higher frequencies, starting from Δf (subcarrier spacing), since the degradation caused by these terms is not corrected in the receiver. However, the lower frequency contributions of phase noise can be corrected together with channel effects.

Figure 4.34 shows the degradation that is obtained in [87] with and without correction of phase noise effects for a given phase noise spectrum. It can be seen that the improvement of common error correction is more noticeable for high phase noise variances, since the performance without correction is very bad in these cases. For lower values of the phase noise variance there is still a considerable improvement. It can be as high as 5 dB for E_b/N_0 values around 10 dB.

For this correction to be possible, the ratio of phase noise bandwidth compared to the OFDM intercarrier spacing $\Delta f = B/N$ must be small, where N is the number of OFDM subcarriers and B the total available bandwidth. Taking into account the phase noise characteristics of commonly used oscillators, this fact leads us to the choice of $\Delta f > 10$ kHz.

Let us consider a channel bandwidth of 10 MHz, high enough to provide high data rates but low enough to achieve low-cost designs. With this value, a number of subcarriers equal to or less than 10 MHz/10 kHz = 1,000 will satisfy these requirements. Besides, for an SNR degradation lower than 1 dB in an E_b/N_0 range of 5 to 10 dB, a total phase noise power lower than 0.1 will be required.

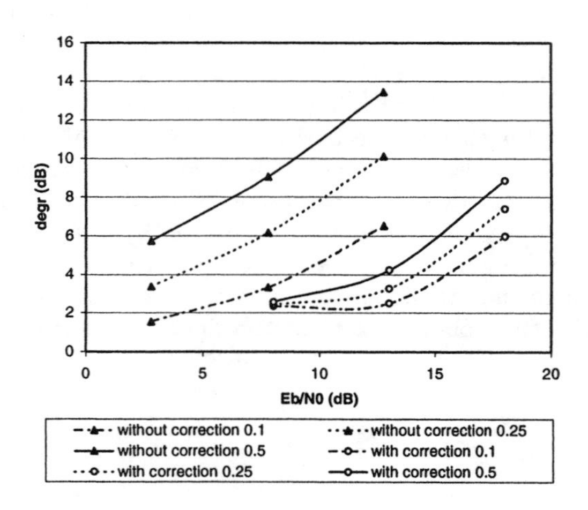

Figure 4.34 SNR degradation due to phase noise.

4.6.2.6 Scalability Considerations and Choice of Parameters

Two possibilities arise when considering how much the range between the minimum and maximum data rates must span. The first possibility, the most exigent, is that we approach the whole range from LDR to HDR. The second possibility is to restrict us to MDR/HDR users, as explained in the beginning of this chapter. In this case the bit rate required by a given transmitter (*Rb*) will be restricted to the range 10 Kbps to 10 Mbps since we are dealing with MDR/HDR users (see Table 4.6).

In any case, the minimum and maximum bit rates ($Rb_{\min}$ and $Rb_{\max}$) can be obtained from $R_{\min}$ and $R_{\max}$ (*R* is the number of bits that a user conveys in an OFDM symbol) as

$$Rb_{\min} = \frac{R_{\min}}{(T + T_{cp})L} \tag{4.25}$$

These numbers do not take into account the inefficiency due to the use of certain OFDM symbols as preambles for channel estimation and synchronization. We will deal with these later.

The following tables summarize the conclusions of the former sections and the proposed values for the main physical layer parameters depending on the scalability choice.

From the comparison of the values of Tables 4.7 and 4.8, it can be concluded that the restriction of scalability to MDR/HDR leads us to a simpler design that is more adequate for a low-cost and low-power goal.

Table 4.7

H-OFDM Parameters for the Whole Range of Scalability

	Limits	Proposed
Number of subcarriers (N)	< 1000	256
Number and distribution frequency guards		16 at each boundary
Number and distribution of pilots		Full-pilot preamble; eight scattered in every symbol
Number of useful subcarriers		216
Bandwidth		10 MHz
Duration of cyclic prefix	> 200 ns	800 ns
Duration of useful OFDM symbol	> 1,600 ns	25.6 μs
Number of slots per frame (L)	$> 10^5/N$	400
Slot length	< 20 ms (< 5 ms conserv.)	3.96 ms ($N_{OFDM} = 150$)
R		1 to 1296
R_b		94.7 bps to 49.1 Mbps

This means that we will use 64 subcarriers with 16 slots per TDMA frame and 150 OFDM symbols per slot.

Table 4.8

H-OFDM Parameters for MDR/HDR Devices

	Limits	proposed
Number of subcarriers (N)	< 1,000	64
Number and distribution frequency guards		8 at each boundary
Number and distribution of pilots		Full-pilot preamble; four scattered in every symbol
Number of useful subcarriers		44
Bandwidth		10 MHz
Duration of cyclic prefix	> 200 ns	400 ns
Duration of useful OFDM symbol	> 1,600 ns	6.4 μs
Number of slots per frame (L)	$> 10^3/N$	16
Slot length	< 20 ms (< 5 ms conserv.)	1.02 ms ($N_{OFDM} = 150$)
R		1 to 264
R_b		9.2 Kbps to 38.8 Mbps

4.6.3 Channel Estimation

In [63] several ways to accomplish channel estimation in OFDM are described: pilot-based, blind, and semi-blind. A pilot-based system has been chosen for H-OFDM not only for channel estimation but also for synchronization. The broadcast preamble of HIPERLAN type 2 [69] has been taken as a reference. This preamble has four OFDM symbols that are transmitted at the beginning of every packet. Only the two last symbols of this preamble will be used for channel estimation purposes. Therefore, in this section we will focus on these last two symbols. An analysis of the convenience of using one or two symbols will be discussed based on the mean square error (MSE) of the estimator and transmission efficiency.

Pilot symbols on this preamble will be BPSK modulated and they will be known at both transmitter and receiver sides. The channel estimation will be performed at the beginning of the packet, and it has to be valid for the whole packet; therefore, the packet duration has to be shorter than the coherence time as explained in section.

As H-OFDM is an OFDMA-based system, the different subchannels are allocated among the totality of the users on the system. This is also a challenge for the method used for channel estimation, since J different channels have to be estimated for each of the J different users instead of only one channel as in the traditional OFDM system.

Two possibilities for using the preamble have been evaluated in order to perform channel estimation. In the first one we divide the preamble in frequency and perform the estimation only on frequencies associated to each user. In the second one we apply a strategy used in MIMO systems in which they deal with a similar problem: the estimation of many different propagation channels.

4.6.3.1 Channel Estimation Based on the Frequency-Division of the Preamble

H-OFDM is a hybrid combination of OFDMA and TDMA. Thus, the N subcarriers are allocated among the different users depending on the rate that they require. In order to offer a large scalability, this frequency allocation is time-varying depending on the needs at a given instant.

Since a specific user only transmits information in a subset of the total subcarriers, there is only a need to perform channel estimation on those subcarriers. The received signal in the frequency domain for each of the N H-OFDM subchannels is

$$R(i) = S(i)H(i) + W(i) \qquad \forall i \in \{0,\ldots,N-1\} \tag{4.26}$$

In order to recover the symbol transmitted in the ith subchannel ($S(i)$), only the channel response on that frequency ($H(i)$) is necessary. Then each user will transmit its preamble independently on each time slot, but it will only transmit pilots on the subcarriers assigned for its transmission. The distribution of the subchannels and the configuration of the preambles will be accomplished at every time slot.

On the other hand, at the receiver side the sum of every transmitted preamble affected by a different channel response is available. Because each transmitter loads with pilots only a subset of subcarriers, and all these subsets are disjoined, the receiver is able to perform the estimation separately for each one of the users.

In order to carry out this task, the least square (LS) estimator [88] is used for several reasons. First, the other estimators with better performance than LS estimator need to know the channel statistics, and in a general scenario this information is unknown. For example, a linear minimum mean-square error (LMMSE) estimator minimizes the mean-square error but it needs to know the correlation matrix for the channel. Second, the LS estimator is very simple and is a way to achieve low-cost, low-power design goals. Its low computational cost causes no additional delays at the receiver. The estimator in frequency has the following expression:

$$H_{LS} = \hat{H}(i) = \frac{R(i)}{P(i)} = \frac{P(i)H(i)}{P(i)} + \frac{W(i)}{P(i)} = H(i) + \Delta H(i) \qquad \forall i \in \{0,...,N-1\} \quad (4.27)$$

where $P(i)$ is the transmitted pilot by the user at ith frequency, $\hat{H}(i)$ is the channel estimation for ith subcarrier.

4.6.3.2 Channel Estimation Based on MIMO Theory

H-OFDM can be viewed as a MIMO system at least from a theoretical point of view. Different techniques have been developed for channel estimation in MIMO using pilots. However, there are some differences between both systems. In H-OFDM L different channels have to be estimated for each user with J different antennas, whereas for MIMO systems, J different channels have to be estimated for the same transmitter.

The LS estimator has been used again here [89]. After evaluating the performance by using the MSE as a measure, the optimal training sequences have been found.

Let J be the number of users in the system, equal to the number of channels to estimate, and $h[n]$ the concatenation of the J channel impulse responses

$$h[n] = \left[h^{(1)}(n)\cdot,......,\cdot h^{(J)}(n)\right] \quad (4.28)$$

Applying the LS method, the estimation of the multiple channels in the time domain will be:

$$\hat{h}(n) = A^{+}(n)preamble(n) \quad (4.29)$$

where $A^{+}(n)$ is a matrix that is composed of pilot sequences transmitted by the J transmitters, and *preamble*(n) is the OFDM symbol received in the time domain that contains information about the J pilot sequences.

Let us denote by K the length of the channel impulse response (CIS); the estimation of the channel in the frequency domain is straightforward, applying the Fourier transform to the impulse response, subject to the following conditions: $N \geq KJ$ where N is the number of OFDM subcarriers, K is the length of the channels impulse response, and J is the number of channels estimated.

For WPAN scenarios, assuming an ISM-band channel model with $K = 3$, an OFDM system with 64 subcarriers can easily fulfill these constraints.

Also in [89] the optimal sequences for channel estimation are proposed:

$$X_k^{(p)} = (-1)^{\left\lfloor \frac{k}{2^{(p-2)}} \right\rfloor} X_k^{(1)}; \qquad p = 1,\dots,J \quad k = 1,\dots,K$$

$$\left| X_k^{(1)} \right| = const \quad \forall k \tag{4.30}$$

It is necessary to remark that a last restriction should be included if one wants to make sure that one is able to reach the inferior bound of the MSE with these optimal sequences

$$N \geq 2^{J-1} K \tag{4.31}$$

Therefore, for the case of 256 subcarriers and ISM-band channel models, there is an upper limit of seven users in the WPAN sharing the H-OFDM signal, which for a first approach seems to be enough. However, the limit for 64 subcarriers is harder (five users). Taking into account that the HDR/MDR devices used in the system use 64 subcarriers, this estimator has been discarded.

4.6.4 Synchronization

Synchronization is a critical problem in OFDM and, as it will be shown later, it is even more important in H-OFDM. The time offset produces an uncertainty of arrival time at the receiver. Multicarrier systems are especially sensitive to this issue because it generates a phase rotation of the received data. Besides the mismatch between the transmitter and receiver, oscillators generate ICI. At the receiver, an initial step of acquisition is needed to determine the correct data decoding time in order to overcome these problems.

Concerning the frequency offset (ε), its value will determine the degree of interference among the subcarriers. ICI can be split into two different effects. The first one is the loss in amplitude of the main lobe, and the second effect is the influence on the rest of subcarriers.

The problem of synchronizing the different users in an OFDM multiuser environment is not trivial. This problem has already been widely discussed in the literature leading to several studies, but it has almost always covered different cases than the ones we are dealing with. Most of them apply to broadcast or downlink scenarios. In these cases the signal transmitted by the base station is used by the totality of users to synchronize in time and frequency, so that this scenario has many similarities with a typical single-user synchronization scenario. The signal transmitted by the base station does not have any interference produced by another terminal.

For example, in [49] the scenario is totally different from the one described above. In order to have a better insight into these differences, a brief description of the scenario in [49] is given here in terms of synchronization procedure. The scenario considers the common situation that several ad hoc advanced terminals (MDR/HDR) are active at the same time. There could exist many different communications among different terminals, but there is not a base station that centralizes communication so the synchronization procedure is more complicated. Moreover, if there are several users sharing the same symbol at the same slot, the ICI will be very important if they are not synchronized.

From this point of view it is clear that every terminal that shares resources in time or frequency must be synchronized to each other. It will therefore be necessary to look for other alternatives for the synchronization of multiple users. The proposed solution assumes some procedures that imply interaction with the MAC layer.

It will rely on a terminal that can be taken as a time reference for synchronization of the rest of the devices. It will be called the leader terminal. Once the leader terminal has been established, the process of synchronization will be reduced to the single-user case [90], since all users will be synchronized with the leader, in this way avoiding the problems triggered by the coexistence of multiple users in the system.

4.6.4.1 Synchronization Algorithm

After initial acquisition with the help of the leader terminal, a synchronization algorithm [90] is proposed here which can be used to estimate the time offset (finding the frame and symbol start) and the frequency offset.

This synchronization algorithm is simple and it maintains a high transmission efficiency, since the ratio between the number of used bits in the header and the number of bits dedicated for the data transmission is very small (2/150).

The algorithm is based on the transmission of a synchronization header at the beginning of every slot, in which two OFDM symbols will be inserted for this purpose. Each of them will be constructed in a way that some given sequences are repeated several times in the same symbol (in the time domain).

At the receiver a correlation will be performed in order to determine the packet start time. Further details can be found in [90].

4.6.4.2 Preamble Definition

Similarly to the HIPERLAN 2 standard, the preamble is designed with two OFDM symbols (besides those for channel estimation) for synchronization purposes. Basically, we have used the HIPERLAN 2 header as defined for the broadcast mode with some differences, as it is shown in Figure 4.35.

Symbols A and B, which are different, contain eight samples each. The symbol IA is the negative reply of A. The blocks of symbols A and B can be generated by 64-point IFFT of a symbol in the frequency domain with eight useful subcarriers, keeping only the first one of the eight exact copies that are generated in the time domain, that is, the first eight samples.

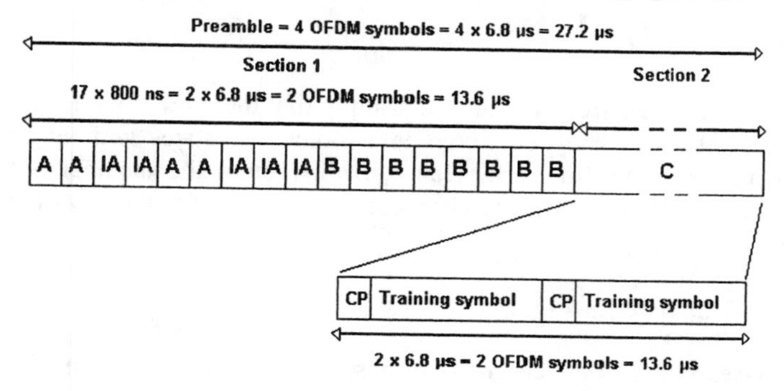

Figure 4.35 Structure of the preamble as considered in the IST project PACWOMAN. (*From:* [49].)

The synchronization algorithm will obtain a peak correlation value when the A IA A IA sequence is detected. This is the reason why the window size value has been chosen to be 32 samples (i.e., the length of that sequence). In the same way, a second peak of correlation will be obtained when a B B B B sequence is detected, which will allow us to better adjust the estimation. Obviously, in real scenarios the peak of the correlation will have different values than those in the ideal AWGN case, and therefore, we need to find a threshold that is suitable for most situations.

The C part represents the two OFDM symbols used for channel estimation, which comprise two training sequences, each one using 48 subcarriers, which are distributed among all users, in addition to the cyclic prefix.

4.6.4.3 Leader Procedure

As it has been pointed out before, in order to carry out synchronization it is necessary that a terminal takes the roll of the leader. This terminal will be the reference for the oscillators and time frame of all devices that share the same signal. When a terminal is switched on it has to find the leader terminal or assume that role if there no leader at that time. A brief description about the whole process is explained in the following.

Sensing Channel

When a new terminal is activated, it needs to know if other terminals already exist or not. The new terminal will sense the channel during a time period guaranteeing the listening of established communications if these exist. In a first approach the duration of two frames (2 ×16,864 ms) has been chosen as sensing time.

There are two possibilities when the terminal senses the channel, namely, the channel is free or it is busy.

Free Channel

If the channel is found free, that does not imply that there are no other terminals in the ad hoc network. They could be in a nonactive state, without established communications at that moment. Hence, even though the channel is free, it is necessary to accomplish a re-search request for the leader terminal (*request for leader packet*).

Request for Leader

For leader-searching (i.e., control) purposes a logical channel *leader channel* (LCH) is needed. This channel will be mapped into the first slot of every frame and it will be used for the communications between the leader and the other terminals.

Since the leader will sense the channel waiting for synchronization requests only at the first slot of every frame (minimizing the leader's power consumption), a new terminal has to send the request (*request for leader*) during two consecutive frames, guaranteeing in this way that the leader listens to the request. The request is therefore 33,728 ms (double the frame duration).

Leader Answer. I'm the Leader

Once the leader has listened to the new terminal request (the leader is sensing the channel at the LCH), the leader must answer the request with an *I'm the Leader* packet in the LCH, also in the first slot of the following frame.

This answer will be the one used by the new terminal to synchronize with the leader. The preamble that precedes the answer will be used for this purpose. This preamble is transmitted by every terminal at the beginning of a slot when this terminal is going to use resources, and it is required by the synchronization algorithm to acquire the slot and frame start and the frequency offset.

At a later time, once the new user has been synchronized, it will inform the rest of the users about its active presence in the network with a new control message (*I'm here*) at the first slot of the next frame.

Answer Is Not Received

Once the petition has been transmitted to locate the leader, the terminal will wait for the answer during two frames (33,728 ms). If no response is received in this time, the new terminal will assume that it is the only one and it becomes the leader. At that point the new terminal will determine the frame time structure, and it will begin to sense the channel only at the first slot of every frame expecting the new user's synchronization request. Also, the new terminal will send in the LCH the message (*So I'm the leader*) to make sure that no one claims the leader role.

Busy Channel

If the terminal detects another transmission on the channel, it will use that communication for synchronization purposes.

This procedure guarantees that all users get the synchronism taking the leader as a reference when they access the network. All terminals that are using the medium will be perfectly synchronized. Any new terminal will listen to the already established communications, and using the preamble transmitted by every user at the beginning of every slot, it will get the synchronism with the algorithm previously referred [90].

But the new terminal still does not know the time structure of the frame. In order to find the frame beginning, the new terminal will send a *Request for Leader* during a frame once the previously established communications have finished. Therefore, complete synchronism is achieved following the same procedure as in the free channel scenario.

Definition of the Control Slot

Besides the LCH, a new logical control channel for MAC is needed. For simplicity and efficiency we have decided to package these three logic channels in the same slot, which will be used as a control slot and will be at the first position in every frame.

This slot, just like the rest of the slots, will consist of a preamble, composed of four H-OFDM symbols used for synchronization and channel estimation, and 150 symbols intended to transmit control information.

For MDR/HDR devices, we have 44 subcarriers and 150 useful symbols in each slot. Therefore, in the first slot of every frame we have to map these three control channels established among the 44 available carriers and 150 symbols. In order to accomplish the mapping coherently, we must closely study each one of the control channels with the aim of finding out the physical resources that will be needed. The three control channels are:

- Random access channel (RACH);
- Paging and access grant channel (PAGCH);
- Leader channel (LCH).

Since there are three logical channels, a three-frequency division is chosen for mapping, as depicted in Figure 4.36:

- RACH: 16 carriers;
- PAGCH: 16 carriers;
- LCH: 12 carriers.

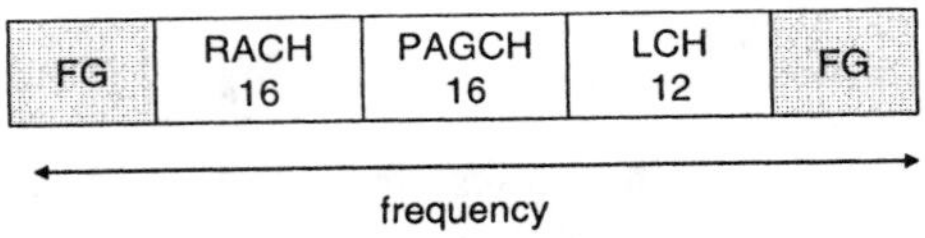

FG	RACH 16	PAGCH 16	LCH 12	FG

frequency

Figure 4.36 Structure of the control slot.

Using the 150 symbols in the slot with a robust modulation BPSK gives:

- RACH: 150 symb. × 16 carrier/symb. × 1 bit/carrier = 2,400 bits;
- PAGCH: 150 symb. × 16 carrier/symb.× 1 bit/carrier = 2,400 bits;
- LCH: 150 symb. × 12 carrier/symb. × 1 bit/carrier = 1,800 bits.

Random Access Channel

This is the logical channel used by the different terminals to access the free resources of the network (both subcarriers and time slots). The H-OFDM time-frequency grid indicates the totality of resources used. The H-OFDM structure will be mapped: a 15 × 44 matrix = 660 bits (the control slot is not signaled). Considering that a CRC of 4 bits is added for error detection purposes, and a convolutional code with rate Rc = 1/2, we have 1,328 bits in total to transmit in the channel. RACH can transmit this number of bits perfectly, using for this purpose only 87 of the 150 slot symbols available.

Since this channel is mapped into the first slot and terminals must listen to it, every terminal has knowledge about the free resources being used.

Paging and Access Grant Channel

A terminal that listens to the PAGCH for a request must answer in the following control slot indicating whether the communications can be established or not. Then, the first terminal has to request resources using the RACH.

Leader Channel

This is the logical channel for communications from and to the leader. As it has been shown, there are many kinds of messages sent on this channel: the first 4 bits of the logical channel are used to specify the type of the message. Possible messages are:

- *I'm here.* This is used to give information about the active presence on the network of a new terminal once the synchronization with the leader has been reached. It is also used to send the MAC address for a new terminal.

- *I'm the leader.* This is the leader answer to a request for leader. This message contains the time-frequency grid indicating the resources that are being used. It is the biggest channel of the LCH, and it will occupy 1,336 bits, taking the CRC and the convolutional code into account.
- *So I'm the leader.* This is the message that a new terminal sends when no other terminal answers the request for terminal message. It is to inform that it is assuming the leader role.
- *Request for leader.* This is the leader search request.
- *Abandon the leader.* This is the message that is transmitted by the leader when it gets ready to abandon the network.

4.6.5 PAR Reduction

A major obstacle in the transmission of OFDM signals is the fact that its probability density function (pdf) behaves like a complex Gaussian function, and therefore, the multiplex signal has a large peak-to-average power ratio (PAR), also named peak-to-mean envelope power ratio (PMEPR), occasionally exhibiting very high peaks. On the contrary, many single-carrier systems yield a constant or bounded envelope.

When passed through a nonlinear device, such as a high power amplifier (HPA) (there are several types of HPAs, including solid-state and TWTA), nonlinearities may get overloaded by high signal peaks [e.g., nonconstant envelope implies higher sensitivity to nonlinear distortions (clipping) and undesired sidelobes].

The HPAs characterization is given by an input-output power relationship, which is linear at low input powers, but reaches a maximum output power, named saturation level, at a certain input power value (see Figure 4.37).

Thus, ideal amplifiers present a straight line as the output power approach of the saturation level, leading to envelope clipping: this is referred to as AM/AM distortion. Similarly, a phase shift depending on the power level will also occur, generating what is denoted as AM/PM distortion. Therefore, in real amplifiers there are two kinds of distortion, amplitude and phase. The final effect of AM/AM and AM/PM distortion is the generation of unwanted spectral energy both in-band and out-of-band. The in-band energy will cause distortion of the transmitted signal and out-of-band energy will cause ACI.

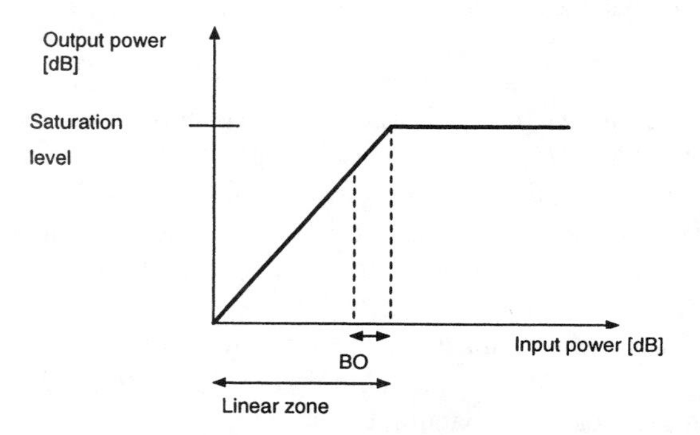

Figure 4.37 Characterization of a high power amplifier response. (BO: back-off.)

Consequently, the signal may suffer significant spectral spreading, in-band distortion and, more critical, undesired out-of-band radiation. The distortion introduced by the HPA is worsened in the OFDM systems by the relatively long-tailed distribution (Rayleigh) of the signal amplitude [91, 92].

The conventional solutions to this problem are to use a linear amplifier or to back off the operating point of a nonlinear amplifier (to prevent saturation and intermodulation products spilling into adjacent subchannels). This leads, respectively, to expensive transmitters and power inefficiency, so that it is aimed to reduce high peaks on the signal, and consequently to diminish the PAR, which is a measure of envelope excursions.

With the increased interest in OFDM, overcoming this problem is a very active area of research. It is highly desirable to reduce the PAR to avoid the effect of the nonconstant envelope of the OFDM signal coupled with the transmitter nonlinearity. As we have already stated, this combination can cause intermodulation products, which may interfere with adjacent channels; also, nonlinearities may also cause ISI and ICI [93].

Typically, the PAR, denoted with χ, is used to quantify the envelope excursions of the signal. Typically, to analyze the PAR in OFDM practical systems, it is more convenient to work with the discrete-time definition. The discrete-time PAR associated with the lth OFDM symbol is a random variable defined as

$$\chi^l = PAR\{s^l[n]\} = \frac{\max\left(\left|s^l[n]\right|^2\right)}{E\left\{\left|s^l[n]\right|^2\right\}} \quad ;n = 0,\ldots,N-1 \tag{4.32}$$

where $\max\left(\left|s^l[n]\right|^2\right)$ denotes the maximum instantaneous power of the time-domain OFDM signal, $E\left\{\left|s^l[n]\right|^2\right\}$ denotes the average power of the signal (where $E\{\cdot\}$ denotes the expected value), and n is the index for the interval over which the PAR is evaluated. The well-known Crest Factor (CF) is preferred by some authors to the PAR; both are simply related since χ is the square of the CF.

4.6.5.1 Methods to Reduce the PAR

Several alternative solutions have been proposed in the literature to face high-PAR signals in OFDM. To organize all these contributions, they can be classified in three main categories:

- The first approach, and the simplest, is to deliberately clip the OFDM signal before amplification. However, clipping is a nonlinear process and may cause even greater distortion [91, 94, 95]. This can be referred to as a *clipping method*, and its main feature is that it causes a predistortion on the OFDM signal prior to amplification, in contrast to the second type of method.
- The second approach, with two subcategories, is a distortionless method with no loss in performance. It is based on the insertion of a treatment block, whose aim is the identification of the signals with the lowest PAR of the various available, which represent the same information. These schemes can be named *distortionless methods*. In this case, the reduction of the PAR can be attained by way of:

    1. Introducing little redundancy (as *selective scrambling* [96], SLM [97], and PTS [98, 99];
    2. A joint block coding and modulation scheme, with Golay and Reed-Muller codes (e.g., [100–102]);
    3. The use of pilot symbols to reduce PAR, as tone reservation [103] and orthogonal pilot sequences [104].

- A final proposal is *clustered OFDM* [104], which divides the signal in different branches with separate HPAs, avoiding high-PAR signals. It is also in fact a distortionless method, but it has been considered as a separate category because it is a special method to combat high-PAR signals. These schemes have not been widely employed, since an increase of the number of HPAs makes this proposal prohibitive; an HPA is the most expensive element in a transmitter, and the increment in the cost it is usually not affordable, especially in the case of a large value of N.

In the following the most relevant of these methods are described in some detail to understand the potential solutions available in the literature to solve the problem of high PAR in OFDM systems.

Partial Transmit Sequences

In the partial transmit sequences (PTS) method, [98, 105] the transmitter constructs its transmit signal with low PAR by coordinated addition of appropriately phase rotated signal parts. In this scheme, the frequency-domain vector $\mathbf{S}^l$ is partitioned into V disjoint subblocks, denoted as $\mathbf{S}^l{}_v$, $1 \le v \le V$. This partition ensures that all subcarrier positions in $\mathbf{S}^l{}_v$ that are already represented in another subblock are set to zero in this one, and therefore,

$$\mathbf{S}^l = \sum_{v=1}^{V} \mathbf{S}^l{}_v \tag{4.33}$$

This method consists of multiplying all subcarriers in each subblock v by the same phase rotation $\varphi_v^l \in [0,2\pi)$, $1 \le v \le V$, $\forall l$. This φ_v^l is the angle of the complex-valued rotation factor for that subblock, $b_v^l = e^{j\varphi_v^l}$. Now this yields a modified frequency-domain vector $\tilde{\mathbf{S}}^l$ given by

$$\tilde{\mathbf{S}}^l = \sum_{v=1}^{V} b_v^l \mathbf{S}_v^l \tag{4.34}$$

which represents the same information as $\mathbf{S}^l$, if the set of V rotation factors $b_v^l, 1 \le v \le V$, is known for each lth OFDM-symbol; these rotation factors are the side information.

A schematic of this method is shown in Figure 4.38, where it must be taken into account that one PTS can always be left unrotated. The rotation factors $b_v^l, 1 \le v \le V$, may be chosen with continuous-phase angle, but it is more efficient for practical systems restricting the possible values to a finite set of W allowed phase angles; in some analyses carried out in [106], they adopt the choice $b_v^l \in \{\pm 1 \pm j\}$, and then W = 4, which is an efficient and interesting implementation.

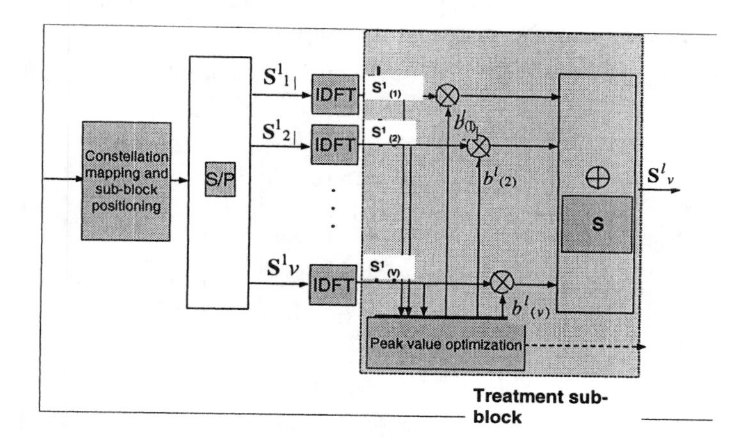

Figure 4.38 PTS scheme for PAR reduction. In this technique, a specific assignment of subcarriers to subblocks (subblock partitioning) has not been given.

Selective Mapping

In the selective mapping (SLM) method [97], the transmitter selects one favorable transmit signal from a set of sufficiently different signals, where all represent the same information. The lth frequency-domain subcarrier vector $\mathbf{S}^l$ is multiplied subcarrier-wise with each one of U vectors, $\mathbf{P}_u$, resulting in a set of U different subcarrier vectors $\mathbf{S}_u{}^l$ with components given by

$$\mathbf{S}_{k,u}^{l} = \mathbf{S}_{k}^{l}\mathbf{P}_{k,u}; \quad 0 \le k \le (N-1); \quad 1 \le u \le U \tag{4.35}$$

where kth is the subcarrier index. The predefined set U is composed of vectors $\mathbf{P}_u$, which are significantly different, distinct, and pseudo-random but fixed; this vector $\mathbf{P}_u = \lfloor P_{0,u} \quad \mathrm{K} \quad P_{k,u} \quad \mathrm{K} \quad P_{(N-1),u} \rfloor$, with $P_{k,u} = e^{j\varphi_{k,u}}$, $\varphi_{k,u} \in [0,2\pi)$ and $0 \le k \le (N-1); \quad 1 \le u \le U$.

After the multiplication, all U alternative frequency-domain vectors $\mathbf{S}_u{}^l$ are transformed into time-domain to get $\mathbf{s}_u{}^l = IDFT\{\mathbf{S}_u{}^l\}$. Therefore, U statistically independent alternative transmit sequences $\mathbf{s}_u{}^l; 1 \le u \le U$, represent the same information. Finally, the transmit sequence with the lowest PAR is selected for transmission. A block diagram of this scheme is shown in Figure 4.39.

Because of the selected assignment of binary data to the transmit signal, this principle is called selective mapping; it must be noted that one of the alternative frequency-domain vectors $\mathbf{S}_u^{\;l}$ can be the unchanged original one.

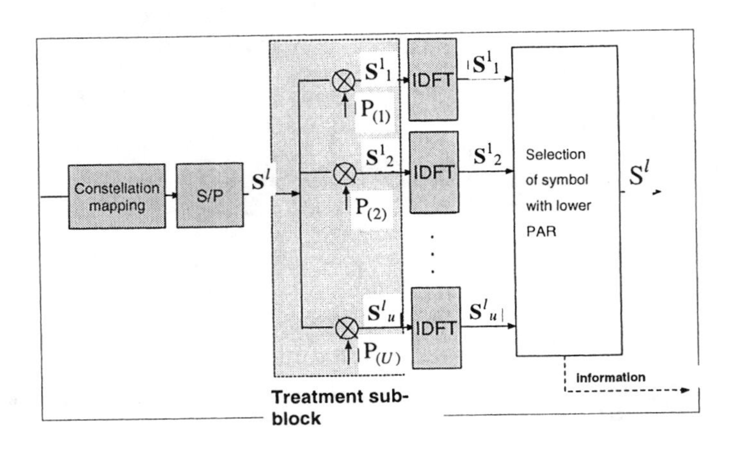

Figure 4.39 Block diagram of SLM scheme for PAR reduction.

Orthogonal Pilot Sequences

The orthogonal pilot sequences (OPS) technique is based on the use of pilot symbol assisted modulation (PSAM) signaling to perform peak power reduction. The idea consists of using certain subcarriers to convey pilot symbols that can be used simultaneously for PAR reduction and channel estimation. The OPS technique is based on the use of a predefined set with a certain number of pilot sequences. Then, the transmitted pilot sequence for lth OFDM symbol, $\tilde{\mathbf{P}}^l$, will be the one that provides the lowest PAR, compared to the other ones in the set. The orthogonality condition is imposed among the different sequences of the set, where orthogonality condition means that the correlation matrix is diagonal.

It is proposed a set S composed of M elements, $0 < \mu \leq M$, where each $\mathbf{P}^{\mu}$ element is a frequency-domain pilot-vector of $N\times 1$ dimensions defined as $\mathbf{P}^{\mu} = \begin{bmatrix} 0 & P_1^{\mu} & \text{K} & P_k^{\mu} & \text{K} & P_{N_p}^{\mu}\text{K} & 0 \end{bmatrix}^T$, with N_p the number of subcarrier positions devoted to pilot symbols, where $P_k^{\;\mu}$ is a complex pilot value and subcarrier positions devoted to data contain zeros.

If orthogonality condition is guaranteed, this will allow blind detection at the receiver. Figure 4.40 shows the treatment subblock employed in the OPS method. In this case, since blind detection will be possible, there will be no need to transmit the little additional information, $B = \log_2 M$ bits accommodated in maybe B subcarriers, required to inform about which element of the set has been chosen.

The orthogonality condition between the elements of the set is determined with the correlation matrix. To ensure orthogonality, the elements of the main diagonal are not zero and the values of the elements that are not placed in the main diagonal have been chosen to equal zero.

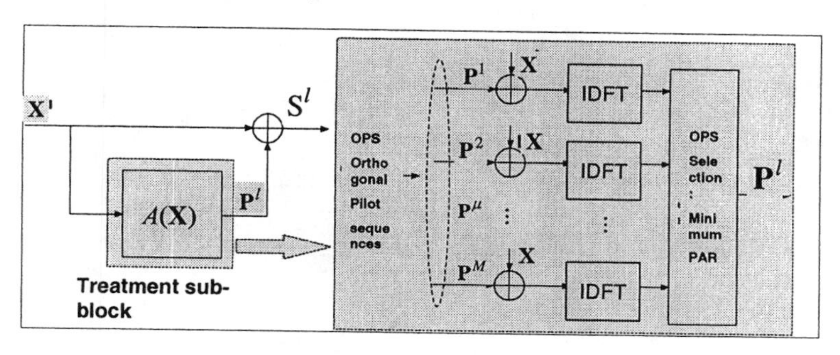

Figure 4.40 Treatment subblock employed in the OPS method.

4.6.6 Performance Evaluation

In this section the performance of the system with the described algorithms for the receiver is analyzed.

4.6.6.1 BER Performance in AWGN

We first start analyzing the performance of H-OFDM in terms of BER in AWGN. Since H-OFDM is a particular case of OFDM, if no multipath channel is present, it is reasonable to obtain the same results as in a traditional OFDM system using the same coding structure. The main advantage of the H-OFDM system is its flexibility, which can be seen in Figures 4.41 and 4.42.

In these figures the performance of a few different *modes* that achieve the same bit rate is shown. The average SNR is the SNR needed to achieve the bit rate measured during the whole frame; that is, the average SNR is calculated as

$$\overline{SNR}(dB) = SNR(dB) + 10\log(L_u) - 10\log(L_{\max}) \tag{4.36}$$

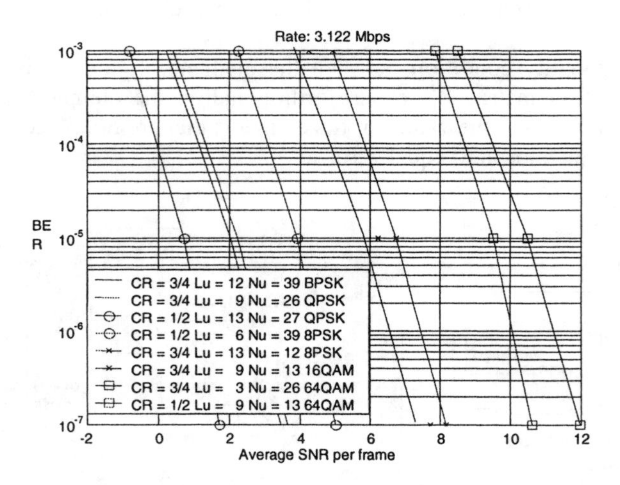

Figure 4.41 Average SNR per frame versus BER for 3.122 Mbps.

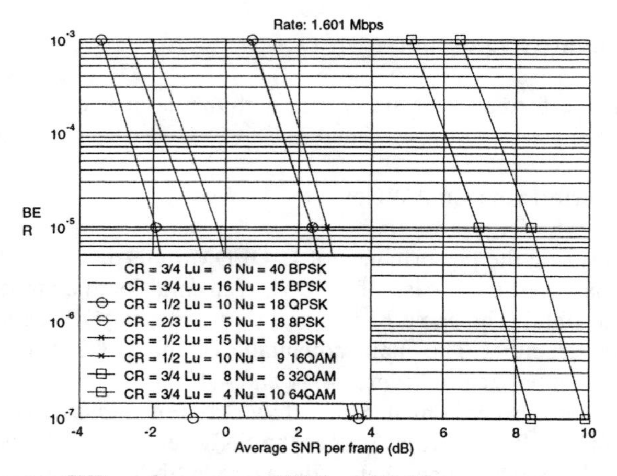

Figure 4.42 Average SNR per frame versus BER for 1.601 Mbps.

Here CR denotes the code rate and L_u and N_u are the number of slots and the number of subcarriers used by the user, respectively. Using these figures together, it is possible to optimize the power consumption. Depending on the terminal power requirements, the specific application, and the system scenario, there is a variety of possibilities to use in order to reach the objectives. Depending on the situation, it may be needed to use a peak SNR or the average SNR (shown in

Figures 4.41 and 4.42). As it is shown in these figures, sometimes it is better (from an average SNR point of view) to use more subcarriers instead of more slots. In other scenarios it is better to change the modulation or the code rate. That kind of decision can be made using the Pareto approach.

4.6.6.2 Channel Estimation

Comparison Between the Two Estimators

The performance of the channel estimation schemes is evaluated in this section by simulation. The MSE has been chosen as a measure of the error committed in estimation.

Due to the different nature of the estimators, MSE cannot be evaluated in the same way in both of them. Nevertheless, it should be noted that the error is always evaluated in frequency domain. In general, the MSE in the frequency domain is calculated as

$$MSE = \frac{1}{Nu} \sum \left\{ \left\| \hat{H}(i) - H(i) \right\|^2 \right\} \tag{4.37}$$

where $\hat{H}(i)$ is the channel estimation of the carrier i, and $H(i)$ is the frequency response of the real channel for the carrier i.

For the channel estimation based on the frequency-division multiplexing of the preamble, we will consider for each subcarrier (i) the frequency response (estimated or real) of the subchannel of interest on that frequency (i.e., the channel associated to the user that uses that subcarrier for transmission). The MSE must be calculated in this way, since only those subcarriers are estimated and not the whole bandwidth. In this case N_u is the number of subcarriers used by the user.

For the estimator based in the multiple users system, we will accomplish the estimation of the L existent channels. Hence, in order to get an idea about the global error, the MSE has to be calculated taking into account the whole bandwidth. In order to do this, H and $\hat{H}$ are considered as the concatenation of the frequency responses of every subchannel. N_u is N times the number of subcarriers of the H-OFDM system, because estimations are accomplished in this case for all the frequencies.

It is foreseeable that the use of more than one symbol in the preamble for channel estimation will improve the performance of estimation, since it allows accomplishing an independent estimation for each preamble symbol and allows, at the end, an average to be taken. However, the more symbols used in the preamble, the fewer symbols may be used for the data transmission in the packet (i.e., the efficiency decreases). Therefore, a trade-off is needed.

A relevant issue should be pointed out concerning the LS estimator based on the theory of multiple antennas. There are optimal pilot sequences for each one of the different user preambles that minimize the MSE in the estimation of the channel:

$$X_i^{(p)} = (-1)^{\left\lfloor \frac{i}{2^{(p-2)}} \right\rfloor} X_i^{(1)}; \qquad p = 1,....,L \quad k = 1,....,N \tag{4.38}$$

where $\left|X_i^{(1)}\right| = cte \quad \forall i$.

It can be deduced that for each subchannel (i) H-OFDM we obtain a BPSK symbol, which will act as a pilot for the estimation. However, frequency guards are ignored, and therefore, this equation is not valid for a real system.

Evaluation of the MSE with guard frequencies is done by the simulation results below. It can be seen in Figures 4.43 and 4.44 that the MSE increases if pilot sequences are used instead of the optimal sequences proposed in [89].

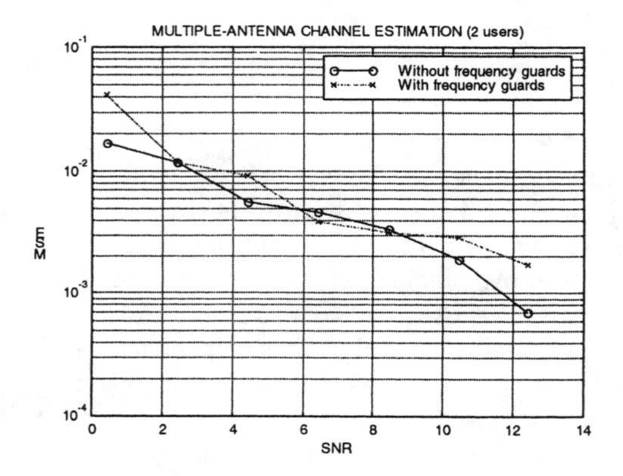

Figure 4.43 Performance with two users.

More users lead to more differences in the results, even though the increase of the MSE is usually quite small, and sometimes it does not even increase. Frequency guards are inserted into the optimal sequences according to their need. Reasonably good results are obtained, even close to the optimal situation.

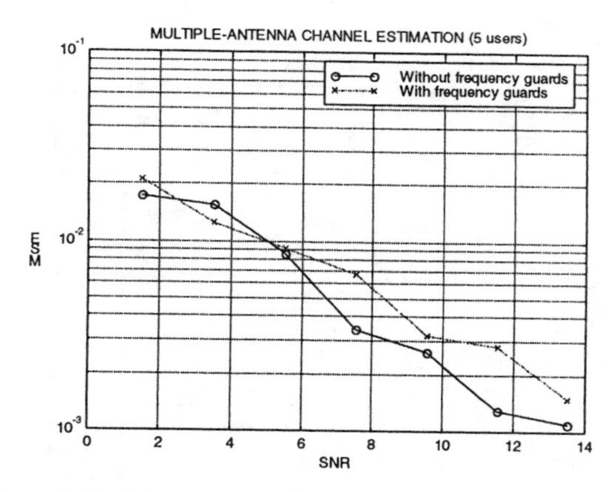

Figure 4.44 Performance with five users.

The use of more than one symbol for channel estimation purposes was also evaluated. Figures 4.45 through 4.48 show that situation. There exists a gain of approximately 3 dB using two pilot symbols.

The simulations show that the use of two symbols in the preamble improves considerably the characteristics of the channel estimation, reducing the MSE, which implies the consequent reduction of the error probability at the receiver. This improvement is produced for all the situations presented, as well as for the two estimators presented, independently of the number of users present in the system.

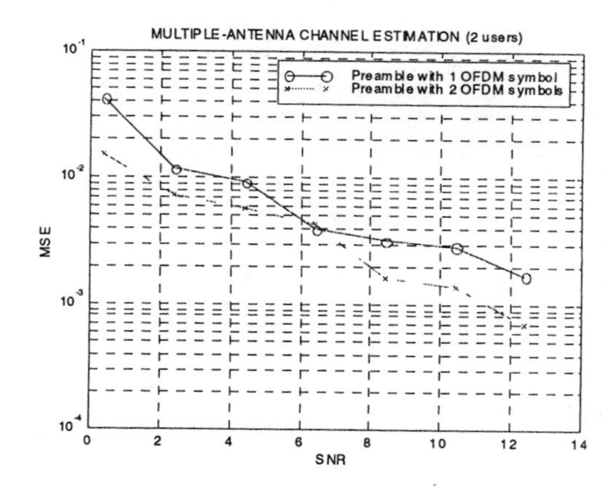

Figure 4.45 Performance with two users.

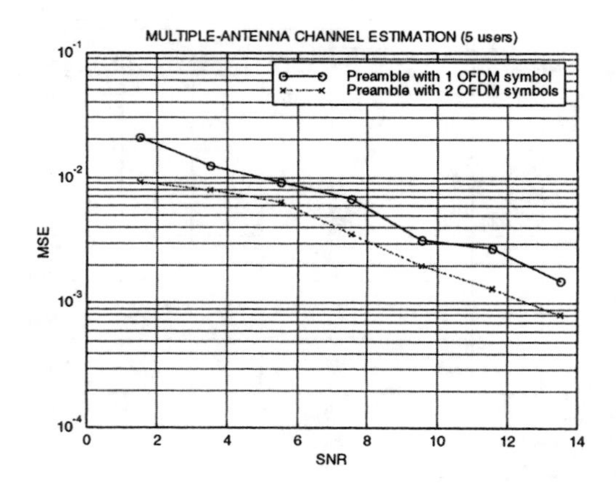

Figure 4.46 Performance with five users.

It has been explained earlier that there are some disadvantages associated with an increase in the number of pilot symbols used for the estimation, due to an efficiency loss in the transmission of information. One of them is that the number of slot symbols effectively used to transmit data decreases. According to the parameters that were fixed earlier for the described H-OFDM system (where 150 OFDM symbols for each packet are transmitted), the use of two symbols for channel estimation can be considered acceptable, maintaining efficiency within more than the acceptable values.

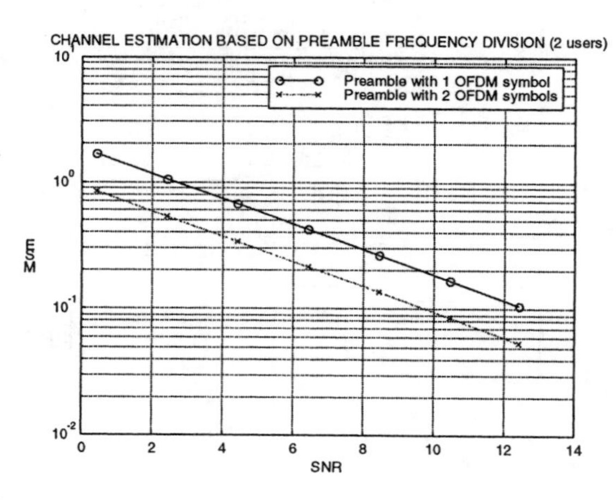

Figure 4.47 Performance with two users.

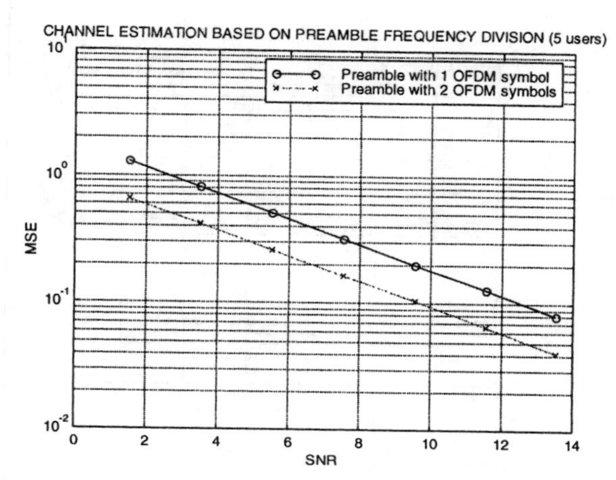

Figure 4.48 Performance with five users.

It can be summarized that the MIMO estimator performs better. A comparison between the schemes is shown in Figure 4.49 and Figure 4.50 for a different number of users.

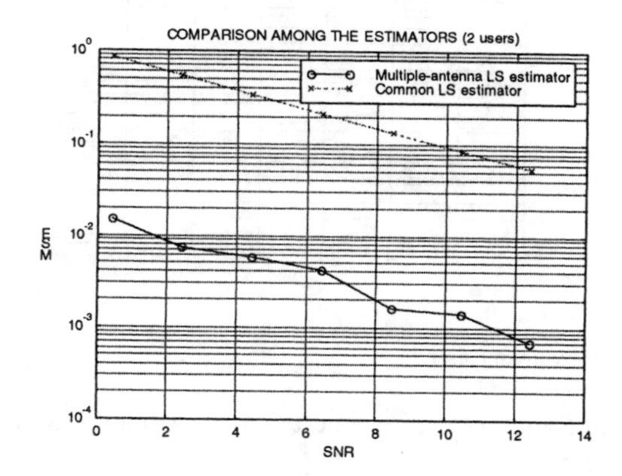

Figure 4.49 Performance comparison with two users.

It should be noted that the MIMO estimator has a limit on the number of users. Figure 4.51 shows the results for only one user.

It is easy to see that we obtain similar results. The MSE committed by both estimators is approximately fixed, despite the variation of the number of users, so that differences are intrinsic to the estimators and independent of the number of users. The efficiency of the MIMO LS estimator when the number of users does not reach the limit (five users for 64 subcarriers) is larger.

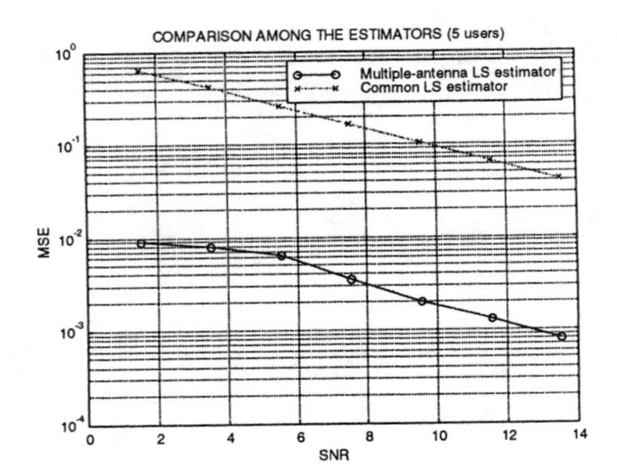

Figure 4.50 Performance comparison with five users.

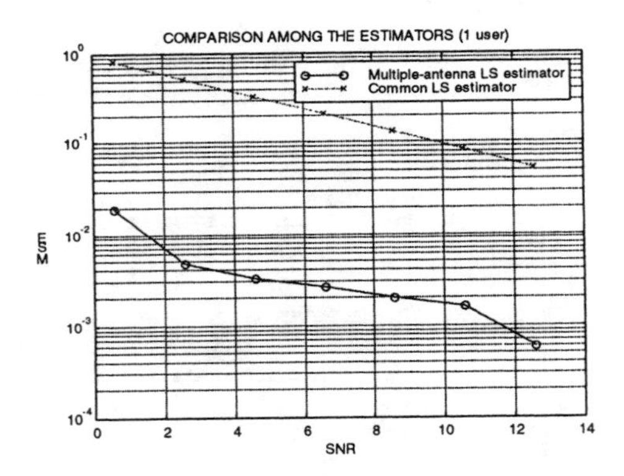

Figure 4.51 Performance with one user.

4.6.6.3 Study of Synchronization-Fault Effects Among Users in Channel Estimation

It is well known that an important issue for an OFDM system is the synchronization. In this section the performance of channel estimation is analyzed when the synchronization is not perfect in order to study its impact.

The aim is to determine the maximum delay among users that results in a probability of error at an acceptable level. These results are shown in Figures 4.52 and 4.53 for the MIMO LS estimator and in Figures 4.54 and 4.55 for the other estimator.

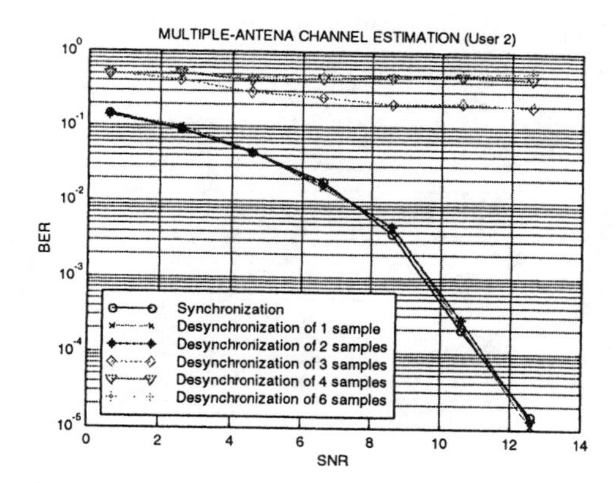

Figure 4.52 Performance with two users.

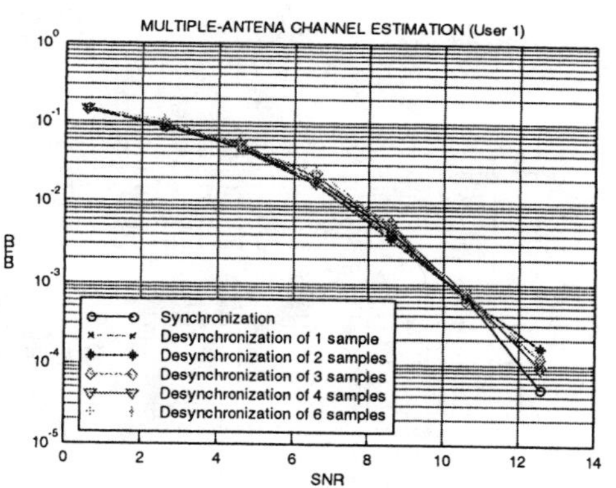

Figure 4.53 Performance with one user.

As we can observe in Figures 4.52 and 4.54, the effect of the desynchronization on user 1 is null because the system is synchronized with user 1. The main effect is therefore on user 2. In Figure 4.53 it can be seen that a delay of more than two samples is enough to ruin the performance of the estimator. That means that the MIMO LS estimator is too sensitive to the delay (i.e., to the synchronization fault). On the other hand, in Figure 4.55 we show the performance of the LS estimator based on the frequency division of the preamble.

Certain degradation can be seen when the delay is more than four samples, which depends on the cyclic prefix length (here, four samples).

In summary, although the MIMO LS estimator is more accurate than the LS estimator based on the frequency division of the preamble, it is very sensitive to the delays (more than two samples ruin the performance of the estimator). The LS estimator must be chosen for channel estimation unless the synchronization algorithm is able to always keep the synchronization within less than one sample.

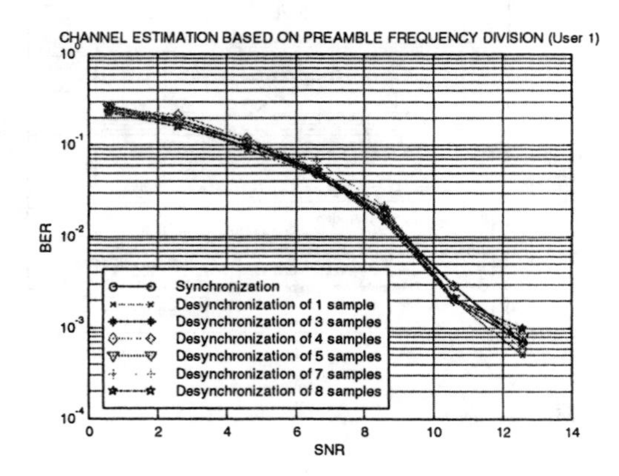

Figure 4.54 Performance with one user.

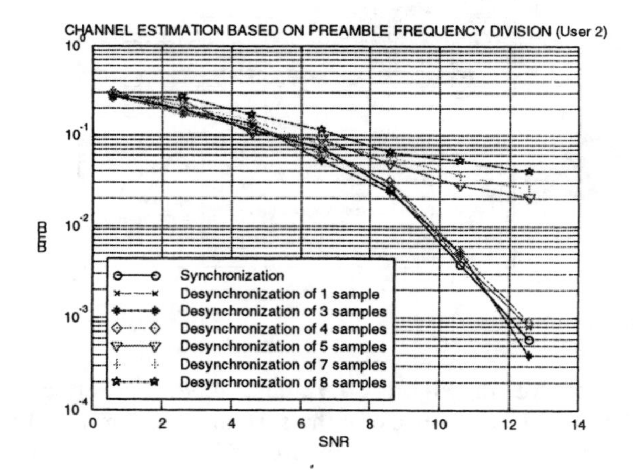

Figure 4.55 Performance with two users.

4.6.7 Synchronization

The synchronization algorithm described in the previous section is the same one that is used for single-user scenarios, and so similar results as for that case could be expected. Indeed, the same results as for the ideal Gaussian scenario were obtained.

With respect to the algorithm, a metric M has to be evaluated in order to determine the beginning of the packet and the estimation of the frequency offset. Ideally that metric will be equal to one, but for ISM channel models [quite similar to an additive white Gaussian noise (AWGN) scenario but with a slight frequency selectivity] that metric must be determined. Two different thresholds are needed: one for the time synchronization and the other for the frequency offset estimation. The optimal threshold values for the ISM channels can be obtained by simulation. Their values are around 0.9 for time synchronization and around 0.6 for frequency offset estimation.

The simulations show that the frequency offset estimation error is lower than 5% of the subcarrier spacing more than 95% of the time for SNR values higher than 10 dB, and the error is lower than 2% more than 80% of the time for the same SNRs.

4.6.8 PAR Reduction and Linearity Requirements

This section evaluates the performance of the PAR reduction methods described in the previous section with the system parameters defined for the project PACWOMAN [49]. In particular, the H-OFDM design handles $N = 64$ subcarriers of which 44 are data subcarriers, and the 16 remaining null subcarriers are employed as guard bands. The channel bandwidth employed for this transmission is of $W = 10$ MHz. Tables 4.9 and 4.10 provide the simulation results for QPSK modulation on each subcarrier.

Table 4.9

Simulation Results for Different PAR Reduction Methods for OFDM Symbols (PAR Threshold > 6 dB)

Method	PAR Original [dB]	PAR Method [dB]	PAR Reduction [dB]	% Application	% Reduction
OPS	7.11	6.06	1.05	60.58	86.58
Clipping	7.20	6.34	0.86	60.58	86.08
OPS + clipping	7.13	6.04	1.09	60.58	92.46
SLM	7.10	5.65	1.45	60.58	94.27
PTS	7.11	5.68	1.43	60.58	93.21

At each trial 10,000 OFDM-symbols were evaluated using the three main methods described earlier: the OPS, the SLM, and the PTS. Additionally, clipping methods have also been analyzed, with a clipping ratio (CR) of 5 dB. Finally, results are given for a two-step method consisting of OPS followed by a clipping subblock that eliminates the remaining peaks after applying the OPS method.

Table 4.10

Simulation Results for Different PAR Reduction Methods for OFDM Symbols (PAR Threshold > 7 dB)

Method	PAR Original [dB]	PAR Method [dB]	PAR Reduction [dB]	% Application	% Reduction
OPS	7.74	6.46	1.28	19.65	94.65
Clipping	7.73	6.39	1.34	19.65	100
OPS + clipping	7.73	6.35	1.38	19.65	100
SLM	7.73	5.71	2.02	19.65	99.74
PTS	7.11	5.71	2.02	19.65	99.84

It makes sense to use a threshold to apply a PAR reduction method since the OFDM-symbols with higher peaks are the ones whose PAR should be more desirably reduced. In these tables, the results are given in five columns. The first column provides the values of PAR without applying any reduction methods. The second column shows the results of PAR after applying the corresponding reduction method. The third column then gives the obtained PAR reduction, which is the difference of the first two columns. The last two columns provide the percentage of OFDM symbols over which we apply the method (considering a threshold), and the percentage of OFDM symbols that reduce the PAR after applying the method, respectively. These results show how all these methods perform similarly in terms of PAR reduction. It must be noted, however, that clipping methods introduce a distortion in the overall system performance. OPS, SLM, and PTS are all distortionless methods; however, SLM and PTS require side information to be known at the receiver side. On the other hand, OPS can perform blind detection at the receiver without any need of additional information.

4.6.9 Phase Noise

In OFDMA the received composite signal can be viewed as an OFDM signal in which each subcarrier is generated by a different user (in the most general case, some subcarriers may come from the same user). The receiver may send some synchronization information so that all transmitters derive the appropriate carrier frequency, symbol rate, and time alignment needed for the orthogonality of the subcarriers. Each of the generated sinusoidal carriers of the users, however, will exhibit phase jitter from a different source. Therefore, it is necessary to consider the effects of phase noise when it comes from different (uncorrelated) sources in an OFDM signal.

Phase noise effects in OFDMA are analyzed in [108] for the general case of different received energies in the different subcarriers (different users may experience different channel attenuation) and different phase noise spectra generated in the transmitters as well as for the particular case of equal energies and phase noise spectra. In both cases the degradation of the SNR due to phase noise is analyzed without attempting to correct the phase noise effects (i.e., the correction of a common error component is not considered).

It is shown in [108] that the complex envelope of the received OFDMA signal is given by

$$r(t)=\sum_{m}\sum_{n=0}^{N-1}a_{m,n}\sqrt{\frac{E_{s,n}}{N}}\sum_{i=0}^{N-1}\exp\left(j2\pi\frac{nl}{N}\right)\cdot p\left[t-(mN+l)\frac{T}{N}\right]\exp[j\phi_n(t)]+n(t) \tag{4.39}$$

where $a_{m,n}$ denotes the mth data symbol (with unit energy) transmitted by the nth user on the subcarrier with frequency n/T, N is the number of subcarriers, $1/T$ is the symbol rate per subcarrier and $p(t)$ is the transmitter pulse. The signal $\phi_n(t)$ constitutes the phase jitter from the nth user and is modeled as a stationary zero-mean process whose bandwidth is much smaller than N/T.

With this notation, the input to the decision device $z_n(0)$ corresponding to the symbol $a_{0,n}$ is shown in [108] to be given by:

$$z_n(0)=\sqrt{E_{s,n}}a_{0,n}E|I_{n,0}|+\sqrt{E_{s,n}}a_{0,n}[I_{n,0}-E(I_{n,0})]+\sum_{\substack{m=0\\m\neq 0}}^{N-1}\sqrt{E_{s,n}}a_{0,m}I_{m,n-m}+W_n(0) \tag{4.40}$$

where $I_{m,k}$ is the discrete Fourier transform of $\{\exp[(j\phi_m(lT/N)], l=0, \ldots, N-1\}$ evaluated at the frequency k/T, and $W_n(0)$ is an additive noise term.

The first term of (4.40) is the useful component, while the second term is a zero-mean disturbance due to the fluctuation of $I_{n,0}$ with respect to its mean. The third term is the ICI, and the fourth term is caused by additive noise.

The SNR degradation caused by phase noise is found in [108] using the above described terms and it is found to be dependent on the received energies $E_{s,m}$ and phase jitter spectra. However, when all received energies are the same and the jitter spectrum is equal for all users, the degradation simplifies to [108]

$$D_n(dB)=10\cdot\log\left\{1+\frac{E_s\sigma^2{}_\phi}{E\left[|W_n(0)|^2\right]}\right\} \tag{4.41}$$

where σ^2_ϕ is the phase noise variance.

This means that in the particular case of equal received energies and phase noise spectra, the degradation depends only on the phase noise variance, as in single-user OFDM. Figure 4.56 shows the SNR degradation in this particular case for different values of the phase noise variance. These results are very close to those obtained by simulating the PACWOMAN system for eight users with identical noise phase. The results are shown in Figure 4.57.

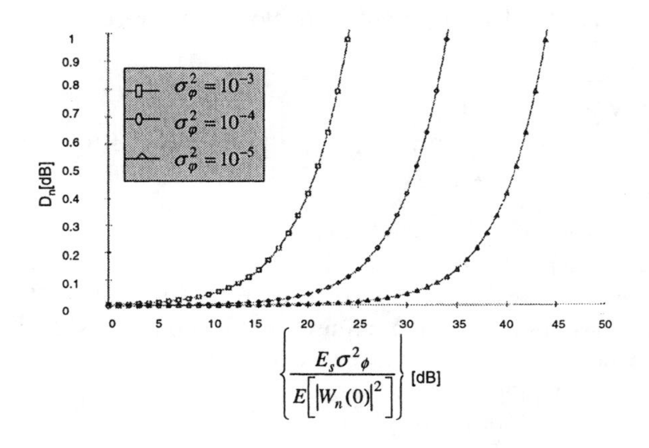

Figure 4.56 Degradation due to phase noise in OFDMA. (*From:* [108].)

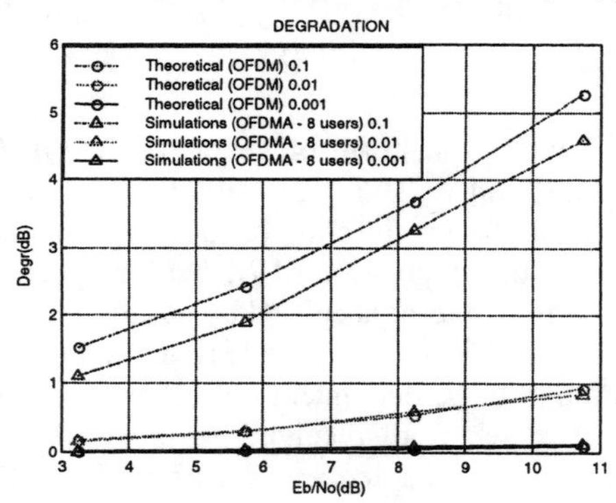

Figure 4.57 Degradation due to the phase noise in OFDMA with eight users with power control.

That situation is, however, not very realistic. The degradation due to the phase noise is shown in Figure 4.58 for the case when different users exhibit different powers. In this figure it can be seen that the lower the power of the user, the more degradation to that user. In the simulations, some users transmit with a given power P and some others with 90%, 80%, or 70% of that power. Given these results, we can conclude that a power control is highly recommended.

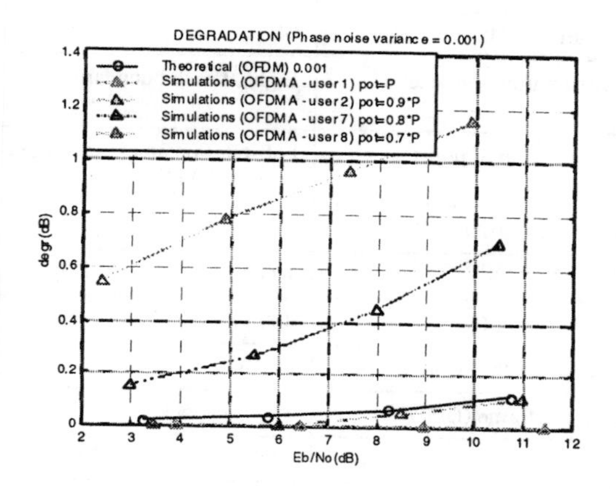

Figure 4.58 Degradation due to the phase noise in OFDMA with eight users, different powers, and noise variance 10^{-3}.

4.6.10 Resource Assignment

In H-ODFM, a given device must be assigned a certain number of OFDM subcarriers and a certain number of time slots. Moreover, the coding and modulation schemes for the used subcarriers can also be assigned, fulfilling the requirements of a high degree of scalability.

In the following section, several transmission modes are identified that can be assigned to the devices according to their transmission needs. This is done by a joint PHY/MAC design taking these modes and the Pareto optimization procedure. Bit loading is considered as a means of assigning physical layer resources to the different devices.

4.6.10.1 Transmission Modes

The main parameters of the H-OFDM physical layer as defined in earlier sections are summarized in Table 4.11.

Although the number of total subcarriers per OFDM symbol and time slots per frame is fixed, each device is allowed to transmit using a variable number of subcarriers and time slots. Moreover, the modulation and coding scheme used in every subcarrier can be chosen according to the channel state. Table 4.12 summarizes the parameters that are selectable depending on the physical layer mode.

Table 4.11

Main Parameters of H-OFDM

Parameter	Value
Number of subcarriers (N)	64
Number and distribution frequency guards	Eight at each boundary
Number and distribution of pilots	Full-pilot preamble; 4 scattered in every symbol
Number of useful subcarriers	44
Bandwidth (B)	10 MHz
Duration of cyclic prefix (CP)	400 ns
Duration of useful OFDM symbol	6.4 μs
Number of slots per frame (L)	16
Slot length	1.02 ms
Number of OFDM symbols per slot (S)	150
Duration of preamble (P)	Five OFDM symbols (at least two for channel estimation)

If the number of bits per symbol that are conveyed in every subcarrier (r_k) of an OFDM symbol is made constant for a given terminal, the total number of bits per symbol for that terminal is

$$R = \sum_{k=1}^{s} r_k = s \cdot r \tag{4.42}$$

The physical layer bit rate achieved with each mode can be obtained as

$$R_{PHY} = (l \cdot s \cdot r \cdot R_c) \cdot \frac{(S-P)}{(N/B+CP) \cdot S \cdot L} \tag{4.43}$$

where the first term in brackets ($l \cdot s \cdot r \cdot R_c$) reflects the values that characterize the different modes (Table 4.12) and the right-hand side parameters are fixed after the signal structure has been designed (see Table 4.11). Substituting the minimum and maximum values of the parameters of Table 4.12, the minimum physical layer bit rate is found to be R_{PHYmin} = 4.44 Kbps and the maximum value is R_{PHYmax} = 35.18 Mbps, fulfilling the flexibility and bit rate requirements. The number of physical layer modes is: 44 · 6 · 15 · 4 = 15,840 modes.

Table 4.12

Parameters that Configure the Physical Layer Modes

Parameter	Description	Range
s	Number of subcarriers	1 to 44
r	Number of bits/symb per subcarrier	1 to 6
R	Total number of bits per symbol	R = s×r
l	Number of TDMA slots	1 to 15
R_c	Convolutional code rate	1/2, 2/3, 3/4, no code

These bit rates are the values achieved at the physical layer level. However, in order to calculate link throughput, the probability of error must be taken into account. In our design, the length of each packet in bits is not constant, but defined by the parameters of physical mode:

$$N_{pack} = (S - P) \cdot s \cdot R_c \cdot r \tag{4.44}$$

For this reason, the PER depends both on the SNR and the physical layer mode (see Figure 4.59).

In Figure 4.59, the PER for a given bit error probability (P_b) and packet length (N_{pack}) is approximated by the upper bound:

$$PER \leq 1 - (1 - P_b)^{N_{pack}} \tag{4.45}$$

The throughput is obtained from the physical layer bit rate (R_{PHY}) and PER, assuming a perfect ARQ scheme, as given by

$$Th = R_{PHY} \cdot (1 - PER) \tag{4.46}$$

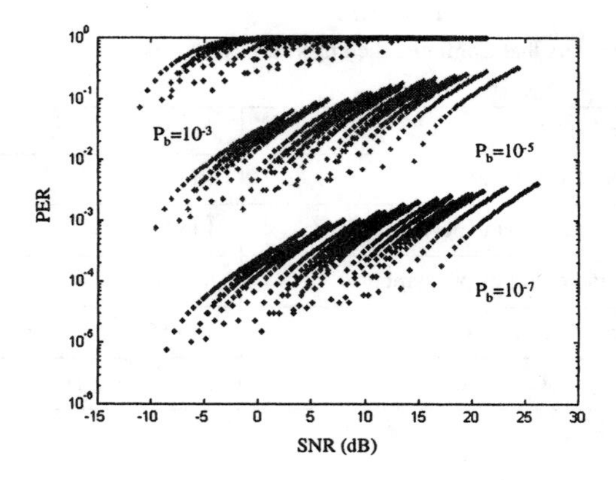

Figure 4.59 PER of the different PHY modes depending on the SNR and the P_b.

4.6.11 Throughput Optimization with Multiple Transmission Modes

Once the achievable throughput is known for every physical layer mode, there are several approaches to choose the best mode to use in a given situation. Looking at a single-user scenario, we may select at every time instant the mode that maximizes throughput. Figure 4.60 shows the maximum achieved throughput when the channel is just corrupted by AWGN and when there is a penalization of 3 dB and 10 dB due to imperfect channel estimation and noise enhancement introduced by the channel correction in the OFDM signal.

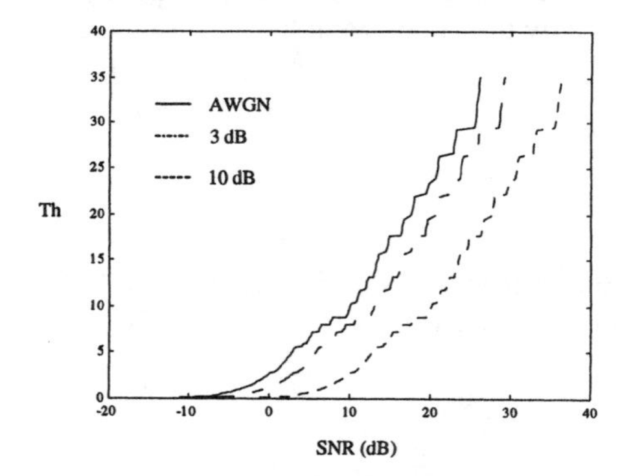

Figure 4.60 Maximum throughput with AWGN and a penalization of 3 dB and 10 dB due to imperfect channel estimation/correction.

This imperfect channel estimation does not only produce degradation in the SNR. The incorrect selection of the proper mode will also cause the throughput to decrease, explaining the deeper difference in throughput for the 10 dB case compared to the others.

The maximization of the individual performance is one possibility. When multiple users are sharing the channel, however, the maximization of individual throughputs does not necessarily lead to maximization of joint throughput [110]. Therefore, it is important to have several configurations (modes) that achieve the same throughput involving a different number of subcarriers or time slots, so that the selection criterion includes the ability to allocate as many users as possible fulfilling their individual bit rate requirements. As an example of the flexibility of this design, Figure 4.61 shows the throughput obtained from multiple combinations of the time slots and subcarriers when every subcarrier is BPSK-modulated and $P_b = 10^{-3}$.

At this point we have a wide pool of modes that will be valuable in order to be able to adapt the transmission to the quality demanded by the users and the instantaneous channel conditions. The number of modes, however, is difficult to manage. Bit loading algorithms appear as a possible solution. Another possibility is the Pareto optimization.

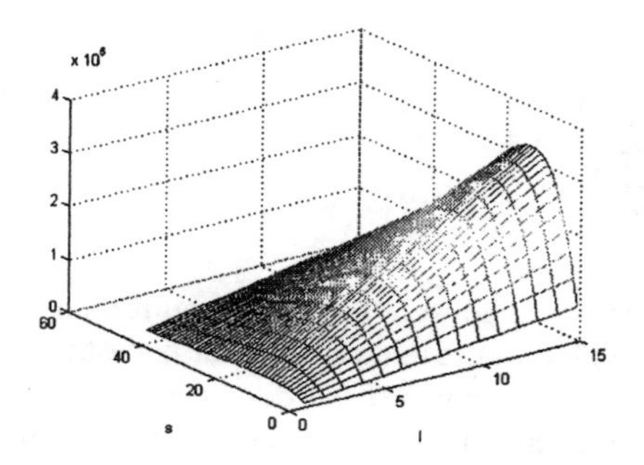

Figure 4.61 Throughput with BPSK-modulated subcarriers and $P_b = 10^{-3}$.

4.6.12 Bit Loading for PHY/MAC Interaction

The energy distribution to achieve capacity in a single-user spectrally shaped Gaussian channel is found by the water-pouring principle [110]. This fact has motivated several optimal and suboptimal bit-loading algorithms that have been suggested for single-user multicarrier systems [111, 112] and later extended to multiuser scenarios [113–117].

Bit-loading algorithms select the number of bits per symbol modulating each of the OFDM subcarriers so that some optimization criterion is fulfilled: generally the energy is minimized and the rate or the margin maximized. To achieve this goal, most bit-loading algorithms impose that a given performance is achieved in every subchannel and use the so-called SNR gap approximation [118, 119].

If γ_k denotes the required SNR to achieve the target error probability in a given kth AWGN subchannel when it carries $R_k = \log_2(M)$ QAM-modulated bits per symbol, the relationship between R_k and γ_k is expressed in a straightforward manner by

$$R_k = \log_2\left(1+\frac{\gamma_k}{\Gamma}\right) \tag{4.47}$$

with the SNR gap is defined as

$$\Gamma = \frac{1}{3}\left[Q^{-1}\left(\frac{SER}{4}\right)\right]^2 \tag{4.48}$$

where the Q function is

$$Q(x) = \int_x^{\infty} \frac{e^{-u^2/2}}{\sqrt{2\pi}} du \tag{4.49}$$

Once the modulation is chosen, the only requirement in order to perform bit-loading is that we are able to estimate SNR, that is, both the channel response and the noise variance must be found. If we denote the noise variance as σ^2 and the frequency response in each subchannel as H_k, the first equation can be rewritten as:

$$R_k = \log_2\left(1+\frac{\gamma_k}{\Gamma}\right) \tag{4.50}$$

where E_k is the energy transmitted over the kth subchannel and G_k is the channel-to-noise ratio given by

$$G_k = \frac{|H_k|^2}{\sigma^2} \tag{4.51}$$

The channel estimation must be implemented in any case in OFDM systems in order to perform coherent demodulation. Additionally, bit-loading techniques require the channel to be estimated too.

In previous works in the literature devoted to bit loading, channel estimation was assumed to be ideally accomplished: in [120] a practical and simple solution using CSMA combined with OFDMA was introduced for uncoordinated WLAN environments, and its performance was compared to optimum multiuser bit-loading and conventional CSMA in the two-user case and for frequency selective fading channels. The performance of this algorithm in the channel models defined for HIPERLAN 2 was analyzed in [121]. However, in both papers the channel was assumed to be perfectly estimated.

Figure 4.62 shows the impact on the symbol error rate (SER) measured in % when erroneous channel estimation is used to perform bit loading.

In Figure 4.62, "error in channel estimation (%)" gives an idea of the magnitude of the estimation error, and it is given as the percentage of the real channel value; the parameter "impact on SER (%)" means the variation in SER obtained at a certain channel estimation error, given as the percentage of the targeted SER value selected as a parameter. Similar results can be expected when using many other bit-loading algorithms that use the SNR gap. It can be seen that as the error increases, the impact on the SER experiences a fast increase. Also, it is shown that for larger E_b/N_0 (lower SER), the impact is greater than for lower ratios. It should be noted that, even for lower errors, the effect is important.

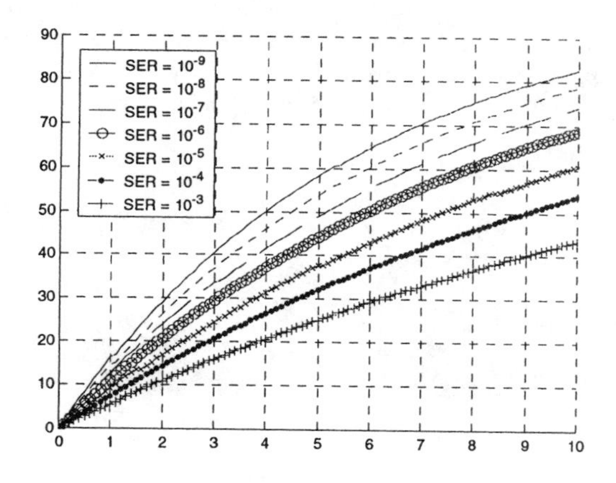

Figure 4.62 Error in channel estimation versus impact on SER.

The % of estimation error committed for a given SNR depends on the chosen estimation scheme. In order to translate this error % to SNR, let us assume that the SNR is 30 dB. From [122] it follows that the MSE equals 10^{-3} using two symbols that is, approximately 3% (note that the MSE is squared). Looking at Figure 4.62 an error of 3% leads to a margin of 42, or 15 % of impact on the SER depending on the targeted SER value. From this point of view, we can conclude that in some cases a better channel estimation may be needed for bit-loading techniques. One possibility is to average over more than two symbols: using 8 symbols results in an MSE of $3 \cdot 10^{-4}$, that is 1.7% error and less than 25% impact, which can be considered enough.

4.6.13 Pareto Approach for PHY/MAC Interaction

Exploiting the trade-off that exists between energy and performance according to the design-time/run-time meta-Pareto framework requires the system to depict sufficient energy scalability [123, 124]. The energy scalability can be defined as the range in which the energy can vary when the performance requirements (e.g., the user data rate) or the external constraints (e.g.,, the propagation conditions) vary from worst to best case. In this section, we analyze the energy scalability of the H-OFDM solutions. Considering a typical energy consumption breakdown (Figure 4.63, also detailed in [125]), we can state that the traditional link adaptation techniques that adapt the modulation order, the code rate, and the transmit energy [126] should affect the total energy mainly via the contribution of the transmit power amplifier (PA) and via the required receiver sensitivity that constrains the receiver digital signal processing and forward error correction (FEC).

However, when considering state-of-the-art implementations of OFDM transceivers [127, 128], it appears that the impact of link adaptation knobs on the total system energy is rather limited. Indeed, with OFDM, the PA power efficiency is highly dominated by the stringent linearity requirements.

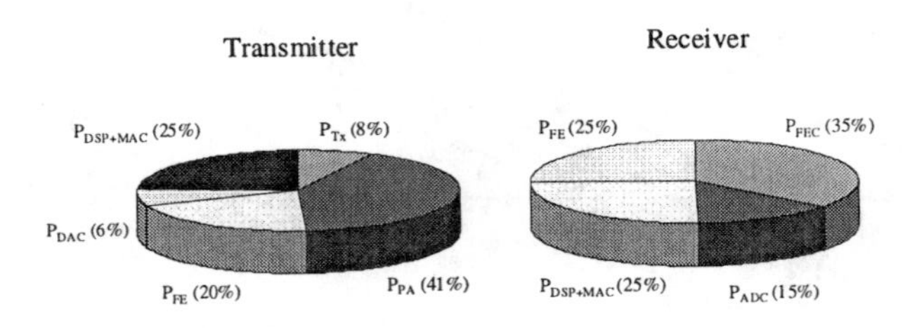

Figure 4.63 Energy consumption breakdown in typical wireless transceivers. (From the compilation of published energy data for state-of-the-art 802.11a WLAN chipsets [127, 128].)

At the transmitter, a reduction of the output power does not result in a significant reduction of the effective PA power consumption (Figure 4.64). Only the duty cycle affects the PA energy, so that the highest modulation order that goes through the channel with reasonable PER (1%) is always the best. Hence, performance downscaling does not bring effective energy gain.

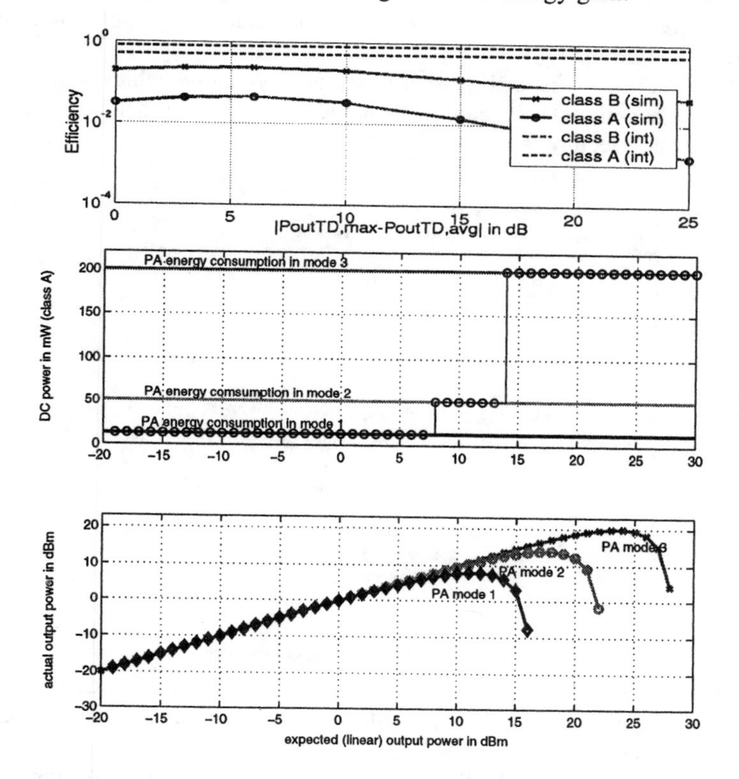

Figure 4.64 Adapting the PA gain compression characteristic allows the translation of a transmit power or linearity reduction into an effective energy consumption gain.

From the receiver perspective, traditional architectures do not really allow translating a reduction in required sensitivity into a real energy benefit. This is mainly due to the lack of flexibility in the DSP and channel decoding. Typically, adapting the code rate does not bring energy gain if typical Viterbi decoders are used. Hence, in order to get energy benefit from link adaptation, the energy scalability of OFDM-based WLAN transceivers has to be enhanced.

To increase the energy scalability of the transmitter, the most effective way is to allow adapting the gain compression characteristic of the power amplifier together with its working point on this characteristic. Indeed, Class A power amplifiers typically used in OFDM-based systems due to the stringent linearity requirements have a gain compression characteristic such that reducing the output power (e.g., to adapt to the lower requirement of lower order subcarrier modulations [129]) translates into an increased back-off and hence into a decreased power efficiency (Figure 4.64). This neutralizes the benefit in terms of total power consumption. Therefore, it makes sense to vary independently the average transmit power and the back-off, which requires the adaptation of the gain compression characteristic. This can be done, for instance, by applying dynamic voltage scaling [130].

From the receiver perspective, energy scalability can be achieved by making the sensitivity adaptive. This can be achieved on the one hand by adopting scalable DSP architecture and, on the other hand, by adopting iterative FEC schemes like turbo codes [131].

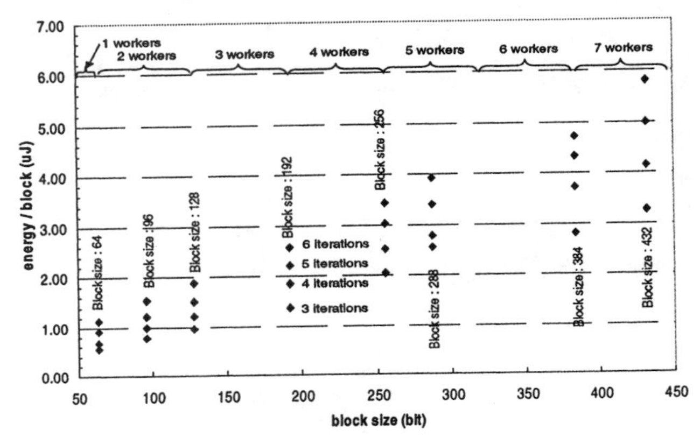

Figure 4.65 Turbo-coding brings energy scalability in wireless receivers.

It has been shown that turbo codes can be implemented with low energy consumption [132] and that they provide a wide freedom to trade-off energy consumption versus code gain [133] (Figure 4.65).

Finally, a last parameter that can be exploited to improve the energy scalability is the packet size. Indeed, for the same bit error rate, the smaller the packet the lower the chance it is corrupted. However, smaller packets have a higher relative overhead due to the headers and channel access mechanisms. From this point of view, H-OFDM allows the flexibility of a variable number of subcarriers and/or time slots. The packet to be transmitted may be fragmented to achieve a better energy performance trade-off, given the propagation conditions. The impact of those flexibility enhancements is assessed in [125].

4.6.14 RF Specifications

This section will cover the implementation view of the H-OFDM approach, fulfilling all the presented requirements by allowing the transmitter to reach any receiver, within the maximum range, with enough power margin and frequency stability to guarantee the expected QoS.

The system architecture is depicted in Figure 4.66.

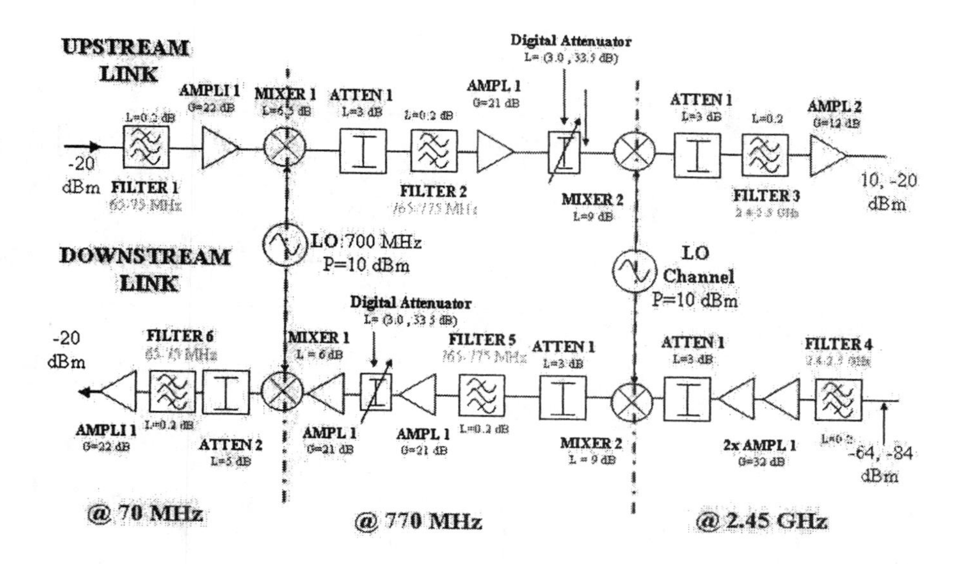

Figure 4.66 RF architecture.

Following a cost-reduction policy, wherever it has been possible, the transceiver will be implemented using the same components in both uplink and downlink chains. This way, a single oscillator can be employed per synthesis, minimizing the jitter influences as well.

4.6.14.1 Synthesizers

As presented in the above scheme, two syntheses are required to move the BB signal up to the radiated frequencies. The first one is at 700 MHz, and the second one takes place in the 1.63- to 1.73-GHz band. Both synthesizers can be controlled from BB through I2C bus, modifying their center frequencies whenever it was necessary. The higher frequency one will be employed for channelling selection, enabling us to locate the 10 MHz of H-OFDM spectrum in one of the 10 available channels between 2.4 and 2.5 GHz.

4.6.14.2 Uplink Chain

Amplifiers

Other important components in this scheme are the amplifiers, whose main characteristics are shown in Table 4.13.

Table 4.13
Amplifiers Characteristics

Model	Freq (GHz)	Gain dB Typical GHz 0.1	1	2	3	4	Output (1 dB Comp) Typ	Min	NF (dB)
AMPLI-1	DC-3	22.1	21.0	18.7	16.8	—	12.5	9	3.5
AMPLI-2	DC-4	12.6	12.5	12.2	11.7	11.3	17.9	16	4.5

Mixers

Two mixers have been employed in this design for the synthesis processes. The first mix moves the center frequency from 70 MHz to 700 MHz, while the second one makes the same operation from 770 MHz to 1 out of 10 different channels whose center frequencies are in the 2.4- to 2.5-GHz band. Table 4.14 shows the characteristics of these two mixers.

Table 4.14
Upconverter IF Section Mixers

Model	Frequency (MHz) LO/RF	IF	Conversion Loss, max (dB)	LO-RF Isolation(dB) Typ.	Min.	LO-IF Isolation (dB) Typ.	Min.
Mixer-1	50/1000	DC-1000	9	42	25	29	18
Mixer-2	3000/4000	DC-700	8.5	26	18	13	7

Attenuators

The digital attenuator is a 6-bit controlled device, with a 0.5- to 32-dB range. Its electrical characteristics are shown in Table 4.15.

This attenuator, as well as the downconverter one, is controlled via flip-flops following the scheme shown in Figure 4.67.

The flip-flops used in this prototype belong to the common D-type.

Table 4.15

Electrical Specifications of the Digital Attenuator

Parameter	Test Conditions	Frequency	Units	Min.	Typ.	Max
Insertion loss	— —	DC–1.0 GHz DC–2.0 GHz	dB dB	— —	3.1 3.6	3.6 4.2
Attenuation accuracy	Any bit or combination of bits	DC–2.0 GHz	dB	—	—	±(3 +4% of atten)
VSWR	Full range	DC–2.0 GHz	Ratio	—	1,8:1	2:1
Switching speed	50% to 90% to 10% RF 10% to 90% or 90% to 10%	— —	nS nS	— —	75 20	150 50
1dB compression	— —	50 MHz 0.5–2.0 GHz	dBm dBm	— —	+21 +29	— —
Input IP_3	Two-tone inputs to +5 dBm	50 MHz 0.5–2.0 GHz	dB dB	— —	+35 +48	— —
Vcc -Vee	— —	— —	V V	4.75 –8.0	5.0 –5.0	5.25 –4.75
Logic "0"	Sink current is 20 μA	—	V	0.0	—	0.8
Logic "1"	Source current is 20 μA	—	V	2.0	—	5.0
Icc	Vcc min to max, Logic "0" or "1"	—	mA	—	0.2	6
-Iee	Vee min to max, Logic "0" to "1"	—	mA	—	–0.2	–1

Filters

Three different filters are necessary to remove undesired LO and spurious components. Discrete element architectures have been employed to keep production cost-reduced.

Filter-1, whose band is 65 to 75 MHz, has the design depicted in Figure 4.68, whose frequency response is shown in Figure 4.69.

For the second filter (Filter-2, see Figure 4.71), the following scheme has been chosen, leading to the frequency response presented in Figure 4.72.

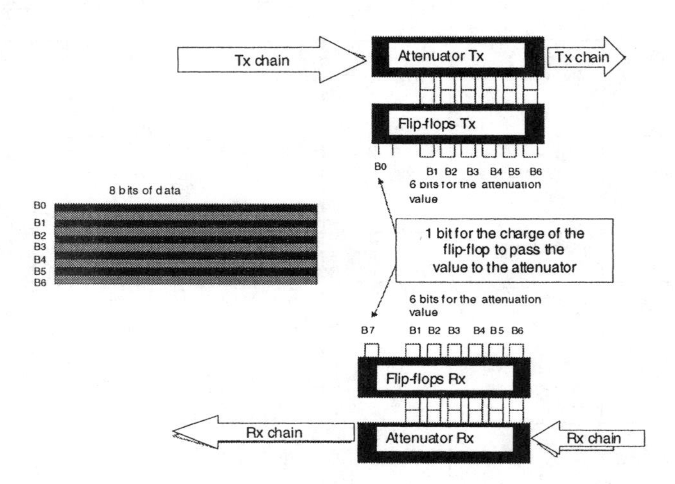

Figure 4.67 Databus for the attenuator controlling.

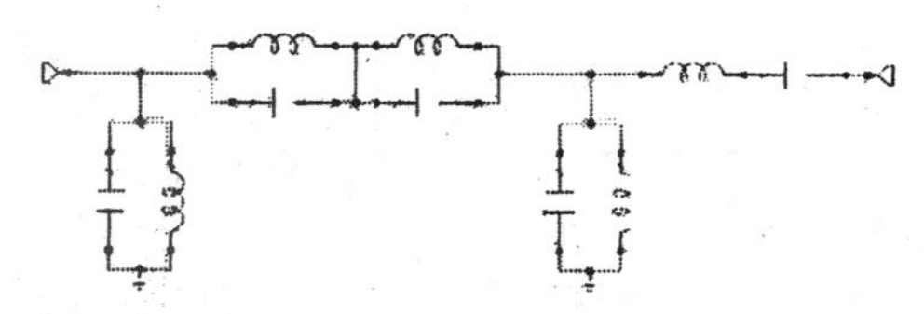

Figure 4.68 Filter-1 architecture.

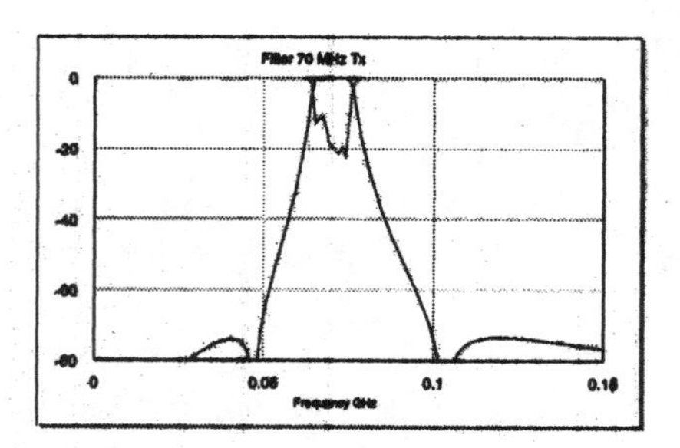

Figure 4.69 Filter-1 frequency response.

4.6.14.3 Downlink Chain

Amplifiers

Only a single model of amplifier is used in this chain, whose characteristics are presented in Table 4.16.

The selected design for Filter-3 appears in Figure 4.73. Figure 4.74 shows its frequency response.

Mixers

Since the mixers are the same as those employed in the uplink, their features have been already presented.

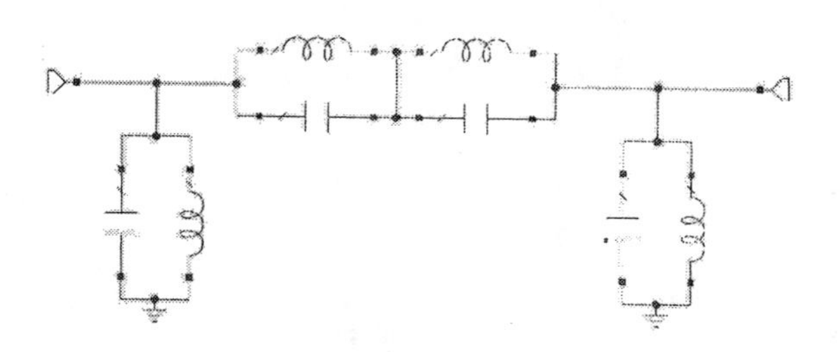

Figure 4.70 Filter-2 architecture.

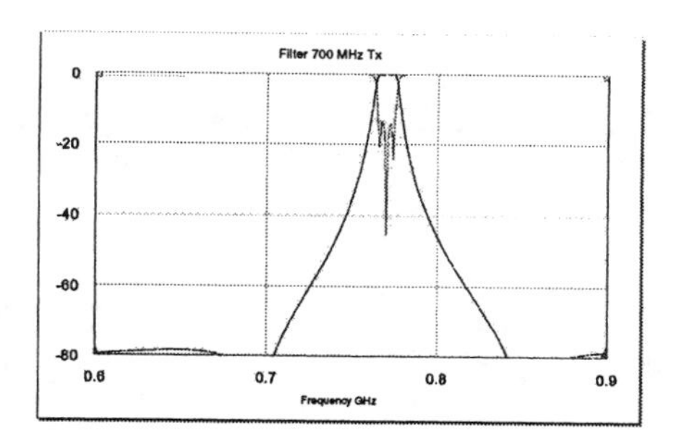

Figure 4.71 Filter-2 frequency response.

Filters

Three more filters were necessary to remove undesired LO and spurious components in the receiver chain. Discrete element architectures were employed to keep the production costs reduced. Since the downlink synthesis has more distanced LO components than for the uplink case, the filter architectures can be relaxed and cost reduced even more.

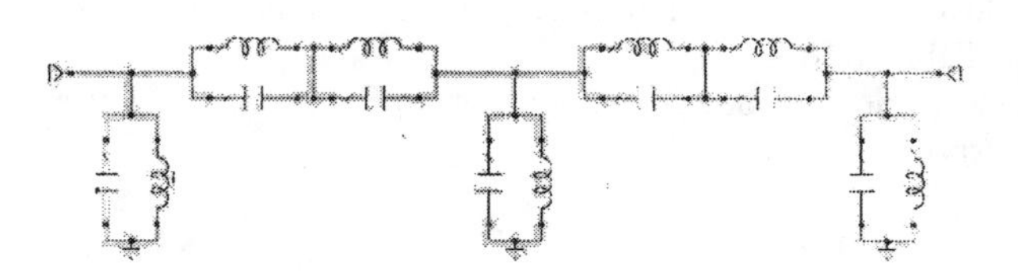

Figure 4.72 Filter-3 architecture.

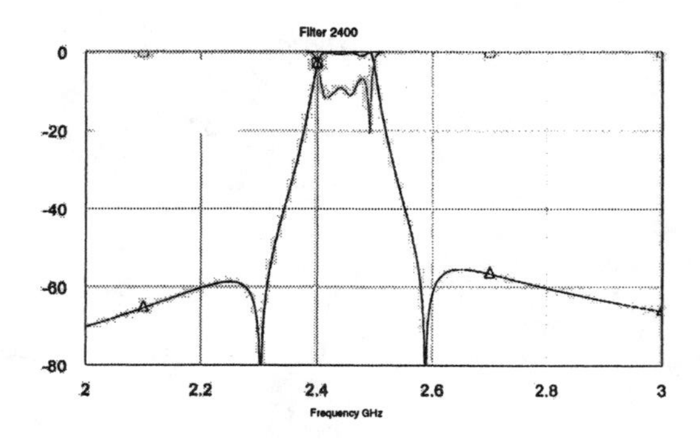

Figure 4.73 Filter-3 frequency response.

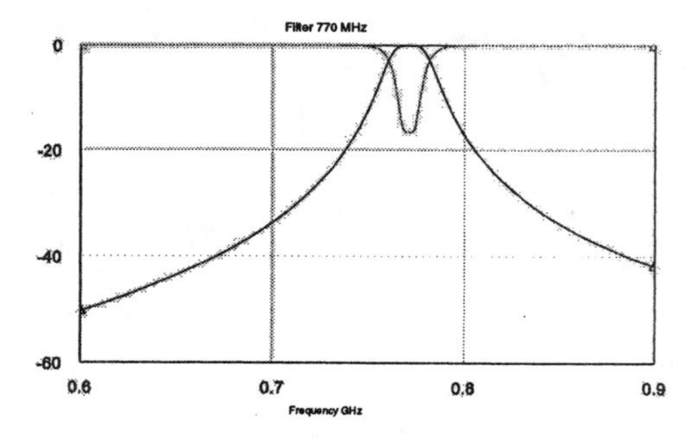

Figure 4.74 Filter-5 frequency response.

Table 4.16

Amplifier Characteristics

Model	Freq (GHz)	Gain dB typical (GHz)					Output (1dB Comp)		NF (dB)
		0.1	1	2	3	4	Typ.	Min.	
AMPLI-1	DC–3	22.1	21.0	18.7	16.8	—	12.5	9	3.5

Filter-4, whose band is 2.4 to 2.5 GHz, had the same design as Filter-3, so no further description is required. Filter-5, centered at 770 MHz, has a different scheme and its frequency response is shown in Figure 4.75.

The selected design for Filter-6 appears in Figure 4.76, whereas Figure 4.77 contains its frequency response.

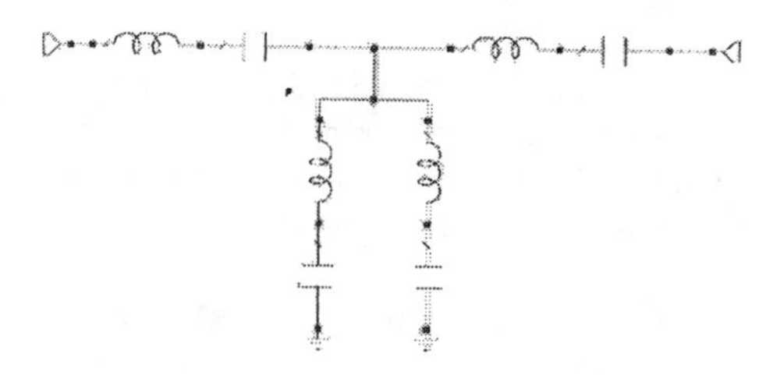

Figure 4.75 Filter-6 architecture.

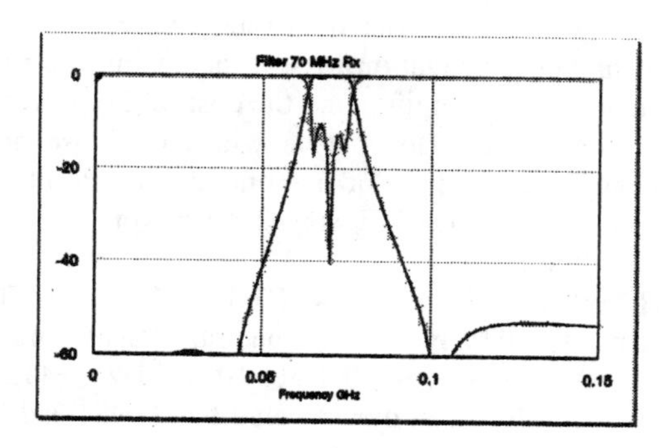

Figure 4.76 Filter-6 frequency response.

4.6.14.4 DC Power Supply

All the presented features will be totally available by providing the following supply values:

- −5v at 10mA;
- GND (0V);
- +5 at 300 mA;
- +12V at 70mA;
- +18V at 10 mA.

A distribution and regulation board can be designed and developed to produce all these supplies from a simplified entry of −5V, GND, and +18V.

4.6.14.5 Choice of Frequency Band

In order to clarify the decision made over the center frequency of the system, some considerations can be presented, taking into account all the advantages and implied features obtained from this choice. Among the most representative reasons for such a decision are the following ones:

- *License free:* The ISM band in the 2.4-GHz band does not require the payment of any kind of taxes or fees for its public utilization.
- *Propagation capabilities:* The selection of higher frequencies is not strictly necessary for our data rates, whereas that selection would have led to higher propagation losses, increasing the required radiated power and thus the consumption and the cost of the power amplifier stage.
- *Component costs:* Due to the widespread use and capabilities of this frequency band, the major manufacturers currently offer a wide range of components for the RF stages at a very low cost with more than reasonable features.
- *UWB not coincident:* New ETSI [134] and FCC [135] regulations covering the UWB power spectral density masks are allowing the maximum values from 3.1 to 10.6 GHz (−41.3 dBM/MHz), having values much lower at our selected band (−51.3 dBm/MHz FCC indoors, −61.3 dBm/MHz FCC outdoors and ETSI). Since the coexistence is not yet totally ensured, is better to avoid these bands. At 900 MHz, FCC masks rises again up to their higher limit (−41.3 dBm/MHz), what could cause any unwanted influence.

It must be taken into account that in IEEE 802.15 working groups looking for solutions for MDR/HDR WPAN devices, several frequency bands are being considered (900 MHz, 2.4 GHz, 3.1 to 10 GHz): 900 MHz is accepted for LDR; 2.4 GHz is the preferred one when low-power, low-cost solutions are targeted [136]; frequencies above 3 GHz are intended for more complex wideband applications, namely those which involve imaging and multimedia.

4.6.15 Complexity

One of the main goals of the IST project PACWOMAN [49] was to make it possible to trade off between high flexibility and low-cost terminals. From this point of view, the system has been designed in such a way that terminals are as simple as possible and every algorithm has been optimized under the constraint of simplicity:

- The simplest of existing channel estimation and synchronization algorithms have been chosen and they have been adapted to the new scenario.
- The method designed for PAR reduction is also simple and it does not reduce transmission efficiency (contrary to most of the existing methods), which is extremely important in order to save power.
- RF has been designed with low power constraints.

On the other hand, the system was built with flexibility and a variety of parameters so that it can be optimized via the Pareto approach.

Finally, efficient architectures are able to perform IFFT/FFT in a power-saving way when not all subcarriers are being used by a transceiver. Indeed, many transmission modes are possible to achieve a given error probability, and it is a design choice to select the mode with the smallest number of subcarriers. The choice will depend on the balance of power consumption between baseband and RF considered in a multiuser environment.

4.6.16 Summary

A hybrid OFDMA-TDMA scheme—H-OFDM in which users are not only assigned different subchannels (OFDMA) but also different time slots (TDMA)—can be devised in order to comply with the requirements of MDR/HDR devices. The signal and physical layer design carried out for an H-OFDM system considering channel estimation, phase noise sensitivity, and scalability issues was described for the scenario adopted by the IST project PACWOMAN [49]. The system was designed in such a way that the terminals are as simple as possible and every algorithm was optimized under the constraint of simplicity. On the other hand, the system has flexibility and a variety of parameters so that it can be optimized via the Pareto approach.

Finally, to evaluate the performance of the different techniques selected, alternatives for resource assignment and RF design were presented.

4.7 CONCLUSIONS

This chapter addressed new radio technologies and their impact on spectrum utilization, including technologies for improving spectrum efficiency, those using multiple antennas such as adaptive antennas and MIMO, and those for handling traffic asymmetry.

To meet the strong demand for broadband multimedia services to both nomadic and mobile users, it is necessary to increase the maximum information bit rate of systems beyond 3G. To enhance capacity, novel technologies or new concepts for improving bandwidth efficiency are indispensable. Advanced RRM algorithms will be beneficial for maximizing the resource utilization. In addition, antenna and coding technologies such as smart antenna, diversity techniques, coding techniques, space time coding, and combined technologies will be necessary for systems beyond 3G to improve the wireless link quality under multipath Rayleigh fading channels. Furthermore, efficient multiple access schemes, adaptive modulation, adaptive downlink modulation, and multihopping technology will be needed to improve the bandwidth efficiency of the system.

Adaptive antennas improve the spectral efficiency of a radio channel, and in so doing, greatly increase the capacity and coverage of most radio transmission networks. This technology uses multiple antennas, digital processing techniques, and complex algorithms to modify the transmitted and received signals at a base station and at a user terminal. In addition, MIMO techniques can provide significant improvements in the radio-link capacity by making positive use of the complex multipath propagation channels found in certain terrestrial mobile communications. MIMO techniques are based on establishing several parallel independent communication channels through the same space and frequency channel by using multiple antenna elements at both ends of the link.

In broadband multimedia communications, asymmetric traffic is envisaged to be dominant. Due to uncertainties in future traffic asymmetry, future mobile communication systems should be adaptable to different ratios of asymmetry especially at the personal-area and the user-access levels in order to deliver the offered traffic asymmetry while simultaneously maintaining high spectrum efficiency. TDD is one of the techniques suitable to support asymmetric high data rate services while providing flexible network deployment including busy urban hotspots and indoor environments as well as wide area applications. TDD systems do not require a duplex frequency pair since both the uplink and downlink transmissions are on the same carrier within the same spectrum band.

Technologies discussed in this chapter included the following:

- *Ultra-wideband*: The basic concept of UWB is to develop, transmit, and receive an extremely short duration burst of RF energy. The resultant waveforms are extremely broadband (typically some gigahertz).
- *Adaptive modulation and coding*: Adaptive modulation and coding schemes adapt to channel variation by varying parameters such as modulation order and code rate based on channel status information (CSI).
- *Flexible frequency sharing:* Sharing of frequency carriers between different operators is a method to optimize the use of spectrum resources.
- *Advanced antenna solutions.*

References

[1] Webb, W. T., and R. Steele, "Variable Rate QAM for Mobile Radio," *IEEE Transactions on Communications,* Vol. 43, July 1995, pp. 2223–2230.

[2] Goldsmith, A. J., and S.-G. Chua, "Variable-Rate Variable-Power MQAM for Fading Channels," *IEEE Transactions on Communications*, Vol. 45, October 1997, pp. 1218–1230.

[3] Goldsmith, A. J. and S.-G. Chua, "Adaptive Coded Modulation for Fading Channels," *IEEE Transactions on Communications*, Vol. 46, May 1998, pp. 595–602.

[4] Matsuoka, H., et al., "Adaptive Modulation System with Variable Coding Rate Concatenated Code for High-Quality MultiMedia Communication Systems," *IEICE Transactions on Communications,* Vol. E79-B, March 1996, pp. 328–334.

[5] Pursley, M. B., and J. M. Shea, "Adaptive Nonuniform Phase-Shift-Key Modulation for Multimedia Traffic in Wireless Networks," *IEEE Journal on Selected Areas in Communications,* Vol. 18, August 2000, pp. 1394–1407.

[6] Van Nee, R., and R. Prasad, *OFDM for Multimedia Communications*, Norwood, MA: Artech House, 2000.

[7] ACTS project SUNBEAM, Smart Universal Beam Forming, Overview Contributions to Standards Results, http:// www.cordis.lu/infowin/acts/rus/projects.

[8] Prasad, R., (ed.), *Towards a Global 3G System, Volumes 1 and 2*, Norwood, MA: Artech House, 2001.

[9] IST-1999-11729 project METRA, Multielement Transmit and Receive Antennas, at htpp://www.ist-metra.org.

[10] IST-1999-11729 project METRA, Schumacher, L., et al., "MIMO Channel Characterization," Deliverable 2, February 2001.

[11] IST- 2000-30148 project I-METRA, Intelligent Multielement Transmit and Receive Antennas, at http://www.ist-metra.org.

[12] IST-2000-30148 project I-METRA Doc.: IST-2000-30148/AAU-WP2-D2-V1.2.doc Title: I-METRA D2: Channel Characterization, 11/7/2002.

[13] Mogensen, P. E., et al., "2D Radio Channel Study," ACTS project SUNBEAM AC347 Deliverable 411, July 1999.

[14] Federal Communications Commission, "Revision of Part 15 of the Commission's Rules Regarding Ultra-Wideband Transmission Systems," *First Report and Order*, ET Docket 98-153, FCC 02-48, April 2002.

[15] Yomo, H., et al., "Medium Access Techniques in Ultra-Wideband Ad Hoc Networks," *Proceedings of Wireless Conference*, Ohrid, Macedonia, 2003.

[16] IEEE Working Group for WPAN, http://www.ieee802.org/15.

[17] IST 2001-34157 Project PACWOMAN, Power Aware Communications for Wireless Optimized Personal Area Networks, http:// www.pacwoman-ist.org.

[18] IST project STRIKE, http//:www.ist-strike.org.

[19] Kermoal, J. P., "Measurement, Modelling, and Performance Evaluation of the MIMO Radio Channel," Ph.D. dissertation, Aalborg University, Denmark, 2002.

[20] Stewart, W., "Mobile Phones and Health," Independent Expert Group on Mobile Phones, 2000.

[21] Pedersen, K. I., "Antenna Arrays in Mobile Communications: Channel Modeling and Receiver Design for DS-CDMA Systems," Ph.D. dissertation, Aalborg University, Denmark, 2000.

[22] Ertel, R. B., et al., "Overview of Spatial Channel Models for Antenna Array Communication Systems," *IEEE Personal Communications*, February 1998, pp. 10–22.

[23] Holma, H., and A. Toskala, *WCDMA for UMTS Radio Access for Third Generation Communications*, Revised Edition, New York: Wiley, 2001.

[24] Telatar, I. E., "Capacity of MultiAntenna Gaussian Channels," AT&T-Bell Labs Internal Tech. Memo, http://mars.belllabs.com/cm/ms/what/mars/index.html, June 1995.

[25] http://www.wlan.org/.

[26] Andersen, J. B., "Array Gain and Capacity for Known Random Channels with Multiple Element Arrays at Both Ends," *IEEE Journal on Selected Areas in Communications-Wireless Communication Series*, Vol. 18, No. 11, 2000, pp. 2172–2178.

[27] Fleury, B. H., "First- and Second-Order Characterization of Direction Dispersion and Space Selectivity in the Radio Channel," *IEEE Transactions on Information Theory*, Vol. 46, No. 6, September 2000.

[28] Petrus, P., J. H. Reed, and T. S. Rappaport, "Effects of Directional Antennas at the Base Station on the Doppler Spectrum," *IEEE Communications Letters*, Vol. 1, No. 2, March 1997.

[29] Lee, W. C. Y., "Effects on Correlation Between Two Mobile Radio Base-Station Antennas," *IEEE Transactions on Vehicular Technology*, Vol. 22, No. 4, November 1973, pp. 130–140.

[30] Adachi, F., et al, "Cross-Correlation Between the Envelopes of 900-MHz Signals Received at a Mobile Radio Base Station Site," *IEE Proceedings*, Vol. 133, Pt. F, No. 6, October 2000, pp. 506–512.

[31] Salz, J., and J. H. Winters, "Effect of Fading Correlation on Adaptive Arrays in Digital Mobile Radio," *IEEE Transactions on Vehicular Technology*, Vol. 43, No. 4, November 1994, pp. 1049–1057.

[32] Pedersen, K. I., P. E. Mogensen, and B. H. Fleury, "Spatial Channel Characteristics in Outdoor Environments and Their Impact on BS Antenna System Performance," *Proceedings of VTC '98*, Ottawa, Canada, May 1998, pp. 719–724.

[33] Gesbert, D., et al., "MIMO Wireless Channels: Capacity and Performance Prediction," *Proceedings of IEEE Global Telecommunications Conference, GLOBECOM '00*, San Francisco, California, Vol. 2, December 2000, pp. 1083–1088.

[34] Chizhik, D., G. J. Foschini, and R. A. Valenzuela, "Capacities of Multielement Transmit and Receive Antennas: Correlations and Keyholes," *Electronics Letters on Communications*, Vol. 36, No. 13, June 2000, pp. 1099–1100.

[35] Chizhik, D., et al., "Keyholes, Correlations, and Capacities of Multielement Transmit and Receive Antennas," *IEEE Transactions on Wireless Communications*, Vol. 1, No. 2, April 2002, pp. 361–368.

[36] Andersen, J. B., "Constraints and Possibilities of Adaptive Antennas for Wireless Broadband," *Proceedings of the International Conference on Antennas and Propagation, ICAP*, Manchester, United Kingdom, Vol. 1, April 2001, pp. 220–225.

[37] Gans, M. J., et al., "Outdoor Blast Measurement System at 2.44 GHz: Calibration and Initial Results," *IEEE Journal on Selected Areas in Communications,* Vol. 20, No. 3, April 2002, pp. 570–583.

[38] Andersen, J. B., "Antenna Arrays in Mobile Communications-Gain, Diversity, and Channel Capacity," *IEEE Antennas and Propagation Magazine*, Vol. 42, No. 2, April 2000, pp. 12–16.

[39] Shannon, C. E., "Communication in the Presence of Noise," *Proceedings of the IRE and Waves and electrons*, January 1949, pp. 10–21.

[40] Shiu, D. S., et al., "Fading Correlation and Its Effects on the Capacity of Multielement Antenna Systems," *IEEE Transactions on Communications*, Vol. 48, No. 3, March 2000, pp. 502–513.

[41] Haykin, S., *Communication Systems*, Third Edition, Englewood Cliffs, NJ: Prentice Hall International, 1994.

[42] Kyritsi, P., et al., "Effect of Antenna Polarization on the Capacity of a Multiple Element System in an Indoor Environment," *IEEE Journal on Selected Areas of Communications*, August 2002.

[43] Mandke, K., et al., "The Evolution of Ultra Wide Band Radio for Wireless Personal Area Networks, High Frequency Electronics," September 2003.

[44] IST-2000-26459 Project PRODEMIS, Hutter, A., "UWB Standardization Technology Watch," Deliverable No.D2.4, available at http://www.prodemis-ist.org.

[45] Federal Communications Commission (FCC), at http://www.fcc.org.

[46] Win, M. Z., and R. A. Scholtz, "Ultra-Wide Bandwidth Time-Hopping Spread-Spectrum Impulse Radio for Wireless Multiple Access Communications," *IEEE Transactions on Communications*, Vol. 48, No. 4, April 2000, pp. 679–691.

[47] Foerster, J., et al., "Ultra-Wideband Technology for Short- and Medium Range Wireless Communications," *Intel Technical Journal*, May 2001, http://developer.intel.com/technology/itj.

[48] Win, M. Z., and R. A. Scholtz, "Impulse Radio: How It Works," *IEEE Communication Letters*, Vol .2, No.1, January 1998, pp. 36–38.

[49] IST project PACWOMAN, at http://www.imec,be/pacwoman.html.

[50] IST Project FLOWS, Deliverable 17, "Analysis of the Market for Multiple Standard Wireless Communications," October 2004, at http://www.flows-ist.org.

[51] IST Project FLOWS, Flexible Convergence of Wireless Standards and Services, at http://www.flows-ist.org.

[52] IST Project FLOWS, Deliverable 15, "Convergence Manager: Mapping Services to Standards," October 2003, at http://www.flows-ist.org.

[53] Rosenberg, J., et al., "SIP: Session Initiation Protocol," Request for Comments 3261, Internet Engineering Task Force, June 2002.

[54] 3GPP TS 23.228, V5.8.0, IP Multimedia Subsystem (IMS), Stage 2 (Release 5), March 2003.

[55] 3GPP TS 24.228, V5.4.0, Signaling Flows for the IP Multimedia Call Control Based on SIP and SDP, Stage 3 (Release 5), March 2003.

[56] 3GPP TS 22.934 V6.2.0, Feasibility Study on 3GPP System to Wireless Local Area Network (WLAN) Interworking (Release 6), September 2003.

[57] IST 1999-10025 Project WIND-FLEX at http://www.vtt.fi/ele/research/els/projects/windflex.

[58] ETSI TR 101 031 v1.1.1 (1997-07) Radio Equipment and Systems (RES); High Performance Radio Local Area Networks (HIPERLAN); Requirements and Architectures for Wireless ATM Access and Interconnection, 1997.

[59] CEPT T/R 22-06 Harmonized Radio Frequency Bands for High Performance Radio Local Area Networks (HIPERLAN) in the 5-GHz and 17-GHz Frequency Range.

[60] ITU Study Groups, Documents 8A-9B/58-E, 28 "Spectrum Aspects of Fixed Wireless Access," September 1998.

[61] Torres, R. P., "CINDOOR: An Engineering Tool for Planning and Design of Wireless Systems in Enclosed Spaces," *IEEE Antennas and Propagation Magazine*, Vol. 41, No. 4, August 1999.

[62] Lobeira, M., et al.,"Channel Modelling and Characterization at 17 GHz for Indoor Broadband WLAN," *IEEE Journal on Selected Areas in Communication—Channel and Propagation Models for Wireless Design*, Vol. 20, No. 3, April 2002.

[63] IST 2000 34157 Project PACWOMAN, D 4.2.2 "State-of-the Art of OFDM and OFDMA with Respect to PHY Layer Requirements," June 2002, at http://www.imec,be/pacwoman.html.

[64] Hirosaki, B., "An Orthogonally Multiplexed QAM System using the Discrete Fourier Transform," *IEEE Transactions on Communications*, Vol. COM-9, July 1981, pp. 982–989.

[65] Kalet, I., "The Multitone Channel," *IEEE Transactions on Communications*, Vol. 37, No.2, February 1989, pp. 119–124.

[66] Bingham, J. A. C., "Multicarrier Modulation for Data Transmission—An Idea Whose Time Has Come," *IEEE Communications Magazine*, Vol. 14, May 1995.

[67] Le Floch, B., R. H. Lassalle, and D. Castelain, "Digital Sound Broadcasting to Mobile Receivers," *IEEE Transactions on Consumer Electronics*, Vol. 35, No. 3, August 1989, pp. 493–503.

[68] EBU/CENELEC/ETSI JTC, "Digital Broadcasting Systems for Television, Sound, and Data Services, Framing Structure, Channel Coding, and Modulation for Digital Terrestrial Television," TM 1545 Rev. 2, January 1996.

[69] ETSI TS 101 475 V1.1.1, Broadband Radio Access Networks (BRAN); HIPERLAN Type 2; Physical (PHY) Layer, April 2000.

[70] ANSI T1.413, "Asymmetric Digital Subscriber Line (ADSL) Metallic Interface."

[71] Weinstein, S. B., and P. M. Ebert, "Data Transmission by Frequency-Division Multiplexing Using the Discrete Fourier Transform," *IEEE Transactions on Communication Technology*, Vol. COM-19. No. 5, October 1971, pp. 628–634.

[72] Chow, P. S., "Bandwidth Optimized Digital Transmission Techniques for Spectrally Shaped Channels with Impulse Noise," Ph. D. Dissertation, Stanford University, 1993.

[73] Rhee, W., and J. M. Cioffi, "Increase in Capacity of Multiuser OFDM System Using Dynamic Subchannel Allocation," *Proceedings IEEE Vehicular Technology Conference*, Vol. 2, May 2000, pp. 1085–1089.

[74] Armada, A. G., "A Simple Multiuser Bit Loading Algorithm for Multicarrier WLAN," *Proceedings IEEE International Conference on Communications, ICC'01*, June 2001.

[75] Wong, C. Y., et al., "A Real-Time Subcarrier Allocation Scheme for Multiple Access Downlink OFDM Transmission," *Proceedings IEEE Vehicular Technology Conference*, Amsterdam, The Netherlands, Vol. 2, September 1999, pp. 1124–1128.

[76] Armada, A. G., M. Calvo, and L. de Haro, "Analysis of an OFDM Access Scheme for the Return Channel on Interactive Services," *Proceedings of the 5th ESA International Workshop on Digital Signal Processing Applied to Space Communications*, Sitges, Spain, 1996, pp. 63–68.

[77] Wei, L., and C. Schlegel, "Synchronization Requirements for Multiuser OFDM on Satellite Mobile and Two-Path Rayleigh Fading Channels," *IEEE Transactions on Communications*, Vol. COM-43, February/March/April 1995, pp. 887–895.

[78] Gallager, R. G., *Information Theory and Reliable Communication*, New York: John Wiley, 1968.

[79] Beek van de, J., et. al., "A Time and Frequency Synchronization Scheme for Multiuser OFDM," *IEEE Journal on Selected Areas in Communications*, Vol. 17, November 1999, pp. 1900–1914.

[80] Vandenameele P., et. al., "A Combined OFDM/SDMA Approach," *IEEE Journal on Selected Areas in Communications*, Vol. 18, November 2000, pp. 2312–2321.

[81] Hara, S., and O. Prasad, "Overview of Multicarrier CDMA," *IEEE Communications Magazine*, December 1997, pp. 126–133.

[82] Saarinen, I., et. al., "Main Approaches for the Design of Wireless Indoor Flexible High Bit Rate WINDFLEX Modem Architecture," *Proceedings of IST Mobile Communications Summit*, Galway, Ireland, 2000, pp. 51–56.

[83] Prasad, R., *OFDM*, Norwood, MA: Artech House, 2004.

[84] Fernandez-Getino Garcia, M. J., J. M. Paez-Borallo, and S. Zazo, "Pilot Patterns for Channel Estimation in OFDM," *Electronic Letters of Communications*, Vol. 36, No 12, June 2000, pp. 1049–1059.

[85] IEEE 802.11a-1999, "Wireless LAN Medium Access Control (MAC) and Physical Layer (PHY) Specifications: High-Speed Physical Layer in the 5-GHz Band."

[86] Rappaport, T. S., *Wireless Communications: Principles and Practice,* Englewood Cliffs, NJ: Prentice Hall, 1996.

[87] Armada, A. G., "Understanding the Effects of Phase Noise in Orthogonal Frequency Division Multiplexing (OFDM)," *IEEE Transactions on Broadcasting*, Vol. 47, No. 2, June 2001, pp. 153–159.

[88] Edfors, O., et al., *OFDM Channel Estimation by Singular Value Decomposition,* Research Report TULEA 1996:18, Division of Signal Processing, Lulea University of Technology.

[89] Tung, T.-L., Y. Kung, and R. E. Hudson, "Channel Estimation and Adaptive Power Allocation for Performance and Capacity Improvement of Multiple-Antenna OFDM Systems," *EURASIP Journal on Applied Signal Processing*, No.3, 2002.

[90] Schmidl, T. M., and D. C. Cox, "Robust Frequency and Timing Synchronization for OFDM," *IEEE Transactions on Communications*, Vol. 45, No. 12, December 1997, pp. 1613–1621.

[91] Di Benedetto, M.-G., and P. Mandarini, "An Application of MMSE Predistortion to OFDM Systems," *IEEE Transactions on Communications*, Vol. 44, No. 11, November 1996, pp. 1417–1420.

[92] Costa, E., M. Midrio, and S. Pupolin, "Impact of Amplifier Nonlinearities on OFDM Transmission System Performance," *IEEE Communications Letters*, Vol. 3, No. 2, February 1999, pp. 37–39.

[93] Berné Martínez, J. M., "Reduction of the Peak-to-Average Power Ratio in OFDM—A Comparative Study," Ms. Thesis, Lund University, Lund, Sweden, June 1998.

[94] Li, X., and L. J. Cimini, Jr., "Effects of Clipping and Filtering on the Performance of OFDM," *IEEE Communications Letters*, Vol. 2, No. 5, May 1998, pp. 131–133.

[95] Rinne, J., and M. Renfors, "The Behavior of Orthogonal Frequency Division Multiplexing Signals in an Amplitude Limiting Channel," *Proceedings of IEEE International Conference on Communications, ICC'94*, Vol. 1, May 1994, pp. 381–385.

[96] Van Eetvelt, P., G. Wade, and M. Tomlinson, "Peak to Average Power Reduction for OFDM Schemes by Selective Scrambling," *IEE Electronic Letters*, Vol. 32, No. 21, October 1996, pp. 1963–1964.

[97] Bäuml, R. W., R. F. H. Fischer, and J. B. Huber, "Reducing the Peak-to-Average Power Ratio of Multicarrier Modulation by Selected Mapping," *IEE Electronic Letters*, Vol. 32, No. 22, October 1996, pp. 2056–2057.

[98] Müller, S. H., and J. B. Huber, "OFDM with Reduced Peak-to-Average Power Ratio by Optimum Combination of Partial Transmit Sequences," *IEE Electronic Letters*, Vol. 33, No. 5, February 1997, pp. 368–369.

[99] Cimini, Jr., L. J., and N. R. Sollenberger, "Peak-to-Average Power Ratio Reduction of an OFDM Signal Using Partial Transmit Sequences," *IEEE Communications Letters*, Vol. 4, No. 3 March 2000, pp. 86–88.

[100] Davis, J. A., and J. Jedwab, "Peak-to-Mean Power Control in OFDM, Golay Complementary Sequences and Reed-Muller Codes," *IEEE Transactions on Information Theory*, Vol. 45, No. 7, November 1999, pp. 2397–2417.

[101] Tarokh, V., and H. Jafarkhani, "On the Computation and Reduction of the Peak-to-Average Power Ratio in Multicarrier Communications," *IEEE Transactions on Communications*, Vol. 48, No. 1, January 2000, pp. 37–44.

[102] Wilkinson, T. A., and A. E. Jones, "Minimization of the Peak to Mean Envelope Power Ratio of Multicarrier Transmission Schemes by Block Coding," *Proceedings of IEEE Vehicular Technology Conference, VTC'95*, Vol. 2, July 1995, pp. 825–829.

[103] Tellado-Mourelo, J., "Peak to Average Power Reduction for Multicarrier Modulation," Ph.D. Dissertation, Stanford University, Stanford, CA, September 1999.

[104] .Fernández-Getino García, M. J., J. M. Páez-Borrallo, and O. Edfors, "Orthogonal Pilot Sequences for Peak-to-Average Power Reduction in OFDM," *Proceedings of IEEE Vehicular Technology Conference, VTC'01-Fall*, Atlantic City, New Jersey, October 2001, CD-ROM.

[105] Cimini Jr., L. J., B. Daneshrad, and N. R. Sollenberger, "Clustered OFDM with Transmitter Diversity and Coding," *Proceedings of IEEE Global Telecommunications Conference, GLOBECOM'96*, Vol. 1, November 1996, pp. 703–707.

[106] Müller, S. H., and J. B. Huber, "A Novel Peak Power Reduction Scheme for OFDM," *Proceedings of IEEE International Symposium On Personal, Indoor and Mobile Radio Communications, PIMRC'97*, Vol. 3, September 1997, pp. 1090–1094.

[107] Müller, S. H., and J. B. Huber, "A Comparison of Peak Power Reduction Schemes for OFDM," *Proceedings of IEEE Global Telecommunications Conference, GLOBECOM'97*, Phoenix, Arizona, Vol. 1, November 1997, pp. 1–5.

[108] Steendam, H. M., and H. Sari, "The Effect of Carrier Phase Jitter on the Performance of Orthogonal Frequency-Division Multiple-Access Systems," *IEEE Transactions on Communications*, Vol. COM-46, April 1998 pp. 456–459.

[109] Cheng, R. S., and S. Verdú, "Gaussian Multiaccess Channels with ISI: Capacity Region and Multiuser Water-Filling," *IEEE Transactions on Information Theory*, Vol. 39, No. 3, May 1993, pp. 773–785.

[110] Gallager, R. G., *Information Theory and Reliable Communication*, New York: John Wiley, 1968.

[111] Hughes-Hartogs, D., "Ensemble Modem Structure for Imperfect Transmission Media," U.S. Patent 4 833 796, May 1989.

[112] Chow, P., S., J. M. Cioffi, and Bingham, J., A., C., "A Practical Discrete Multitone Transceiver Loading Algorithm for Data Transmission over Spectrally Shaped Channels," *IEEE Transactions on Communications,* Vol. 43, February/March/April 1995, pp. 773–775.

[113] Yu, W., J. M. Cioffi, "FDMA Capacity Region for Gaussian Multiple Access Channels with ISI," *Proceedings of IEEE International Conference on Communications (ICC)*, Vol. 3, June 2000, pp. 1365–1369.

[114] Rhee, W., J. M. Cioffi, "Increase in Capacity of Multiuser OFDM System Using Dynamic Subchannel Allocation," *Proceedings of IEEE Vehicular Technology Conference VTC'00*, Tokyo, Japan, Vol. 2, May 2000, pp. 1085–1089.

[115] Wong, C. Y., et al., "A Real-Time Subcarrier Allocation Scheme for Multiple Access Downlink OFDM Transmission," *Proceedings of IEEE Vehicular Technology Conference, VTC'99*, Amsterdam, The Netherlands, Vol. 2, September 1999, pp. 1124–1128.

[116] Hoo, L. M. C., J. Tellado, and J. M. Cioffi, "Dual QoS Loading Algorithms for Multicarrier Systems Offering Different CBR Services," *Proceedings of IEEE International Symposium On Personal, Indoor and Mobile Radio Communications, PIMRC*, Vol. 1, September 1998, pp. 278–282.

[117] García-Armada, A., "CSMA Multiuser Bit Loading Algorithm for Multicarrier Wireless Local Area Networks," *Proceedings of IEEE Vehicular Technology Conference, VTC'01—Spring*, Rhodes, Greece, Vol. 2, May 2001, pp. 1099–1103.

[118] Cioffi, J. M., et al., "MMSE Decision-Feedback Equalizers and Coding—Part II: Coding Results," *IEEE Transactions on Communications*, Vol. 43, No. 10, October 1995, pp. 2595–2604.

[119] Starr, T., J. M. Cioffi, and P. J. Silverman, *Understanding Digital Subscriber Line Technology*, Englewood Cliffs, NJ: Prentice Hall, 1999.

[120] Armada, A. G., "A Simple Multiuser Bit Loading Algorithm for Multicarrier WLAN," *Proceedings of IEEE International Conference on Communications*, Vol. 4, 2001, pp. 1168–1171.

[121] Armada, A. G., and J. M. Cioffi, "Performance of Single-User and Multiuser Constant Energy Bit Loading in HIPERLAN Channels," *Proceedings of XI European Signal Processing Conference (EUSIPCO)*, Toulouse, France, Vol. III, September 2002, pp. 187–190.

[122] Gil Jiménez, V. P., M. J. Fernández-Getino García, and A. García Armada, "Channel Estimation for Bit-Loading in OFDM-based WLAN," *Proceedings of IEEE International Symposium on Signal Processing and Information Technology (ISSPIT)*, Marrakech, Morocco, December 2002, pp. 581–585.

[123] Min, R., et al., "A Framework for Energy-Scalable Communication in High-Density Wireless Networks," *Proceedings of the International Symposium on Low Power Electronics and Design*, Monterey, August 2002, pp 36–41.

[124] Bougard, B., et al., "Energy-Scalability Enhancement of WLAN Transceivers," *Proceedings of IEEE International Conference on Communications, ICC'04*, Paris, May 2004.

[125] IST Project PACWOMAN, "How to Generalize the Pareto-Curve Approach to a Communication Node/Link Optimization Problem," Deliverable 4.3., PACWOMAN IST-2001-34157, 2002.

[126] Goldsmith, A. J., and S.-G. Chua, "Variable-Rate Variable-Power MQAM for Fading Channels," *IEEE Transactions on Communications*, Vol. 45, No. 10, October 1997, pp. 1218–30.

[127] Zargari, M., et al., "A 5-GHz CMOS Transceiver for IEEE 802.11a Wireless LAN Systems," *IEEE Journal of Solid-State Circuits*, Vol. 37, No. 12, December 2002, pp. 1698–1694.

[128] Belnad, A., et al., "Direct-Conversion CMOS Transceiver with Automative Frequency Control for 802.11a WLANs," *Proceedings of the International Solid State Circuit Conference*, San Francisco, California, February 2003, pp. 356–357.

[129] Côme, B., et al., "Impact of Front-End Nonidealities on BER Performance of WLAN-OFDM Transceivers," *Proceedings of IEEE RAWCON*, September 2000, pp. 91–94.

[130] Hanington, G., et al., "High-Efficiency Power Amplifier Using Dynamic Power Supply Voltage for CDMA Application," *IEEE Transactions On Microwave Theory and Techniques*, Vol. 47, No. 8, August 1999, pp 1471–1476.

[131] Berrou, C., et al, "Near Shannon Limit Error Correcting Coding and Decoding: Turbo Codes," *Proceedings of IEEE International Conference on Communications*, Geneva, Switzerland, Vol. 2/3, May 1993, pp. 1064–1071.

[132] Bougard, B., et al., "A Scalable 8.7-nJ/bit 75.6-Mb/s Parallel Concatenated Convolutional (Turbo-) Codec," *Proceedings of the International Solid State Circuit Conference*, San Francisco, California, February 2003, pp. 152–153.

[133] Bougard, B., et al., "A Class of Power Efficient VLSI Architectures for High-Speed Turbo-Decoding," *Proceedings of IEEE Global Telecommunications Conference, GLOBECOM*, Taipei, Taiwan, Vol. 1, November 2002, pp. 549–553.

[134] European Telecommunications Standards Institute, (2003) EN 302 065—Ultra Wide Band for Short Range Devices.

[135] Federal Communications Commission (2002) Revision of Part 15 of the Commission's Rules Regarding Ultra-Wide Band Transmission Systems, FCC 02-48, ET docket 98-153, adopted February 14, 2002.

[136] TG3 PHY: 802.15.3 Channel Plan, IEEE P802.15 Working Group for Wireless Personal Area Networks (WPANs), July 2001.

Chapter 5

Advanced Mobile Satellite Systems

5.1 INTRODUCTION

Efficient support of Internet-based applications to mobile/nomadic users is a key feature of third generation networks. In light of the shortage and the high cost of the terrestrial-UMTS (T-UMTS) spectrum, operators are looking into the provision of integrated broadcast/multicast services through hybrid broadcast-UMTS systems. Satellite UMTS (S-UMTS) is one way to ensure the efficient delivery of some UMTS services. These services include, for example, broadcasting and multicasting applications such as audio/video, e-newspaper, and live stock exchange data [1].

Second generation mobile satellite systems failed to grab the mobile market for a number of reasons including cost and that they were stand-alone systems, raising a lot of concerns about the future of commercial satellite systems in general. However, the multimedia concept, strongly embedded within UMTS, introduced a new perspective for the mobile satellite systems as collaborative parts of T-UMTS rather than stand-alone systems [1].

The satellite component of the UMTS system architecture was extensively studied in a number of European projects, such as the projects within the fourth research framework on Advanced Communication Technology and Services (ACTS), Satellite-UMTS Multimedia Service Trials over Integrated Testbeds (SUMO), and Satellite Integration into Networks for UMTS Services (SINUS) [3, 4], and some basic principles were established. The requirement for interoperability and integration with T-UMTS was one of the main drivers of these studies, while the concept of discriminating between radio-independent and radio-dependent functions in the system design, as it was first coined in the ACTS project Radio Access Independent Broadband on Wireless (RAINBOW) [3] for T-UMTS, seems to have achieved wide acceptance.

Recent developments in the T-UMTS architecture, however, generate some new challenges for the S-UMTS architecture design. The introduction of packet-mode into the system definition and the ever-increasing penetration of the Internet protocol into the system functions constitute the basic reasons for a reexamination of the system architectures proposed so far and their modification/optimization [1].

Satellite systems may offer a complementary solution, with a possible long-term impact to the way in which multicast data is delivered over the Internet. The broadcast nature of satellites and its ubiquitous coverage offer a natural way to multicast data over a large cell. Their wide footprint coverage enables mobility and flexibility. In particular, with just one satellite in geostationary Earth orbit (GEO), for example, it is possible to cover one-third of the Earth's surface by having one downlink stream. Even in the cases of low Earth orbit (LEO) satellites, it is possible to spread information streams among more expanded regions, compared to a cell with a radius of just a few kilometers, as it is the case for a standard UMTS cell [5].

Thus, satellites may constitute a content delivery network (CDN); that is a multicast layer over the Internet, which is in charge of distributing any kind of large content to the edge of the IP network, as close as possible to the user. In this way, mobile satellite broadcast systems may become a very efficient complement to terrestrial mobile networks, removing their asymmetric load and providing them with far more point-to-point equivalent capacity for far less investment cost [5].

By combining satellite and terrestrial component deployment, infrastructure cost could be significantly reduced, providing the system with a global and complete coverage and allowing distribution cost to really fit price constraints of multimedia business.

A number of mobile and satellite IST projects within the Fifth Program framework focused on research and development of the S-UMTS. The project Satellite-UMTS IP-Based Network (SATIN) [1], for example, defined and evaluated efficient S-UMTS access schemes (using as much as possible the UTRA access scheme model to allow maximum commonality of terminals) based on packet-based protocols. At the same time the project maintained close harmonization activities with other ACTS or IST projects, such as the previously mentioned projects, SINUS [3], SUMO [3], and RAINBOW [3], and the projects Multicast Over Geostationary EHF Satellites (GEOCAST) [6], Virtual Home UMTS on Satellite (VIRTUOUS) [7], and Functional UMTS Real Emulator (FUTURE) [8]. The continuity and harmonization of the S-UMTS research and development activities is shown in Figure 5.1.

SATIN was an IST project focusing on the particular implications of the IP-based packet mode on the S-UMTS design. The requirements dictated by the UMTS core network for the S-UMTS access network were a fundamental drive for the architecture design paving the way for full integration with the T-UMTS for efficient delivery of a series of packet-based services.

The project FLOWS [5], on the other hand, was motivated by a vision of future communications in which IP plays an increasing role and where there is a convergence of wireless systems such as 3G and WLANs. FLOWS identified the flexibility and improved performance offered by MIMO antenna techniques as a means of enabling convergence between standards for the user. FLOWS investigated and developed the antenna, radio, processing, systems, and network convergence techniques to demonstrate this. FLOWS studied the benefits that the adoption of a converged approach using multiple standards can bring to the service provider.

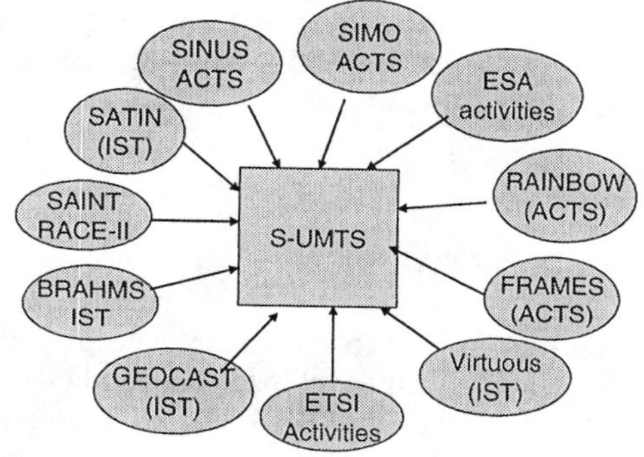

Figure 5.1 Continuity and harmonization of the S-UMTS research and development activities. (*After:* [1, 9].)

An important concept adopted by the IST project FLOWS was to use a common access network based on IP. Onto this common network a variety of wireless access points were connected using, for example, GSM, UMTS, or HIPERLAN/2. These will provide the consumer with access to a range of services depending on factors such as location, mobility, and the characteristics of the service (e.g., required data rates). DVB- and DAB-based broadcast networks are also becoming increasingly available and can augment the future mobile services with entertainment and information services. By creating and managing the convergence of these wireless standards, the FLOWS project showed how a consumer with a single mobile terminal can be connected to different kinds of access points to use the extended services.

This chapter is organized as follows. Section 5.2 discusses the technological challenges for satellite IP-based networks. Input to this section is based on the results achieved within the IST project SATIN [1]. Section 5.2 includes the proposal for a new satellite-UMTS air interface.

Section 5.3 discusses some of the challenges related to the development of satellite broadband multimedia systems. The IST project BRAHMS [10] defined a generic user access interface for broadband multimedia satellite services, which is open to different satellite system implementations, including GEO and LEO constellations. The concept is known as the broadband multimedia satellite system (BMSS) and addresses a range of multimedia user groups with data rate requirements up to 150 Mbps. This BMSS approach was seen as a vehicle for convergence between fixed and mobile multimedia networks towards the global multimedia mobility (GMM) architecture by merging service functions derived from the UMTS/IMT-2000 and from fixed broadband access. The use of IP-based satellite transmission was considered as a solution to convergence towards seamless broadband service provision.

Section 5.4 briefly introduces the concept of high-altitude platforms (HAPS). These systems were not a main topic of research of the FP5 IST research program. HAPS, however, play an important role in current mobile communications development. Several projects within the FP5 follow-up, the Sixth Framework Research Program (FP6) [11], investigate the challenges related to their deployment.

5.2 SATELLITE IP-BASED NETWORKS

In the past the satellite portion of UMTS was developed in isolation from the terrestrial portion [12]. The latter is well developed and standards are agreed upon as part of IMT-2000. S-UMTS, on the other hand, had no international agreement and is in the stage of having several candidate submissions being considered by standards bodies (3GPP, ETSI-TC-SES, ITU-WG).

Initially, the priority with the operators was to roll out the 3G terrestrially without a role for satellite and, therefore, satellite communications standardization did not receive a lot of attention.

Based on the success of traditional mobile satellite systems (MSS) (INMARSAT) and the failure of 2G MSS (Iridium, Globalstar), it is possible to conclude that conventionally, satellites have the following unique advantages:

- Quick to provide services (time niche);
- Broadcast (wide area) coverage;
- Coverage in areas where terrestrial infrastructure cannot economically provide the service;
- Supplementary to terrestrial services;
- Satellite is part of an integrated UMTS service.

From the 2G terrestrial and satellite systems, it is obvious that the use of satellite for voice applications is very small (30,000 users for Iridium and around 40,000 for Globalstar) compared to GSM. New predictions suggest that 200,000 subscribers is a possible number for S-UMTS as a stand-alone system [12]. This figure is not enough to support a satellite business. The satellite community believes that the only way to target the mass market for S-UMTS is to combine the cooperative services (broadcast and multicast) with other interactive services (such as voice) through the terrestrial system [1]. It is also important to mention here that the mass market is concentrated in and around built-up areas. Hence, satellite coverage must be augmented by terrestrial "gap fillers" to retransmit the satellite signal to be received in urban areas and inside buildings.

IP multicast defines an architecture that allows IP applications to send data to a set of recipients (a multicast group) specified by a single IP address. Audio/video conferencing, push-based data dissemination, and remote education are examples of applications that make use of this architecture. Their efficiency lies in the fact that only a single datagram traverses a link rather than a number equal to the sum of receivers involved in such applications. The main functional elements of this architecture are the multicast routing and the multicast group management functions.

5.2.1 S-UMTS Architecture Scenarios

The following elements—RNC, node-B, radio network subsystem (RNS), interface, Uu interface from 3GPP specifications, and the elements specific to the satellite systems, network control center (NCC), and fixed Earth station (FES)—are used here to define different system architectures for S-UMTS under two main categories: coverage-oriented and broadcast-oriented. The architecture scenarios selected within SATIN are shown in Figure 5.2.

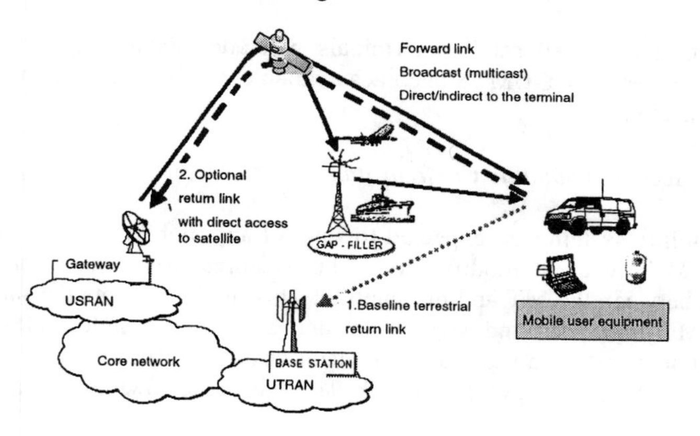

Figure 5.2 SATIN architecture [1].

Two scenarios, a baseline and an optional, were selected to address the service requirements. In the baseline scenario, a handheld mobile terminal receives data through the satellite and/or the intermediate module that features one-way repeater functionality. The satellite path would be the preferred communication link, but if the satellite path of the user were blocked, the communication link would be sustained via the intermediate module repeater (IMR) stations. The introduction of IMR modules can help to overcome the inability of satellite-only systems to offer in-building and in-urban areas coverage (where the mass market resides) and support the moderate and high bit rate multicast/broadcast (MC/BC) services envisaged within SATIN. The return path is provided via the T-UMTS network (baseline case). A satellite receive-only terminal may well serve a given subset of services (pure broadcast and nonhighly interactive multicast). Alternatively, the terminal may also support direct transmission (in the return path) to the satellite (optional case), leading to the more conventional system configuration that allows a stand-alone system to be built at the expense of a more expensive and complex terminal [13].

5.2.1.1 Coverage-Oriented Scenario

There are two ways to provide coverage, either through a direct link between the MT and the satellite or indirectly using intermediate equipment.

Direct Access to Satellite Configuration

The services supported by this type of access are basically the same as those provided by the T-UMTS. Due to link budget constraints, operation in indoor conditions is limited. Therefore, additional techniques need to be adapted to cover this case. The cost for the usage of the S-UMTS will remain higher than that of T-UMTS.

Consequently, all satellite terminals will additionally support T-UMTS as well. Whenever the T-UMTS becomes available, the bimode terminal will restore to terrestrial mode.

Indirect Access to Satellite Configuration

Here satellite systems are expected to support any mobile terminal compatible to the T-UMTS without modification. This requires insertion of an intermediate module between the MT and the satellite. This module adapts the satellite signals to the MT interfaces and vice versa, and enables full independence from the terminal segment. The satellite component ensures traffic transportation between local networks and the public network. This has the following advantages:

- Reduced investment and delay in the development due to a possible reduction in complexity/constraints on the terminal design because the system is compatible with existing terminals, and thus enables early introduction of services.
- The user does not have to learn the usage of another terminal with a different man-machine interface. The environment of the user is not affected. This is increasingly important as the number of features in a terminal grows.
- The subscribers are only faced with a small additional fee for the services delivered through satellite.

- The S-UMTS may be improved and optimized for capacity as well as bandwidth performance provided that the booster accommodate new features or S-UMTS evolutions.

Two system configurations may then be envisaged, collective and individual. These are shown in Figures 5.3 and 5.4, respectively. A system supporting both can also be envisaged.

In the collective configuration, the satellite-based system is inserted within a RAN of the T-UMTS. The system is used in a trunking mode and transports the traffic exchanged between the terrestrial network and the local network. The intermediate module constitutes an entry point for a local network. It consists of a part of the radio access network or a single BS. It provides UMTS services to all terminals within the coverage area. Rapid installation of the intermediate module could be an advantageous feature.

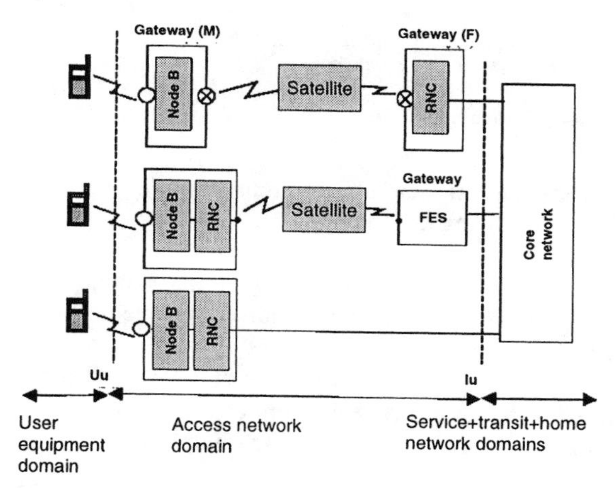

Figure 5.3 Collective configuration. (*From:* [12].)

Installation on a building roof or terrestrial mast for Earth fixed coverage and on-board a vehicle transporting passengers, as well as maritime and aeronautical applications, can be foreseen for this type of coverage. The approach of individual configuration is similar to that of direct access to satellite systems except that it is based on a distributed terminal concept (mobile station). It consists of booster equipment and a standard terrestrial terminal. The booster converts the satellite signals into a format compatible to the short-range wireless interface of the terrestrial terminal. It relies on the assumption that mobile stations will support such short-range wireless interfaces to connect phone accessories as well as computing devices.

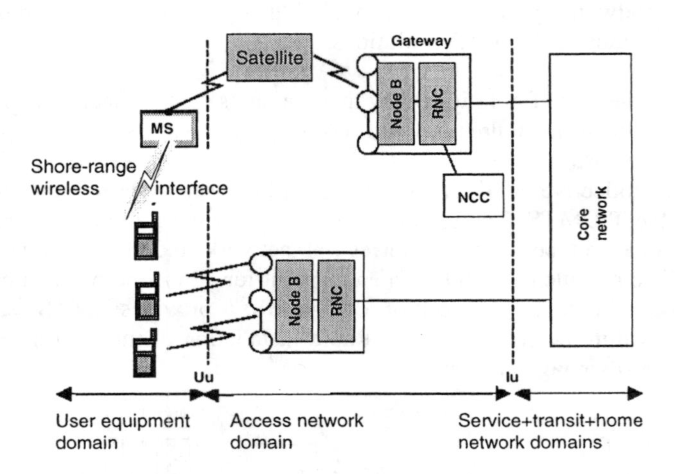

Figure 5.4 Individual configuration. (*From:* [12].)

To reach the largest market, different kinds of booster may be envisaged according to the following criteria:

- *Mobility capability criteria:* The transportable or nomadic types are longer in size but can be installed in a vehicle or easily carried in a suitcase.
- *Service capability criteria:* Voice and low rate data only; video, voice, and high data rate; traffic asymmetry for video, voice, high data rate on downlink and voice, low data rate on uplink.

Basically such systems can address nearly the same market as the configuration of direct access to a satellite because most of the market segments identified can be targeted with a terminal in a distributed configuration (several parts). In most cases, a nomadic terminal is able to satisfy the needs of the users. It can either be a transportable terminal or a terminal installed on-board a vehicle.

Broadcast Oriented

The S-UMTS is based on similar transport capabilities provided by the DAB and/or DVB technology.

The end user benefits from T-UMTS services and can simultaneously access services offered by the S-UMTS terminal configurations in two modes:

- *Indirect access to the satellite or distributed terminal configuration:* an external module enables a terrestrial terminal to benefit from broadcast services offered by the S-UMTS. The inter-connection between the terminal and the external module can be realized using short-range wireless technology (see Figure 5.5).
- *Direct access to the satellite or integrated terminal configuration:* the terrestrial terminal contains embedded functions to benefit from the broadcast services (Figure 5.6).

Figures 5.5 and 5.6 show that the user benefits from broadcast/multicast services either with an integrated or a distributed terminal. Another configuration could be envisaged which provides transport for broadcast/multicast traffic towards a type of base station, typically known as the high rate packet node. This enables lower congestion within the UMTS RAN (UTRAN). Such a system supports all unidirectional services. Broadcast can be supported for public information and multicast for value-added services (VAS).

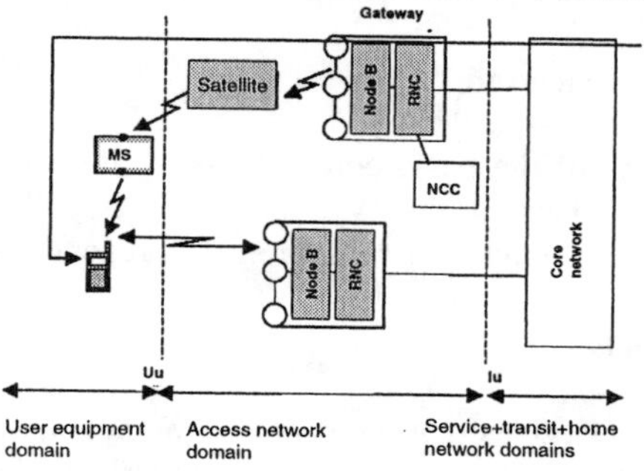

Figure 5.5 Indirect reception from the satellite. (*From:* [1].)

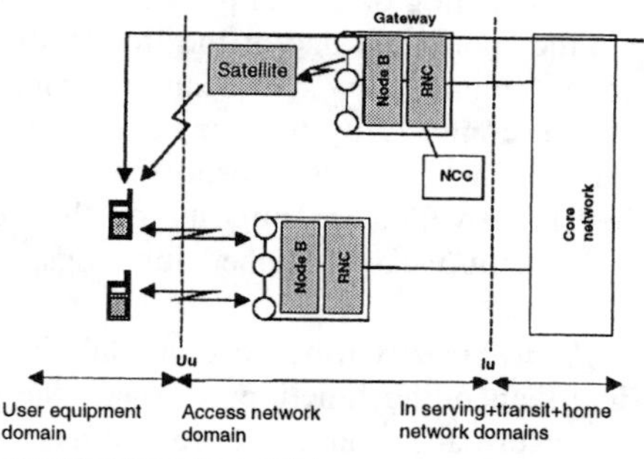

Figure 5.6 Direct reception from the satellite.

Terminals equipped with positioning devices can filter useful broadcast/multicast information according to their geographical location, their subscriber profile, or other criteria.

Gap Filler Concept

Broadcast and multicast are considered as promising candidates for S-UMTS services, and the mass market for them is in and around built-up areas (urban areas). But the two broadcast scenarios shown in Figures 5.5 and 5.6 are not suitable for urban areas because of the following reasons:

- There is no direct satellite reception inside the built-up area because of the high blockage.
- Users tend to use their mobiles inside buildings.

A gap filler is thought to be the solution to solve the problem of urban area satellite coverage. The gap filler acts as a repeater in both directions or in one direction depending on the services. Because voice is not considered here, the services are mainly asymmetric. The nature and the position of the gap fillers were investigated in detail in the IST project SATIN [1].

5.2.2 Space-Segment Selection

The space segment is one of the basic components of any satellite system. Two choices must be made regarding the type of satellites (GEO or a non-GEO constellation) and the type of payload (transparent and regenerative).

5.2.2.1 Payload Aspects

The main functional added values that the space segment can bring, in addition to its traditional signal amplifying function, can be summarized as follows:

- *Connectivity:* Transparent analog or digital processors can offer a layer 1 connectivity between the spot-beam and/or the frequency channels, thus allowing either regional (among a few spot-beams) or global (through the whole satellite coverage) connectivity in a single satellite hop; this is of particular interest in the case of a multibeam coverage. Regenerative payloads can further improve the granularity of the connectivity by implementing layer 2 circuit/packet-switched functions, or multiplexing functions.
- *Link performance enhancement:* On-board demodulation/re-modulation and possible decoding/reencoding functions alleviate the constraints put on the transmission performance, and as a consequence, on their cost. Low-cost satellite terminals are usually characterized by low-level transmitting power and poor local oscillator stability that could prevent meeting the link budget through a transparent payload.
- *Flexible use of the satellite resource:* This aspect is closely related to the level of granularity that the satellite payload possesses (i.e., on-board

access to, for example, layer 2 packets), and also to resource management techniques (which allows allocation of a given resource—channel, carrier, slot—to a specific demand more or less dynamically, and care for minimizing the congestion occurrence). A digital regenerative on-board processor enables on-board buffering and statistical multiplexing of layer 2 resources. These two functions contribute to the improvement of the resource management.

From the above argument, a hybrid payload looks promising to handle the mixture of needs in a balanced and cost-effective manner. A space segment based on digital bent-pipe with multibeams will surely not meet all the service requirements (but still meet most of them) and will probably be less effective. On the other hand, it is open to waveform upgrades, is less complex, and is based on well-tried techniques and technology.

5.2.2.2 QoS and Resource Management

Although the IP QoS framework does not come directly into play in the RAN procedures, ultimately these procedures have to serve the QoS requirements of the IP [transport control protocol (TCP) and the real-time transport/user datagram protocol (RTP/UDP)] flows. There are a number of points that can be made regarding the issues arising here. These are summarized as follows:

- *TCP-adaptive flows:* In a satellite environment, when crossing wireless links, error-prone links are aggravated, mainly because of the higher propagation delay. Quite evidently, the larger the propagation delay, the more adverse the effects on the TCP throughput, implying that in this context a LEO constellation is preferred over a GEO. On the other hand, a LEO constellation introduces variable RTTs, which impose difficulties on the TCP congestion control mechanisms (timers, ACK sequence). Things become worse in cases where the connection path involves more than one LEO satellite. This is the case when a LEO regenerative payload is under consideration, since the variation of the connection's RTT increases given an increased probability of a handover within the satellite constellation.
- *Real-time flows:* Regarding the RTP/UDP flows that will bear the real-time services, the conclusions are in a sense similar; that is, a GEO satellite introduces more difficulties for interactive services such as voice (the UMTS conversational class of service) especially when the payload does not have on-board switching functionality. In this case, direct single-hop, mesh connectivity is not feasible and the end-to-end delay for two users of the satellite network exceeds the upper limit of 400 ms end-to-end delay. On the other hand, a LEO constellation might introduce

more jitter and affect jitter-sensitive services (e.g., impose a larger playback buffer for streaming applications).

- *Call admission control (CAC):* The resource management procedures and the CAC function in S-UMTS face extra challenges due to the time-varying capacity available (particularly LEO/MEO) to the calls/connections served by the network. The task of preserving the required QoS/GoS while making an efficient utilization of the network resources becomes even more difficult as a direct consequence of the increased mobility of the satellites and the variability of the network [particularly if interactive satellite links (ISLs) are assumed]. There are ways around these difficulties [14, 15] relying on the periodicity of these orbits which allows some prediction of the future capacity changes. However, they demand efficiently accurate positioning methods and introduce significant complexity to the system.

Session Initiation Protocol in S-UMTS

The S-UMTS is regarded as an access network to the all IP-based core network, and the SIP functionality is mainly found in the IP multimedia (IM) subsystem in the core network. Thus S-UMTS only relays SIP messages to the respective SIP servers (i.e., the call state control functions). Therefore, the SIP messages are transparent to S-UMTS. The type of constellation used for S-UMTS does not have much impact on the performance of SIP. However, the only impact (foreseeable) of S-UMTS is the longer propagation delay.

Entities that are able to parse SIP messages are those that are SIP-enabled (in UMTS, these are the UEs and CSCFs). S-UMTS with bent-pipe configuration will not parse SIP messages. Likewise, a satellite with on-board processing will not parse SIP messages (this is because the protocol stack will go up to the IP layer at the most, but to parse SIP, it needs to go up to the application layer). Based on the capability of the satellite, this has no impact at all on the SIP.

5.2.2.3 Mobility

If the GEO constellation is used, there will not be big differences between cellular IP and IP-based S-UMTS in terms of mobility management except for the long delay which will increase the handover delay. For non-GEO cases, three types of handover are possible. The effectiveness of each handover depends on the capability of each node in the network and IP adaptation to each node.

- *Spotbeam handover:* It can be considered as micromobility. The rate is very high. There should be a very efficient handover mechanism to guarantee required QoS with packet mode.

- *Satellite handover:* This can lead to micromobility or macromobility depending on the connectivity of the satellite with the ground station.
- *FES handover:* This comes under macromobility.

Location management is not that crucial compared to handover but, still, location management with GEO is much easier than with non-GEO. It does not directly contribute to QoS improvement of the system except in saving some of the signaling load.

5.2.2.4 Multicast

Multicast applications fit better in the GEO paradigm. Their wide coverage in combination with the complexity related to the implementation of multicast (IP or not) within a satellite constellation renders GEO more appropriate for this purpose. In particular, regenerative satellites with on-board switching capabilities can be beneficial for the system capacity budget—a packet only needs to be transmitted once at the uplink and can be copied multiple times on-board.

5.2.3 Multiplexing Mode (FDD/TDD) Selection

There are a number of aspects to be taken into consideration, and for each aspect there are advantages and disadvantages for any technique. However, it may be appropriate to critically review some of the claims usually put forward to sustain the TDD case, with reference to the satellite application. These are outlined below.

TDD allows significant terminal size reduction. It is obviously true that for TDD a frequency duplexer is not needed; however, to limit adjacent band interference, a high-quality RF filter (with its insertion loss) is still a necessity. Furthermore, for FDD, duplexers of the same size as the ones used in T-UMTS terminals may be used in S-UMTS terminals provided that the satellite is sufficiently large to compensate for the low G/T figure yielded by these duplexers. For the sake of fairness, it should be added that some circuit reuse is possible for TDD and that the CPU never operates simultaneously over forward and reverse link bursts, thus saving battery power.

TDD efficiently handles asymmetric traffic. It is not possible to accommodate the asymmetric traffic requirements of every single user, but only those of the aggregate traffic, which in general is less unbalanced (particularly for multicast services). Second, even considering the aggregate traffic, it should be clear that in a multibeam scenario the necessary degree of intra- and interbeam switching point synchronization will largely reduce the flexibility of TDD.

TDD efficiently exploits channel reciprocity. Unfortunately, at the FES and on-board the satellite, channel reciprocity is spoiled by the fact that the propagation delay is large with respect to the channel coherence time. Channel reciprocity may be exploited at the MT.

These facts certainly do not rule out the TDD option, but somewhat reduce its appeal at least for the satellite scenario.

The choice of the FDD option becomes a necessity when the most relevant issues in selecting the duplexing mode are as follows:

- Spectrum allocation;
- Opportunity of exploiting commonalities with T-UMTS;
- Flexibility in adapting to ever-changing market and traffic scenarios.

5.2.3.1 Summary

A hybrid type payload with some switching functionality would best serve the S-UMTS service requirements.

Selection of multiplexing modes for S-UMTS shows FDD as the best candidate for S-UMTS considering the terminal and payload complexity and interoperability with the T-UMTS.

5.2.4 Proposals for S-UMTS Air Interface

For S-UMTS, it is necessary to establish the critical mass of customers needed to provide affordable service. It is a common belief that the inability of satellite systems to provide urban and indoor coverage has prevented the MSS industry from achieving its potential success as a provider of competitive services to all areas, including rural and remote areas [17].

In order to overcome that problem, ground repeaters/intermediate module repeaters in urban and rural areas, and on highways can be introduced as a solution [17].

This proposed solution allows S-UMTS operators to extend multimedia services to indoor and urban areas, this way addressing a mass market in terms of coverage. It also makes S-UMTS terminals (SMT) more consumer-friendly and affordable. The attractiveness (in terms of cost, mainly) of the SMT is anticipated to be a decisive factor for the S-UMTS success since the potential UMTS customer is not willing to pay much more than what he pays now for a 2G/2G+ terminal.

This section explains the SATIN approach on IMR to provide multicast and broadcast services as a terrestrial UMTS complement in a more efficient way to the mass market. The IMR concept is explained for different environments and with its functionalities.

5.2.4.1 IMR Module Concept

This section explains possible IMR scenarios, which can target the mass market and a variety of services. The following issues may be different for different scenarios or may be the same.

- IMR functions (e.g., just like a booster);
- Interfaces SAT-IMR and IMR-SMT.

Urban and Suburban Environments

Figure 5.7 shows the arrangement of an IMR capable of satellite reception inside the built-up area and inside the buildings. There are two possible service scenarios, broadcast and multi-cast only services via satellite to the local users and full services via satellite to international roamers. However, the IMR may also be just a repeater without incorporating any functions of RNC or node-B.

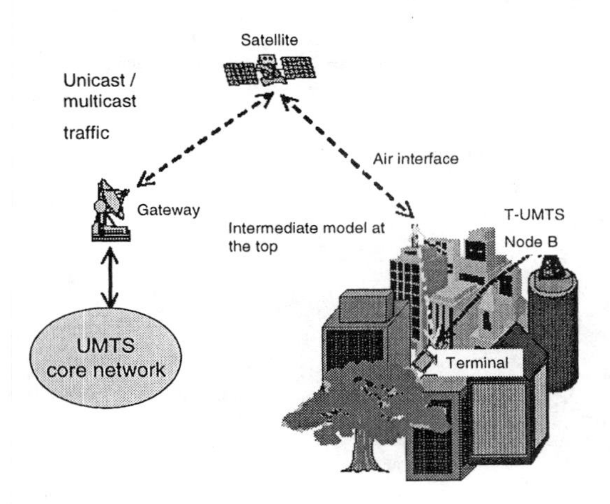

Figure 5.7 IMR in urban environment. (*After:* [1, 17].)

Vehicular and Highway Environment

IMR positions for the in-car application and the respective configurations are shown in Figure 5.8. The IMR can be just a repeater, and, hence, the terminals use the satellite mode or the IMR can translate the signal into terrestrial form so that the terminal can use the terrestrial mode.

Ship, Plane, and UMTS Islands Case

In this scenario (except UMTS islands), the IMR may feature a node-B or a simple repeater functionality. In the UMTS islands case, the satellite link represents the interface between the UTRAN and the CN (Iu). This scenario is shown in Figure 5.9.

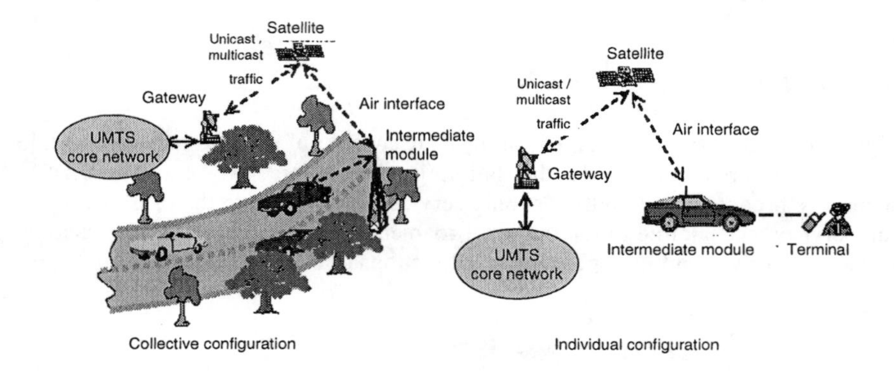

Figure 5.8 Vehicular or highway environment. (*After:*[17, 18].

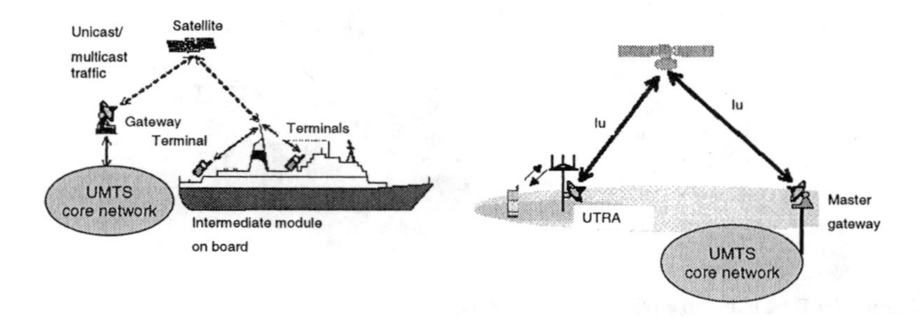

Figure 5.9 Remote environment (ship, plane, and UMTS islands). (*From:* [17].)

IMR Possible Functionalities

There are different possibilities for defining the functionalities of IMR and the interfaces SAT-IMR and IMR-SMT considering the following points:

- Multicast and broadcast services can be well served by satellite.
- Terminal complexity should not increase significantly due to the introduction of the IMR.

A big constraint experienced by the terrestrial system was placing the base stations in a cost-effective and environment-friendly way for demonstration purposes [1]. Therefore, the satellite industry may also experience the same problem in installing the IMRs.

In the case of Figure 5.10, the IMR acts as a simple repeater.

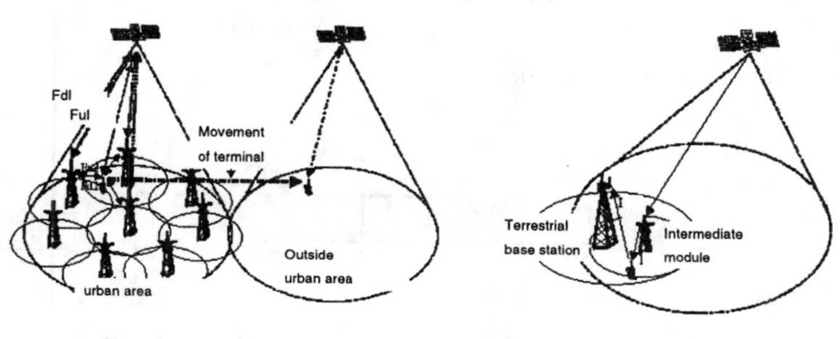

Figure 5.10 Simple repeater case.

The IMR receives the signal in the S-UMTS band from the satellite, then amplifies and retransmits it towards the terminal. Similarly, it receives the signal from terminals and transmits it towards the satellite. The same frequency band may be used for both links, namely the SAT-IMR link and the IMR-SMT. Alternatively, different bands may be used for each link; in the latter case the IMR features frequency conversion capability. Therefore, the terminal can receive the same signal from two or more IMRs as shown in Figure 5.11 similar to multipath propagation. When the terminal moves out of coverage of the IMR, it can directly communicate with the satellite since the signal attenuation is very low outside the built-up area. Hence, the S-UMTS mode can be used at the terminal inside and outside the built-up areas.

Contrary to the terrestrial case where the signal received from other cells is considered as interference, the signals transmitted by other IMRs can be considered as multipath signals except when the IMRs are located in different spot-beam coverage areas. Here a trade-off exists between the IMR system cost and the terminal complexity.

The multipath arrival delays of the signals coming from different IMRs will mostly be larger than the arrival delays of the multipaths caused by reflections of the signal coming from the IMR closest to the terminal. Extending the RAKE search window (larger delay line) implies on one hand a more costly terminal, but on the other hand a similar amount of signal code power can be received with lower power IMRs or less densely distributed IMRs.

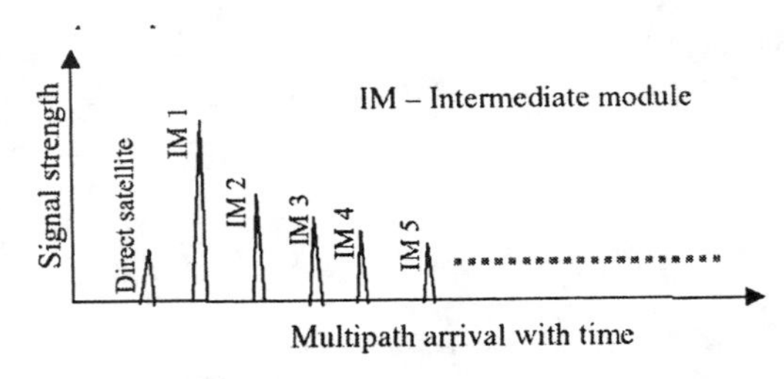

Figure 5.11 Same signal through different IMRs.

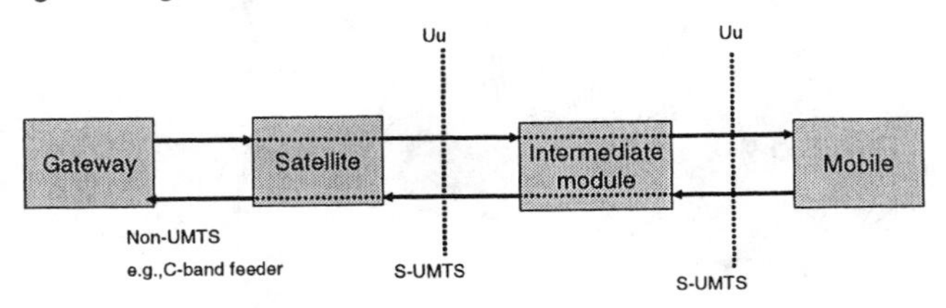

Figure 5.12 A bidirectional repeater.

In the IST project SATIN [1], two types of repeaters are considered based on the SATIN architecture concept [18]: a bidirectional (see Figure 5.12) and a unidirectional (see Figure 5.13) simple repeater.

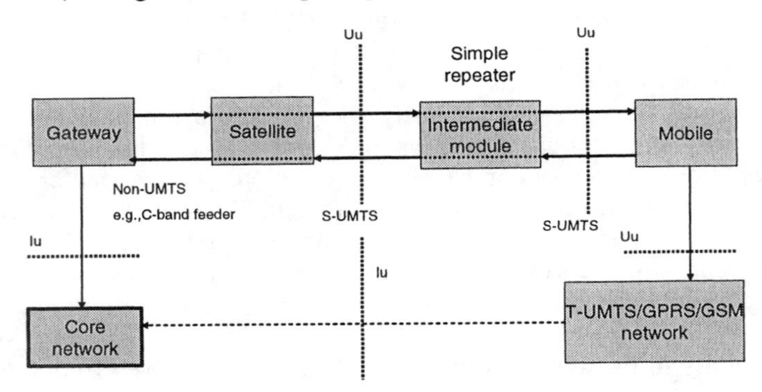

Figure 5.13 A unidirectional simple repeater.

In the bidirectional case, both the downlink and the uplink will use S-UMTS frequency bands. The positive aspects of this approach can be summarized as follows:

- The creation of a multipath environment becomes possible; a RAKE receiver in the terminal can exploit this and enhance the signal-to-noise-interference ratio (SNIR) of the signal. Note that this is limited to urban areas; in rural environments the channel still has a Ricean/ LOS character.

- Effectively, everywhere/anytime coverage becomes possible, because the terminal can communicate pseudo-directly to a satellite in an urban environment and directly in an open environment.

The negative aspects are summarized as follows:

- Only slow inner loop power control is possible due to a large propagation delay between the IMR and the satellite. (Note that the implied comparison is made with a case where the intermediate module features some node-B or an RNC functionality.) Power control instructions will be given on a frame-to-frame basis (100 Hz instead of 1,500 Hz as in T-UMTS). This will result in a serious decrease in the ability to compensate for fading channels.
- There is no possibility to implement any form of PC that will regulate the transmit level of the IMRs to mitigate intraspot-beam interference.
- Terminals will have to be dual-mode for both the Tx and Rx chains, and hence more expensive.

Terminal Considerations

It seems difficult to design low-cost power effective-handheld terminals that can handle the full-rate uplink straight to the satellite, as is the case for rural areas not covered by the IMRs. This does not necessarily mean that the receive-only scenario is the only option left. For low data rates the processing gain can be high enough to boost the uplink signal sufficiently at the satellite receiver. Hence, an asymmetrical link scenario (multicast/broadcast) seems feasible for handheld terminals.

S-UMTS handheld terminals, when used for the geographical complement concept, will only be able to provide low uplink bit rates. A possible way to alter the uplink bit rate is to use an extension module with enough transmit power connected to the terminal (laptop or PDA) with a short-range wireless link or a cable or to use the nomadic terminal. The highest performance will probably be reached when using a vehicular IMR that can either be a simple repeater or a short-range wireless interface/S-UMTS converter, because in this case available Tx power will be highest.

The unidirectional case has the following advantages compared to the bi-directional case:

- The IMR complexity (and cost) will be greatly reduced because the RF front-end must only be capable of receiving from a satellite and transmitting to the mobiles.

- The terminal complexity (and cost) can be made considerably lower because it must only be able to receive S-UMTS. The most cost-saving factor in that case is the considerably reduced complexity of the RF/IF part. Power consumption will be considerably lower because no S-UMTS has a Tx in the terminal. Other benefits related to the terrestrial uplink infrastructure features that are feasible are the (fast) uplink PC and RAKE combining (T-UMTS).

If T-UMTS is selected for the uplink, the geographical complement concept is violated. If the uplink is GSM/GPRS, the geographical complement is in a way achieved, since not many areas are outside the GSM/GPRS coverage, but the uplink capabilities will of course be insufficient to support full T-UMTS services. Additionally, the T-UMTS/GSM/GPRS uplink might get overloaded. This is, however, rarely the case, since the targeted set of services are broadcast/multicast.

This scenario seems to be the most interesting when the geographical complement is not the main objective.

Although different types of terminals (in terms of T/S-UMTS capabilities) will probably be available on the market soon, customers will need to pay more to get terminals with uplink S-UMTS capabilities (low rate).

IMR Simple Repeater with Some Functionalities of Node-B

Depending on the extra cost involved, some node-B functionalities could be implemented in the IMRs. The functions of interest are summarized as follows:

- Power control;
- Multipath reception (RAKE Rx).

As indicated in Figure 5.14, the IMR will need to be able to communicate with the mobile in a direct and independent manner. One or more control channels per mobile user will need to be present to manage the envisaged functionalities. Inherently, the IMR must be capable of performing demodulation and remodulation of the control signal.

This will place demand on the digital part, its complexity depending on the functionalities to be included. The analog part (RF) will also be more complex because some additional filtering, frequency conversion, amplification, and A/D-D/A conversion will be necessary.

It is important to note that the IMR is still a repeater, so the interface with the satellite is the same S-UMTS interface as between the IMR and the mobile. For most of the signals the IMR will be transparent.

PC is an essential feature of any CDMA-based cellular system. The mechanism to be considered in this scenario is the inner loop PC (both for the uplink and the downlink). It continuously adjusts the terminal transmit levels in order to meet a specified SNR (depending on needed QoS) set by the outer loop PC. Open loop PC involves the RNC and is certainly not to be implemented in the IMR (see Figure 5.15).

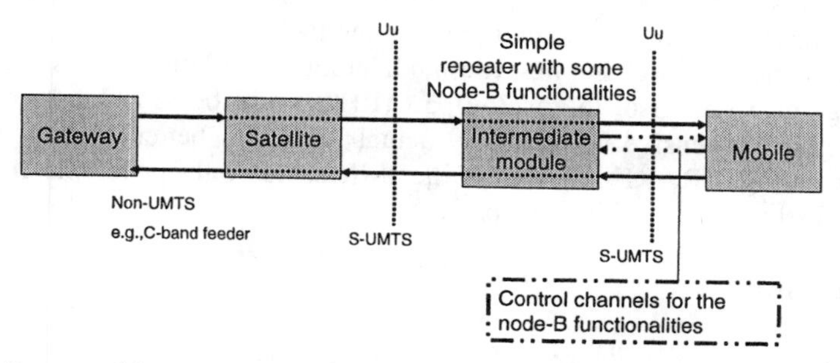

Figure 5.14 Repeater with some node-B functionalities.

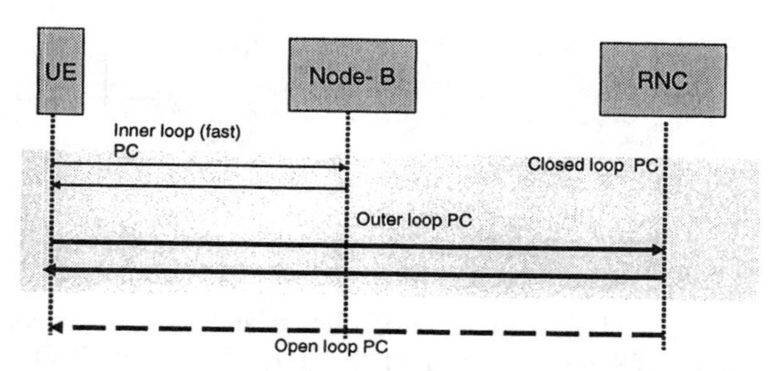

Figure 5.15 WCDMA power control mechanisms.

The main reasons for implementing PC are the near-far problem, the interference-dependent capacity of the WCDMA system, the limited power source of the terminal, and the presence of fading channels. But only the latter would really require fast inner loop PC (1,500 Hz). A frame-based PC (100 Hz) should be sufficient to effectively handle the other drawbacks. If the PC signal would only have to travel the distance between mobile and IMR, a T-UMTS, like the PC mechanism, could be implemented and fading would be effectively mitigated.

The channel between the mobile and satellite does not usually have Rayleigh multipath characteristics. The satellite environment with the IMR, however, can be seen as a multipath environment, the same way as in T-UMTS. The multipath characteristics can be exploited by incorporating a RAKE receiver into the module. An advantage of putting a RAKE receiver into the IMR as opposed to having a RAKE only in the gateway is that the IMR would demand less transmitted power.

Putting a RAKE receiver in the gateway makes it possible to exploit macro-diversity (from different IMRs). If the IMR uplink transmitted power is not really an issue, there is probably no considerable benefit since the path between IMR and the gateway should not really distort the signal (only path loss), so the multipath characteristics of the signal prevail and can be exploited by a RAKE in the gateway. The large increase in complexity, however, raises the cost and this outweighs the gain in implementing PC in the IMR.

Implementing PC implies (de)modulating capabilities and some decision-making software. Also, the analog part (RF/IF) will become more complex and thus more expensive. The only actual gain is better fading mitigation. Implementing a RAKE receiver in the IMR seems only beneficial if the IMR's uplink Tx power is a critical factor.

IMR as a Node-B

This case is shown in Figure 5.16.

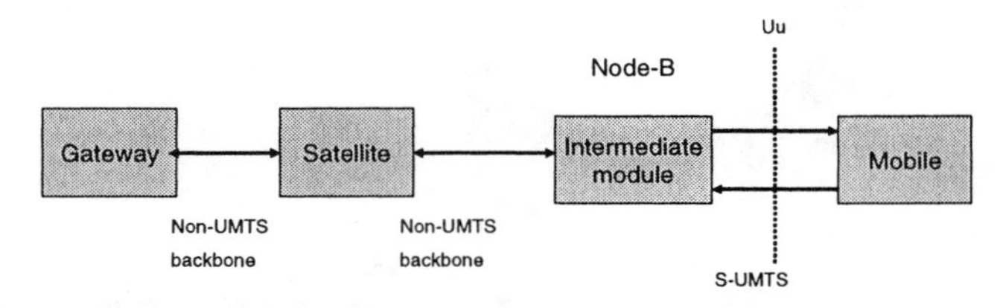

Figure 5.16 IMR with node-B functionality.

This case is similar to the T-UMTS island case, with the difference that it uses the S-UMTS band instead of the T-UMTS band. ICO proposed a system similar to the one explained to handle the coverage in the urban environment. It is known as the Ancillary Terrestrial Component (ATC). It is important to note that the satellite only acts as a backbone network, or, as shown in Figure 5.17, there is no satellite involvement at all except when both satellite and ATC share the frequency band. It has been mentioned in the ICO proposal that there should be a single entity responsible for fully integrated operation of the MSS network in order to reduce the interference and share the spectrum.

IMR as Node-B and RNC

This setup could be interesting for the UMTS island scenario in the sense that the satellite link is responsible for the interface between the island and the UMTS core network. In some cases this could be much cheaper than connecting the island to the CN with cables. An island can be a remote, although relatively densely populated, area, a ship, or a stadium. It is shown in Figure 5.18.

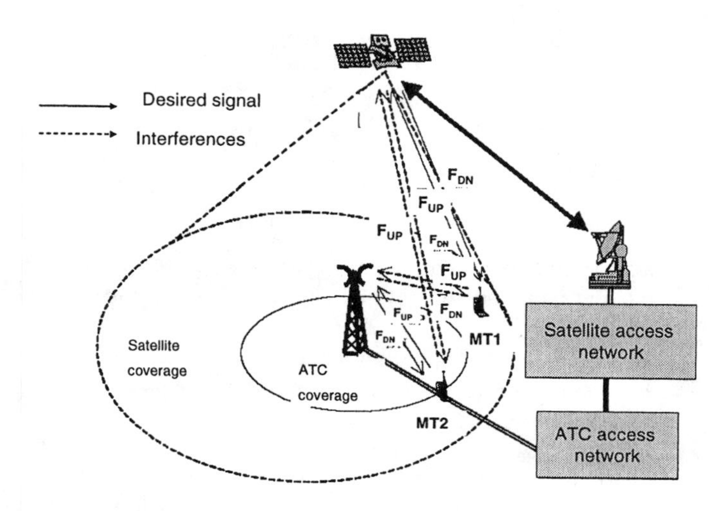

Figure 5.17 ICO forward band sharing mode.

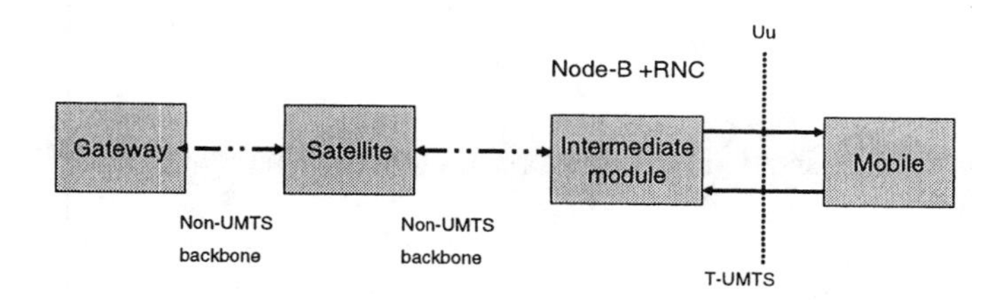

Figure 5.18 IMR with node-B and RNC functionality.

5.2.4.2 IMR Architecture

The overview of some basic architecture issues for the unidirectional simple repeater is given next.

IMR Functional Elements

The IMR's functions are limited to receiving, amplifying, and retransmitting the signal coming from the satellite towards the mobile. Therefore, the entire module can be kept analog, since only RF-related functions have to be implemented.

Figure 5.19 displays a simple model of a possible architecture. The components of this repeater type are limited:

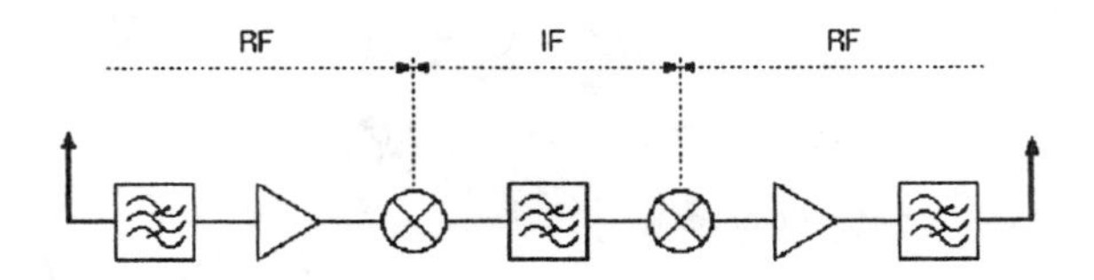

Figure 5.19 Architecture for unidirectional simple repeater.

- *Donor antenna:* The repeater antenna directed towards the satellite, picking up the downlink signal. This antenna should be highly directional.
- *Service antenna:* Omnidirectional antenna to cover the service area.
- *RF bandpass filters:* These determine the frequency range for operational configuration.
- *IF bandpass filter:* Defines the actual pass band and is a determining factor in important issues like out-of-band gain, delay, or error vector magnitude (EVM), for which a compromise will have to be made.
- *Mixers.*
- *Local oscillators.*
- *Low noise amplifier (input).*
- *Power amplifier (output).*

Typical characteristics of this repeater (partially based on T-UMTS repeaters) are as follows:

- Gain: 70 to 90 dB;
- Maximum output power: 30 dBm;
- Rx antenna gain: 28 dBi;
- Noise figure: 3.5 dB (G/T = 2.5 dB).

Most repeaters feature auto limit control (ALC) or automatic gain control (AGC), an adjustable limit for the output power to be able to inhibit out-of-band gain and emissions, and to prevent self-oscillation. Some mechanical characteristics are as follows:

- Size in cm: 40H×35W×30D;
- Weight: < 20 kg.

High power repeaters used for large coverage areas might imply a frequency separation of both links, because the antenna isolation requirement can become too strict. So for high power repeaters a trade-off exists between extra effort in antenna isolation (shielding, highly directional Rx antenna pattern) and the need for extra spectrum. The increase in repeater complexity as a consequence of the need for frequency conversion will be negligible.

Using a GEO constellation, highly directional antennas are possible, and because the majority of repeaters will probably not be large coverage area oriented, the most favorable option with respect to efficient spectrum usage seems to be a single-frequency repeater.

Indoor reception will already be greatly improved by the outdoor repeaters. Still, coverage dead zones might exist (e.g., in tunnels underground parking lots). The repeater will take the outside-received signal and retransmit it inside a building. These repeaters will be similar to the outdoor ones but will need less gain and less output power.

Another remark is to be made concerning moving IMRs (on a ship or a train). The presence of Doppler frequency shifts due to the relative motion with respect to the satellite will most likely not ask for a different type of IMR, since the shifts will be very small with respect to the signal bandwidth.

Some extra attention should be given to the IF filter characteristics. In the case of moving IMRs, it can be better to use a wider although steeper filter characteristic to tolerate a slightly frequency shifted signal with a minimum amount of distortion. The Doppler frequency shift will be removed in the user equipment.

5.3 SATELLITE BROADBAND MULTIMEDIA SYSTEMS

Traditionally, separated worlds such as telecommunications, computing, and broadcasting have shown in the recent years a tendency to converge towards common objectives. This had led to a scenario that is characterized by instability and continuous changes. Research is engaged in identifying common guidelines for this chaotic dynamic in order to define a universal communication framework [20].

These should be capable of accommodating, on one hand, the needs of various types of customers, and on the other hand, the exigency of network operators to invest capital in network infrastructures that will be obsolete only in the long term. The heterogeneous character of future mobile communications will be clearly supported by the Internet technology [21] that has become a real standard for interconnecting remote applications in the last few years, thanks to the independence of the supporting underlying technology.

In such a scenario, satellite systems can play a key role because of the following advantages:

- Wide-area coverage;
- Fast deployment;
- Intrinsic broadcast capabilities.

ATM technology [22] has been considered for a long time an attractive solution for the support of multimedia applications, and a great deal of research has been engaged in the past few years in investigating efficient solutions for the support of ATM in satellite systems. Yet, the unexpected booming of Internet technology in recent years has led the satellite industry to consider IP over satellite platforms. There is the problem, however, of how to exploit the investments made up until now in the satellite technology, and possibly, to find a way of converting it, in the light of the new emerging scenarios. The IST project BRAHMS [10] provided a solution to all the previous issues by harmonizing the majority of the common satellite access network functions, while allowing flexibility by the development of a "module" able to interwork with optimized or proprietary air interfaces for the space segment. BRAHMS developed a new concept for generic broadband satellite access networks providing high-speed multimedia services to a range of users, known as BMSS. This concept is one option for a future common platform for satellite access networks connecting to core networks such as the Internet and B-ISDN.

The BMSS platform was demonstrated among the maritime society: several European ferry line operators recommended that a selected ferry along its maritime route serve as the mobile platform for the demonstration. As maritime conditions are very different from land, a special digital satellite TV measurement system had to be used for the performance analysis [24]. The terminal for the reception from the ship also required special consideration [25].

In the long term, medium delay applications requiring high bandwidth will be served via a start configuration using transparent satellites (Ku/Ku, Ka/Ku, Ka/Ka). For applications requiring a short delay and a high bandwidth, a mesh structure is possible, using in this case on-board-processing (OBP) satellite transponders (similar to the existing Skyplex transponders) [26]. Having OBP will allow data to be routed from any of the uplink coverage footprints onto any combination of downlink coverage footprints [27].

While many wireless systems can be used to deliver interactive broadcasting (e.g., LMDS, UMTS) [28] (see Chapter 6 of this book), satellite systems have a major advantage of being able to serve a large customer base in a large service area including rural environments once the satellite is in orbit. Figure 5.20 shows the reference model of a satellite interactive network.

Generally, in the satellite interactive network, the NCC provides the monitoring and control functions. The traffic gateway (TG) will process the data from the return channel satellite terminal (RCST) [29] and reroute the data. The feeder will broadcast the forward link signal, which may include user data and/or the control and timing signals.

The forward traffic to the user is multiplexed into a conventional DVB/MPEG-2 broadcast stream at a broadcast center (at the hub) and relayed to the RCST. The broadcast stream is transmitted in the Ku band using QPSK modulation and concatenated convolutional and Reed-Solomon coding (providing a maximum forward data rate of approximately 45 Mbps). Turbo coding can also be used. The return channel path will be operated as part of a digital network, with the hub station providing the gateway to other (satellite and terrestrial) networks.

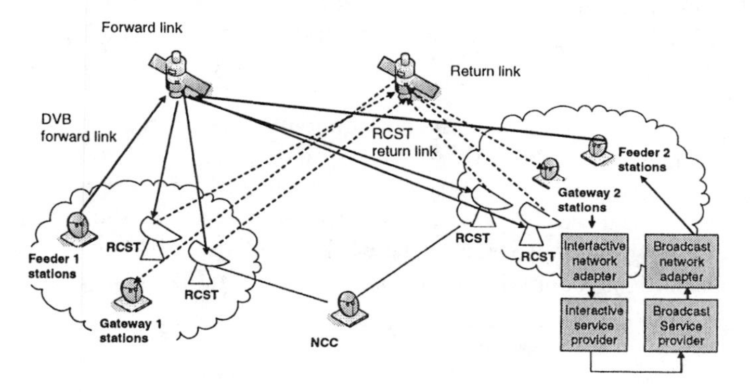

Figure 5.20 Reference model for the satellite interactive network. (*From:* [28].)

A return path from the individual user can be provided from an RCST with a fixed, small antenna (0.5 to 1.2m) and an end user terminal (i.e., easy-to-use, small, low-cost equipment) to an interactive server at a hub station using a multiple access scheme. The RCST also acts as a router/multiplexer for different traffic sources. Where applicable, the individual return link components will be routed from the hub to their final destination. To achieve this, a scheduled MF-TDMA scheme to access the network is employed. MF-TDMA was selected over other access schemes such as CDMA because of its high efficiency and relatively low complexity of implementation. MF-TDMA allows a group of RCSTs to communicate with a hub using a set of carrier frequencies, each of which is divided into time slots. The hub allocates to each active RCST a series of bursts, each defined by frequency, bandwidth, start time, and duration.

The return channel air interface specifies how the traffic and signaling data will be transmitted. The design of the air interface is important because it ensures that the network can operate reliably during the communication processes between the RCST and the hub equipment. More details of the frame structure and access techniques are provided below.

5.3.1 Generic Access User Interface

While satellites are increasingly being used for high-speed links in the Internet, this use is presently focused on long distance access to nodes.

The access network, involving the largest number of point-to-point links in a telecommunication network, also requires the largest share of investment. The BMSS as an access network has therefore the greatest potential for cost savings since it can obviate many new "holes in the ground" for cabling. The star topology of an access network where many users are connected to an access node is well suited to a satellite system with a large gateway station acting as the node at the center of the star of small user terminals. The satellite system can also be designed to provide mesh (i.e., multiple user to user) connections in parallel.

The typical asymmetry of multimedia services (especially IP) traffic is also well suited to satellite links, which lend themselves to be engineered with greater return than forward capacity.

The BMSS has features that are common to other terrestrial broadband access networks, such as the ETSI BRAN, xDSL, and WLAN, in order for the user terminals to be used interchangeably between them as far as possible. The BMSS complements these access technologies since their reach is somewhat limited. Interworking between these access networks via core networks, and via the Internet especially, is required. The scenario is shown in Figure 5.21.

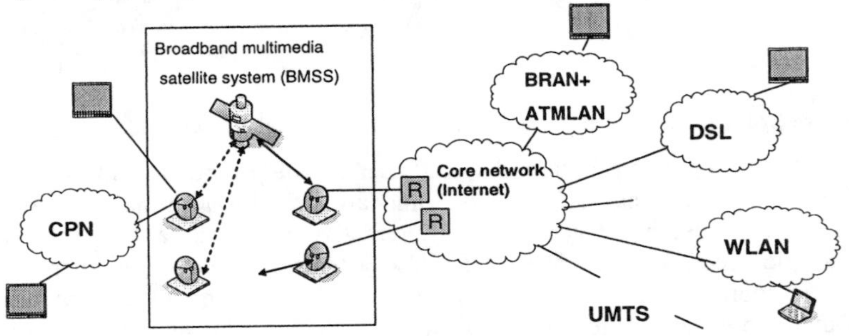

Figure 5.21 End-to-end broadband internetworking scenario. (*From:* [28].)

A number of new broadband satellite systems, such as Teledesic and Skybridge, have been proposed in recent years to ensure direct user access. Each of these systems is a proprietary solution with little or no intercompatibility of satellites or user terminals.

In terms of transmission formats, the main options for support of multimedia services at present are based on ATM or DVB packet types, although in the longer term, a dedicated transport scheme optimized to carry IP datagrams could prove more efficient.

To open up and expand the market for satellite user terminals and for satellite systems in general, a generic solution for satellite access networks is necessary in order to allow harmonization of many access network functions while allowing flexibility for optimized or proprietary air interfaces (e.g., frequency, access type, orbit) to satellite systems.

The BMSS is therefore a system that allows maximum commonality of terminal functions and common interfaces between them. These features should be common to different broadband satellite (and terrestrial) access networks to ensure easier production, cost reduction, and multiple choice of access (i.e., multimode) for terminals. The satellite communications market as a whole will benefit from users not having to worry that an investment in a terminal will tie them to one system and service provider, as is now the case for GSM terminals.

BMSS can offer, where possible, compatibility with current satellite systems at some level. BMSS can serve as a convergence path from current satellite systems to a more generic next generation system with common terminal architectures supported by new standards.

The need for standardization in broadband satellite systems has been recognized by the industry, which has resulted in the establishment of a working group on Broadband Multimedia Satellite (BMS) Systems in ETSI, to which the IST project BRAHMS [10] contributed.

5.3.1.1 System Design Approach

The commonality and flexibility of the BMSS systems architecture (e.g., for frequency, access type, orbit) is obtained by separating physically related functions from common service and access functions, as indicated in Figure 5.22.

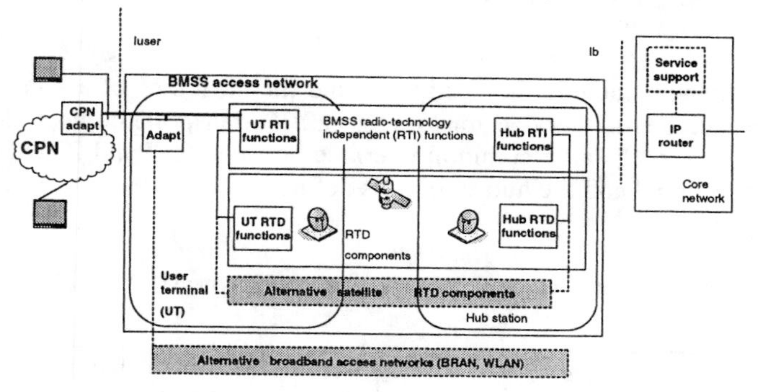

Figure 5.22 The BMSS system architecture.

The separation of higher layer radio-technology independent (RTI) access network functions, which "hide" the lower layer dependent (RTD) functions from the user and the core network, is a key strategy for the BMSS and is known as the generic radio access network (GRAN). The common RTI layers in the user and hub stations in access networks are designed for IP transport of the higher protocol layers across the system, and for connection to alternative customer premises and core networks. The RTI layers should therefore be transparent to a full range of IP-based multimedia services (i.e., broadband and Internet). Different RTD layers are intended for different satellite systems (e.g., LEO, GEO constellations) and different radio access schemes (e.g., TDMA/FDMA/CDMA frequencies) within the access network, and each will support an extensive but different subset of the services of the RTI layer suited to particular markets.

This RTI/RTD concept has similarities to the UMTS access/nonaccess stratum architecture, which although intended for different services may also lead to further commonality between systems and terminals.

The BMSS architecture defined in this way enables easier interworking and roaming of user terminals between alternative satellite (and terrestrial) networks.

A network adaptation layer middleware function at the user terminal side performs service adaptation in order to match the requirements of an application service to the functions and QoS available from a selected access network.

The separation of network functions between the RTI and RTD layers is one of the first considerations to be made in the system design. These functions fall into user, control, and management planes. An initial separation of some of the main functions of interest for these layers is summarized in Table 5.1.

The use of the IP protocol in BMSS is considered in terms of its ability to ease interworking and speed convergence towards seamless broadband service provision; that is practicality in the satellite access network and compatibility with the RTI/RTD approach.

RTI Layer Requirements

The main challenges include the need to define the attributes of a common RTI layer, which includes all service functions (e.g., IP transport) compatible with a range of RTD layers. Next, a common interface between RTI and RTD layers (in both the user terminal and the hub station) is defined.

Table 5.1

Separation of Main Functions in the BMSS System [30]

Function	Plane	RTI Layer	RTD Layer
Call setup/release	Control	Yes	No
Bearer setup/release	Control	?	Yes
Location	Management	Yes	No
Resource management	Management control	Yes (network resources)	Yes (radio bearer)
Handover (nonGEO)	Management	Yes	Yes
Encryption	User	?	?
Congestion	User	?	?
IP packet	User	Yes	No
QoS management	Management	Yes	No

The RTI part includes all end-to-end service control functions with interworking towards core networks such as B-ISDN and Internet. Mobility is included for user roaming and intelligent network (IN) functionalities such as virtual home environment (VHE) in which the user's desired operating system environment is downloaded from his home service provider, allowing services to be tailored flexibly to each end user independently of his location. This part will also address interworking between the BMSS and other fixed/mobile terrestrial/satellite segments, and thus issues of fixed-mobile and satellite-terrestrial convergence.

RTD Layer Requirements

The RTD layer is not unique, but some examples suitable for high-speed multimedia services are considered based on Ku- and Ka-band GEO and LEO satellites and with selected access schemes (e.g., TDMA, FDMA, DVB), which are well suited to packet transport. Different forward/return link combinations are envisaged according to interactivity requirements for two main user types to be addressed: direct-to-office (DtO) and direct-to-home (DtH) multimedia services. MAC protocols [31] based on those of wireless terrestrial networks or on generic wireless IP schemes investigated in the IST Program can be adopted where possible. The DVB (-RCS) multiple access scheme needs to be included as an option in view of its increasing importance for DtH services and in terms of the population of terminals installed. Turbo coding schemes are being examined as a means to increase capacity whilst maintaining error rates compatible with packet transport.

The satellite itself is included in the RTD layer. Several types of satellite configuration will be considered including transparent and regenerative, GEO and nonGEO. For regenerative satellites, on-board IP routing and multicasting will be considered.

5.3.1.2 Common Transmission Medium

IP is attractive as a common transport medium, and ideally it should be transmitted over the radio medium in a form as close to its "native" format as possible in order to retain its full features and to avoid the inefficiency of overheads.

Radio-link resources in general (including satellites) are limited and precious and need to be employed and shared efficiently in a public or shared network between the users requesting services.

Therefore, radio resources are best used by the establishment of connections for each user, which is at odds with the connectionless nature of IP services. Typically, radio capacity is shared on time-multiplexed channels, and on these channels fixed-length slots or packets (e.g., ATM) are preferred for easier resource control and allocation. IP packets, however, are of variable length, which is not easily matched to efficient capacity sharing. Also IP headers alone can reach uncomfortable lengths, which in the case of RTP/UDP/IP, typical for real-time multimedia, are 40 bytes for IPv4 and 60 bytes for IPv6. Lossless header compression is one solution. Header stripping/reinsertion is another. Both of these methods involve processing of IP packets in the BMSS type of architecture.

Furthermore, special support or adaptation of IP is therefore needed over satellite links to ensure the following features:

1. QoS for multimedia services;
2. Fair access;
3. Efficient use of radio resources for bursty traffic.

A challenge is therefore to conceive a more IP-dedicated transport scheme for shared, QoS enabled systems compared to existing satellite transport schemes such as IP/ATM (or IP/ATM-like), IP/MPEG/DVB, and IP/DVB, and to evaluate relative efficiency and cost-effectiveness.

Other issues are delay and bit errors (over a GEO satellite in particular) which may limit throughput especially at the TCP layer. Solutions are available (e.g., use of RTP, spoofing) which have been addressed previously [32], and which also involve processing of TCP/IP packets in the BMSS.

The concepts described are embodied within the overall BMSS protocol architecture that is shown in Figure 5.23.

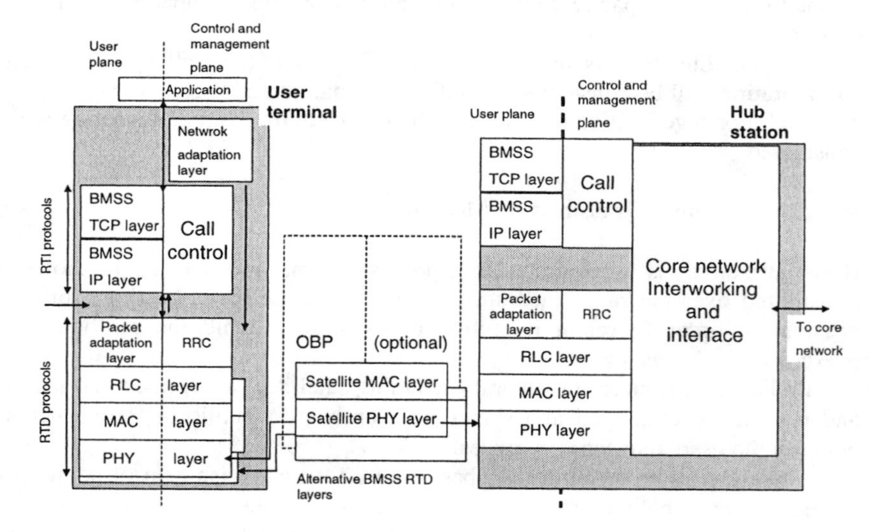

Figure 5.23 Overall BMSS protocol architecture.

5.3.2 Security in Satellite Networks

The IP-dedicated satellite access scheme is a new access solution intended to enhance the compatibility between satellite systems and the Internet [33–35]. This is mainly achieved by providing a connectionless layer 2, based on label switching, that is naturally able to exploit the satellite multicast capabilities, and that has been designed with built-in security functionalities. The design of the IP-dedicated scheme is motivated by the objective to offer a connectionless transfer capability in order to better adapt to the IP connectionless nature. It aims also at alleviating some of the inherent drawbacks of connection-oriented schemes with regard to the Internet: complex protocols, latency, and signaling load.

The current generation of satellite packet technologies, which operate on the basis of "partial security by default" (e.g., DVB-RCS), provide insufficient security support, especially in satellite networks with natural broadcast/multicast capability over large areas.

The use of well-known upper layer security protocols such as secure socket layer (SSL) or IP security/Internet key exchange (IPsec/IKE) [36–38] may be recommended but are dedicated to unicast communications.

New security mechanisms able to secure transparently both unicast and multicast transmissions are introduced by the Sat IPsec protocol [33]. IPsec is the most widely used security protocol in terrestrial networks. It provides strong security mechanisms. It is, however, not well adapted to satellite systems particularly with regard to their native multicast capability.

The term IPsec refers generally to the IPsec data plane protocol and to the IKE control plane protocol [38], which allow strong, secure point-to-point communications, by providing entity authentication, data confidentiality, integrity and message origin authentication services. They can be implemented in multicast environments, but as IKE is only a point-to-point protocol, it is not an optimized solution. The following example illustrates this point.

A group of satellite terminals securely exchange multicast data. For that purpose, they use IPsec (and IKE). They establish IPsec tunnels between themselves with IKE. In order to protect all traffic transmitted on satellite links, an IPSec tunnel (a secure bidirectional link) has to be established between each satellite terminal pair. As a consequence, with a large number N of satellite terminals, the security architecture establishment is very expensive, as it requires $N(N-1)/2$ tunnels.

Besides, the security configuration (i.e., encryption and authentication algorithms, key generation) applied to data is different for each IPsec tunnel, as it is negotiated by both the terminals that have established that tunnel. A satellite terminal sending a multicast packet will therefore have to duplicate the packet, independently encrypt the copies, and send them on their respective satellite link (see Figure 5.24).

If n is the number of terminals, for each packet, $n - 1$ encryption computations will be performed by the sender for securing only one multicast session. This solution does not take advantage at all of the natural broadcast capability of the satellite link.

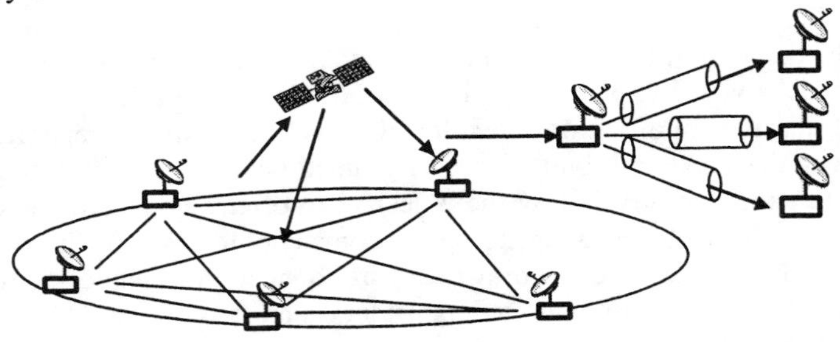

Figure 5.24 IP multicast over IPsec/IKE: duplication of the IP packets. (*From:* [33].)

Finally, in the case of interconnection of several types of networks, additional security mechanisms would be required if other types of traffic (e.g., ATM or ethernet traffic) need to be protected. This represents an additional drawback for IPsec.

Sat IPsec, when implemented at layer 2 (IP dedicated), provides a more efficient alternative to those nonoptimized mechanisms.

In the first place, a security layer-2 solution allows securing transparently any type of traffic (IPv4, ATM, ethernet) without additional security mechanisms.

Second, the native multicast capability of IP-dedicated schemes offers to Sat IPsec the capacity to enable multicast security mechanisms, optimized for the satellite context.

In the previous example of a group of satellite terminals, the Sat IPsec requirements are to limit the number of session establishments and encryption computations and to avoid the duplication of multicast packets over the satellite links. For that purpose, Sat IPsec can be based on a centralized management solution (achieved by a central server) and on a new key exchange protocol called flat multicast key exchange (FMKE). A secure link is established between each satellite terminal and the central server by FMKE, as shown in Figure 5.25.

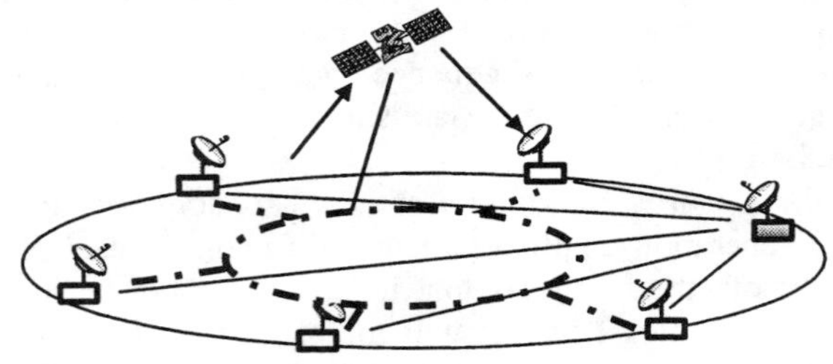

Figure 5.25 Use of Sat IPsec for the exchange of secured multicast data.

The secure links are used to send a common data security configuration to all satellite terminals (sent by the central server). Thus, any terminal having to send a multicast packet encrypts it once according to the common security configuration, and sends it in multicast on the satellite link. The packet can then be decrypted by all terminals. One can notice that the security architecture establishment is therefore less expensive than with IPsec ($N-1$ tunnels in the control plane and 1 in the data plane). In Figure 5.25 the connections in the control plane are given with solid lines, while the connections in the data plane are given with dashed/dotted lines.

The central server is called the Sat IPsec master module. It is integrated in a satellite terminal and is in charge of authenticating the satellite terminals and providing them with sufficient information for securing data transmissions, according to the required security level (cryptographic functions, security services, keys). All this information is defined in tables called security associations (SA). The SA are distributed to each new authenticated satellite terminal if it is authorized to get access to it.

These modules apply security processes (encryption, authentication algorithm) on unicast and multicast IP packets at their entry in the satellite system and send them in tunnel mode.

Upon reception, the client modules have to decrypt these packets before sending them on the terrestrial side. These processes are shown in Figures 5.26 and Figure 5.27 for unicast and multicast transmissions, respectively.

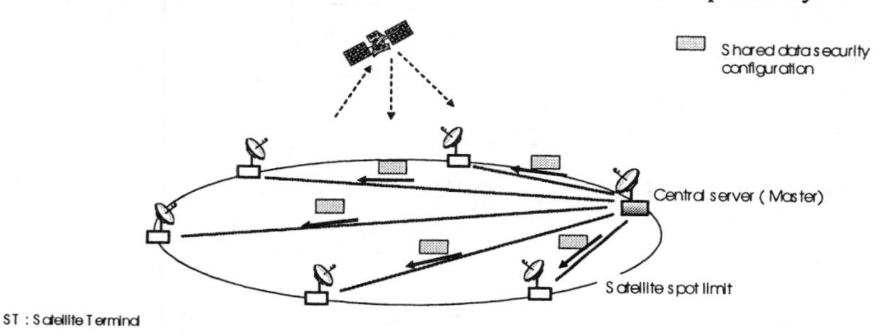

Figure 5.26 Secure transmission of the shared data security configurations in unicast. (*From:* [33].)

The Sat IPsec mechanisms are directly implemented inside the IP stack of the satellite terminals as a module called Sat IPsec Client. The control plane is implemented at layer 3, over the IP layer, while the data plane is the IP-dedicated security data plane, implemented at the DLC IP-dedicated sublayer. The control plane and the data plane are therefore separated, as shown in Figure 5.28.

The Sat IPsec protocol allows the flexible setup of large secure multicast networks in a flat environment (no intermediate routers between the master and a large amount of clients). It is well adapted to satellite or terrestrial mobile systems that want to efficiently offer multicast secure services. This protocol is able to configure the well-known IPsec stacks and it integrates a possible functional extension that optimizes the wireless bandwidth.

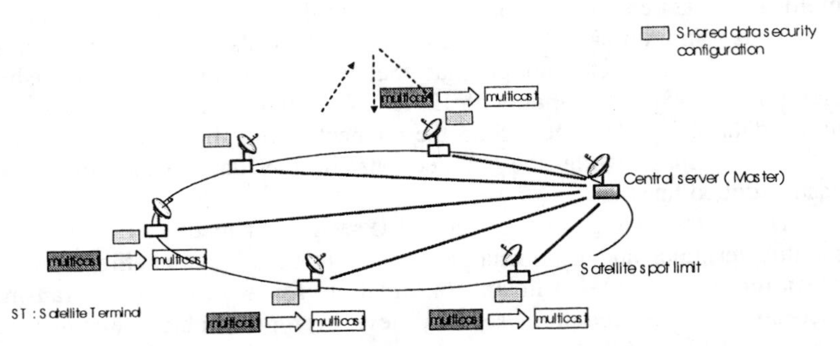

Figure 5.27 Multicast encryption, transmission, and decryption with Sat IPSec. (*From:* [33].)

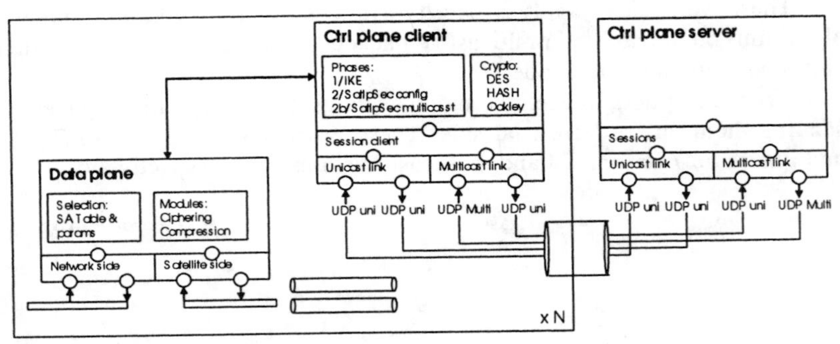

Figure 5.28 Sat IPsec functionalities.

5.3.3 Summary

The previous sections described a modular satellite access network architecture for the BMSS for multimedia service provision and optimized IP transport. Such an architecture allows standardization of broadband satellite terminals while allowing flexibility of the underlying physical satellite transmission. The BMSS architecture was defined in terms of the split of RTI/RTD functions.

IP-dedicated transport schemes for shared, QoS-enabled systems are suitable to increase efficiency and cost-effectiveness compared to existing satellite transport schemes based on segmentation into fixed packets (e.g., ATM, DVB).

Next generation networks will allow for uncoupling services and transport provision in packet-based (IP), multimedia/multiservices/multinetworks, QoS, and security.

5.4 HIGH-ALTITUDE PLATFORMS

The increasing demand for higher data rate wireless mobile communications services has accelerated the need to develop more innovative communications infrastructures. Terrestrial ground-based systems and satellite systems are the existing well-established ways of providing mobile communication services. At the present time, technologies and services are emerging indicated by an increasing interest in combining two or more technologies and packing them to offer a new form of applications and services. It is not only the matter of higher and higher data rate services; the main issue is moving towards a user-centric approach that will actually answer the question: what kind of services do users really require, including the data rate, coverage, mobility, security, identity, and location services. In this context, HAPS systems emerge having enormous potential for support of future synergic convergence wireless systems by offering:

- *Customizability:* Most of the radio transmission technologies can be implemented on HAPS systems to provide wide additional coverage in the region. The range of available platforms nowadays and in the future can actually support the implementation of regional to global telecommunication services starting from the light information class of service such as telemetry, sensor network, and location services.
- *Adaptability:* Sometimes people see less endurance of any type of platforms as weakness of the system. However, this less endurance characteristic (compared to satellite endurance) can be taken as a positive value of a HAPS system as the system can then be maintained and upgraded on the ground. This poses less strict redundancy components on-board and also improves its adaptability to the requirements driven by technical and/or business reasons. For instance, initially HAPS can easily be deployed to support interactive text messaging services, which is already included by the capability of delivering regional coverage of sensor network, telemetry, and location services. If the demands are for additional services (for example, radio broadcasting via DAB), the system can be upgraded accordingly and ready in the next flight. With that, then, the service provider can now think more of providing a new form of services by combining interactive text messaging and radio broadcasting (which becomes an interactive radio broadcasting service). An additional source of revenue is then available.
- *Scalability:* Not only can the system support most radio transmission technologies available today, its coverage and capacity can also be upgraded following the demands.

An example of capacity evaluation of HAPS for 3G systems is presented in [39]. HAPS could bring more modularity to 3G, thanks to its easy setup and high capacity support. Its fast deployment rate, for instance, would allow it to accommodate emergency and events coverage. The capacity studies show HAPS as a real asset in some areas as a complement to any existing infrastructures.

Enormous efforts in HAPS development have been carried out for almost a decade. The ITU has also accepted the approach of delivering the IMT-2000 services in the frequency ranges 1,885 to 1,980 MHz, 2,010 to 2,025 MHz, and 2,110 to 2,170 MHz in Regions 1 and 3, and 1,885 to 1,980 MHz and 2,110 to 2,160 MHz in Region 2. ITU has also approved the use of 47/48-GHz and 28/31-GHz (Region 3 only) bands for the delivery of fixed wireless services using HAPS.

At WRC 2003, the 28- to 31-GHz bands were approved to be used by HAPS for fixed broadband applications, HAPS initiatives were successful at WRC-03 in increasing the number of countries with access to the HAPS allocations in these bands from the existing 12 Asian countries established at WRC-00 to more than 50 countries in North and South America, Europe, Africa, and Asia. This broad support was enabled not only by the projected benefits from utilizing HAPS technologies, but partly because of recently completed technical studies submitted to the ITU that show that sharing in these bands with satellite and terrestrial systems on a noninterference basis is feasible. The ITU HAPS allocations now consist of:

- 27 to 28 and 31 GHz (300 MHz in each direction shared on a nonharmful interference, nonprotection basis with fixed satellite and terrestrial services);
- 47 to 48 GHz (300 MHz in each direction, shared on a coprimary basis with fixed satellite);
- 2.1 GHz IMT-2000 (up to 50 to 60 MHz total bandwidth, to be used as an alternative to terrestrial towers).

The most challenging element of the concept to be proven is the platform development. Up to now, only one demonstration with limited environmental setup has been done. The demo was conducted by SkyTower [40]. The other projects are still under development.

The design of the platform and its station keeping mechanism has a direct impact on the performance of the HAPS communications system. Two types of HAPS platforms are currently proposed: the fixed wing flying aircraft (manned or unmanned) and the lighter-than-air airship. During operation, the fixed wing aircraft will fly in a tight circle, whereas the airship will be kept stationary at the stratospheric layer above the coverage area.

From an operational point of view, airspace above Flight Level 650 (19.8 km) is uncontrolled (i.e., there is no need to coordinate airship movements with air traffic authorities). There is very little civil traffic above Flight Level 450 (13.7 km). This level used to be the upper ceiling of controlled airspace (and still is in some countries) until the arrival of the Concorde in 1976, which pushed the ceiling to FL 650. Both fixed wing aircraft and airships can operate satisfactorily at these altitudes, but only the airship can theoretically remain geostationary. The fixed wing aircraft would have a latitude limit due to the requirement of subsonic flight over land but would also have a very large operational orbit as the margin between high and low speed stall decreases with the turn rate. A HAPS communications system consists of platform and ground-based equipment. A HAPS system is shown in Figure 5.29.

5.4.1 Comparisons Between Terrestrial/Satellite Systems and HAPS

The IST Fifth Framework program did not fund any project research on HAPS as this is an emerging technology. However, for the completeness of this book, HAPS are compared to the terrestrial/satellite systems.

Terrestrial tower-based systems and satellite systems are two well-established methods of providing wireless communication. One of the main reasons why HAPS-based communications systems are highly favored is the free-space like path loss characteristic. For example, for a HAPS located at 22 km above the ground, the path loss is comparable to a location at the edge of a small terrestrial ground-based cell with a radius of 2 km [41]. Also, the HAPS propagation channel is characterized by Rician distributed fading, while the ground-based terrestrial channel has deeper (i.e., Rayleigh distributed) fading. Other main advantages of HAPS are as follows:

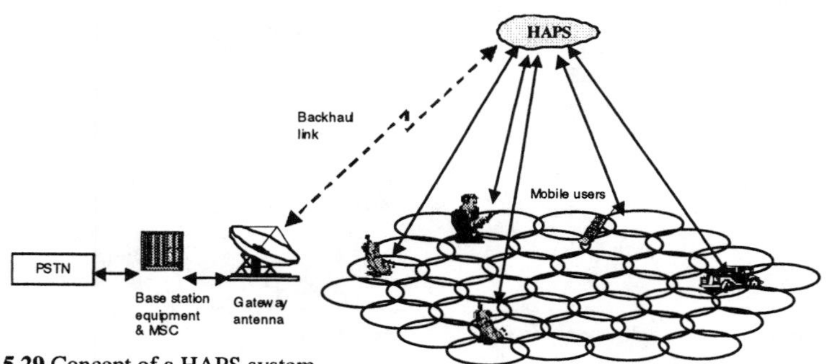

Figure 5.29 Concept of a HAPS system.

- Centralized architecture can be implemented, which improves efficiency in resource allocation and channel utilization.
- Synchronization among different cells is possible, because a single timer can be implemented. Hence, handover between cells can be softer and faster.
- System capacity can be increased by reducing the cell size through antenna beam shaping.

Unlike satellite systems, HAPS can be brought down for servicing and upgrading easily. Also, equipment upgrading can be easily done at a central location. Table 5.2 summarizes the main advantages of HAPS over terrestrial and satellite systems.

Table 5.2

Summary Comparison of HAPS, Terrestrial, and Satellite Systems

Issue	Terrestrial Wireless	Satellite	High-Altitude Platform
Availability and cost of mobile	Huge cellular/PCS market drives high volumes resulting in small, low-cost, low-power units	Specialized, more stringent requirements lead to expensive, bulky terminals with short battery life	Terrestrial terminals applicable
Propagation delay	Not an issue	Causes noticeable impairment in voice communications in GEO (and MEO to some extent)	Not an issue
Health concerns with radio emissions from handsets	Low-power handsets minimize concerns	High-power handsets due to large path losses (possibly alleviated by careful antenna design)	Power levels like in terrestrial systems (except for large coverage areas)
Communications technology risk	Mature technology and well-established industry	Considerable new technology for LEOs and MEOs; GEOs still lag cellular/PCS in volume, cost, and performance	Terrestrial wireless technology, supplemented with spot-beam antennas; if widely deployed, opportunities for specialized equipment (scanning beams to follow traffic)
Deployment timing	Deployment can be staged; substantial initial build-out to provide sufficient coverage for commercial service	Service cannot start before the entire system is deployed	One platform and ground support typically enough for initial commercial service
System growth	Cell-splitting to add capacity, requiring system reengineering; easy equipment update/repair	System capacity increased only by adding satellites; hardware upgrade only with replacement satellites	Capacity increase through spot-beam resizing and additional platforms; equipment upgrades relatively easy.

System complexity due to motion of components	Only user terminals are mobile	Motion of LEOs and MEOs a major source of complexity, especially when intersatellite links are used.	Motion low to moderate (stability characteristics to be proven)
Operational complexity and cost	Well understood	High for GEOs, and especially LEOs due to continual launches to replace old for failed satellites	Capacity increase through spot-beam resizing and additional platforms; equipment upgrades relatively easy
Radio channel quality	Rayleigh fading limits distance and data rate; path loss up to 50 dB/decade; good signal quality through proper antenna placement	Free-space like channel with Ricean fading; path loss roughly 20 dB/decade; GEO distance limits spectrum efficiency	Free-space like channel at distances comparable to terrestrial
Indoor coverage	Substantial coverage achieved	Generally not available (high-power signals in Iridium to trigger ringing only for incoming calls)	Substantial coverage possible
Breadth of geographical coverage	A few kilometers per base station	Large regions in GEO; global for LEO and MEO	Hundreds of kilometers per platform
Shadowing from terrain	Causes gaps in coverage; requires additional equipment	Problem only at low angles	Similar to satellite
Communications and power infrastructure; real estate	Numerous base stations to be sited, powered, and linked by cables for microwave	Single gateway collects traffic from a large area	Comparable to satellite
Aesthetic issues and health concerns with towers and antennas	Many sites required for coverage and capacity; "smart" antennas might make them more visible; continued public debates expected	Earth stations located away from populated areas	Similar to satellite
Public safety concern about flying objects	Not an issue	Occasional concern about space junk falling to Earth	Large craft floating or flying overhead can raise significant objections

5.5 CONCLUSIONS

Terrestrial tower-based systems have advantages such as low-cost, low-power user terminals, short propagation delays, and good scalability of system capacity. However, they have various disadvantages as well. The radio signal is subjected to high scattering and multipath effects that affect the delivered QoS. Furthermore, as the base stations are dispersed over a wide geographical area, communications resources cannot be optimally used. In addition, it is expected that more infrastructure is required, as smaller cells will be required to provide high-quality broadband services.

Although satellites can provide similar services with much less infrastructure and higher elevation angles, they have their own limitations. Geo-stationary satellite systems suffer large signal delays due to their large distances from the Earth, while nongeostationary satellite systems are more complex in design. Furthermore, high launching costs and limited orbit space lead to high connection costs.

An innovative way of overcoming the shortcomings of both the terrestrial tower-based and satellite systems is to provide cellular communications via HAPS [41].

References

[1] IST-2000-25030 Project SATIN, at http://www.ist-satin.org.

[2] Karaliopoulos, M., et al., "The Implications of IP on Satellite UMTS," *AIAA 20th International Communications Satellite Systems Conference,* 2002.

[3] Prasad, R., (ed.), *Towards a Global 3G System, Volumes 1 and 2,* Norwood, MA: Artech House, 2001.

[4] http://www.cordis.lu/acts.

[5] IST Project FLOWS, at http://www.flows-ist.org/.

[6] IST project GEOCAST, at http://www.geocast-satellite.com/.

[7] IST-1999-10167 Project VIRTUOUS, at http://www.ebanet.it/virtuous.htm.

[8] IST-2000-25355 Project FUTURE, at http://www.ebanet.it/future.htm.

[9] Evans, B. G., R. Tafazolli, and K. Narenthiran, "Satellite-UMTS IP-Based Network (SATIN)," *AIAA 19th International Communications Satellite Systems Conference*, 2001, available at http://www.ist-satin.org.

[10] IST-1999-10440 Project BRAHMS, at http://brahms.telecomitalialab.com/.

[11] FP6, at http://www/cordis.lu/ist/ka4/fp6.htm.

[12] Evans, B. G., and M. Mazzella, "Satellite-UMTS IP-Based Network," *Proceedings of the IST Mobile Summit*, Barcelona, Spain, September 2001.

[13] Karaliopoulos, M., et al., "The Implications of IP on Satellite UMTS," *AIAA 20th International Communications Satellite Systems Conference 2002*, at http://www.ist-satin.org.

[14] Siwko, J., and I. Rubin, "Call Admission Control for NonGeostationary Orbit Satellite Networks and Other Capacity-Varying Networks," *International Journal of Satellite Communications*, March to April 2000, pp. 87–106.

[15] Mertzanis, I., "ATM QoS Provisioning over Broadband Satellite Networks," Ph.D. dissertation, October 1999.

[16] Wood, L., et al., "IP Routing Issues on Satellite Constellations," *International Journal of Satellite Communications*, January–February 2001, pp. 75–85.

[17] Severijns, T., et al., "The Intermediate Module Concept Within the SATIN Proposal for the S-UMTS Air Interface," *Proceedings of the IST Mobile Summit 2002*, Thessaloniki, Greece, June 2002.

[18] SATIN Project, "SATIN Architecture specifications," Del. No. 3, February 2002.

[19] SATIN Project, "S-UMTS IP-Specific Service Requirements," Del. No. 2, October 2001.

[20] Delli Priscoli, F., T. Inzerilli, and V. Morsella, "BRAHMS Reference Architecture."

[21] "Internet Protocol," IETF RFC 791, September 1981.

[22] http://www.atmforum.org.

[23] Heiman, R., et al., "Optimization of Spectrum Utilisation in Satellite Access to Public Area Kiosk Network," *Proceedings of IST Mobile & Wireless Telecommunications Summit 2002*, Thessaloniki, Greece, June 2002, pp. 278–282.

[24] Papadakis, N., V. Gennatos, and S. Costicoglou, "Designing a Feasible Digital Satellite TV Measurement System for Maritime Scenario," *Proceedings of the IST Mobile & Wireless Telecommunications Summit 2002*, Thessaloniki, Greece, June 2002, pp. 269–273.

[25] Perez, J., et al., "MOBILITY Ku-band Terminal On Board Ships for DVB-S Reception," *Proceedings of IST Mobile & Wireless Telecommunications Summit 2002*, Thessaloniki, Greece, June 2002, pp. 275–277.

[26] "Plain Sailing with DVB-S," DVB News, Edition No. 2, September 2001.

[27] Chancon, S., et al., "Networking over the IBIS System," *Proceedings of IST Mobile & Wireless Telecommunications Summit 2002*, Thessaloniki, Greece, June 2002, pp. 283–287.

[28] Digital Video Broadcasting (DVB), "Interaction Channel for Satellite Distribution Systems," ETSI, EN 301 790 v1.2.2, December 2000.

[29] "Short Overview on Technical Aspects of Integration of Terrestrial and Satellite System," *Extract from Input of Telenor and Space Engineering to the ASMS-TF Commercial Requirements Report*, May 2001.

[30] Mort, R. J., "A Generic Approach to IP-based Broadband Satellite Multimedia Systems," *Proceedings of the IST Mobile Summit 2000*, Galway, Ireland, October 2000, CD-ROM.

[31] Perravi, H., "Medium Access Control Protocols Performance in Satellite Communications," *IEEE Communications Magazine*, March 1999.

[32] Balakrishnan, H., et al., "Comparison of Mechanisms for Improving TCP Performance over Wireless Links," *IEEE/ACM Transactions on Networking*, December 1997.

[33] IST SATIP6 Project, "Sat IPSec: General Presentation," ALCATEL SPACE INDUSTRIES, Toulouse, France, available at http://satip6.tilab.com/.

[34] Josset, S., et al., "Secure Multicast with the Connectionless IP-Dedicated Satellite Access Scheme," *Proceedings of 8th Ka-Band Utilization Conference*, Baveno, Italy, September 2002.

[35] Buret, I., et al., "IP Dedicated: A New Internet Oriented Satellite Transfer Scheme," *Proceedings of 19th AIAA Conference*, Toulouse, France, 2001.

[36] Kent, S., and R. Atkinson, "Security Architecture for the Internet Protocol," *RFC 2401*, IETF, November 1998.

[37] Kent, S., and R. Atkinson, "IP Encapsulating Security Payload (ESP)," *RFC 2406*, IETF, November 1998.

[38] Harkins, D., and D. Carrel, "The Internet Key Exchange (IKE)," *RFC 2409*, IETF, November 1998.

[39] Widiawan, A., and R. Tafazolli, "HAPS Capacity Evaluation for Mobile Communication Systems," *Proceedings of the 6th International Symposium on Wireless Personal Multimedia Communications*, Yokosuka, Japan, Vol. 2, October 2003, pp. 473–477.

[40] http://www.skytowerglobal.com/network.html.

[41] Djuknic, G. M., and J. Freidenfelds, "Establishing Wireless Communications Services Via High-Altitude Aeronautical Platforms: A Concept Whose Time Has Come?" *IEEE Communications Magazine*, September 1997.

Chapter 6

Interactive Broadcasting

6.1 INTRODUCTION

Telecommunications today is evolving towards increased interactivity. Customers want to choose, sort, order, store, and manipulate what they receive on their terminal, and ideally also interact from the same terminal [1]. There is a growing demand for much greater selection of programming on hundreds of channels, control over and customization of television content, on-demand delivery of video, and real-time interaction between people via game playing [2]. Because of increasing interactivity requirements, the traditional distribution infrastructure develops into an asymmetric interactive network, which includes the return path from the customer to the service provider. Evolution towards fully symmetric networks is also envisaged, given the increasing bandwidth requirements of future communications services.

Besides the telecommunications infrastructure, telecommunications applications are also changing to meet the demands of interactivity. The convergence between broadcasting and communication leads to an evolution from traditional broadcasting techniques to multicasting, which makes it possible for service providers to offer contents/services designed for individuals or a group of people with restricted access and billing [1]. There is also a significant growth in the variety of consumer equipment that could support interactive broadcasting applications, such as personal computers, video remote control keypads, keyboards, video game consoles, and information kiosks.

Interactive broadcasting has traditionally been delivered through wired infrastructures. With the growing demand for communications on the move, technologies are being developed to support high bandwidth interactive broadcasting services via the wireless medium. Interactive broadcasting may also be delivered via a combination of wired and wireless infrastructures from the service provider to the end user and vice versa. Here, we are concerned with wireless interactive broadcasting.

Wireless interactive broadcasting can be delivered through various wireless infrastructures, with different architecture and standards, such as terrestrial (GSM,

UMTS, DVB-RCT[1]) and satellite communications. The technological developments in the design of the forward and return wireless links to support interactive broadcasting are covered in this chapter.

Wireless communications can usually be divided into two broad categories: fixed and mobile. Both categories are equally important in delivering interactive broadcasting and both have their own unique markets that suit different needs and requirements.

The ultimate aim of the mobile wireless interactive broadcasting is to allow users to be able to link to the network anytime, anywhere. Although users can access services while on the move, the supportable data throughput is generally lower as compared to that supportable by fixed wireless systems. Examples of mobile wireless systems include GSM, UMTS, and DVB-RCT. Figure 6.1 shows an overview of a mobile interactive broadcasting network.

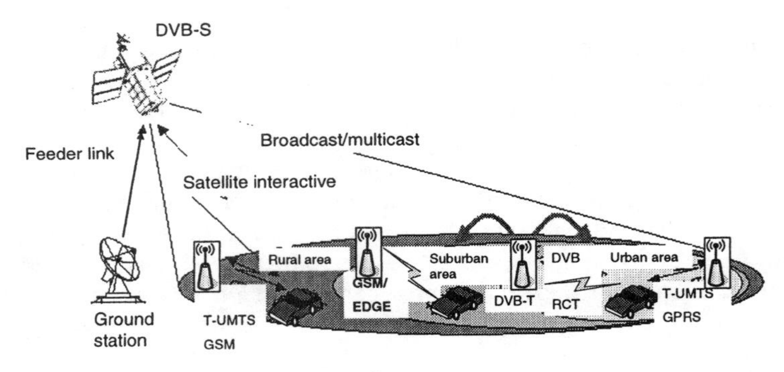

Figure 6.1 Mobile interactive broadcasting network.

Unfortunately, most of these mobile services do not have coverage in the most remote parts of the world and most of the ocean region. Therefore, in order to let the seafarers of the world have a fair share of the advantages offered by interactive broadcasting, satellite systems are set in place. The resulting technology is known as digital video broadcasting via satellite (DVB-S). A detailed discussion on satellite systems and technologies is provided in Chapter 5.

Interactive broadcasting services range from low bit rate applications (text, voice) to bandwidth hungry applications such as video (see Figure 6.2 for an overview). Usually, high data rate applications are provided by fixed wireless systems such as the local multipoint distribution service (LMDS). LMDS is a broadband wireless point-to-multipoint communication system operating above 20 GHz (depending on the country of licensing), which can be used to provide digital two-way voice, data, Internet, and video services. More recent advances in a point-to-multipoint technology offer service providers a method of providing

---

[1] DVB-RCT is a standard for terrestrial interactive digital video broadcasting (DVB-T). It is approved as EN 301 958 ETSI standard.

high-capacity local access that is less capital-intensive than a wireline solution, faster to deploy than wireline, and able to offer a combination of applications [3].

The availability of satellite and terrestrial UMTS allows that broadband interactive broadcasting services can be provided to mobile users on these networks. Although GSM cannot be used for broadband applications, its well-established infrastructure and standards can be used to support narrowband interactive broadcasting services without a large increase in capital cost. The GPRS radio access network (GERAN), which is based on the mature GSM system concept, can be used to support a limited range of higher data rate interactive broadcasting services. GERAN has a peak rate of 59.2 Kbps per time slot and 473.6 Kbps for eight time slots, which meets the 3G urban/suburban requirements of 384 Kbps.

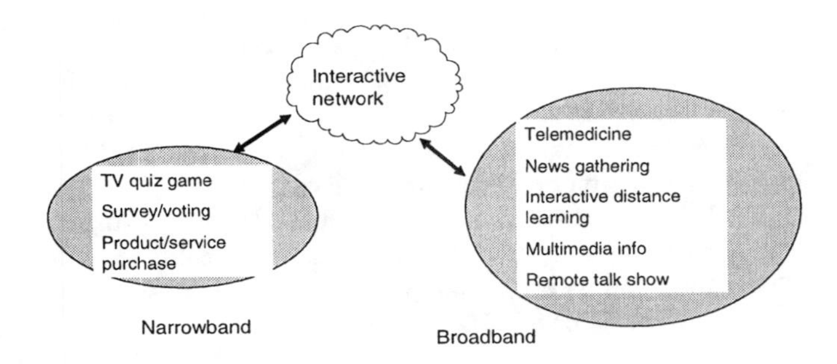

Figure 6.2 Interactive broadcasting applications.

Some new services and features have been recently offered to TV viewers, with no need of a return channel from the user to the service provider: data carousel or electronic program guides (EPG) are examples of such enhanced TV services that implement a rough type of local interactivity [4].

However, real T-commerce services require full interactivity, implying data and commands exchanged between the users and the service providers. The long connection time needed by existing solutions for a return channel based on telecom networks (PSTN, ISDN, or GSM) and the very limited bandwidth limits considerably the category of possible interactive services and the interest from users. Pay-per-view is an example of services that can cope with such long latency connection time. However, to implement real bandwidth demanding interactive service (e.g., Internet) or new services having a strong, real-time relationship with the TV programs (e.g., interactive advertising, televoting, and telequiz), a low latency, high-bandwidth return channel technology is mandatory.

DVB-RCT offers a wireless interaction channel for real-time interactive digital terrestrial television services. Some of the main features of this standard are described for completeness, as follows:

- DVB-RCT is spectrum efficient, low cost, and powerful and provides a flexible wireless multiple-access system based on the OFDM technique that is well suited for transmission in the terrestrial channel.
- DVB-RCT can serve large cells, up to 65 km in radius, providing a typical bit rate capacity of several kilobits per second for each TV viewer, even at the edge of the coverage area. Typically, these large cells closely match the downstream coverage area of the digital television broadcast signal.
- DVB-RCT can handle very large peaks in traffic, as it has been specifically designed to process up to 20,000 short interactions per second in tele-polling mode, this in each sector of each cell.
- DVB-RCT can be employed with smaller cells to constitute denser networks of up to 3.5-km-radius cells providing to the user a bit rate capacity of up to several megabits per second per user.
- DVB-RCT does not require access to spectrum on a primary basis; the system has been designed to use any gaps or under-utilized spectrum anywhere in the bands III, IV, and V without interfering with primary analog and digital broadcasting services.
- DVB-RCT is able to serve portable devices; bringing interactivity everywhere the terrestrial digital broadcast signal is receivable.
- DVB-RCT can be used around the world, which uses the different DVB-T system: 6-, 7-, or 8-MHz channels.
- DVB-RCT does not require more than 0.5-W rms transmission power from the user terminal or settop box to the base station.

A number of European projects focused on research in the area of interactive broadcasting.

The IST project OVERDRIVE (Spectrum Efficient Uni- and Multicast Services over Dynamic Multi-Radio Networks in Vehicular Environments) [5] developed and demonstrated efficient mobile multicast techniques based on a dynamic spectrum allocation scheme. The results of this project are discussed in Section 6.2.

The IST project FIFTH (Fast Internet to Fast Train Hosts) [6] focused on the provision of passengers on high-speed trains with Internet and digital broadcast TV services. In particular, an innovative solution for the QoS support was described, which was based on the adaptation and enhancement of the CDN approach to the mobile environment. The way to simultaneously manage the mobility profiles of passengers (by means of mobile IP) and the mobility of CDN servers mounted on-board the trains [by means of multiprotocol label switching (MPLS)] is presented in Section 6.3.

The IST project ULTRAWAVES (Ultra Wideband Audio Video Entertainment System) [7] developed a solution for wireless home broadband connectivity for audio and video entertainment devices. ULTRAWAVES-based systems support devices in the home requiring multistreaming of high-quality video and broadband media, including DVD, high-definition television (HDTV), PDA, MP3 player downloads, television, or even images from digital camcorders and cameras. As the UWB technology enables high data rate link, one can transfer a favorite home movie from a camcorder to a projector without using any cables. This solution is discussed in detail in Section 6.4.

6.2 DYNAMIC SPECTRUM ALLOCATION FOR VEHICULAR ENVIRONMENTS

The current method of assigning spectrum to different radio systems is a fixed spectrum allocation scheme. With this technique, radio spectrum is allocated to a particular radio standard, and these spectrum blocks are of fixed size and are usually separated by a guard band. The spectrum then remains solely for the use of the radio license owner. This method of spectrum allocation can control interference between differing networks using the spectrum, provided adequate guard bands are maintained, with no coordination between the networks. However, most communications networks are designed to cope with a certain maximum amount of traffic called the busy hour, which is the time of the peak use of the network. If this network uses its allocated spectrum fully during this hour, then the rest of the time the spectrum is not fully used. In fact, almost all services, such as speech, video, Web browsing, and multicast applications, which are envisaged as future mobile services, have distinct time-varying traffic demands. It will also be seen that the demand for different services on different networks depends on location, leading to a spatial variation in the spectrum usage. Therefore, the radio spectrum, while scarce and economically valuable, is often underused or idle at certain times or areas. This is the motivation for a more spectrum efficient technique, called dynamic spectrum allocation [8–14], which was investigated as part of the projects Dynamic Radio for IP-Services in Vehicular Environments (DRIVE) [15, 16] and OVERDRIVE [5, 17], with the objective to ensure delivery of multimedia services to the vehicle.

Although spectrum management bodies have already mentioned issues such as spectrum trading [18], little work had previously been done on the potential for DSA to improve the spectrum efficiency. The goal of DSA is to allocate only the amount of spectrum to a RAN that is required to satisfy the short-term traffic load within a given area for a certain user satisfaction level, thus making spectrum unused by one RAN available to other RANs. The DSA scheme that was investigated in the project DRIVE was called contiguous DSA. This scheme allocates contiguous blocks of spectrum to the RANs, which are separated by suitable guard bands. However, the widths of the assigned spectrum blocks can vary to allow for a changing demand. This technique allows the spectrum partitioning to change at the expense of a spectrally adjacent RAN bandwidth, allowing spectrum to be reallocated to other RANs when it is unused. This can be seen in Figure 6.3. Two aspects of DSA were investigated in DRIVE. One section investigated changing the spectrum allocations over time, known as temporal DSA, and the other over space, called spatial DSA. A complete report on the results and investigations that were performed can be found in [9], and a summary can be found in [8].

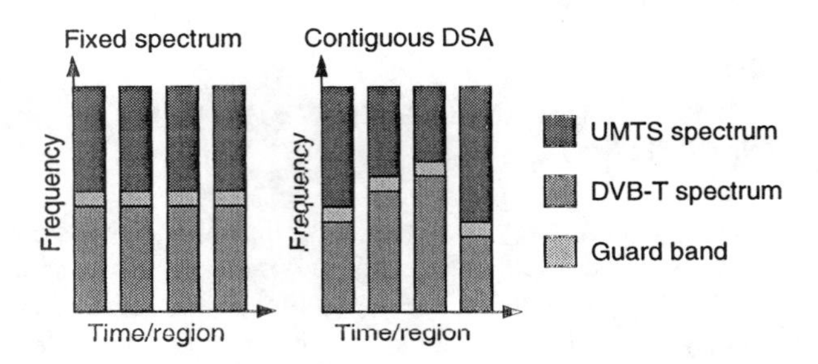

Figure 6.3 DRIVE contiguous DSA scheme.

6.2.1 Basic DSA Concepts and Results

As stated, the DRIVE work investigated the DSA problem on two levels. It considered how to allow the allocations of spectrum to adapt to temporal changes in demand, and also how to dynamically allocate spectrum over space. The concepts and basic results for each of these parts are described below.

6.2.1.1 Temporal DSA

An algorithm has been designed to implement temporal DSA, as described in [9, 10]. The algorithm runs at set periods, and the calculation of the spectrum required by each RAN is based on a prediction of the offered load until the next reallocation. There are two main aspects to the prediction: a load history, which is a database of the loads seen in the past, and a time-series prediction algorithm. If unusual traffic events occur, then the load history alone is not able to adapt, leading to inappropriate spectrum allocations, so time-series prediction is used to estimate the loads. From the load prediction, the RANs estimate the number of carriers they will require for the forthcoming time interval. They declare to the DSA if they have any currently unused carriers that could be reallocated, as a carrier can only be allocated to another RAN if there are no ongoing calls on it. An allocation algorithm decides how the free carriers are distributed. This spectrum allocation then applies until the next DSA run. To ensure that as many carriers as possible are free for DSA, all new calls are supported on carriers farthest away from the guard bands between RANs, and are handed over to these carriers whenever possible, making the ones nearest the guard bands free for reallocation.

The operation of a temporal DSA system relies on the fact that the time-varying traffic patterns that are seen on RANs sharing the spectrum are different, or at best uncorrelated. Typical traffic patterns for a speech service and a video service can be seen in Figure 6.4, as could be seen on RAN such as UMTS and DVB-T. These curves are inspired from [17–19]. The y-axis shows example traffic loads for the services. A fixed assignment scheme would require enough spectrum for each RAN to support its peak loads, shown at (a) and (b). The overall amount of spectrum that would be required would be the sum of these, shown at (c). For DSA, the overall amount of spectrum required is the cumulative value of the UMTS and DVB-T demands, shown at (d). The curve for this can be seen to be below the FSA requirements, leading to a significant gain in spectrum efficiency over the FSA case.

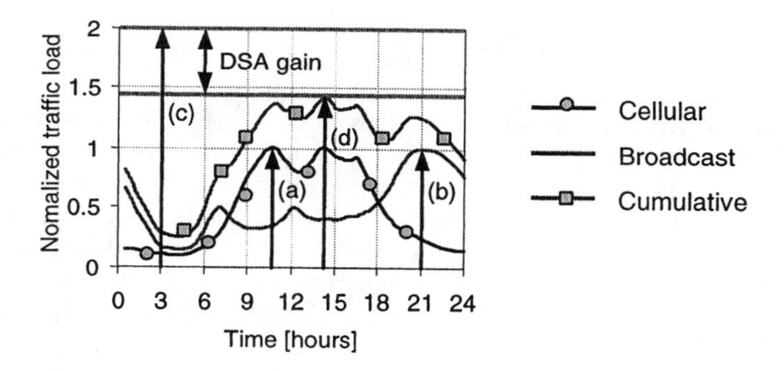

Figure 6.4 Time-varying spectrum demands.

Obviously, this is an ideal case, and real systems are more constrained in their spectrum allocations. For example, the spectrum allocations cannot follow the loads with an infinitely small resolution, because whole radio carriers need to be allocated, and the spectrum allocations cannot be changed continuously in time. Typical performance results are shown below in Figure 6.5. This shows how the use of DSA can increase the traffic load supportable at the same user satisfaction ratio as FSA. This is called the DSA gain and corresponds to the increase in spectrum efficiency. These results show, for this scenario, a gain of 35% for UMTS, 18% for DVB-T, and 29% gain for the overall system.

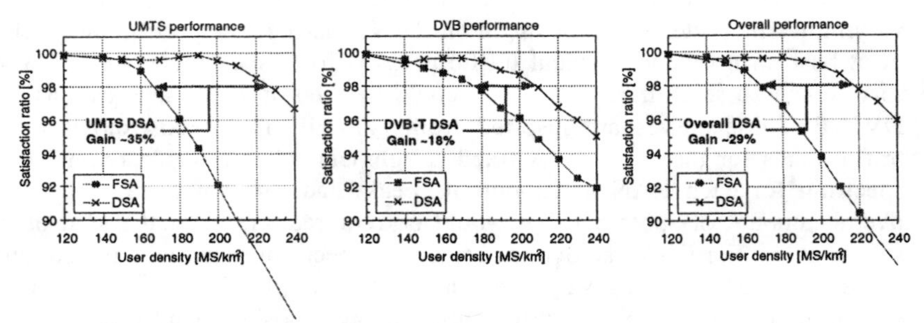

Figure 6.5 Basic performance results of temporal DSA.

Spatial DSA

Apart from the temporal variations, the traffic demand also varies regionally. For example, the demand for broadcast services may be larger in recreational areas, whereas the demand for unicast services may be larger in areas where people desire bidirectional communication, for example, in business areas. This motivates a regional adaptation of the spectrum allocations. It is advisable to confine the areas of uniform spectrum allocation to regions, called *DSA areas*, in which the traffic demands of different RANs are rather constant in space (yet they may be time variant). Figure 6.6 shows an example of spectrum allocations in adjacent DSA areas (linearly arranged), whereby the considered strategy is that spectrum allocations to different RANs belonging to adjacent DSA areas do not overlap. This is a reasonable strategy in the case that the radio network planning of the RANs is uncoordinated (i.e., no further measures are taken to prevent interference between RANs if they operate in the same spectrum).

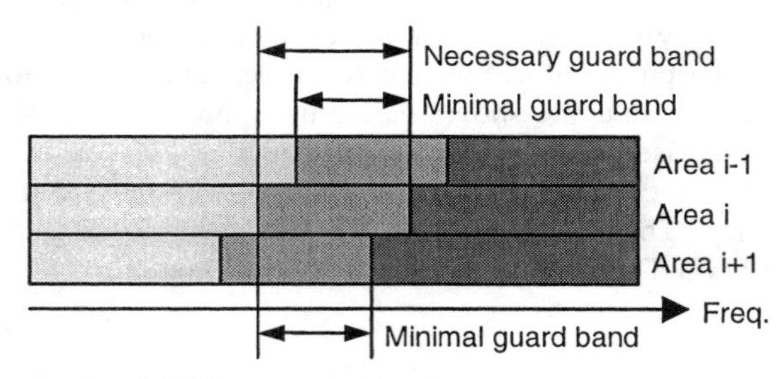

Figure 6.6 Concept of spatial DSA spectrum adaptation.

Spatial DSA is described as a mathematical optimization problem, aiming at allocating spectrum for each RAN in each DSA area in a way that an overlap of spectra allocated to different RANs in adjacent areas is avoided. The theoretical maximum of the achievable DSA gain in GoS considering percentiles is 100% (whereas only 33% when considering the mean GoS), assuming a traffic distribution for which FSA performs worst. The gain in a GoS percentile can be translated into a gain in the total traffic demand that can be accepted with DSA, so that the same GoS percentile as with FSA is maintained, as shown in Figure 6.7(a), for different values of the standard deviation in traffic demand. For the 1% percentile, the gain in acceptable traffic demand increases with the considered value of the 1% percentile [Figure 6.7(b)]. For example, about 44% more traffic demand is acceptable for $\sigma = 0.2$ with DSA, if a GoS of 100% shall be achieved at the 1% percentile (i.e., in 99% of the DSA areas), while only about 38% more traffic can be accepted with DSA if a GoS of only 90% shall be achieved at the same percentile.

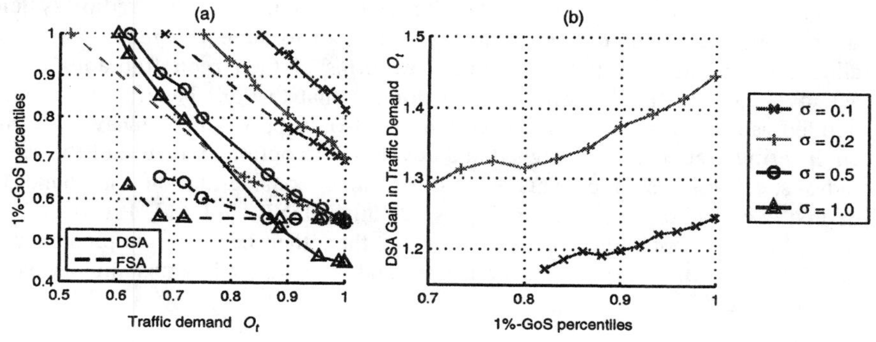

Figure 6.7 (a, b) Typical performance results of spatial DSA.

6.2.1.2 Flexible Fragmented DSA

The project DRIVE concentrated on the evaluation and investigation of the contiguous DSA scheme, where contiguous blocks of spectrum were assigned to different RANs, separated by suitable guard bands. If a RAN (e.g., UMTS) required additional spectrum, it was taken from the RAN in the adjacent spectrum (e.g., DVB-T) and vice versa, whereby the guard band was shifted accordingly, maintaining the same width. This resulted in limited flexibility for spectrum partitioning.

The project OVERDRIVE [19] overcame these limitations by allowing more flexible, albeit more complex, fragmented DSA schemes. Moreover, the employed hierarchical architectures require investigations for regional coexistence of different cell-layer access systems in a single shared frequency band. Based on the requirements for the envisioned operational scenarios, analytical and simulation-based studies were carried out for advanced DSA methods, such as fragmented and cell-by-cell DSA.

Gains of DSA schemes, both contiguous and fragmented, may be enhanced by carefully planning the coexistence of multiple radio access schemes within the same local area. Therefore, part of the research into advanced DSA schemes was dedicated to studying the possible local coexistence of different RAN networks within one frequency band. The goal of these studies was to develop algorithms that can be used for the definition of DSA areas and cell planning, whereby regional traffic demands for the services and RANs involved, were used as limiting parameters.

Dynamic spectrum coordination requires techniques for adaptation of sender and receiver to different spectrum using different physical layer specifications. This approach requires a quite high flexibility of the RF and BB chains necessary to fulfill the specified performance of the individual Rx and Tx requirements. It is therefore necessary to investigate the requirements on the radio systems (reconfigurable terminals and networks) for them to support dynamic spectrum allocation (contiguous and fragmented) in terms of their reconfigurability. It is acknowledged in the reconfigurability domain that the full benefit of software-defined radio and reconfigurable systems will only be reached if new spectrum engineering practices are designed and developed to optimize the use of the scarce radio spectrum resource. This has to be done considering on one hand the diversity of radio access technologies (cellular, DXB, WLAN, PAN) and the multiplicity of potential operators, and on the other hand the implementation feasibility of the developed concepts and the related gains in terms of performance.

6.3 PROVISION OF INTERNET AND DIGITAL TV SERVICES ON-BOARD HIGH-SPEED TRAINS

The railway environment of high-speed trains represents a highly challenging scenario. The possibility of providing services with a guaranteed QoS in such mobility conditions requires that a number of critical issues be solved both at the physical and network layers. The focus of this section is on the solutions developed in the framework of the IST project FIFTH [6] both to manage mobility and to support QoS at the network layer for high-speed train environments.

The FIFTH system was devised with the objective of providing passengers of high-speed trains with two major types of service: Internet and broadcast digital TV. Moreover, the system developed by the project FIFTH communication infrastructure may also be usefully employed by railway operators for fleet management purposes.

In order to describe the solutions designed for the mobility and QoS support in the railway environment, it is useful to highlight some of the main characteristics of the system architecture.

6.3.1 Network Infrastructure

The design of a global, mobile, and integrated communication infrastructure and the evaluation of its performance via both theoretical (simulations) and experimental (trials) analyses were carried out in the IST project SUITED (Multisegment System for Broadband Ubiquitous Access to Internet Services and Demonstrator) [20]. The system, named as the global mobile broadband system (GMBS), consists of multiple (satellite and terrestrial) system components (or segments) that are mutually complementary. Such infrastructure is conceptually comprised of the following elements:

- The Ka-band regenerative satellite system, providing connectivity to ultra-small aperture mobile terminals through a cluster of geo-stationary multibeam continental-wide coverage satellites; the European satellite component of the GMBS system can be represented by the EuroSkyWay Ka-band satellite cluster, envisaged to support mobile service.
- The mobile terrestrial systems such as the GPRS based on the GSM/TDMA network, representing the near-term version of UMTS; the evolution strategy of the GMBS comprised a near-term solution (phase 1) based on GPRS and a final solution based on UMTS.
- Wireless radio access (W-LINK) is a very useful network enhancement for the provision of short-range connectivity for both outdoor and indoor environments in areas where the above-mentioned coverages are poor or not available; the aggregate traffic can be transported by satellite or terrestrial lines.

The network infrastructure supporting the distribution of contents to the passengers of high-speed trains consists of an access portion and a fixed terrestrial portion.

The access part of the architecture is based on a satellite system. A geostationary Ka- or Ku-band regenerative satellite payload is used, which implements on-board switching/routing and a capacity assignment functionality to guarantee full-meshed connectivity; a dynamic bandwidth assignment mechanism is used for the satellite channels. A digital video broadcast–return channel by satellite (DVB-RCS) [21, 22] data format may be adopted eventually to implement the specific solutions necessary to counteract the periodic fading of the satellite link due to the obstacles along the railway routes. An overall architecture based on a full-meshed configuration among a given population of satellite gateways is adopted in this case: each gateway supports Internet traffic over both forward and return links (i.e., the system is characterized by several points of access to the Internet network). Other gateway stations, under the control of TV broadcasters, are used to support TV channel distribution. An example of a commercial satellite able to meet these requirements could be represented by Hot Bird 6 (manufactured by Alcatel and operated by Eutelsat), if it was enhanced at the data-link layer and the network layer with suitable mechanisms to dynamically manage the bandwidth assignment, and at the physical layer with mechanisms able to counteract the link worsening factors of the railroad environment.

The satellite connectivity is bridged in areas where the satellite coverage is not available (e.g., within tunnels, at the railway stations) by means of suitable wireless terrestrial technologies. Several solutions may be adopted to implement these gap fillers (also referred to as bridging segments): (1) WLAN based on the IEEE 802.11 standard [21], or (2) satellite local repeaters. The combination of the gap filler and the satellite segment is referred to as an extended segment. Considering the return direction, in order to reach the Internet network, the traffic generated by the in-train terminal is first transmitted over the radio link provided by the gap filler and then over the satellite link provided by the satellite gateway—referred to as a bridging terminal satellite terminal (BT-ST)—directly connected to the gap filler base station (i.e., either the WLAN access point or the satellite local repeater). In the forward direction, traffic flow follows the reversed path.

The general architecture of the described system is schematically shown in Figure 6.8.

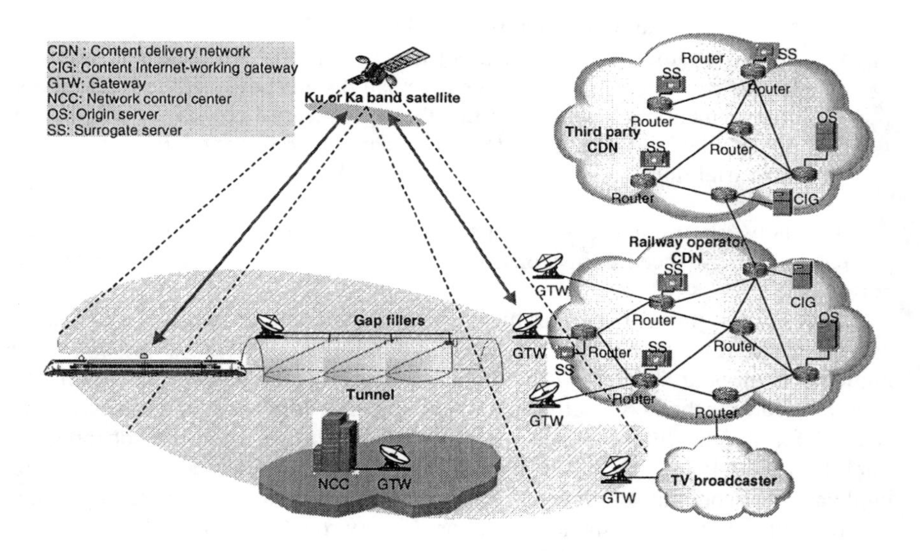

Figure 6.8 System architecture for support of services in high-speed trains [6].

A short-term version of this satellite system is envisaged to reduce the time-to-market for the service deployment. Such an early version is based on a Ku- or Ka-band transparent satellite payload. Also in this case a DVB-RCS data format (eventually enhanced) may be used. A hub-centric configuration is adopted for the overall architecture.

In particular, there are two main configurations envisaged for the access network in charge of bridging the satellite connectivity in the shadowed areas:

- *Single-segment access network:* This configuration, mainly suitable for deployment in the railway tunnel, only consists of an extended segment used to support both the traffic to/from the end users (i.e., the passengers) and the refilling traffic coming from "origin" servers located in the fixed terrestrial network and directed to the proxy server mounted on-board the train (this traffic "refreshes" the content stored in the proxy server).
- *Mixed-segment access network:* This configuration is devised to operate in the train station. It consists of a WLAN-based extended segment supporting end users' traffic and a WLAN hotspot segment for proxy server refilling traffic. End users are unaware of the presence of this direct (hotspot) connection. As it can be noted in Figure 6.9, the same WLAN radio link is used to support both types of traffic, which are then correctly routed to/from the satellite gateway and to/from the origin server (or, more generally, to/from the Internet point of access) by the WLAN distribution system.

The fixed terrestrial portion of the FIFTH system is provided by an Internet network suitably enhanced in some portions to implement the functionalities necessary for the QoS support.

6.3.2 Terminal

A key element of the FIFTH system is the terminal mounted on-board the trains and specially designed fo the purpose. The characteristics of the access network, envisaging that the satellite component is bridged in shadowed areas by means of a wireless terrestrial technology, entails that such a terminal is developed in a multimode configuration whose main constituting blocks are shown in Figure 6.10 (this is the case of a gap filler implemented by a WLAN station[2]). Therefore, the train terminal is also referred to as a multimode terminal (MMT). By means of standard devices, referred to as terminal equipment (TE) (e.g., a TE may be a laptop), the passengers connect to the LANs of the coaches and, via a distribution network, to the terminal interworking unit (T-IWU). The T-IWU, in addition to acting as a router, implements the functionalities for mobility support, QoS support, and protocol conversion to/from the satellite terminal (ST) and the WLAN Station (STA).

---

[2] Optionally, a WLAN hot spot may be used also to support the exchange of fleet management information between the train and the station.

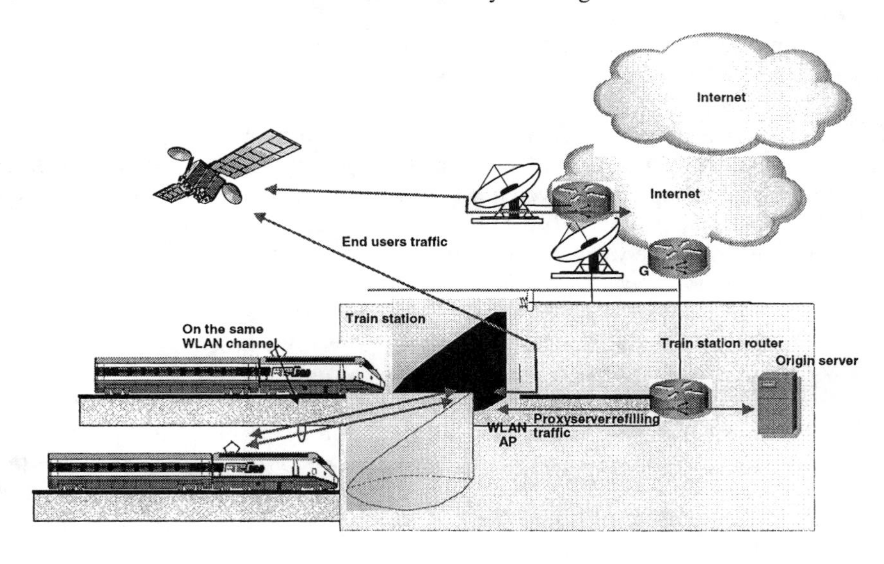

Figure 6.9 Mixed-segment access network at the train station.

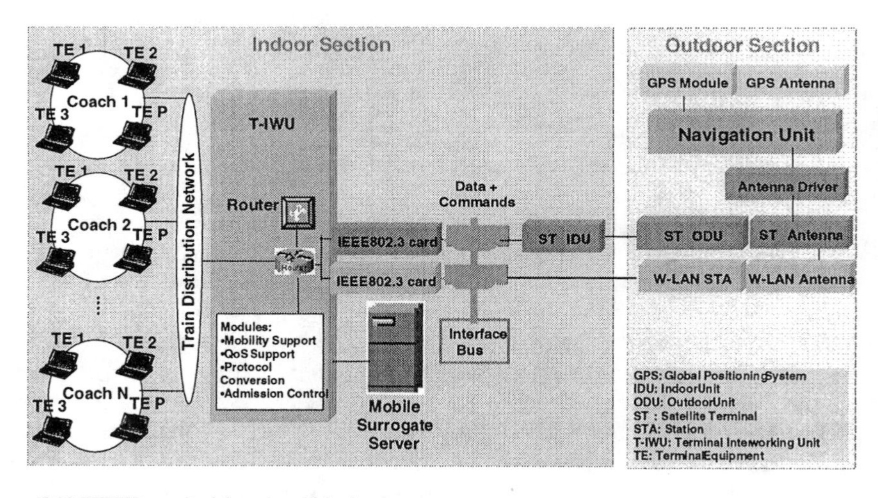

Figure 6.10 FIFTH terminal functional blocks.

Several alternatives can be considered to implement the LANs internal to the train coaches. They can be either wired (based on the IEEE 802.3 standard) or wireless (based on the IEEE 802.11 a/b standard [23]). By means of the wired LANs the services can be provided at the seats where passengers may directly find a user interface (e.g., an LCD touch screen) or they may find a standard LAN connector to plug in their own TE. In the case of an internal WLAN, each TE brought on board has to be provided with an IEEE 802.11 card.

6.3.3 Solutions for QoS Support

The system supports Internet access with a QoS guarantee. The main difference with the access to Internet without any QoS is that the former hosts services provided with an improved QoS level. This is realized by adopting a CDN [24] architecture. A CDN consists of an origin server (OS), where the original contents reside, and a number of surrogate servers (SSs) where the contents are partially or completely replicated. This architecture allows the distribution of both computing and network resources requests, in order to alleviate congestion at servers and at network links, respectively. A suitable routing system is envisaged in order to distribute users' requests, on the basis of contents, server load (load balancing), and network congestion (congestion control). Routing devices (also called decision makers) are in charge of routing requests to the "best" site, according to a given metric.

The application of the CDN approach to the railway scenario (a railway operator CDN is referred to as RO-CDN) entails that new solutions have to be developed in order to cope with mobility. Towards this end, in addition to the fixed SSs envisaged by the usual CDN, a new typology of SS is introduced, the so-called mobile SS (MSS), to be mounted on-board the train. The MSS functionality is implemented in the proxy server mentioned in the previous sections. Moreover, the router functionalities are encompassed in the T-IWU.

The solution designed to manage traffic within the RO-CDN exploits the functionalities of multiprotocol label switching (MPLS) [25]. MPLS is used for updating purposes, to distribute contents from the OS both to SSs and MSSs and for data delivery from OS/SSs to MSSs, or to accomplish service requests, coming from end users on-board the trains.

A number of label switched paths (LSPs) are set up between the router 4 in the train terminal and the fixed SSs in the terrestrial network. Depending on areas where the train is moving, these LSPs may be supported by the satellite segment by an extended segment or by a WLAN hotspot. Whenever a content required by an end user is not available in the MSS, the router in the train terminal selects the best LSP (connecting one of the SS) for delivering the requested service.

In order to exploit the MPLS features, the proposed solution assumes that LSPs are associated with classes of service: a number of LSPs are set up for each class of service. In particular, if one train connected through one given access segment is considered, there will be a number S of SSs that can provide contents to that train, a number M of available service classes, and a number p of LSPs used to route traffic of the same class to the same SS. If N is the number of trains that might be traveling at the same time, the theoretical number of contemporarily active LSPs for data delivery is

$$n_{del} = S \times M \times p \times N \tag{6.1}$$

The LSPs for data delivery are dynamically established and released depending on the requests arriving at the train routers. This is just a maximum value and the number of LSPs for data delivery that are simultaneously active will usually be lower.

When requested contents are not found on the local MSS, the router selects the optimal SS according to predefined administrative policies and network status. The optimum SS is not always the same, due to changing network conditions and server loads. This selection of the best SS is executed taking into account the congestion status of the network. Toward this end, a procedure has been defined which is used to gather network/server status information. Moreover, the metric used to select the best LSP for delivering requested services has been identified. On the basis of the network topology information provided by the Open Shortest Path First (OSPF) [26] routing protocol assumed to run in the fixed terrestrial Internet portion, the number of permanent shortest paths to each SS is calculated. The same approach described above for the definition of the maximum number of LSPs for data delivery is adopted. Measurements are over all these LSPs, according to the procedure described in the following. This discovering algorithm runs in the background in each router and it measures what is the status of each LSP in terms of bandwidth, round-trip time delay, and load of the reached SS. Since LSPs are unidirectional, it is important to note that the direction measured by means of the presented algorithm is the direction SS to router. But, in order to allow the execution of the algorithm, a feedback channel has to be provided for the measurements. A common LSP for each SS can be established. This LSP will have a low loss probability, and it will be used for all M classes and for all p available paths that share the same SS. Taking into account this consideration, the total number of permanent LSPs used for measurements that have to be contemporarily active is

$$n_{mes} = (S \times M \times p + S) \times N \tag{6.2}$$

When a service request for class arrives at a router, the router already knows (as a result of the discovery algorithm) what is the optimal SS and the best route (among the p available routes) to serve this request, and it selects accordingly an LSP. Then it establishes a new LSP from the selected SS to the router, along the chosen route, and with the same service class m, to transfer user contents. Of course, a traffic LSP, created on demand, has to be consistent with the measure LSP, which selects the optimal route and SS for that class.

The procedure designed to measure the available bandwidth on the LSPs and the congestion status of SSs may be seen as an extension of the assured forwarding per hop behavior (AF-PHB) [27] of the DiffServ model [28], where the number of the differentiated service code points (DSCPs) is raised from 3 to

$2n+1$. The number n represents the number of bits dedicated to this procedure in the DiffServ field. In particular, only one DSCP configuration is used for labelling data traffic, while the rest of the $2n$ configurations are used for the discovery of the available bandwidth on a network path and the server status by means of repeated and always more precise approximations through probes, named discovery probe packets (DPPs). The available bandwidth on the LSP between the considered SS (a possible source for data downloading) and the router is encoded by means of these n bits. Since a binary search procedure is employed, the number of necessary trials is equal to n, being able to fix the value of 1 bit in the considered field at each attempt. In particular, to determine the ith bit of the selected set, the ith DPP is used. Within each router on the LSP, there is a measurement module (as in the standard AF), which is in charge of deciding when to forward or drop packets, on the basis of runtime measurements and a packet label (i.e., the DSCP).

Furthermore, when the router wants to start the procedure with a given SS on a selected LSP, it sends a *request packet* to such an SS; upon reception of the request at the SS, the procedure starts and the SS sends the first DPP and activates a timer. The value of the most significant bit of the set of bits X used to encode the bandwidth on that LSP is determined by the outcome of the first DPP, and it is provisionally set to 1, whereas the others are set to 0. The code X corresponds to the level of bandwidth $b(X)$. When arriving at the first router on the LSP, the dropping algorithm will operate according to the rule that the current DPP, with code X, will be forwarded only if the estimated value of available bandwidth is higher than the codified level of bandwidth ($b(X)$), that is, if $b_{ava} \geq b(X)$ where

$$b_{ava} = C_{local} - b_{meas} \tag{6.3}$$

If the DPP reaches the router, this router generates a *feedback packet*. If the SS receives this feedback packet before the time-out expiry, it understands that the available bandwidth on the LSP is greater than the value associated with this DPP. The procedure is repeated for the other bits of the code X. At the end of the process, a binary number, X, representing the minimum amount of available bandwidth as $b(X)$ along the considered LSP, is generated.

In addition, the selected SS inserts in the DPP payload additional information such as (1) its current load status L_s (given by the number of active TCP connections divided by the maximum allowed number of TCP connections) and (2) an estimation of the delay in the download direction, d_{est} (by introducing a suitable timestamp).

Due to this information being available to the router on-board the train, the LSP that is selected is that with the lowest value of the LSP cost:

$$c_{LSP} = f\left(b(X), L_s, d_{est}\right) \tag{6.4}$$

where $f(\cdot,\cdot,\cdot)$ is a suitably defined function.

6.3.4 Solutions for Mobility Support

This section focuses on routing and handover procedures.

6.3.4.1 Routing

The way the routing is performed in the context of the FIFTH system is strongly influenced by the characteristics of the train terminal, which hosts both the end users and the MSS of the RO-CDN. In order to flexibly manage different network layer configurations, it was envisaged that both mobile IP (MIP) [29] and MPLS are supported on top of the classical IP layer. In particular, end users mobility is managed by means of the MIP protocol, while MSS mobility is managed by means of the MPLS protocol.

The T-IWU is configured as a *fixed router*[3] with a number of (physical and logical) wireless interfaces supported by different logical satellite connections towards the Internet network. The LAN in the train is configured as a subnetwork with public IP addresses. The T-IWU has at its disposal a static pool of IP addresses to be assigned to the TE connected to the internal LAN on the train. When a TE connects to the internal LAN, it is assigned an IP address by the T-IWU, by means of dynamic host configuration protocol (DHCP) [30]. If the generic TE implements MIP, then such a TE considers this IP address as its CoA and registers it with its home agent. If the TE supports only IP, no registration is executed.

When the train enters into a shadowed area, such as a train station or a tunnel, a handover (if sessions are active) between the satellite segment and the access network serving the area (and vice versa when the train leaves the shadowed area) takes place. The shadowed area may be served either by a single-segment access network or a mixed-segment access network. In both cases, since the traffic to/from the TE is supported by the extended segment, from the TEs' perspective this handover does not entail any change in the point of access to the Internet

---

[3] In this context the terms mobile router and fixed router are adopted with reference to the possible implementation of the MIP protocol and not with reference to the physical movement of the router. A router is said to be mobile if it can change its point of attachment from one link (i.e., network) to another while maintaining all existing communications and using its (permanent) IP home address. A mobile router is therefore assigned an IP home address and has a home agent in its home network with which it registers the CoA received when it connects a foreign network. Conversely, a router is assumed fixed if it does not present these features. When it connects to a foreign network, it has to be reconfigured and reassigned a new (set of) address(es). The fixed router does not maintain any identity (there is no IP home address) and when it connects to a new network it becomes a part of it.

network. The TEs' CoAs remain unchanged and no new binding registration procedures towards the corresponding home agents are required.

As highlighted in the previous section, the content refilling of the MSS takes place by means of a set of LSPs established between the T-IWU and the terrestrial SSs. In the case of a handover to/from a single-segment access network, because the points of access to the Internet network do not change, no LSP recomputation is required. Conversely, when a handover to/from a mixed-segment access network is executed, the MSS refilling takes place by means of the WLAN hotspot. This demands that the LSPs that are set up between the T-IWU and the different SSs distributed over the terrestrial network, and initially supported by the satellite connections, have to be recomputed. For each active LSP a complete rerouting computation can be performed. This means that a new route has to be computed on the new access segment for the old LSP. Then a new LSP, with the same quality of service, can be established on the new access segment. When the new LSP is set up, the traffic flow can be switched to it and the old LSP is released. This "make-before-break" approach implies that two different LSPs will coexist for the same service during the handover procedure. An important parameter that should be taken into account, in order to reduce the signaling overhead after LSP handovers, is the number of rerouted links. This is an important complexity factor, since, in an MPLS scenario, every rerouted link causes signaling due to the label distribution. After an LSP rerouting, the resulting route is still expected to use a portion of the previous route, and only in some cases does a whole new route have to be set up. In order to reduce the signaling amount, the number of rerouted links should be reduced, thereby causing a reduction of the number of labels distributed after a handover. This allows a resulting LSP which mostly overlaps with the old one and consequently to distribute only label mappings relevant to the new portions of the path. In order to accomplish this aim, a routing algorithm has been designed. This is a modified version of the Dijkstra algorithm. The Dijkstra algorithm computes the shortest path by using link delay d_{link}. The proposed algorithm, instead of d_{link}, adopts the following delay metric:

$$\bar{d} = d_{link}(1 - \alpha X) \tag{6.5}$$

with $\alpha \in [0;1]$ and $X = \begin{cases} 1 & \text{if the link was included in the "old" LSP} \\ 0 & \text{if the link was not included in the "old" LSP} \end{cases}$

If the routers provide the new delay values as inputs for the Dijkstra algorithm, the result is a computation of routes, which takes into account the old paths. Links used by the old path are believed to be "shorter" by the Dijkstra algorithm, which will most likely include them in the new LSP route.

6.3.4.2 Handover

From the perspective of the mobile terminal, there are two types of handover procedures: intersegment handover (ISHO) and intrasegment handover (IASHO). The ISHO occurs when the satellite segment currently supporting the IP traffic towards/from the MMT is no longer available while a different segment (single-segment or mixed-segment access network depending on the environment) is available, and vice versa. Generally, the ISHO procedure will take place within the overlapping region of the satellite and the extended segment/WLAN hotspot. This overlapping region is where the extended segment/WLAN hotspot coverage coexists with the satellite coverage. In some cases this overlapping could be not present. The IASHO occurs when the MMT moves by crossing the boundary of the areas covered by adjacent local satellite repeaters or, alternatively, the boundary between APs in WLAN.

The ISHO procedure consists of the following three phases:

1. *Handover information gathering:* The handover handler module in the T-IWU of MMT collects information from the radio links from the segment-specific terminals; location information from Global Positioning System (GPS)/track sensors; and information regarding the QoS perceived by the user.
2. *Handover decision:* Using the above feedback information the ISHO procedure is triggered.
3. *Handover execution:* Two types of handover are executed: (1) satellite to/from extended segment handover for both end users and MSS refilling traffic (in areas where a single-segment access network is available), and (2) satellite to/from extended segment handover for end users' traffic and satellite to/from a WLAN hotspot segment for MSS refilling traffic (in areas where a mixed-segment access network is available). In both cases, the satellite connections are always active, so:
    - For both types of ISHO the handover module in the T-IWU triggers bridging segment-specific procedures so as to allocate and deallocate resources on the bridging segment (providing both a satellite extension or a WLAN hotspot) when the train enters and leaves the shadowed area, respectively.
    - For both types of ISHO the handover module triggers segment-specific procedures so as to switch the satellite connections from the ST to the BS-ST and vice versa when the train enters and leaves the shadowed area, respectively.
    - The type (1) of ISHO occurs only on the link and physical layers; the network layer functionality is unaffected.
    - The type (2) of ISHO occurs, in addition to the link and physical layers [as for the type (1)], also at the network layer since the

LSPs between the MMT and the SSs are recomputed by the router in the MMT.

Either of the two types of handover mentioned above is executed depending on the environment (and the relevant access network configurations) considered:

- *Tunnel:* The type (1) of ISHO takes place in this environment. The mobile terminal could perform signal measurements to be aware of the radio link availability and also to check for the available resources. The use of location detection procedures (location information is provided to the terminal either by track sensors or by the GPS) and antenna diversity could extend the time duration provided for handover.
- *Train station:* Both type (1) and type (2) of ISHO may take place in this environment depending on the particular access network configuration deployed by the railway operator. Also in this case, the utilization of location detection procedures may significantly help the execution of the handover procedure (see Figure 6.11).

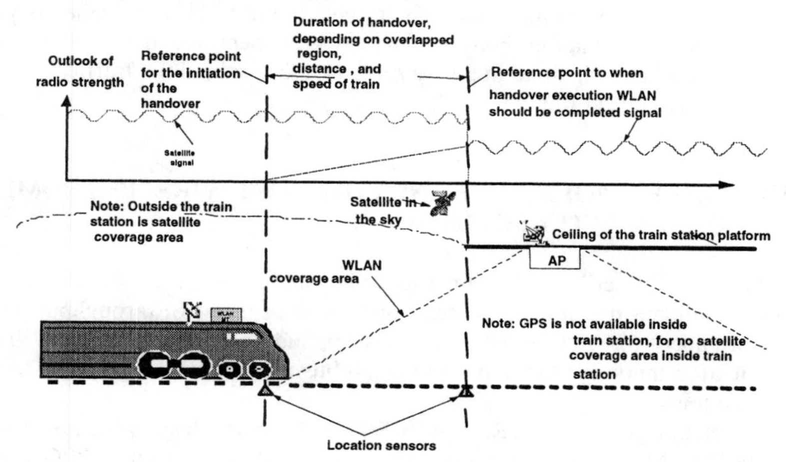

Figure 6.11 Intersegment handover at the train station.

The speed of the train and the overlapped region distance are key factors that have to be considered in both cases. The selected handover strategies can be summarized as follows:

- *Handover controlling scheme:* A mobile controlled handover (MCHO) scheme is adopted. The mobile terminal, in this scheme, is responsible for gathering radio-link measurements from its current and targeted systems. It will also monitor the location of the train with the aid of the location sensors and/or GPS. The mobile terminal is responsible for making the handover decision. This scheme is suitable for fast handover, and in a fast train environment where the speed of the train could be up to 300 km/h, it is necessary for the handover to be as quick as possible.

- *Connection establishing scheme:* A forward handover scheme is adopted. When there is an unexpected failure in the current connection, in this scheme the mobile terminal must promptly establish a new signaling channel with the targeted connection (i.e., the new access segment available). Once the channel is established, the handover signaling exchange would be performed. The location sensors can be used as trigger points during handover. Adopting this scheme, with the train traveling at very high speed, the establishment of the new channel in the extended segment should be completed before the mobile terminal leaves the coverage area of the old channel.
- *Connection transference scheme:* A signaling diversity scheme is adopted. This scheme, similar to soft handover, is a seamless "make-before-break" handover. While the mobile terminal is still utilizing the current traffic link, the signaling procedures are carried out between the mobile terminal and the targeted link. This scheme requires no synchronization between the two connections because the signaling exchange is done in the targeted connection and the traffic link would be the current link.

6.4 HIGH-PERFORMANCE AND LOW-COST WIRELESS HOME CONNECTIVITY SOLUTION

This section describes a system supporting multistreaming of high-quality video and broadband multimedia content, optimized in terms of throughput, range, and performance for indoor environments, including guaranteed QoS. The main application market is the home and small business where a cost-effective solution is a main issue.

The IST project ULTRAWAVES [7] built a wireless system able to deliver up to 100 Mbps data supporting distribution of up to four MPEG-2 channels, bidirectional real-time channels and data services, and providing easy configuration and support for IPv6.

To achieve this, work was carried out towards the development of an UWB air interface including, selection of waveforms, modulation, channel coding techniques, channel estimation, and prediction and detection algorithms. The UWB channel was modeled for home and indoor environments (of linear and spatial behavior) and the interference (statistics and power levels) was analyzed. The analysis focused on intrinsic UWB radio capacity limits and assessment of coexistence capabilities with other narrow and wideband systems, and the results were applied for the development of novel high-performance antennas for UWB communications.

An example of a home wireless network is shown in Figure 6.12.

To support high-quality communications in such a scenario, the project ULTRAWAVES designed and implemented a demonstration platform that enabled a transmission of MPEG-2 video/audio streaming from storage media to three PC projectors using an UWB-based system.

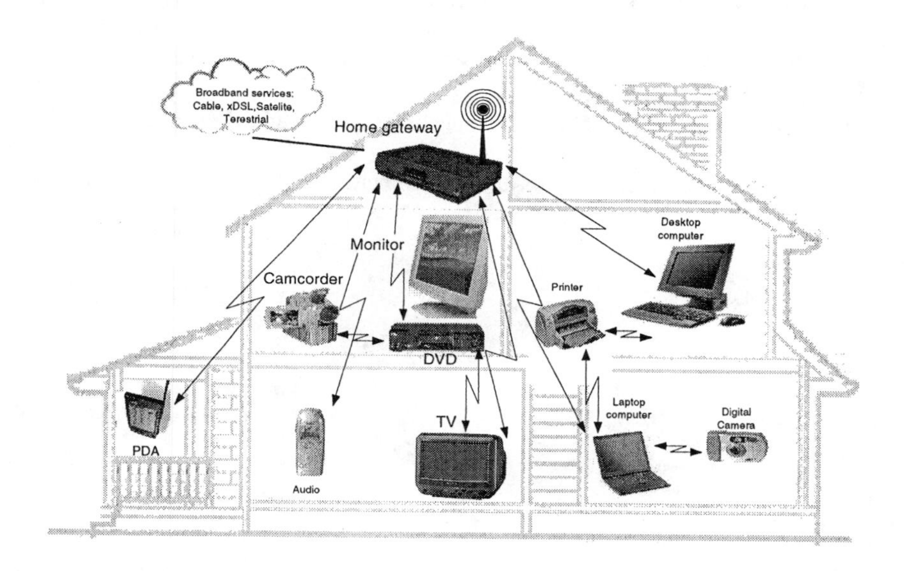

Figure 6.12 A home wireless network scenario. (*From:* [31].)

The application/communication system architecture consists of the following layers (see Figure 6.13):

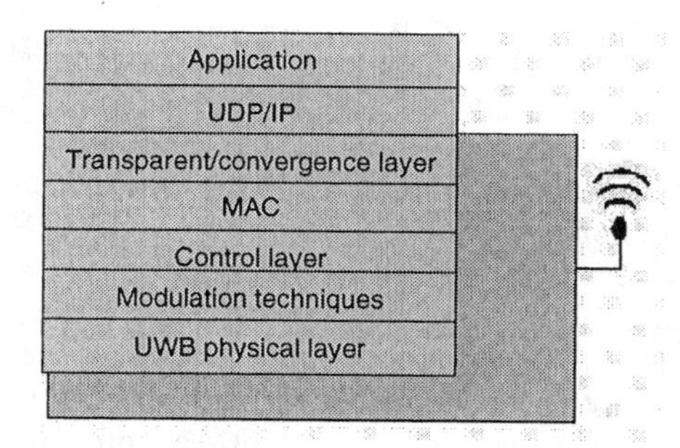

Figure 6.13 ULTRAWAVES protocol layer structure.

- *PHY layer:* ULTRAWAVES-specific UWB physical layer;
- *MAC layer:* ULTRAWAVES-specific MAC layer;
- *Convergence layer:* This sublayer is used to adapt between the ULTRAWAVES link layer to connecting link layer technologies such as ethernet, IEEE-1394, and USB.
- *Network layer:* The network (NWK) layer is performed using the Internet Protocol.
- *Transport layer:* The transport layer is performed using the user datagram protocol (UDP).
- *Application layer:* MPEG-2 Transport stream over UDP/IP.

The designed communication system architecture is shown in Figure 6.14.

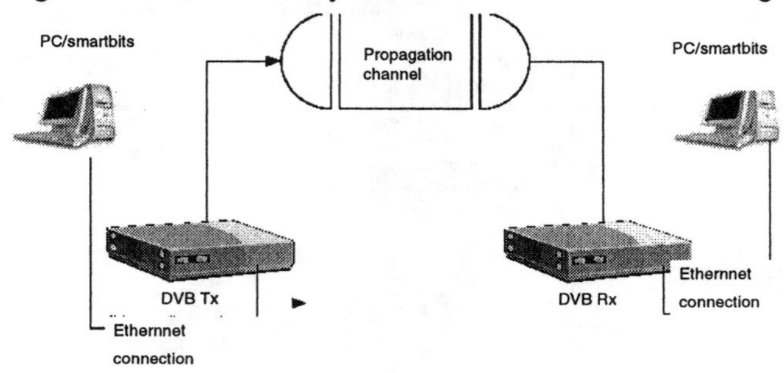

Figure 6.14 Point-to-point application system demonstration.

The communication system provides a link capable of performing an Ethernet bridge. It can operate in point-to-point mode or in point-to-multipoint broadcast mode. The bit rate (PHY) is 60 Mbps.

A detailed description of the actual throughput that can be provided at the Ethernet ports and information about the modulation scheme can be found in [32].

On power-up the DVP functions either as a master and then it will start transmission according to the described frame, or as a slave and then it will start searching for a signal and lock on to it. The master or slave settings are set by dipswitches on the board. A system block diagram of the possible implementation is shown in Figure 6.15.

In order to define the interfacing and control functionality between the application and communication system, the impact on MPEG-2 video with the introduction of errors in a controlled environment was investigated. In order to approve the required QoS concerning the MPEG-2 transmission, some subjective tests were performed to see what those numbers meant in a visual way. These tests were done with a wired LAN setup between a VideoLAN server and a VideoLAN client with a modified version of the VideoLAN software (the choice of software is based on that performed within the project surveys). Figure 6.16 shows the application testing setup.

Use of HDTV MPEG-2 as a source to split up to three different DVD transport-streams to form one complete image on the walls was one type of test.

It was concluded that three MPEG-2 (DVD level) streams might not be the preferred approach. Sometimes, it would be better to achieve a HDTV quality projection using three projectors that reflect three parts of one HDTV stream.

This can be done by splitting one HDTV MPEG-2 stream into three streams.

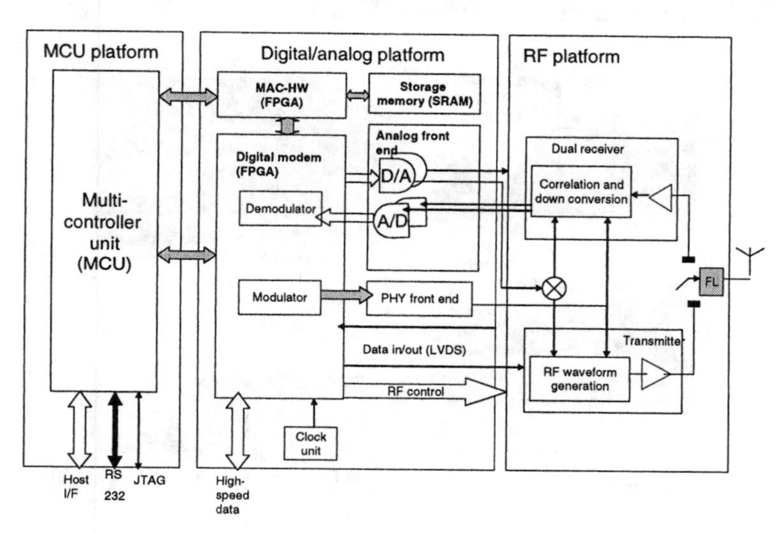

Figure 6.15 System block diagram. (*From:* [33].)

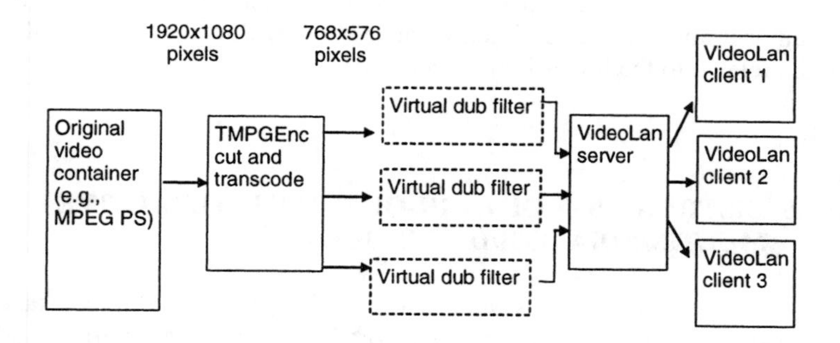

Figure 6.16 Application testing setup. (*From:* [32].)

Such an action can require a redesign of the setup of the demonstrator from opposite-wall projection to a panoramic wall with all projectors placed on a movable rack. This way the demonstrator is much more movable and gives better results regarding visual perception.

Based on the available communication platform (see Figure 6.16), a demonstrator can be built up where the UWB communication platform uses a main part of a multistreaming user scenario for consumer electronics. The hardware and software are then integrated into a working demonstration platform.

The goal of the platform built in within the project ULTRAWAVES was to enable the wireless transfer of a high-quality video from a source to an extra-wide (wall size) panoramic screen as shown in Figure 6.17.

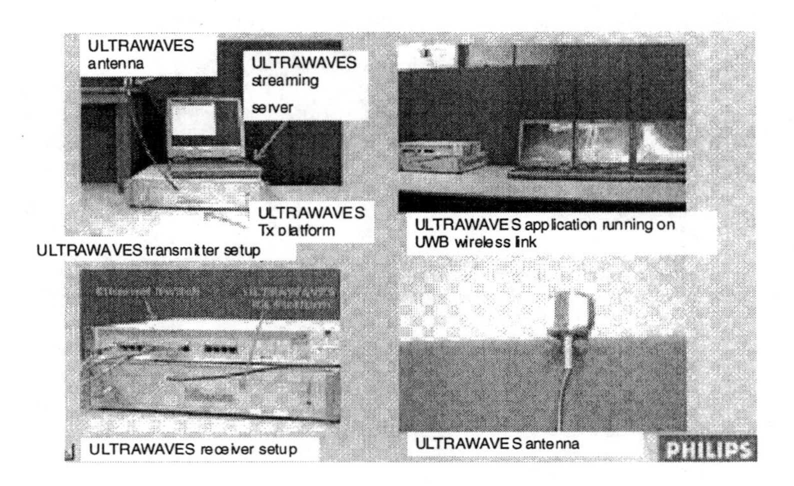

Figure 6.17 Demonstrator setup.

The DVD accompanying this book has complete details about the demonstration and experiments carried out within the project ULTRAWAVES and related to the issues in this section.

6.5 DISTRIBUTION OF HIGH-QUALITY INTERACTIVE MULTIMEDIA DVB/IP SERVICES

A satellite interactive system can be integrated within a standard DVB-S broadcast network, in order to support interactive TV, Internet, and multimedia services. The IST project IBIS (Integrated Satellite Broadcast Network) [34] combined the DVB-RCS standard (on the uplink) and DVB-S standard (on the downlink) into a single regenerative multispot satellite system allowing for full cross-connectivity between the different uplink and downlink beams.

The system was implemented with a fully regenerative OBP designed to provide direct (distributed) DVB-RCS compliant satellite access for individual

Use of HDTV MPEG-2 as a source to split up to three different DVD transport-streams to form one complete image on the walls was one type of test.

It was concluded that three MPEG-2 (DVD level) streams might not be the preferred approach. Sometimes, it would be better to achieve a HDTV quality projection using three projectors that reflect three parts of one HDTV stream.

This can be done by splitting one HDTV MPEG-2 stream into three streams.

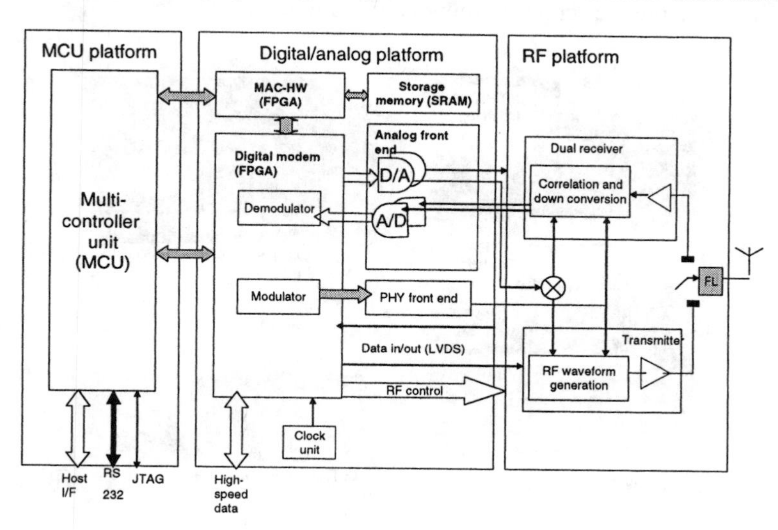

Figure 6.15 System block diagram. (*From:* [33].)

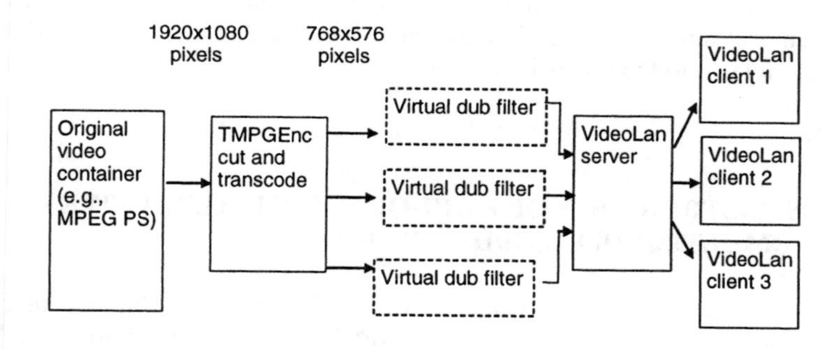

Figure 6.16 Application testing setup. (*From:* [32].)

Such an action can require a redesign of the setup of the demonstrator from opposite-wall projection to a panoramic wall with all projectors placed on a movable rack. This way the demonstrator is much more movable and gives better results regarding visual perception.

Based on the available communication platform (see Figure 6.16), a demonstrator can be built up where the UWB communication platform uses a main part of a multistreaming user scenario for consumer electronics. The hardware and software are then integrated into a working demonstration platform.

The goal of the platform built in within the project ULTRAWAVES was to enable the wireless transfer of a high-quality video from a source to an extra-wide (wall size) panoramic screen as shown in Figure 6.17.

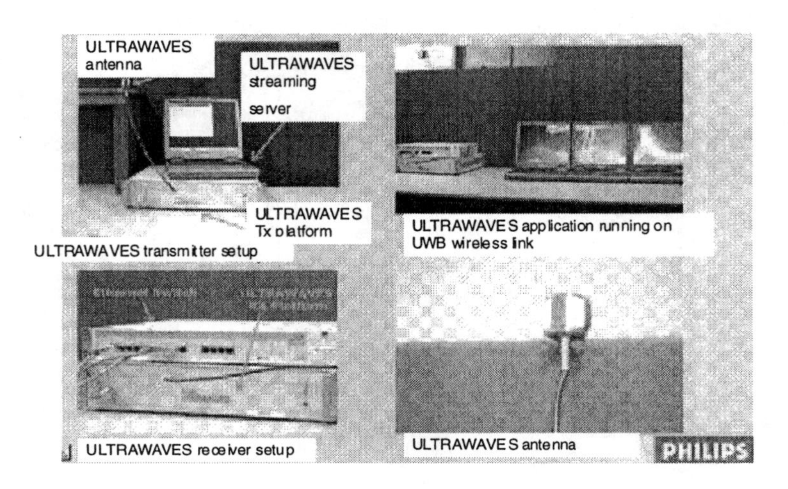

Figure 6.17 Demonstrator setup.

The DVD accompanying this book has complete details about the demonstration and experiments carried out within the project ULTRAWAVES and related to the issues in this section.

6.5 DISTRIBUTION OF HIGH-QUALITY INTERACTIVE MULTIMEDIA DVB/IP SERVICES

A satellite interactive system can be integrated within a standard DVB-S broadcast network, in order to support interactive TV, Internet, and multimedia services. The IST project IBIS (Integrated Satellite Broadcast Network) [34] combined the DVB-RCS standard (on the uplink) and DVB-S standard (on the downlink) into a single regenerative multispot satellite system allowing for full cross-connectivity between the different uplink and downlink beams.

The system was implemented with a fully regenerative OBP designed to provide direct (distributed) DVB-RCS compliant satellite access for individual digital video broadcasters and Internet service providers. This is shown in Figure 6.18.

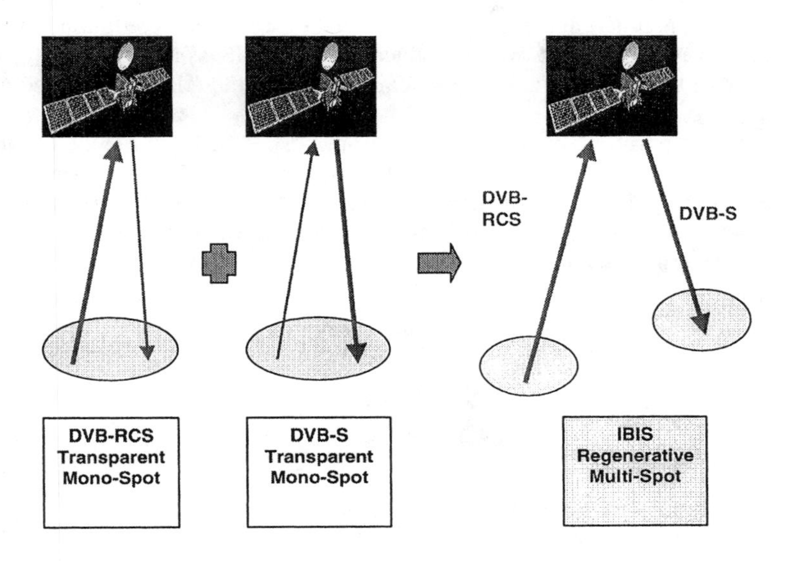

Figure 6.18 Integrated satellite broadcast network combining different standards.

The OBP combines the individual DVB-RCS contributions into one or several DVB-S downlinks as required by users. This approach provides a cost-effective solution to support interactive TV, Internet, and multimedia services, as well as business-to-business (B2B) networking (true mesh networking). IBIS users can take advantage of the economies of scale provided by the use of standard DVB-RCS/S user stations (RCSTs).

The overall system functional architecture was defined based on a set of service and system requirements and on the basis of a satellite multibeam approach and on-board processing techniques.

The various system units of the architecture are as follows:

- A payload unit;
- User units;
- A service provider unit;
- A network control center.

The overall architecture of the proposed broadcast interactive satellite system approach is shown in Figure 6.19.

A main component of the architecture in Figure 6.19 is the broadcast interaction model (BIM, not shown here). This model constitutes a reduced configuration of the payload equipment and the ground system elements (the service provider station, the user station, and the system management).

The objective of the BIM is to perform a complete development, integration, and test, at functional and performance levels, of a reduced configuration of the satellite system. The BIM was introduced in the IBIS system to allow for the validation of the on-board and ground equipment and the related supported digital services.

For the purpose of testing the project specified, a complete ground environment was developed and tested. The BIM performed on the basis of an IF-baseband configuration, in principle, one uplink with several carriers, and two downlinks will be defined for the communication between the space segment and the ground segment stations.

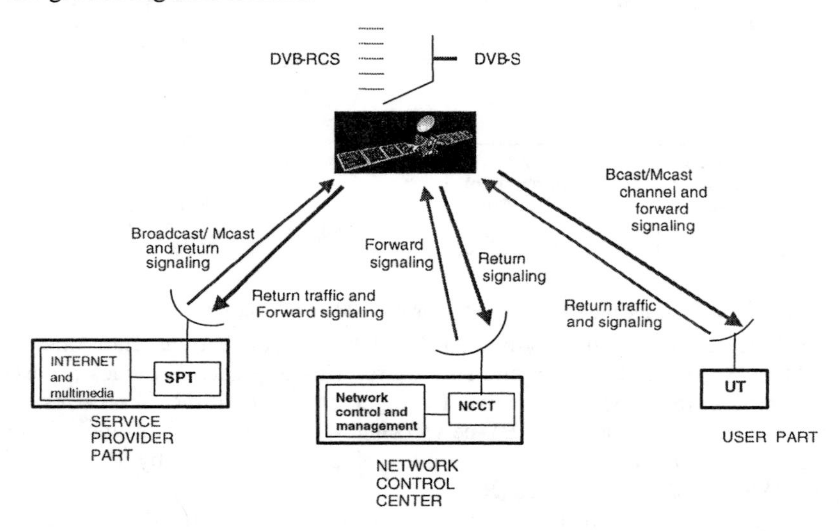

Figure 6.19 The IBIS system architecture. (*From:* [34].)

The MF-TDMA access mode was tested and validated under the nominal performances and operation system requirements. To this end, the configuration of one service provider station and multiple user stations was configured, and the complete set of flexibility features in carrier allocation, bit rate allocation and management were demonstrated for the various channels and different configurations.

A representative set of the multiplexing and routing facilities for the different types of configurations was afterwards validated.

The monitoring functionalities for the various modes of operation could be demonstrated by means of the system management unit.

The complete specifications, as well as an analysis of the standards applicable to the IBIS system (i.e., DVB-S, DVB-RCS), are available on the DVD accompanying this book.

6.6 CONCLUSIONS

The convergence of technologies impacts the way services are delivered and changes the habits of users accessing broadcast services. This has a direct impact on personalization of services. In order to be able to access and enjoy the services easily (no technical knowledge required) and on multiple usage environments and from different access points/devices, and to be able to access any broadcasting services anytime and anywhere, a global solution for interactive broadcasting is necessary. Several wireless technologies are available that can provide such a solution. These are listed in Table 6.1.

The technology trend for interactive broadcasting can generally be divided into four main areas: return channels for broadcast systems, enhancements to UMTS for broad/multicast, the convergence of broadcast and cellular systems into multiradio environments, and the development of new applications and markets for interactive broadcasting.

At the present time, interactive broadcasting uses mostly DVB as the interactive platform. DVB is a family of compatible standards for all television delivery media, including satellite, terrestrial, and cable. It created a harmonious digital broadcast market for all service delivery media. DVB-T uses COFDM technology, which allows the construction of single-frequency networks (SFNs). Adjacent transmitters will use the same transmission frequency, which also permits gap-filling for optimizing coverage in different spots.

The core of the DVB digital data stream is the standard MPEG-2 data container, which holds the broadcast and service information. This flexible carry-all can contain anything that can be digitized, including multimedia data. Future standards, which are based on software (e.g., software radio) rather than hardware, are also likely to be enhanced by this flexibility, which allows immediate downloading of the latest decoders.

Apart from DVB, the other main technology trend is in the enhancement of UMTS to allow for the delivery of broadcast and multicast services. In particular, this is typified by the increased standardization effort that has begun for MBMS by 3GPP. This shows a high level of interest in the concepts of allowing multimedia multicasting and broadcasting over UMTS, which, due to the inherently bidirectional abilities of UMTS, will certainly be used for interactive broadcasting services. Since the standardization progress of MBMS is now progressing well, after a slow start, it can be expected that there will be more interest in the potential that this enhancement has for UMTS.

There are a number of active research areas surrounding the concept of broadcasting over UMTS. These include power control issues, multicast group management, error correction, and the application of schemes such as H-ARQ, security, and mobility management. It can be expected that all of these areas will see increased effort in the near future.

Table 6.1

Technologies for Interactive Broadcasting Services

Technology	Terrestrial/Satellite/ HAPS [T/S/H]	Fixed/Mobile [F/M]	Broadband/Narrowband [B/N]
DVB	T/S/H	M	B
UMTS	T/S/H	M	B/N
GSM/GPRS /EDGE	T/H	M	N(GSM/GPRS)/B(EDGE)
DAB	T/S/H	M	B
LMDS	T/H	F	B
MMDS	T/H	F	B
MVDS	T/H	F	B

Source: [34].

In addition to these aspects, there is a current trend towards the consideration of new downlink bearers for UMTS. There are several potential advantages of a new bearer, not only for the support of interactive broadcasting, but also for the efficient support of other asymmetric services. A significant amount of research is still required on this issue, but it is gaining a significant amount of interest.

References

[1] Paxal, V., "DVB with Return Channel Via Satellite," *Telenor R&D, DVB-RCS200*, DVB White Paper, http://www.dvb.org/dvb_technology/.

[2] Carey, J., "Winky Dink to Stargazer: Five Decades of Interactive Television," Greystone Communication, *UnivEd Conference on Interactive Television*, http://www.columbia.edu/cu/business/courses/download/B9201XX/carey/history_of_interactive_tv.pdf.

[3] LMDS overview, at http://www.wcai.com/lmds.htm.

[4] Faria, G., and F. Scalise, "DVB-RCT: A Standard for Interactive DVB-T," available at http://www.broadcastpapers.com/tvtran/HarrisIntDVBTStandard02.htm.

[5] IST-2001-35125 Project OVERDRIVE, http://www.cordis.lu/ist/ka4/projects.html.

[6] IST-2001-39097 Project FIFTH, http://www.cordis.lu/ist/ka4/projects.html.

[7] IST-2001-35189 Project ULTRAWAVES, http://www.ultrawaves.org.

[8] Leaves, P., and J. Huschke, "A Summary of Dynamic Spectrum Allocation Results From DRiVE," *IST Mobile and Wireless Telecommunications Summit*, June 2002, pp. 245–250.

[9] Huschke, J., and P. Leaves, "Dynamic Spectrum Allocation Including Results of DSA Performance Simulations," DRiVE deliverable D09, January 2002.

[10] Leaves, P., et al., "Dynamic Spectrum Allocation in a MultiRadio Environment: Concept and Algorithm," *Proceedings of IEE 3G2001 Conference,* March 2001, pp. 53–57.

[11] Leaves, P., et al., "Performance Evaluation of Dynamic Spectrum Allocation for Multiradio Environments," *Proceedings of the IST Mobile Summit'01,* September 2001, Barcelona, Spain, CD-ROM.

[12] Huschke, J., and S. Ghaheri-Niri, "Guard Band Coordination of Areas with Differing Spectrum Allocation," *Proceedings of the IST Mobile Summit'01*, Barcelona, Spain, September 2001, CD-ROM.

[13] Leaves, P., et al., "Dynamic Spectrum Allocation in Hybrid Networks with Imperfect Load Prediction," *IEE 3G2002 Conference*, May 2002.

[14] Huschke, J., et al., "Dynamic Spectrum Allocation for Mobile Interactive Multimedia Systems," *Proceedings of European Wireless 2000 Conference*, September 2000.

[15] IST-1999-12515 Project DRIVE, http://www.cordis.lu/ist/ka4/projects.html.

[16] Tönjes, R., et al., "Architecture for a Future Generation Multiaccess Wireless System with Dynamic Spectrum Allocation," *Proceedings of IST Mobile Summit'00*, Galway, Ireland, October 2000, CD-ROM.

[17] Tönjes, R., et al., "OverDRiVE—Spectrum Efficient Multicast Services to Vehicles," *Proceedings of IST Mobile Summit'02*, Thessaloniki, Greece, June 2002, CD-ROM.

[18] Almeida, S., et al., "Spatial and Temporal Traffic Distribution Models for GSM," *IEEE Vehicular Technology Conference*, Amsterdam, the Netherlands, Vol. 1, September 1999, pp. 131–135.

[19] Kiefl, B., "What Will We Watch? A Forecast of TV Viewing Habits in 10 Years," *The Advertising Research Foundation*, New York, 1998.

[20] IST Project SUITED, http://www.suited.it.

[21] ETSI EN 300-421, "Digital Video Broadcasting (DVB); Framing Structure, Channel Coding and Modulation for 11/12 GHz Satellite Services," 1997.

[22] ETSI EN 301-790, "Digital Video Broadcasting (DVB); Interaction Channel for Satellite Distribution System," 2000.

[23] IEEE Std. 802.11b, "Part 11: Wireless LAN Medium Access Control (MAC) and Physical Layer (PHY) Specifications: Higher-Speed Physical Layer Extension in the 2.4GHz Band," *in Information Technology-Telecommunications and Information Exchange Between Systems—Local and Metropolitan Area Network—Specific Requirements*, New York, 1999.

[24] Verma, D. C., *Content Distribution Networks: An Engineering Approach*, New York: John Wiley & Sons, 2002.

[25] Rosen, E., A. Viswanathan, and R. Callon, "Multiprotocol Label Switching Architecture," *IETF RFC 3031*, January 2001.

[26] Moy, J., "OSPF version 2," *IETF RFC 2328 (Std. 54)*, April 1998.

[27] Heinanen, J., et al., "Assured Forwarding PHB Group," *IETF RFC 2597*, June 1999.

[28] Blake, S., et al., "An Architecture for Differentiated Services," *IETF RFC 2475*, December 1998.

[29] Johnson, D., C. Perkins, and J. Arkko, "Mobility Support in IPv6," *IETF Internet Draft*, draft-ietf-mobileip-ipv6-21, February 2003.

[30] Droms, R., et al., (eds.), "Dynamic Host Configuration Protocol for IPv6 (DHCPv6)," *IETF Internet Draft*, draft-ietf-dhc-dhcpv6-28, November 2002.

[31] IST Project ULTRAWAVES, "Wireless Video Links Via UWB Technology—Application Requirements," submitted to doc: *IEEE 802.15-SGAP3a-02/119r0,* http://www.ultrawaves.org.

[32] IST Project ULTRAWAVES, "D10.1—Final Report," http://www.ultrawaves.org.

[33] IST-2001-35189 Project ULTRAWAVES, "D7.2—ULTRAWAVES Communication System," August 2004, http://www.ultrawaves.org.

[34] IST Project IBIS, Integrated Broadcast Interaction System.

Chapter 7

Realization and Management of Future Mobile Systems

7.1 INTRODUCTION

It is difficult to give a definition to the fourth generation (4G) of wireless systems, however, the scope of a 4G system encompasses the integration of broadband networks with a common network layer, thus creating an all-IP network.

The design of such a system has to consider a requirement for greater efficiency to meet future traffic requirements and has to be more cost-effective in the presence of scarce radio resources. In addition, future networks will evolve to provide preferential treatment to QoS traffic; therefore, it is necessary to accommodate QoS mechanisms and to support true intersystem roaming between IP networks. The relevant network-related design issues can be addressed by an IP convergence layer, and this has a direct impact on the design of the link layer. A generic IP convergence layer can provide a degree of transparency between the network and the link layer in line with the IP design philosophy.

A number of projects addressed issues related to realization and management of future mobile systems and networks within the FP5 of the European Commission.

The IST project Trials Realization on an Experimental 3G UMTS Platform (REAL) [1] assessed the 3G technology sufficient to provide mobile users with advanced, high-bandwidth applications by performing advanced traffic experiments on the core part of a well-established 3G platform. Specifically, within the framework of the project, the traffic experiments concentrated on the Iu and Gi interfaces that implement the UTRAN environment and connectivity to backbone networks [e.g., the public switched telephone network (PSTN), GPRS, and wide area networks (WAN)]. The objective was to perform stress and load traffic experiments at the protocol and service levels by simulating thousands of mobile subscribers making use of real 3G services. The traffic experiments covered not only the user (*u*)-plane but also the control (*c*)-plane functionality of the core switches. These experiments are discussed in Section 7.2.

The IST project MAMBO (MultiServices Management Wireless Network with Bandwidth Optimization) [2] addressed the problem of managing audio-video (AV) services distribution on the available bandwidth. The main advantage of using digital transmission for audio-video services is the ability to use less bandwidth and infrastructure for reduced operation expenses per program. The project proposed an innovative concept of a system capable of optimizing the bandwidth allocation for every service as a function of the complexity of the service and based on the perceived QoS. The project focused on the development and implementation of key components of the distributed feedback loop bandwidth allocation system, which is the key to an open and scalable solution.

Section 7.3 focuses on an end-to-end QoS solution across the Internet. This section is based on work performed within the IST project MESCAL (Management of End-to-End Quality of Service Across the Internet at Large) [3]. The project proposed and validated scalable, incremental solutions that enable the flexible deployment and delivery of interdomain QoS across the Internet.

This involved the development of templates, protocols, and algorithms for establishing service level specifications (SLS) between ISPs and their customers, including their peers, and scalable solutions for interdomain traffic engineering based on enhancements to the existing BGP routing protocol and associated route selection logic.

MESCAL considered both unicast- and multicast-based services and ensured that the proposed solutions were applicable to both IPv4 and IPv6.

Finally, a solution for the end-to-end provision of QoS for S-UMTS services is discussed in Section 7.4. One of the projects that researched this topic was the IST project VIRTUOUS [4, 5]. The integration of a satellite UMTS network with a terrestrial UMTS represents one of the most attractive proposals to overcome the coverage limitations of the future 3G cellular mobile networks, together with the possibility of exploiting the intrinsic satellite transmission capabilities.

Section 7.5 concludes the chapter.

7.2 TRAFFIC EXPERIMENTS FOR SIMULATION OF 3G SERVICES

One of the main reasons for developing the UMTS standard was the rapid growth of the Internet and the transfer of its features and capabilities into the mobile world. It was envisaged that, as a result, the switch of manufacturers and network providers would be able to test the switch performance especially regarding the transfer of IP traffic even under very high load conditions [6].

The main focus of the trials performed within the IST project REAL [1] was the data transfer in the uplink from the Iu interface to the Gi interface, and in the downlink from the Gi interface to the Iu interface, with a special focus on the QoS and the packet drop rate. A trial testbed was set up that consisted of the following components (see Figure 7.1):

- Test equipment;
- Serving GPRS node (SGSN);
- Gateway GPRS node (GGSN);
- 100-Mbps hub;
- Monitoring PC for reference and cross-check measurements.

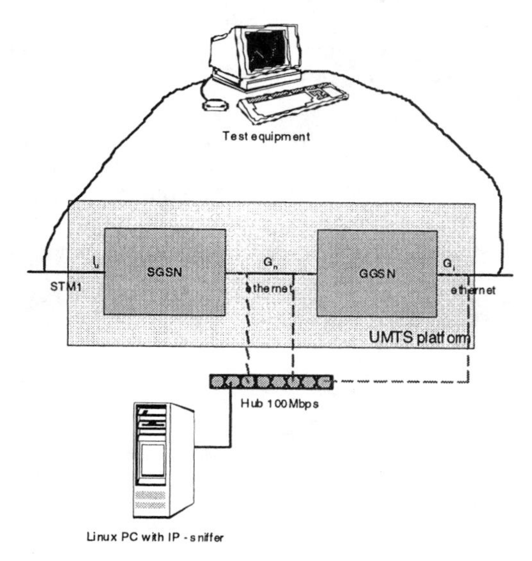

Figure 7.1 UMTS trial platform.

The test equipment supported one Iu interface connected with the SGSN through an STM-1 physical interface, and one Gi interface connected with the GGSN through a 100-baseT ethernet physical connection. These interfaces were chosen because such a solution provides the capability to trace the transferred messages via commercially available IP sniffer software at the Gn as well as at the Gi interface.

The test equipment is connected on the side to the Iu interface of the switch and on the other side to the Gi interface. The aim of this configuration is to verify the signaling capability and the number of IP packets that can be transferred without any packet loss. As a result of this, the performance of the switch concerning the packet drop rate and the obtained QoS could easily be measured.

The focus of the trials was on the performance evaluation of the core 3G UMTS network through its individual components performance and the influence on the overall services provision to the end users.

The performance experiments were carried out in the form of sophisticated traffic scenarios characterizing the 3G UMTS underlying protocols technology. Large-scale experiments were also conducted for the determination of the minimum set of requirements that should be satisfied by the UMTS switching equipment in order to achieve an acceptable performance level.

7.2.1 Background

According to [7], Figure 7.2 shows the basic domains in UMTS as defined by the 3GPP organization body. A basic architectural split is between the user equipment (terminals) and the infrastructure. It results in two domains: the user equipment domain and the infrastructure domain.

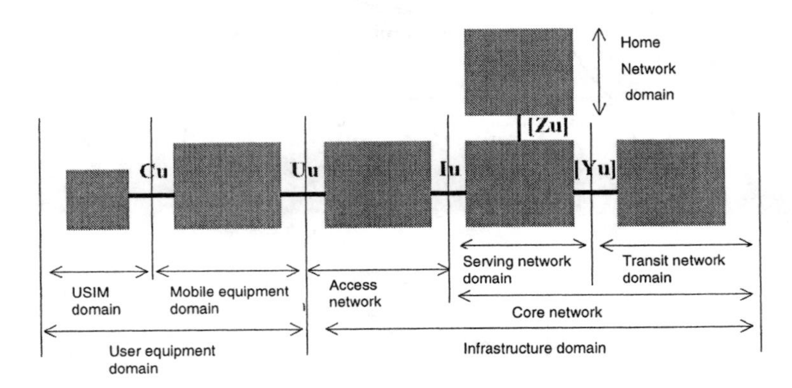

Figure 7.2 UMTS domains and reference points. (*From:* [7].)

The user equipment is the equipment used by the user to access UMTS services. The user equipment has a radio interface to the infrastructure. The infrastructure consists of the physical nodes, which perform the various functions required to terminate the radio interface and to support the telecommunication services requirements of the users. The infrastructure is a shared resource that provides services to all authorized end users within its coverage area.

The reference point between the user equipment domain and the infrastructure domain is termed the "Uu" reference point (UMTS radio interface).

The user equipment domain encompasses a variety of equipment types with different levels of functionality. These equipment types are referred to as user equipment (terminals), and they may also be compatible with one or more existing access (fixed or radio) interfaces (e.g., dual-mode UMTS-GSM user equipment). The user equipment may include a removable smart card that may be used in different user equipment types. The user equipment is further subdivided into the mobile equipment (ME) domain and the user services identity module domain (USIM). The reference point between the ME and the USIM is termed the "Cu" reference point.

The mobile equipment performs radio transmissions and contains applications. The mobile equipment may be further subdivided into several entities, for example, the one that performs the radio transmission and some related functions, such as mobile termination (MT), and the one that contains the end-to-end application (e.g., laptop connected to a mobile phone) or the terminal equipment. This separation is used in the description of the functional communication but no reference point is defined.

The USIM contains data and procedures that unambiguously and securely identify themselves. These functions are typically embedded in a stand-alone smart card. This device is associated to a given user, and as such allows the identification of this user regardless of the used ME.

The infrastructure domain is further split into the access network domain, which is in direct contact with the user equipment, and the core network domain. This split is intended to simplify/assist the process of decoupling access-related functionality from nonaccess related functionality and is in line with the modular principle adopted for the UMTS.

The access network domain comprises roughly the functions specific to the access technique, while the functions in the core network domain may potentially be used with information flows that use any access technique. This split allows for different approaches for the core network domain; each approach specifying distinct types of core networks connectable to the access network domain, as well as different access techniques, where each type of access network is connectable to the core network domain.

The reference point between the access network domain and the core network domain is termed the "Iu" reference point.

The access network domain consists of the physical entities that manage the resources of the access network and provide the user with a mechanism to access the core network domain.

The core network domain consists of the physical entities that provide support for the network features and telecommunication services. The support provided includes functionalities such as the management of user location information, control of network features and services, and the transfer (switching and transmission) mechanisms for signaling and for user-generated information. The core network domain is subdivided into the following domains:

- Serving network domain;
- Home network domain;
- Transit network domain.

The serving network domain is the part of the core network domain to which the access network domain that provides the user's access is connected. It represents the core network functions that are local to the user's access point, and thus their location changes when the user moves. The serving network domain is responsible for routing calls and transporting user data/information from source to destination. It has the ability to interact with the home domain to cater for user-specific data/services and with the transit domain for nonuser-specific data/services purposes.

The transit network domain is the core network part located on the communication path between the serving network domain and the remote party. If, for a given call, the remote party is located inside the same network as the originating UE, then no particular instance of the transit domain is activated.

The home network domain represents the core network functions that are conducted at a permanent location regardless of the location of the user's access point. The USIM is related by subscription to the home network domain. The home network domain, therefore, contains permanent user-specific data and is responsible for management of subscription information. It may also handle home-specific services, potentially not offered by the serving network domain.

A simplified block diagram comprised of the most important elements of the 3G UMTS network is shown in Figure 7.3.

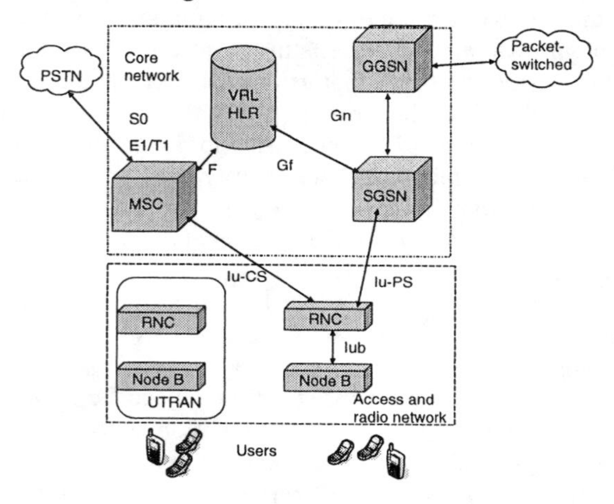

Figure 7.3 Simplified 3G UMTS network model. (*From:* [8].)

In general, a UMTS network consists of three domains: user equipment, UTRAN for 3G or BSS for 2G and 2.5G, and core network. The UE is a mobile terminal that communicates with UTRAN via the air interface. UTRAN provides the air interface access method for the UE. The CN provides switching, routing, and transit for user traffic. It also stores databases and provides network management functions [9].

From a specification and standardization point of view, both the UE and the UTRAN consist of completely new protocols and underlying communication technology, the design of which is based on the needs of the new WCDMA radio technology. Regarding the underlying technology, N-ISDN, formerly used by 2G UMTS (GSM), has now been replaced with the B-ISDN (ATM) technology to cope with higher data rates and improved voice quality.

A UTRAN consists of two distinct elements: the node-B and the RNC.

The RNC (logically corresponds to the GSM BSC) controls the radio resources in its domain. RNC is the service access point for all services that the UTRAN provides to the CN. It also terminates the RRC protocol that defines the messages and procedures between the UE and UTRAN. A UTRAN may consist of one or more RNSs. An RNS is a subnetwork within UTRAN that consists of one RNC and one or more node-Bs. The RNCs, which belong to different RNSs can be connected to each other via the Iur interface. The logical function of an RNC is further divided into controlling, serving, and drift. The controlling RNC administers the node-B for load and congestion control. It also executes admission control and channel code allocation for the new radio links to be established by the node-B [9].

The serving RNC is the RNC that terminates both the Iu and Iub links from the core network and user equipment, respectively. It performs L2 (MAC layer) processing of data to/from the radio interface. Mobility management functions such as power control and handoff decision are also handled by the serving RNC. The drift RNC complements the serving RNC by providing diversity when the UE is in the state of inter-RNC soft handoff (which requires two RNCs). During the handoff, the drift RNC does not perform L2 processing; rather, it routes data transparently between the Iub and Iur interfaces [9].

There is a clear difference between the user plane protocols that have been used to provide the means of support in parallel voice and the protocols used to provide the means of support in data calls. This is shown in Figure 7.4.

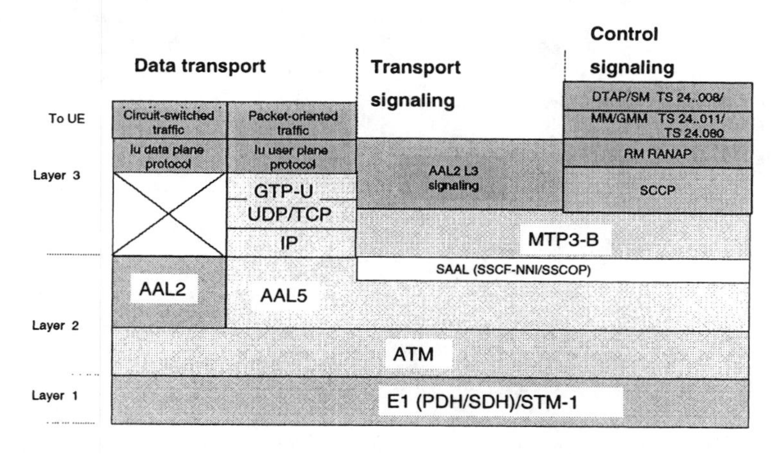

Figure 7.4 Iu interface protocol architecture.

Voice calls are serviced through the use of AAL2 comprising the requirement for small packet handling in order to avoid delays. Packet calls are inherently mapped onto the principle of IP (all versions) and handled by tunneling protocols

(GTP) over AAL5. Both user-plane stacks require the initiation of signaling procedures, which must be maintained throughout the life cycle of the user calls and finally be released.

To achieve that, 3G UMTS technology has adopted the concept of using separate signaling stacks, both associated with specific user services (data or voice) and each other through radio access bearers.

Though it is beyond the scope of this section to explain the functionality of RAB signaling, it can be shown (see Figure 7.4) that signaling is separated in RANAP, AAL2 signaling (ALCAP), and NAS. RANAP and AAL2 signaling are associated to the circuit (Iu-CS for voice) and the packet (Iu-PS for packets) switched services, respectively, while NAS is associated to a unique user call context used to identify, in a one-way manner, a communicating user.

7.2.2 Trials Environment

The 3G UMTS network trial environment set up by the IST project REAL is shown in Figure 7.5.

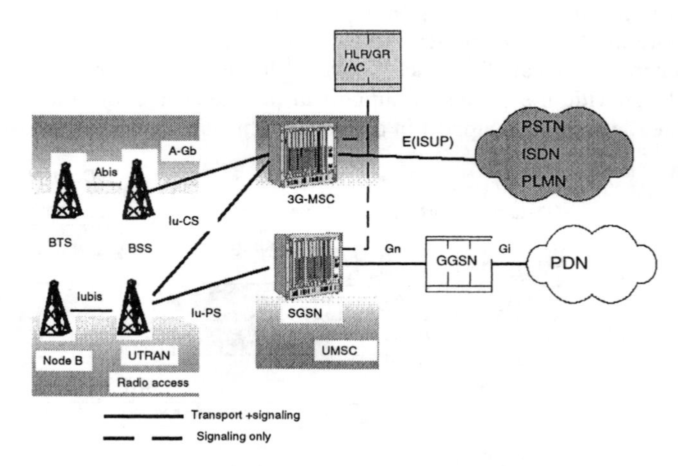

Figure 7.5 3G UMTS core network environment. (*From:* [8].)

The technical objectives of the trials can be summarized as follows:

- To assess the status of 3G UMTS technology by performing traffic experiments and simulating real users on the Iu and Gi interfaces. The Iu and Gi identify interfaces that are located at the communication ends of the 3G UMTS CN.
- To investigate, clearly describe, and finally adopt QoS measurements for the assessment of the performance of UMTS networks on the Iu and Gi interfaces.

- To develop 3G UMTS-specific conformance testing scenarios for the Iu and Gi interfaces.
- To perform stress load tests on both CS and PS signaling and data domains.

The interfaces A and Gb were beyond the scope of the REAL trials as they represent communication interfaces handling the adaptation of 2G and 2.5G terminal equipment (TE-GSM, GPRS) to the 3G UMTS core network infrastructure (UMSC).

The trials were conducted by performing traffic experiments in both the CS and PS domains, thus generating results related to the assessment of both voice and packet services

The 3G UMTS core network architecture setup for the trials is shown in Figure 7.6.

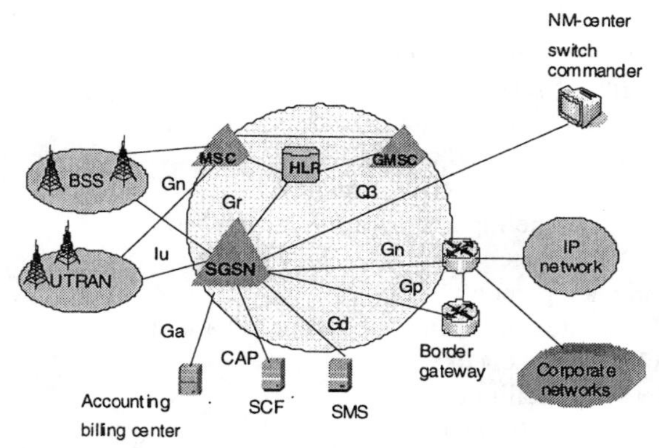

Figure 7.6 3G UMTS trials environment.

The main components of the SGSN are:

- ATM mobile platform, including modular multiprocessor design subscriber database, packet control, signaling control, and SS7 interfaces;
- IP mobile platform, including IP packet transport, the interfaces to RAN, the interface to the packet backbone;
- A LAN switch, including an interface to the NM-center and to the billing center.

In general, the SGSN hosts the following main features:

- Mobility management, including the following functionalities:
  - Hosting of the subscriber database (SLR);
  - Security functions (i.e., authentication, data integrity, and user confidentiality);

  - Paging for packet-switched and circuit-switched services;
  - GTP-MAP conversion.

- Handover control, including the following functionalities:
  - Control of the SGSN change;
  - Control of the RNC relocation;
  - SMS handling;
  - SMS over UMTS.

- CAMEL handling:
  - CAMEL phase 3 support (PDP context model);
  - Prepaid service;
  - Support of Location services based on cell ID technology;
  - VPN service for SMS and change of access point name.

- Packet routing and transfer:
  - Quality of service support;
  - IP header compression;
  - Flow and overload control;
  - Dynamic routing RIPv2 and OSPF;
  - GPRS Tunneling Protocol (GTP) V0/V1.

- Session management:
  - PDP context handling (PDP type IPv4 and PPP);
  - Subscription check;
  - Network facility check;
  - Local exception handling.

- Resource management:
  - Management of internal resources for admission control;
  - Resource management-specific RANAP signaling.

The network structure of the mobile networks with respect to the GGSN is shown in Figure 7.7.

The main tasks of GGSN are as follows:

- Network access and subscription control;
- Interworking functionality between PLMN and public data networks;
- Mediation for accounting.

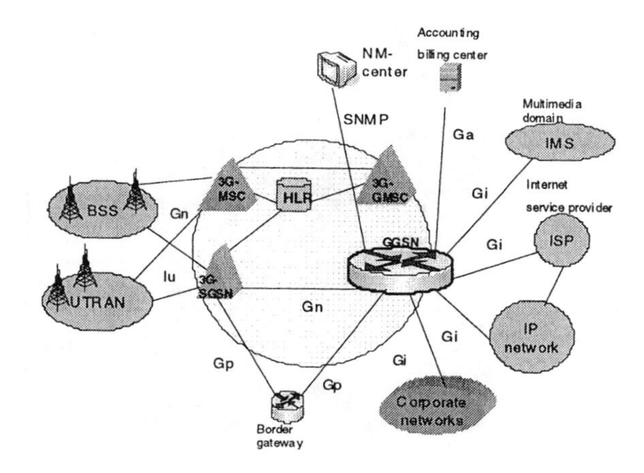

Figure 7.7 Network structure of the mobile networks with respect to the GGSN.

In order to simulate the behavior of real users, the test equipment functionality should implement the *u*-plane and the *c*-plane logic of the whole communication branch towards the Iu interface (see Figure 7.3) that is, towards the UE, the node-B, and the RNC. This is the Iu-Gi format of the system, called the UTRAN simulator.

In terms of protocols functionality, the Iu version of the test equipment implements the protocol stack shown in Figure 7.8.

In comparison to the original Iu stack defined by 3GPP, the stack implemented in the test equipment additionally incorporates a nonaccess stratum layer (NAS), which carries the information of the mobile user. This information includes the identities of the mobile users (IMSI), the terminals (LAC, RAC, PLMid), and authentication and billing messages.

With the amended stack of Figure 7.8, the test equipment acquires the ability of not only performing traffic generation and analysis at the user plane, but also in simulating the behavior of real mobile users as the system allows transportation of user-specific information associated to both the Iu signaling and the user plane links, towards the core network.

The Gi and Gn version of the test equipment implements the standard protocol stacks of GPRS [10].

By using comprehensive GUI, test engineers may activate users for specific services and can perform the following traffic experiments:

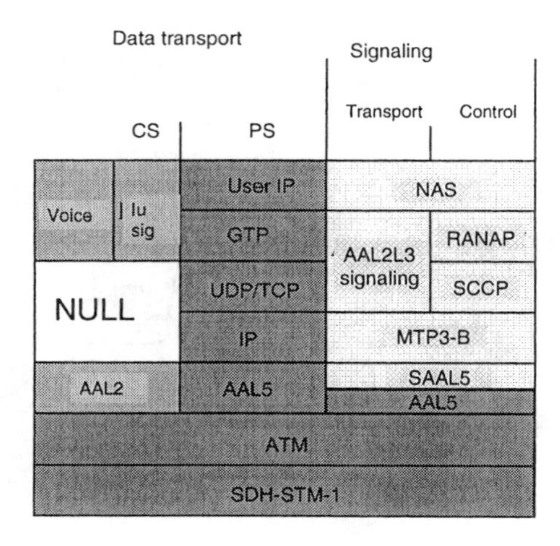

Figure 7.8 Stack implemented in the test equipment for the REAL trials.

- *A conformance test:* The test equipment serves to verify the Iu and the Gi protocols and how these comply against the existing standards.
- *Simulation:* Simulation of mobile users accessing voice and packet services at several rates.
- *A stress test:* The test equipment provides the ability of generating traffic exceeding the signaling capacity of the Iu and Gi nodes in order to enable monitoring of the core network behavior under flood conditions, which may occur when many mobile subscribers attempt to get connected to the UMTS network.
- *A duration test:* The test equipment provides the ability of testing the stability of the core network for long-duration runs by circulating traffic on the Iu and Gi interfaces.
- *A performance test:* This test includes measurements related to QoS aspects and other requirements related to real-time performance.

The logic of activating, monitoring, and releasing users is associated with the concept of creating and running scenarios (see Figure 7.9) with individual traffic profile settings.

Selection of traffic profiles is done in the same way for both circuit-switched and packet-oriented traffic (see Figure 7.10).

After checking the Send Data check box, the operator selects from one of the available traffic files (extension: .evt) and selects the source that will produce this profile's data. Traffic files are created through the Link Profile dialog box.

Once the individual user profiles are created, the operator may invoke the dialogs explained below to create a testing scenario incorporating many users of different or the same traffic profiles.

Three parameter groups are defined for each scenario. Some of these are automatically calculated based on other ones.

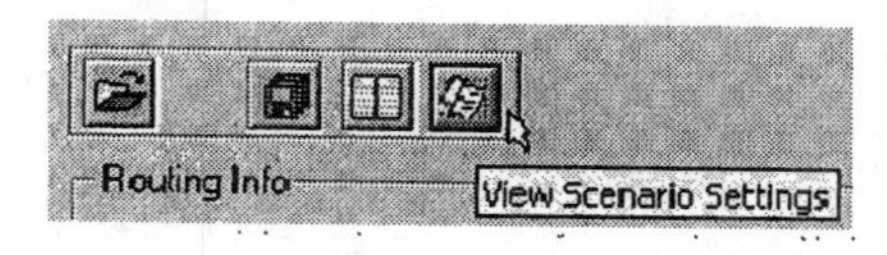

Figure 7.9 Invoking the Scenario Settings dialog box.

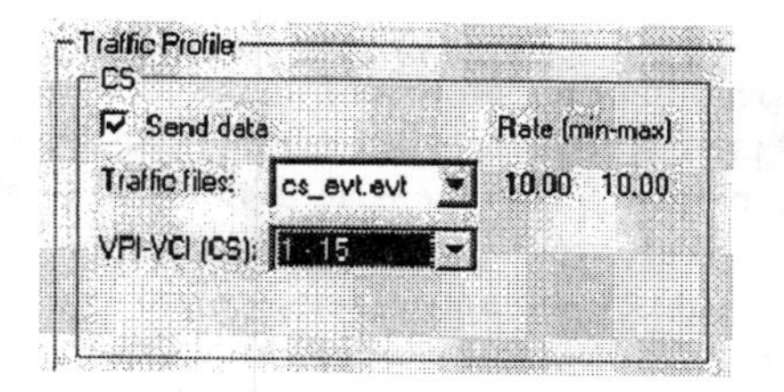

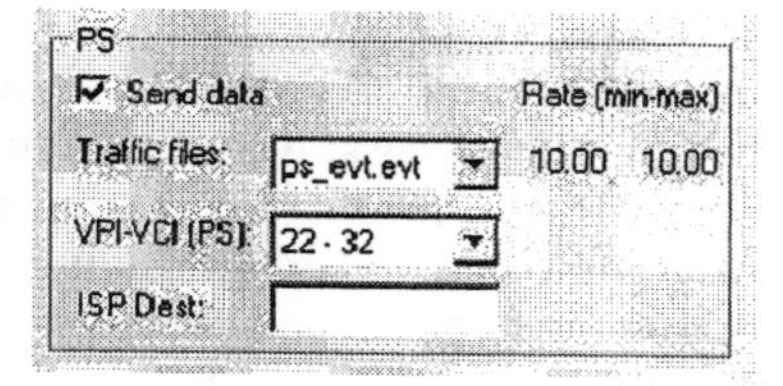

Figure 7.10 Selection of Traffic Profiles.

The first group is the Message Flow: it allows the operator to select the Message Cycle performed by the tester (for each user) in this scenario. Note that some of the dialog's fields are only enabled in certain message cycles. The edit boxes contain the time (in seconds) that the tester will wait in between messages. For example (see Figure 7.11), after a successful ATTACH the tester will wait 580 seconds before issuing a SRV_REQ. When the SRV_REQ is completed, a conversation will be performed and then the tester will wait another 580 seconds before DETACHing.

These intermessage times however, are influenced by the second group, the Traffic Parameters. In the above example, the operator has specified that 10 users are going to perform this scenario, and each user will perform two busy hour call attempts (BHCA). This means that the tester will perform 20 BHCA for this scenario. Since the conversation time is 60 seconds, the time intervals computed are 580 seconds [(580+580+60+580)×2 = 3,600 seconds = 1 hour]. Note that these intervals are checked automatically. If the operator modifies one of them, the others are recalculated.

The operator can also choose whether all users (of this scenario) will send requests at the same time or not. In the example above, if the traffic shape is set to bursty, all users will ask for SRV_REQ as soon as the 580 seconds pass. The minimum and maximum delay values in the start delay field specify a common delay value for all users. This means that all user requests will be sent in bursts.

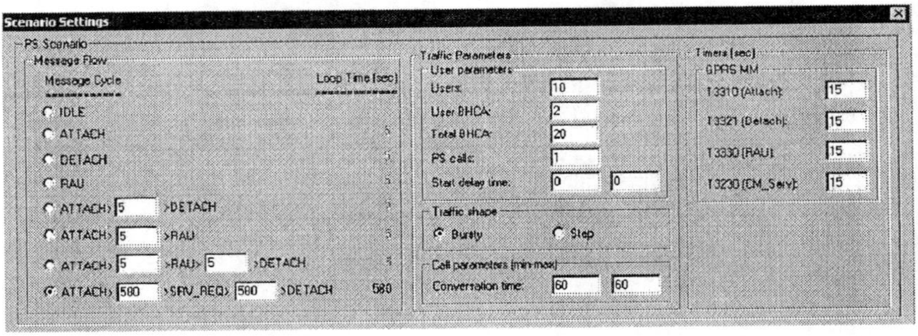

Figure 7.11 Example of a test for a packet-oriented scenario. (*From:* [8].)

The Step option will allow the tester to randomly distribute the calls (with a random interval time chosen between the operator-provided minimum and maximum). The third group allows modification of the protocol timers.

After the successful configuration of the User Profiles, the operator can create a Test Scenario for the test equipment to perform. This scenario is in effect a set of user profiles. In other words, this final step groups the user profiles together.

Similar steps are followed to complete the configuration settings and to start generating the traffic data. For details, the reader is kindly referred to [8].

Besides the basic statistics, the gathered statistics can be displayed in detail. An example of c-plane gathered statistics and their display is shown in Figure 7.12.

The statistics window allows for graphical reviews of the gathered data, as well as for saving data for later review (through the Save button). The operator can also select the messages of interest through the Message Selection buttons. A specific message can be added or removed from the list through the list control buttons.

For each protocol displayed, the operator can also double-click on the protocol window's title bar (on the main window), and see specific per-protocol statistics.

7.2.3 Measurement Results

7.2.3.1 Signaling Procedures

In general, the signaling procedures required for the authentication and registration of users to the core 3G UMTS network can be identified by the signaling flows as shown in Table 7.1.

Mobile and signaling call (MOC) is the procedure of activating a CS or PS call (from the user side). For the PS this procedure is associated with the PDP context activation (deactivation). For the CS this procedure is associated to an actual voice call activation.

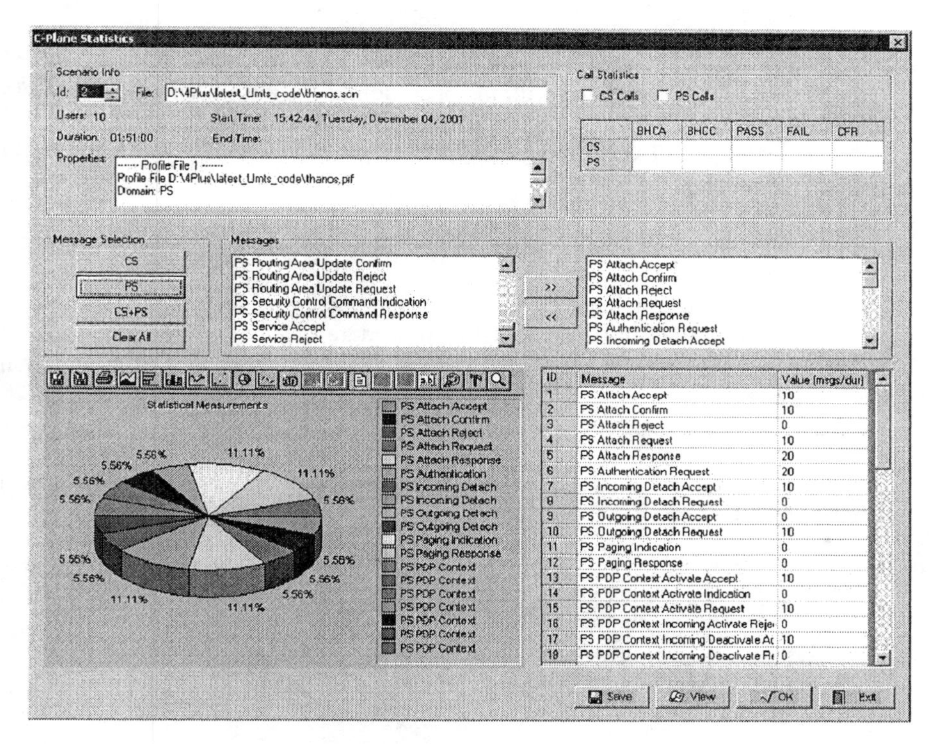

Figure 7.12 Display of c-plane statistics. (*From:* [8].)

Table 7.1

Signaling Flows for the REAL Experiments

Signaling Flow	Domain
GPRS ATTACH	PS
GPRS DETACH	PS
Location update (LUP)	PS, CS
Mobile originating call (MOC)	PS, CS

Specifically for the case of PS MOC, the GPRS ATTACH (DETACH) procedure is required to take place before the actual call, so that the core network can get the parameters of both user and terminal and assign back temporary identities relevant to identify the user and the terminal within a specific location area.

LUP is the procedure of updating the assigned parameters (TMSI) of the user and the terminal when moving from one location area to another (cell coverage area).

All the above procedures can be further disassembled into message flows of internal procedures associated with the operation of the individual protocol layers and stacks.

The main goal of performing experiments on these signaling procedures is to measure or monitor the following parameters:

- The core network processing time in completing each signaling procedure;
- To monitor the behavior of the switch when a wide number requires the invocation of the same signaling procedure;
- To monitor the behavior of the switch in the long term (long duration tests);
- To measure the success/failure ratio.

Some indicative diagrams of the above-discussed procedures are shown in Figures 7.13 to 7.16.

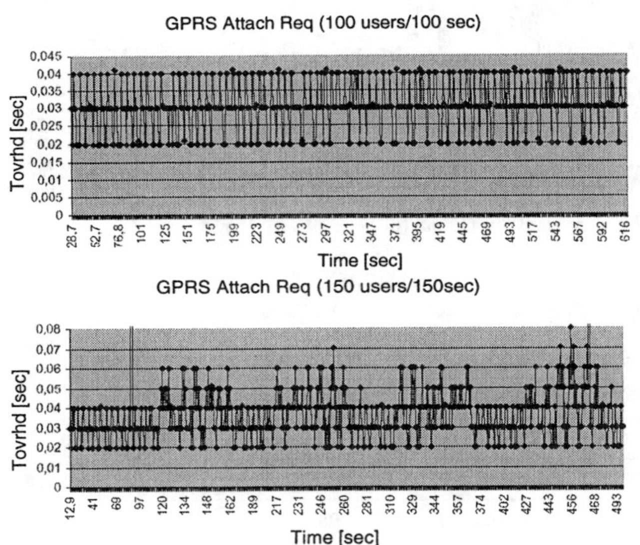

Figure 7.13 Diagram for the GPRS ATTACH procedure for: (a) 100 users and (b) 150 users.

T_{proc} and T_{ovrhd} are values representing the time in seconds required by the switch (SGSN or 3GMSC for PS and CS, respectively) to accomplish a given signaling procedure.

For the case of GPRS ATTACH [see Figure 7.13], the SGSN processing time constantly fluctuates between 0.02 and 0.04 second for a number of up to 100 users (per UTRAN); but for higher numbers the SGSN processing time increases considerably, also showing failures in accomplishing the given procedure.

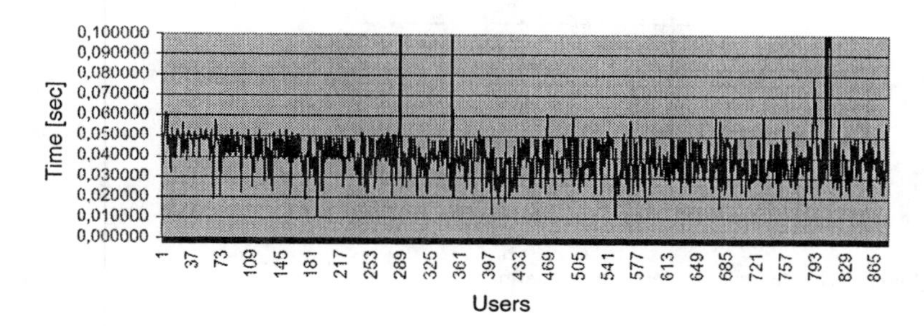

Figure 7.14 Diagram for the GPRS DETACH procedure.

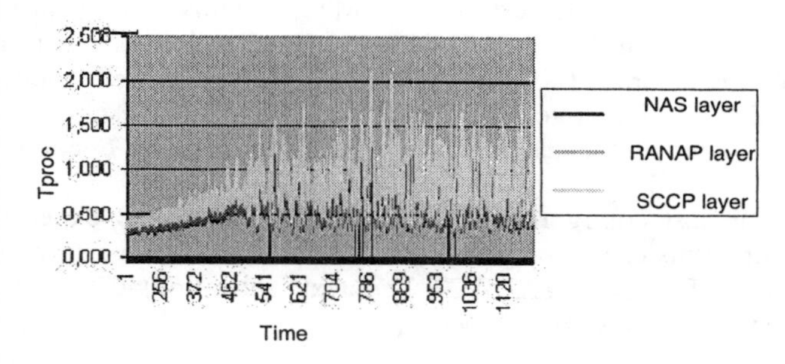

Figure 7.15 Diagram for the LUP scenario.

The GPRS DETACH procedure (see Figure 7.14) is evaluated by activating the DETACH request message flow for 100 users per second. Also, while the time required by the switch to accomplish the procedure is fluctuated around a constant value, as the number of users per UTRAN increases, failures in accomplishing the procedure emerge. These failures do not occur periodically. This means that the internal signaling algorithms of the switch need to be reinvestigated.

The LUP procedure (see Figure 7.15) is evaluated by invoking the LUP message flows for 500 users, progressively. Here, the T_{proc} value increases progressively with the number of users without failures.

The MOC procedure (see Figure 7.16) is evaluated by initiating 100 users per second. Here the T_{proc} value t fluctuates around a certain range up to 4 seconds, meaning that the internal MOC implementation algorithms are well tuned.

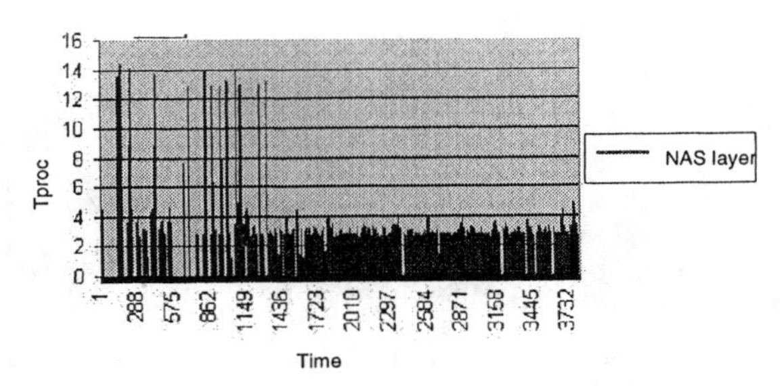

Figure 7.16 Diagram for the MOC scenario.

7.2.3.2 PS U-Plane Procedures

The most demanding feature of 3G UMTS systems is the data load that can be transferred through the system without any packet loss or other errors caused by this. Within the project REAL [1], several experiments were conducted to provide data load at high data rates.

The setup of the testbed was described earlier in Section 7.2.2; see also Figures 7.1 and 7.6.

In the test configuration of Figure 7.17, a typical experiment performed during the trials execution phase consisted of 400 users performing the signaling scenario: ATTACH–CTX ACT (context activation)–CTX DEACT (context deactivation)–DETACH.

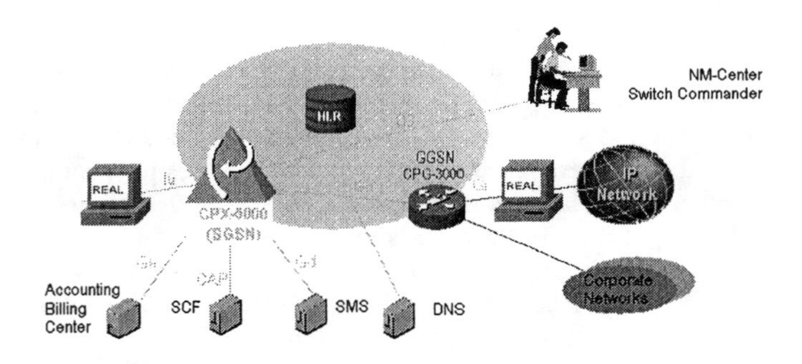

Figure 7.17 Trial configuration for REAL.

While the context (user plane link per user with specific destination and source IP) is active, the test equipment produces and analyzes data at rates of 20 Mbps per user. The test duration for this experiment was 65 hours.

After the test was accomplished, several counters at the switch side (SGSN and GGSN) as well as at the side of the test device had to be compared and a call failure rate and a packet drop rate were calculated.

Another important issue to be considered while performing very demanding traffic experiments is the performance counters provided by the switch per main processor type.

These counters are provided by the switch (SGSN-GGSN) periodically and can be retrieved after the test execution. The obtained raw data is then loaded into a certain post-processing tool and distinct data can be extracted. For instance, one can obtain time-correlated information of the processor loads for every processor type that is involved. Out of these measurements, a prediction of the load of the different main processors at certain signaling and date load is possible. The processor load over a period of 60 hours is shown in Figure 7.18.

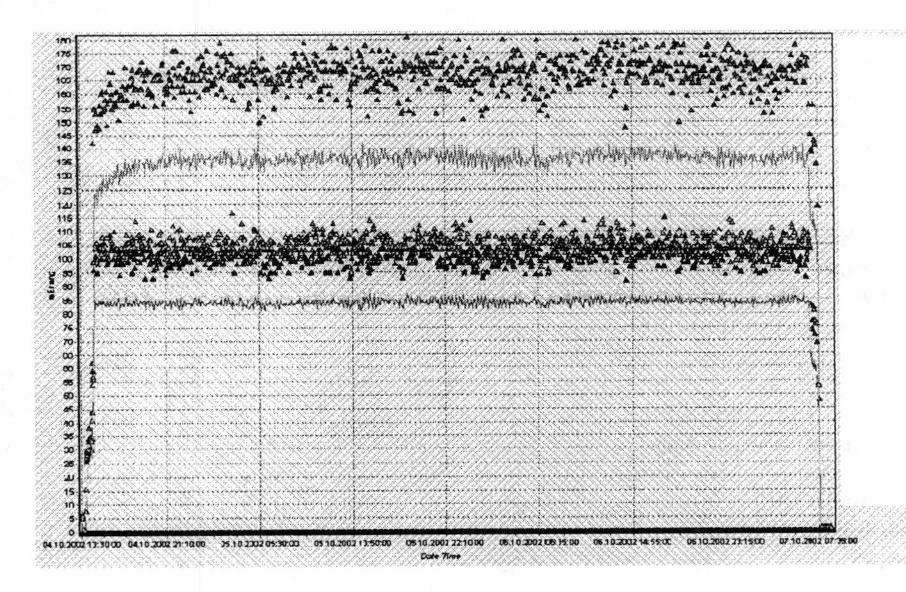

Figure 7.18 Processor load over a period of 60 hours.

In the following plots, the small dots represent the maximum measured values, while the continuous line graphs represent the average values produced from those actually measured.

Another type of measurement that was performed during the trials was the delay measurement. For that reason the packet is marked with a certain label and then its round-trip delay is measured from the Iu to the Gi interface and vice versa.

By modifying the data load, the relationship between the cell transfer delay and the data load routed through the switch is detected. A series of measurements was performed for 10,000 and 15,000 packets per second in the up- and downlink directions. The results showed a fluctuation of packets interarrival period between 300 μs and 1.3 ms, which is caused by the processing load on the SGSN and GGSN nodes. This fluctuation may worsen as the number of users increases [8].

Though considerably high, packets interarrival period may be considered as acceptable since it copes with the general requirements of IP connectivity to external networks or the Internet. The only factor that is affected by such behavior is the time required for IP users to get connected to the Internet.

7.2.3.3 PS C-Plane Measurements

The following table shows the assumed subscriber behavior, which is used for the dimensioning of the GSN. The definition is a copy from the requirement specification.

The traffic model assumes an attach ratio of 60%. For a worst-case scenario all subscribers can be attached (in the SLR). However, the system supports only the load according to the average subscriber behavior as defined by the traffic model. The 60 users perform a scenario in the following way:

- ATTACH – 8 times routing area update – DETACH;
- ATTACH – context activation – context deactivation – DETACH.

This leads to a traffic rate of about 100,000 BHCAs. The evaluation of the performance counters of the involved main processors is shown in Figures 7.19 and 7.20.

After the start of the scenario the load on the MPs reaches a certain value, and this value remains constant until a problem occurs in the switch. The graph representing the load of the CALLP processors starts fluctuating and produces some peaks up to higher performance values.

At the same time the processors responsible for handling the context load of the system can experience remarkable changes in their processor load. Based on this observation the error notebook of the switch for that time interval was investigated and the reason for that problem was detected to be caused by certain indication procedures. The consequence of this test, for instance, was the development of a patch to overcome this problem.

The importance of the performance counters is shown in Figure 7.21.

In the beginning two test devices are running; after some time one is switched off and, therefore, only the load of one test device must be handled by the switch. This is reflected in the reduced load on the involved processor types. In this test the following scenario was run:

- 50 subscribers performing ATT-RAU-DET scenario;
- 400 subscribers performing ATT-CTX-DET scenario.

Compared to the previous test, the number of users is higher, and therefore the load on the related processors is also increasing as expected. This increasing of the load on the related processors is used as a prediction for the number of users that can be handled in parallel by the switch.

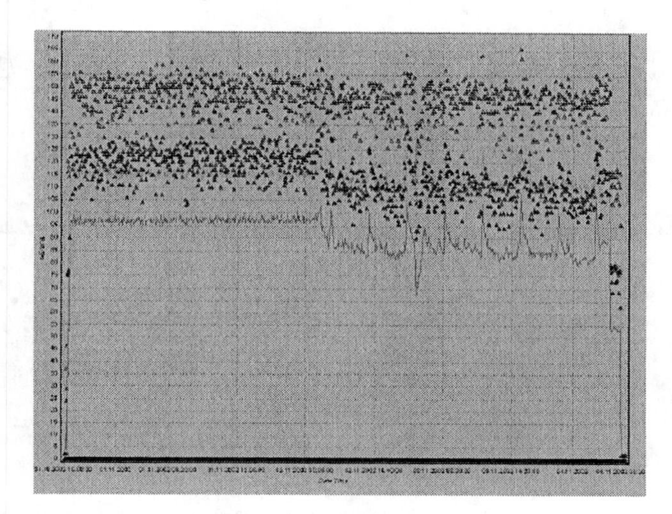

Figure 7.19 Call processing load over the observed time period.

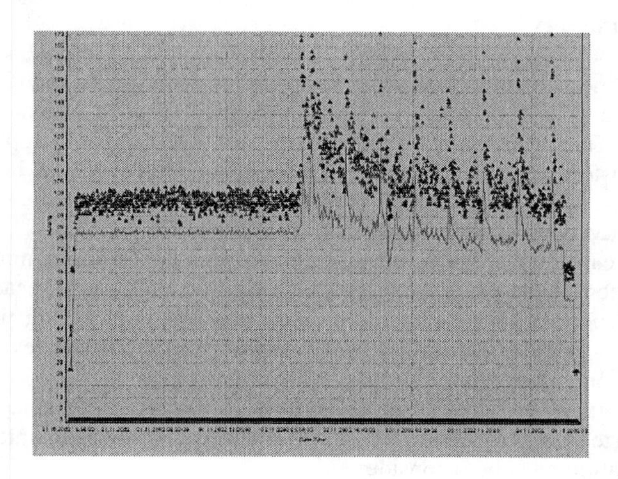

Figure 7.20 Load of the MP:MM and MP:PD.

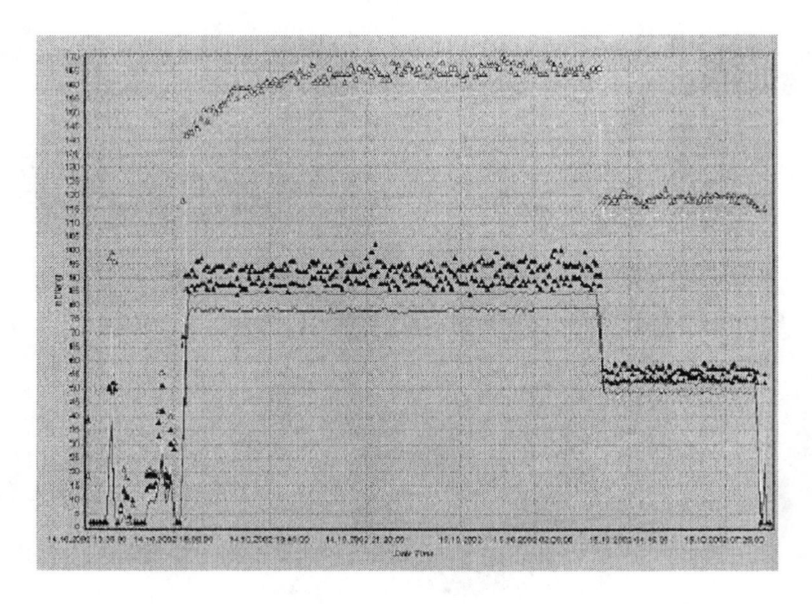

Figure 7.21 Performance values for the involved processor types.

7.2.4 Summary

One of the most important parameters for the evaluation of a mobile network and its components is the overhead introduced with the performance of signaling operations. This overhead assumes even greater significance in applications when many short-term channels should be opened and closed at a high rate. Such requirements arise in the context of Internet applications, distributing computing, and interactive video transactions.

These call for the dynamic establishment of a considerable number of short-lived transport channels. It is evident that a significant time overhead would result in network performance degradation. Time-consuming signaling procedures can be identified as those associated with database operations (retrieval of user IDs, authentication, billing).

These procedures are associated to NAS and RANAP layers operation. To minimize this side effect, VLR and HLR databases with internal structure implementation must be reconsidered.

Regarding user plane data transactions, UDP is a good choice for the fast transfer of user information, tolerable to a minimum packet loss since no time-consuming functions, such as flow control, reordering, or retransmission functions, are activated. In its turn, UDP packets will be encapsulated in IP packets where the IP source and destination address indicate the static address of the respective GSN (in the case of a GPRS network). To keep the delay variation of the data transactions at a minimum level, TCP on the Iu should be replaced by UDP.

The trials performed by the project REAL deduced some important issues related to QoS critical parameters. An impairment of those may affect the overall core network performance. Among the most critical are the following:

- *Busy hours call attempt:* A nominal value should always be defined per MSC, SGSN, and GGSN. During network operation, impairment of this value is associated with the increase of loss ratio in signaling packets due to protocol malfunctions, which is likely to happen when much more than the expected number of users attempts to get connected (congestive condition).
- *Packet delay variation across Iu-Gi/Iu-PSTN paths:* Packet delay should always comply with the service features that are associated for each user application (PDP context). For IP services related to data retrieval, a delay of 35 ms is usually acceptable. For IP services related to point-to-point communications (e.g., video telephony) a maximum delay of 6 ms is acceptable.

The impact of vendor-specific options on the Iu interface may be related to the number of supported users per port. For instance, the use of ATM over E1/T1 (2 Mbps) links reduces the number of supported users by 77% per port if combined to that of ATM over STM-1 (155 Mbps).

BHCA nominal value is also strictly related to vendor implementation options. BHCA determines the number of supported users per "box," that is, how many users a SGSN or MSC may accept.

Transport IP and its assignment to port or to logical entities determines how many GPRS nodes a core network may have. If, for instance, a core network presents GGSN and SGNS nodes supporting more than one transport IP, then the GPRS part of the network is said to have the ability of servicing more than one GPRS path.

The signaling protocols implementation version constitutes a factor that indicates what services a core network may support. RANAP is the protocol mostly subjected to several versions related to the granularity of the supported QoS.

7.3 MANAGEMENT OF END-TO-END QUALITY OF SERVICE ACROSS THE INTERNET AT LARGE

This section describes the proposal and validation of scalable, incremental solutions that enable the flexible deployment and delivery of interdomain QoS across the Internet. This work was performed within the scope of the IST project

MESCAL [3]. The project validated its results through prototypes, and evaluated the overall performance through simulations and prototype testing.

7.3.1 Initial Specification of Protocols and Algorithms for Interdomain QoS Delivery

This section specifies algorithms and protocols that enable interdomain QoS across the Internet, including the following [11]:

- Algorithms and protocols that enable SLS establishment between peers, including domain advertisement of QoS capabilities and QoS capability discovery;
- Algorithms and protocols for invocation of service instances across domains;
- Off-line interdomain and intra-domain traffic engineering algorithms. Two interdomain provisioning cycles are described: a longer timescale cycle in which pSLS requirements are determined by the traffic engineering algorithms and then negotiated with peer domains, and a shorter term cycle in which interdomain bandwidth is invoked within the framework of existing pSLSs;
- Dynamic interdomain traffic engineering algorithms and protocols, including QoS enhancements to BGP and a path computation server (PCS) communications protocol (PCP) for MPLS traffic engineering;
- Algorithms for interdomain traffic forecast;
- Algorithms that integrate inter- and intradomain SLS management with traffic engineering, defining the data that needs to be passed between the SLS handling functional blocks, traffic forecast, and traffic engineering components;
- Algorithms for on-line SLS invocation handling (i.e., interdomain admission control);
- Multicast SLS definitions and multicast traffic engineering algorithms.

7.3.1.1 Meta-QoS-Class Concept

Although there has been much work done in the QoS field over the last decade, little has been undertaken to provide guidance on how to deploy QoS throughout the whole Internet. This section presents a new concept: the meta-QoS-class concept. This concept, if it succeeds in raising general interest and agreement, is likely to dramatically help to achieve a QoS-enabled Internet [11].

A meta QoS-class is a standardized set of qualitative QoS transfer capabilities with the scope of a single provider domain, understood to meet common application requirements.

Based on current best practices, we can hardly say that QoS (if over-provisioning is not considered as a part of the QoS management) has been currently deployed interdomain and even intra-domain in the networks of the service providers.

The Internet remains an interconnection of best-effort networks. The only worldwide transport service usable throughout the Internet is the best-effort service. For instance, there is no means for a video content provider to make it possible for their willing-to-pay customers to access the service via a performance guaranteed transport on a large scale.

An interdomain QoS delivery solution should take into account some requirements that would prevent QoS techniques and architecture from impairing the spirit in which the Internet has been devised since its early days. The idea is, of course, not to refuse any evolution in the Internet paradigm just because the Internet is as it is. The intention is to keep the features the majority of users can agree on rather than emphasize on technical and financial considerations. We should preserve the facility to spread Internet access, the facility to welcome new applications, and the possibility to communicate from any point to any other point.

From this angle, the list of requirements can be summarized as follows:

- Networks should be ready to convey interdomain QoS traffic before customers can initiate end-to-end SLS negotiations (just like interdomain routing is).
- The solution must not, to the greatest extent possible, preclude unanticipated applications.
- A best-effort route must be available when no QoS route is known.
- Best-effort delivery must survive QoS.
- The solution should not rely only on the existence of a centralized entity that has knowledge and control over the Internet.

Based on the meta-QoS concept and considering the above-listed requirements, an Internet QoS model can be built [11]. The purpose of this model is to build a QoS-enabled Internet, which keeps, as much as possible, the openness of the existing best-effort Internet and more precisely conforms to the requirements expressed above. In this model, the resulting Internet appears as a set of parallel Internets or meta-QoS-class planes. Each Internet is devoted to serve a meta-QoS-class.

Each Internet consists of all the local QoS classes (l-QCs) bound in the name of the same meta-QoS-class. An l-QC is a basic network-wide QoS transfer capability that can be provided by means employed in the provider domain itself. Evidently, the domain boundaries appearing in the topological constraints of an l-QC should belong to the boundaries of the provider domain. When an l-QC maps several meta-QoS-classes it belongs to several Internets. The user can select the Internet that is the closest to his/her needs as long as there is currently a path available for the destination. A meta-QoS-class is an abstract concept, not a real local QoS-class (l-QC) implemented in a real network. Providers may map the l-QCs they have engineered to meta-QoS-classes if they wish to support standardized QoS characteristics.

In order to comply with the Internet paradigm, it can be assumed that in a meta-QoS-class plane all paths are to a reasonable extent, born equal. Therefore, the problem of path selection amounts to finding at least one path for each selected meta-QoS-class plane. This is similar to the traditional routing system used by the Internet routers. Therefore, a BGP-like protocol can be used for the path-selection process. By destination, q-BGP selects and advertises one path for each meta-QoS-class plane. When, for a given meta-QoS-class plane, there is no path available to a destination, the only way for a datagram to travel to this destination is to use another meta-QoS-class plane. The only meta-QoS-class plane available for all destinations is the best-effort meta-QoS-class plane (also known as the Internet). There is no straightforward solution to change from one plane to another on the fly. So there is no straightforward way to span a meta-QoS-class plane hole by a best-effort bridge.

This solution gives only loose administrative guarantees; however, as long as all actors (especially all service peers involved in the QoS AS path) do their job properly, the actual level of guarantee will be what is expected. This solution stands only if l-QC meta-QoS-class-based binding is largely accepted and processed.

7.3.1.2 The MESCAL Functional Architecture

This section introduces the functionality required for the provision of interdomain QoS services from the perspective of a single provider. The functional architecture analyzes the overall problem of providing interdomain QoS and decomposes it into a set of finer grained components. One of the objectives of this exercise is to aid the development of the MESCAL solutions by breaking it down into manageable entities while maintaining a holistic view of the overall issues to be solved. The detailed MESCAL functional architecture is available in [12]. Here, the architecture is presented in order to describe the implemented algorithms and protocols for each functionality. The functional architecture is shown in Figure 7.22.

The data plane is responsible for per-packet treatment within packet arrival epochs. The control plane covers intra- and interdomain routing, SLS invocation handling, including authentication, authorization, and admission control, and dynamic resource management, including load distribution and capacity management functions. Typically, the control plane functions are embedded within the network equipment although they are not involved in packet-by-packet decisions.

The management plane is an off-line functionality, typically located outside of the network elements in the management servers. The management plane functions are responsible for planning, dimensioning, and configuring the control and data planes and interacting with customers and service peers to negotiate contracts. While management plane functions are not as dynamic as control and data plane functions, they are by no means static. Within the MESCAL system there is a continual background activity within the management plane at the epochs of the so-called resource provisioning cycles (RPCs).

Two RPCs can be distinguished: the intradomain RPC, which involves off-line intradomain traffic engineering, and the interdomain RPC, which involves off-line interdomain traffic engineering.

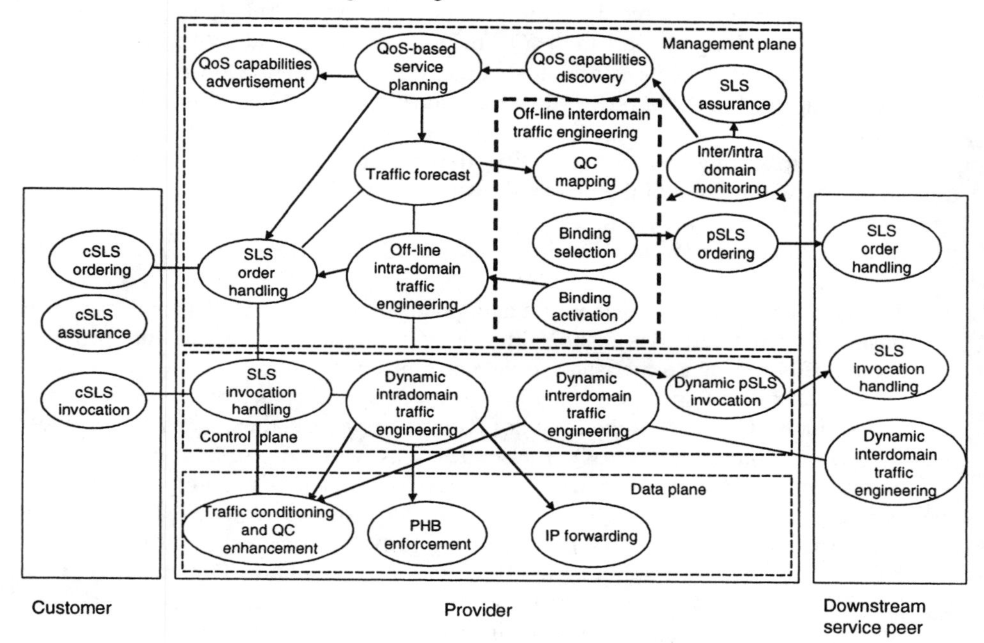

Figure 7.22 Initial functional architecture. (*From:* [11].)

The latter may be further decomposed into a binding selection cycle and a binding activation cycle (see [11]). The RPCs aim at proactively optimizing network resources to meet the predicted demands and to build in sufficient spare capacity to avoid the burden of reconfiguring the network for each and every SLS subscription or renegotiation, without the inefficiencies and costs associated with massively over-provisioning resources.

The architecture describes the full set of functions required for a provider to participate in the end-to-end provision of QoS-based IP services. It does not, however, prescribe the implementation means by which they will be realized within the network equipment or in the external management servers, with automated or manual processes. This is a matter for each provider. While the full set of functional blocks, (or their equivalent) are expected to be in place in the downstream providers, the proposed architecture does not assume that automated processes will always implement all blocks. Here, the algorithms suitable for deployment as automated processes in the traffic engineering and service management functional blocks are presented, but it is also possible to deploy much of the management plane through manual processes, at the cost of reduced responsiveness or flexibility.

For some of the service options, the algorithms or manual processes required to implement the functionality might be trivial. For instance, the loose guarantees service option does not require explicit admission control functionality in the SLS invocation handling block, and the QC mapping, binding, and activation processes are simplified due to its adoption of the well-known meta-QoS-classes and the restriction to bindings only with the same meta-QoS-class in service peer domains.

The following sections identify the major aspects of the functionality contained within each of the blocks shown in Figure 7.22.

QoS-Based Service Planning and QoS Capabilities Discovery and Advertisement

The QoS-based service planning functionality encompasses all the higher-level business-related activities responsible for defining the services that the provider should offer to its customers and service peer providers.

These are specified according to the business objectives of the provider, and include l-QCs within the scope of its own network and extended QoS-classes (e-QCs) combining its local QoS-based services with those offered by its service peers. An e-QC is a basic network-wide QoS transfer capability that can be provided by means employed not only in the provider domain but also using appropriate means in other (service-peering) provider domains. In other words, an e-QC is provided by combining the QoS transfer capabilities (QoS-classes) of the provider domain with appropriate capabilities (QoS-classes, l-QC or e-QC) of other provider domains. The domain boundaries appearing in the topological constraints of an e-QC could be outside the boundaries of the provider domain, thus extending the topological scope of the QoS transfer capabilities of the provider domain.

Prior to any SLS established between two providers (pSLS) agreement with a neighboring provider, a provider discovers the QoS capabilities, capacities, destination prefixes, and costs of potential service peer providers thanks to the QoS capabilities discovery functional block. Once l-QCs and e-QCs have been defined and engineered (by intra- and/or interdomain TE), the QoS capabilities advertisement block is responsible for promoting the offered services so that its customers and service peer providers are aware of its offerings.

Off-Line Traffic Engineering

The traffic forecast block is responsible for aggregating and forecasting traffic demands. During a provisioning cycle, the set of subscribed SLSs established between customers and providers (cSLSs) and pSLSs are retrieved from the SLS order handling and an aggregation process derives a traffic matrix with the demand per ordered aggregate between the ingress and the egress points of the domain (ASBRs). The demand matrix is used by the intra- and interdomain traffic engineering processes to calculate and provision the local and interdomain resources needed to accommodate the traffic from established SLSs as well as those anticipated to be ordered during the provisioning cycle.

Binding selection is the process of combining l-QCs of the local domain with o-QCs of other domains, learned through QoS capabilities discovery, to construct potential e-QCs that meet the service requirements defined by QoS-based service planning. It should be noted that binding selection might result in a number of QoS-bindings for a given e-QC. QoS-bindings with the same service-peering provider may differ in the l-QC and subsequently in the o-QC they use. Alternatively, QoS-bindings may differ when established with different service-peering providers.

The binding activation functionality is responsible for mapping the predicted traffic matrix to the interdomain network resources (once pSLSs have been established), satisfying QoS requirements while optimizing the use of network resources. Binding activation decides which of the established QoS-bindings will be put in effect in the network for implementing an e-QC together with the associated routing constraints for those e-QCs. The QC-bindings in effect will be enforced through routing decisions as well as configurations of the traffic conditioning and QC enforcement block (e.g., configuring the egress ASBR to perform DSCP remarking for realizing a QC-binding). The latter configuration can be made directly to the egress router or passed through dynamic interdomain traffic engineering.

Figure 7.23 shows a change in the above-described decomposition. First, because the functionality of the QC mapping functional block can be too lightweight to justify a single functional block, it can be incorporated in the binding selection block as the first step of its algorithm.

Moreover, both binding selection and binding activation have to run an optimization algorithm, which will decide on the most optimal resource allocation in terms of inter- and intradomain cost in order to satisfy a predicted traffic demand. This resource allocation could be either for the establishment of the pSLSs for the next binding selection period or for the allocation of the interdomain resources for the next provisioning cycle.

Consequently, the traffic engineering system includes an interdomain resource optimization block, which realizes the algorithm described above and is called both by binding selection and binding activation. Of course, the input to the algorithm will be different when called by binding selection and when called by binding activation, since in the first case a traffic demand for a longer period will be passed as input to the algorithm, while in the latter case a shorter-term prediction of the traffic demand will be the input.

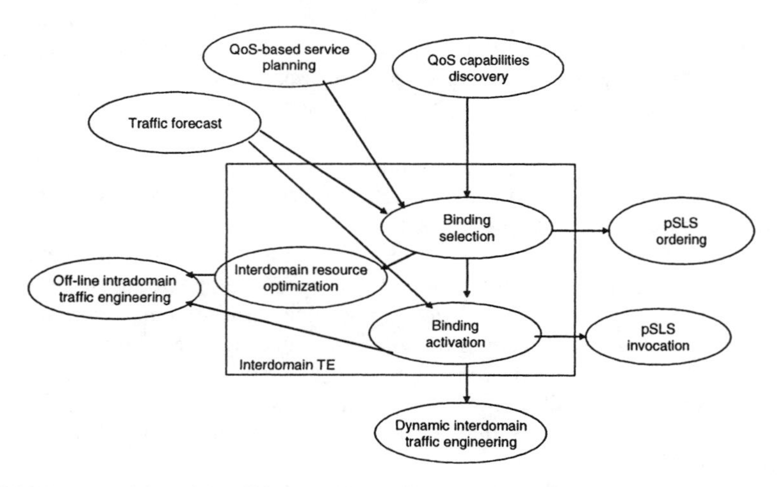

Figure 7.23 Decomposition of the off-line interdomain TE.

Moreover, when binding activation triggers the interdomain resource optimization algorithm, the allocation of resources is constrained by the already established pSLSs, while binding selection has to consider different hypothetical scenarios of pSLSs in order to decide which one of them leads to a more optimal solution in terms of resource use and at the same time satisfying the traffic demand.

Off-line intradomain traffic engineering computes the intradomain network configuration in terms of routing constraints and PHB capacity requirements in order to satisfy the predicted traffic demand at intra-domain RPC epochs.

The off-line intradomain TE block has been further decomposed into two subcomponents: resource optimization and network reconfiguration scheduler. The network reconfiguration scheduler is the control system for the offline intra-TE block. It has two main purposes, handling computation requests to resource optimization (resource provisioning cycles, interdomain traffic engineering "what if" queries) and scheduling the reconfiguration of the network using link weight settings computed by the resource optimization block. The resource optimization block contains the OSPF link weight optimization algorithm. It is a passive block, until called by the network reconfiguration scheduler at which point it collects a traffic demand matrix and a network topology and computes an optimal set of link weights. Computed weights are deposited in a link weight database inside the offline intradomain traffic engineering block, until they are put into operation in the network by the network reconfiguration scheduler.

The interactions required between off-line inter- and intradomain TE and the options for coupling/decoupling the inter- and intradomain RPCs are analyzed in detail in [11].

Dynamic Traffic Engineering

Dynamic interdomain traffic engineering runs within an interdomain RPC and is responsible for interdomain routing [e.g., QoS-enhanced BGP (qBGP) advertisement, qBGP path selection], and for dynamically performing load balancing between the multiple paths defined by the static component based on real-time monitoring information changing appropriately the ratio of the traffic mapped on to the interdomain paths. (qBGP is an enhanced BGP that takes into account QoS information it carries in its messages as an input to its route selection process.)

Dynamic intradomain traffic engineering is the dynamic management layer as defined in [13]. This includes the intradomain routing algorithms (e.g., QoS-enhanced OSPF), together with other dynamic algorithms, to manage the resources allocated by off-line intradomain traffic engineering during the system operation in real time, in order to react to statistical traffic fluctuations and special conditions arising within an intradomain RPC. It basically monitors the network resources and is responsible for managing the routing processes dynamically as well as ensuring that the capacity is appropriately distributed among the PHBs.

SLS Management

The SLS management functionality can be split into two parts, namely:

- The part responsible for the contracts offered by the provider to its customers (i.e., the end-customers and interconnected providers);
- The part responsible for the contracts requested by the provider from its peer providers. The resulting functional components are named SLS order handling and SLS ordering, respectively.

While the ordering process establishes the contracts between the peering providers, the invocation process is required to commit resources before traffic can be exchanged, with SLS invocation handling and pSLS invocation providing the necessary functionality.

SLS order handling is the functional block implementing the server side of the SLS negotiation process. Its job is to perform subscription level admission control. The off-line intradomain traffic engineering block will provide SLS order handling with the resource availability matrix (RAM), which indicates the available capacity of the engineered network to accept new SLS orders. SLS order handling will negotiate the subscription of both cSLSs and pSLSs; they will be (largely) treated in the same way. SLS order handling maps incoming SLS requests onto the o-QCs; it can offer and investigate whether there is sufficient intra- and interdomain capacity, based on the RAM for that o-QC.

pSLS ordering is the client side of the pSLS negotiation process. During an interdomain RPC, binding selection may identify the need for new pSLSs with service peers. pSLS ordering implements the decisions of the binding selection algorithms and undertakes the negotiation process.

The pSLS invocation function block is responsible for invoking pSLSs with peer domains. The pSLSs have already been subscribed through an ordering process between pSLS ordering and SLS order handling. Optionally, pSLS invocation may be directly invoked by dynamic interdomain traffic engineering to cater for fluctuations in traffic demand, which are significantly different to those forecasted and used by binding activation for the current RPC. Whether or not this should trigger a new binding activation cycle by involving binding activation and interdomain resource optimization is a topic for further study.

Admission control is needed to ensure that the network is not overwhelmed with traffic when the network adopts a policy of overbooking network resources at the subscription level. SLS invocation handling, containing the admission control algorithm, receives signaling requests from customers or peer providers for cSLS and pSLS invocations, respectively. SLS invocation handling checks whether the invocation is conformant to the subscribed SLS and whether there is sufficient capacity in the local AS and also on the interdomain pSLSs in the case of SLSs that are not terminated locally.

Monitoring is responsible for both node and network level monitoring through both passive and active techniques. It is able to collect data at the request of the other functional blocks and asynchronously notify the other functional blocks when thresholds are crossed on both elementary data and derived statistics.

SLS assurance compares monitored performance statistics to the contracted QoS levels agreed in the SLSs to confirm that the network or service peer-networks are delivering the agreed service levels.

Traffic conditioning and QC enforcement is responsible for packet classification, policing, traffic shaping, and DSCP marking according to the conditions laid out in previously agreed SLSs and the invocation of those SLSs. At ingress routers the traffic conditioning function is responsible for classifying incoming packets to their o-QC, and subsequently marking them with the correct DSCP for the required l-QC. At the egress router the QC enforcement function may need to remark outgoing packets with the correct DSCP as agreed in the pSLS with the service peer. In other words, QC enforcement is responsible for implementing the data-plane binding from l-QC to o-QC of the service peer. Note that QC enforcement is not responsible for selecting the correct peer AS: this is decided by the q-BGP (part of the dynamic traffic engineering blocks in Figure 7.22), therefore, QC enforcement does not implement the full QC mapping/binding process in the data plane.

PHB enforcement represents the queuing and scheduling mechanisms required to be present in order to realize the different PHBs with the appropriate configuration as defined by the TE related blocks.

IP forwarding represents the functionality needed to forward IP datagrams based on the information maintained in the corresponding FIBs. Optionally, IP forwarding may also include mechanisms to perform multipath load balancing.

Interactions Between SLS Management and Dynamic Interdomain Traffic Engineering

This section describes the relationship between the SLS management and the qBGP when there is an agreement either for a new or an updated pSLS. Note that these interactions are not only interactions between the management functions and qBGP; for example, the traffic engineering decisions will also control and influence the qBGP machinery.

A pSLS contains the following elements that have been agreed between two autonomous systems (ASs) as part of the SLS order handling function:

- A defined offered QoS class, o-QC (required for all solution options);
- Reachability information: a set of destination addresses to which this o-QC is valid (required for statistical and hard solution options; not required for loose solution options);
- A bandwidth (i.e., a data rate, in units of bits/second) that defines the rate at which traffic may be sent within the terms of this pSLS, possibly including a traffic profile (required for statistical and hard solution options; not required for loose solution option);
- Time schedule (required for all solution options).

In the case of multiple links between two ASs, for each link there will be different values for some of the constituent parameters listed above. For example, the bandwidth may be different, or the reachable address prefixes may be different for different peering links. This can be addressed by assigning separate pSLSs to each link.

The pSLS/qBGP interaction is illustrated with the pair of ASs in Figure 7.24.

Each AS contains a management node, denoted X and Y (one per AS). For the pSLS agreement between AS1 and AS2, X is responsible for performing the pSLS ordering function, and Y is responsible for the SLS order handling function. Nodes X and Y are thus responsible for agreeing upon the pSLS (or pSLSs) between AS1 and AS2.

The other entities here are:

- Upstream AS ingress node(s) (i.e., A in Figure 7.24);
- Upstream AS egress node(s) (i.e., B in Figure 7.24);
- Downstream AS ingress node(s) (i.e., C in Figure 7.24).

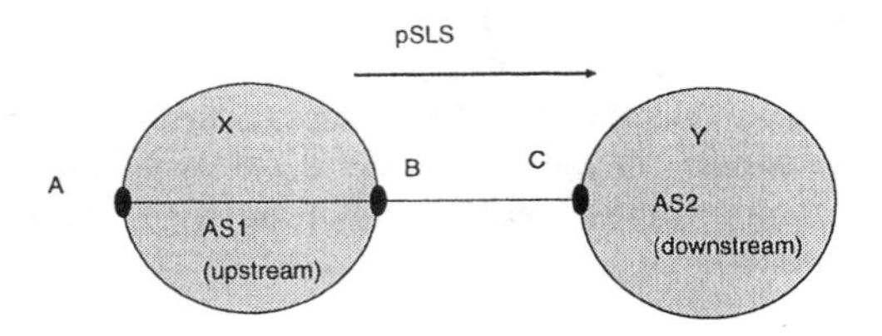

Figure 7.24 Two adjacent autonomous systems.

Between B and C there is an exterior qBGP protocol flow (e-qBGP), and between A and B there is an interior qBGP (i-qBGP) session. Let us assume that a pSLS has just been agreed (either a new or a revised old one) between AS1 and AS2. The interaction required between the management functions and qBGP will be based on where this interaction is foreseen, what information is included in that interaction, when this interaction happens, and finally who is responsible for performing that interaction.

Routing advertisements are propagated from AS2 to AS1 using e-qBGP. These advertisements must include some QoS information that is part of the agreed pSLSs between the two ASs. In general, the two domains will filter out any other advertisement that is not part of an agreement. Thus, after a pSLS is agreed, whether new or revised, both parties should enable the exchange of such advertisements.

Interaction between pSLS information and qBGP is required at the following locations:

- At the ingress nodes of the downstream AS, to implement a policy that enables the related qBGP advertisements towards the upstream AS;
- At the egress nodes of the upstream AS (i.e., B in Figure 7.6), to implement a policy that allows (stops filtering out) the related qBGP advertisements.

Reachability information (i.e., specific address prefixes), is required both as part of the policy filter information and also for injection into qBGP. For the former reason, it is therefore required at both upstream AS egress nodes and at downstream AS ingress nodes. If the information about specific address prefixes is not part of the pSLS agreement, then it is assumed to be a "wild card"; that is, it equals all the address prefixes to which there is reachability with the best-effort class.

Bandwidth availability on the egress link for a particular QC is required for TE functions within the upstream AS (i.e., AS1 in Figure 7.24). One of the TE functions that requires this information is the egress path selection process of the ingress nodes of the upstream AS (e.g., node-A). In the rest of this section it is assumed that the preferred alternative is qBGP, and it is discussed how and where this bandwidth must be injected into qBGP.

If qBGP is used to propagate pSLS bandwidth within the upstream domain, the scope of this propagation is only between the ingress node of the downstream AS (i.e., node-C), and all the ingress nodes of the upstream AS (e.g., node-A), (see Figure 7.25).

There are two principal alternatives as to where bandwidth is injected into qBGP if this policy is adopted. One is at the egress point of the upstream AS of the agreement (i.e., node-B in the example), and the other alternative is at the ingress node of the downstream AS (i.e., node-C).

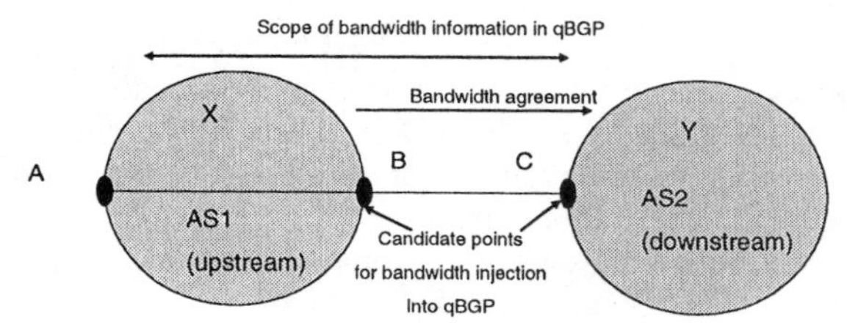

Figure 7.25 The case of bandwidth in qBGP.

For the described case, the latter alternative is chosen for two reasons. First, the node-C already is responsible for setting the QoS attributes of the qBGP advertisements towards the node-B. Second, this alternative gives the ability to perform dynamic TE with e-qBGP at node-B, in addition to the TE for egress selection with i-qBGP at node-A (see Figure 7.26). Figure 7.26 extends the model to the case of multiple links between the upstream and downstream AS: the node-B can use bandwidth information propagated using e-qBGP to select either path BC or BE.

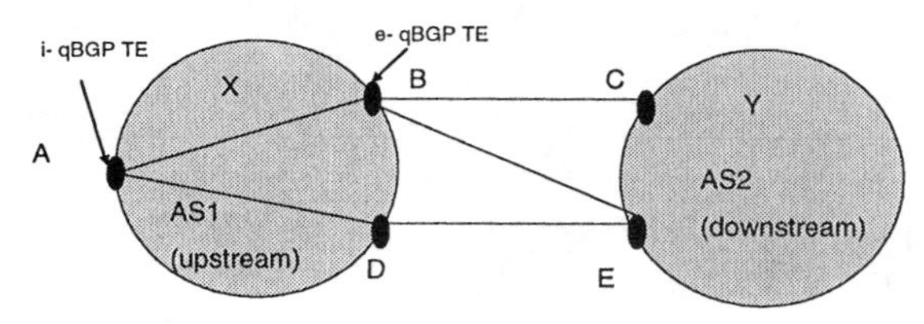

Figure 7.26 Illustration of i-qBGP and e-qBGP dynamic traffic engineering.

Other Functions and Capabilities

The functional architecture covers those capabilities necessary for deploying and operating interdomain QoS services. A provider may need other more general support functions such as fault and configuration management, but as these are not an explicit part of the interdomain QoS provision problem they are not covered in this architecture.

Policies are expected to cover the SLS management and traffic engineering functional blocks. There are no explicit functional blocks shown to handle multicast services. It is assumed in [3] that the multicast functionality is distributed over several of the blocks and only two additional blocks can be identified: dynamic group management and RPF checking. For most providers, an important aspect of providing service differentiation is the means for charging appropriate rates for different service levels. The issues associated with financial settlements according to the various business models for interactions between network providers are analyzed in detail in [13, 14].

Following the previous discussions, Figure 7.27 shows an updated view of the functional architecture in Figure 7.22.

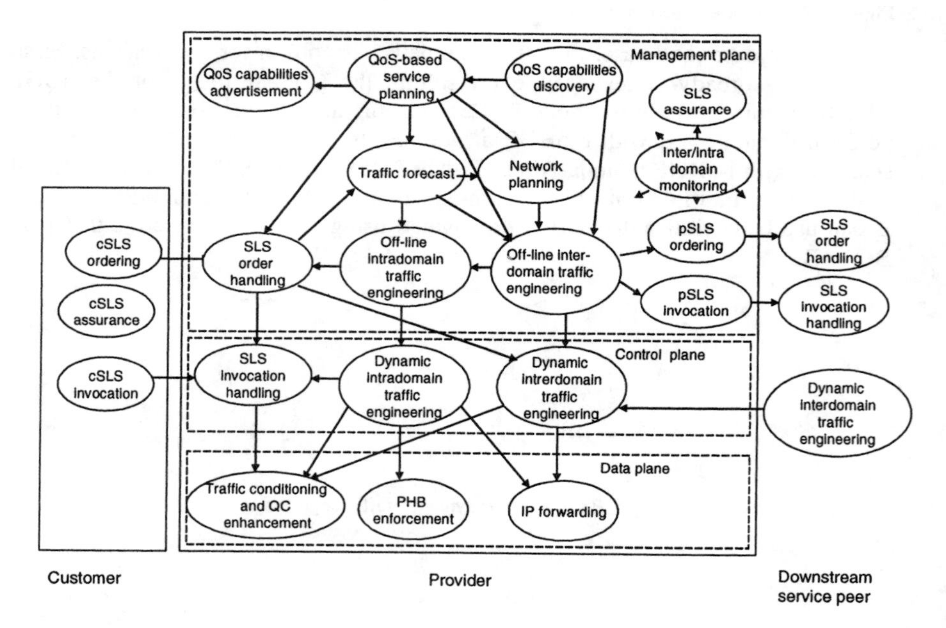

Figure 7.27 Updated functional architecture. (*From:* [11].)

As a summary, the main changes are the following:

- pSLS invocation has been clearly positioned in the management plane as the component implementing the decisions of binding activation.
- Additional interactions have been specified between SLS order handling and the qBGP processes contained within dynamic interdomain TE.
- A network-planning block has been included to position the role of dynamic network provisioning within the MESCAL functionality.

Off-line interdomain TE and off-line intradomain TE have both been further decomposed as highlighted in the previous discussions. For simplicity, both groups of functions are shown as a single functional block in the overall functional architecture.

7.3.2 Bidirectionality

The MESCAL solutions allow QoS-based IP delivery service between end points spanning a substantial number of domains. The general requirements of providing bidirectional services with possibly different QoS assurances in the forward and reverse paths should be considered.

In the cascaded approach adopted by the project, each network provider (NP) or ISP forms pSLS contracts with adjacent NPs. Thus, the QoS peering agreements are only between BGP peers. This process is repeated recursively to provision QoS to reachable destinations that may be several domains away. Figure 7.28 shows an example for end-to-end unidirectional QoS service implementation using the cascaded approach.

Each NP/ISP administers its own domain and the interconnection links that it is responsible for. For example, in Figure 7.28, ISP1 is responsible for the network provisioning and resource allocation in AS1 including the configuration of both "*a*" and "*b*" interfaces.

7.3.3 Summary

From the above-described investigations it can be summarized that any solution for QoS provision needs to fulfill the requirement for service/application level signaling between the communicating parties. In the context of the above studies, this means to find out how this requirement will be satisfied for the meta-QoS-class plane for the reverse direction, for the information billing and admission control, to specify the desired link for the return traffic for the *destination AS* and the *l-QC*/*e-QC* for the return traffic, and to define the information flow for billing and admission control. Service level communication can also be required for obtaining the necessary information for authentication and billing purposes.

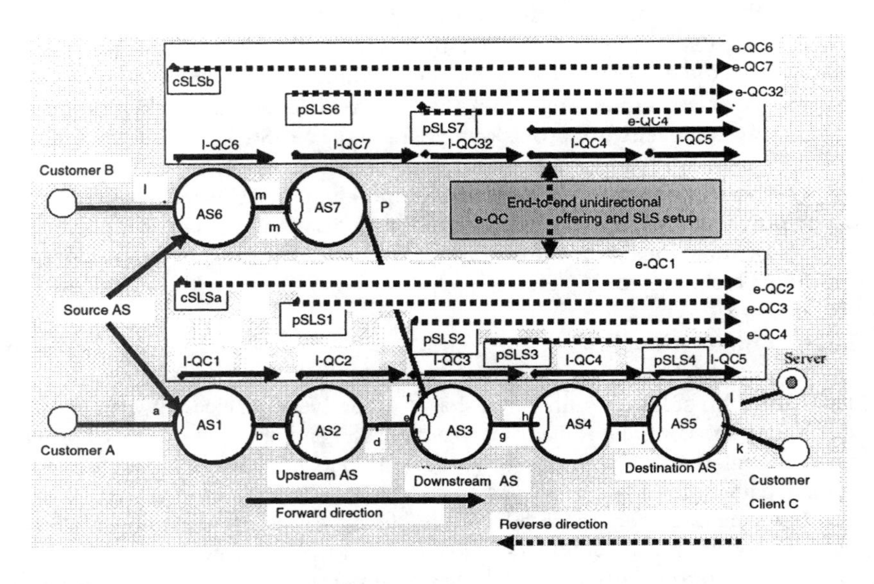

Figure 7.28 End-to-end unidirectional QoS implementation.

7.4 END-TO-END QOS PROVISION STRATEGY FOR PACKET-ORIENTED S-UMTS SERVICES

The integration of a satellite UMTS network with a terrestrial UMTS one represents one of the most attractive proposals for overcoming the coverage limitations of the future 3G cellular mobile networks, together with the possibility of exploiting the intrinsic satellite transmission capabilities. An objective of some IST projects (e.g., VIRTUOUS [4] and FUTURE [5, 17]) was to study and define a suitable QoS management concept for an integration scenario with two 3G systems (i.e., T-UMTS and S-UMTS [18]). A part of the investigation in the scope of the project VIRTUOUS focused on the scheduling algorithm on the mobile terminal side to support many kinds of sources (voice, FTP, Web).

The IST project FUTURE was the logical continuation of the IST project VIRTUOUS and focused on a QoS management concept towards the access and core network level to complete end-to-end QoS provision for the target UMTS. Focus was on the integration of multimedia services and on the development of enhanced QoS strategies for a packet-based UMTS to fulfill the requirements of the integrated services.

For the satellite segment, the general consensus of the industry is towards a geostationary Earth orbit-based solution for S-UMTS, which means that the system has to deal with high delays and BERs.

The topic of this section is QoS concepts as proposed by the above-mentioned projects. The proposed concepts were applied to a satellite channel and in compliance with UMTS specifications [19–23]. Both concepts are described by adopting suitable QoS architectures. After the description of the overall algorithms, each implemented QoS supporting function is introduced. The section includes a short investigation of the CN within an integrated terrestrial-satellite UMTS environment in order to give an end-to-end QoS approach.

7.4.1 QoS Architectures

The demonstrator shown in Figure 7.29 is based on the architecture of a classical mobile network, separating the elements into several domains: user equipment, radio access network, and core network, plus the external ISP domain. This demonstrator has the following additional components:

- A user mobility server (UMS),
- Call and session control function (CSCF);
- A feature server, where the S-UMTS part implements QoS mechanisms.

7.4.1.1 VIRTUOUS QoS Implementation

In the scope of the project VIRTUOUS, a general target QoS management system architecture was proposed, and its implementation focused onto the radio access stratum (RAS) on the terminal side (see Figure 7.29).

The RAS protocols were enhanced through the design of appropriate QoS modules/devices [24, 25].

The following describes the functionalities of the implementation:

- *Radio resource control:* According to the 3GPP standard, the RRC layer provides a group of functionalities for the general management of the already admitted connections. The implementation of RRC here is for ensuring dynamic resource management and an algorithm for setting the best way to manage connections.
- *Radio link control:* The RLC layer inserts the data packets belonging to different flows into the relevant queues, also performing the usual traffic shaping/policing through a set of dual leaky buckets (DLB). Based on the inputs derived from the MAC scheduling algorithm implemented in VIRTUOUS, the RLC is in charge of performing the segmentation of the IP datagrams in order to adapt their format to the MAC PDU length.

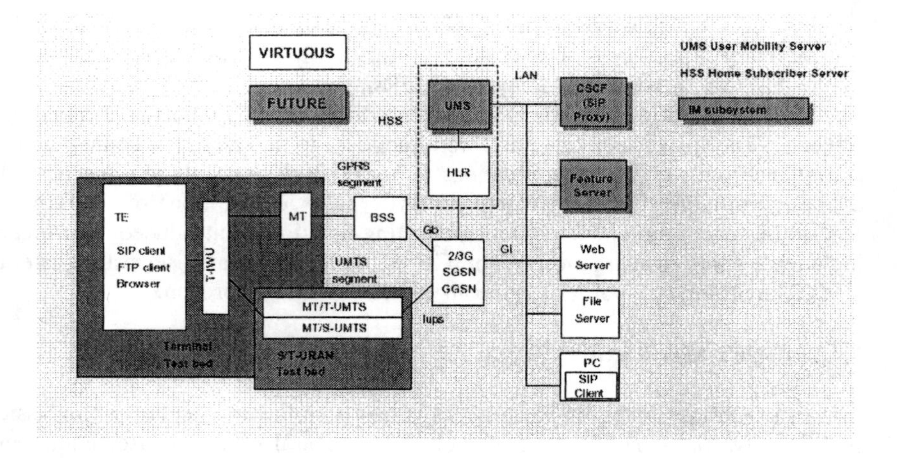

Figure 7.29 VIRTUOUS/FUTURE demonstrator.

- *Medium access control:* The MAC scheduling algorithm selects both the most suitable transport format for transmission (this decision is taken during each TTI (i.e., each 10 ms in the above implementation), and the packet to be actually transmitted among those is stored in the RLC queues. This combined selection is performed through an innovative VIRTUOUS-specific algorithm.

7.4.1.2 QoS Implementation in the Project FUTURE

The QoS guaranteeing algorithm is amplified/completed to the VIRTUOUS one where its implementation is more focused on the RAS of the RAN.

As shown in Figure 7.30, in FUTURE the RRC and MAC layers are extended out of the state of VIRTUOUS by means of a connection admission control (CAC), measurement control (MC), active set handling (ASH), and an additional scheduling algorithm [26, 27].

The separate entities are described as follows:

- *CAC:* Apparently no admission request could be satisfied if the electromagnetic coverage conditions, at the point where the requesting MN resides, are not adequate. In FUTURE, three algorithms for CAC were investigated, namely the wideband power-based admission control strategy, the throughput-based admission control strategy, and the CAC-based on the signal-to-noise-plus-interference ratio, any of which can be adopted in order for the CAC process to be realized at the radio bearer layer.

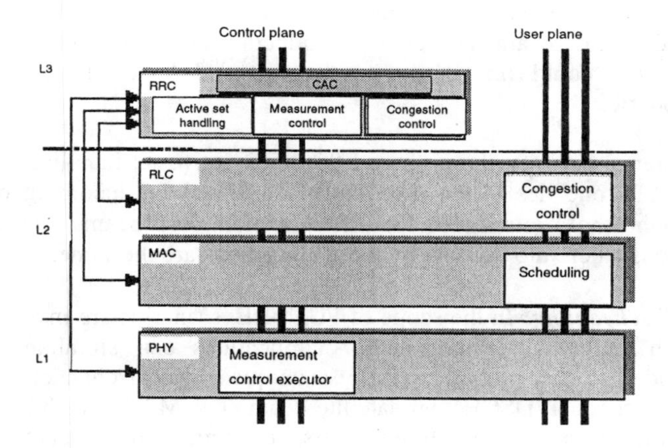

Figure 7.30 FUTURE QoS implementation.

- *Measurement control:* To support the provision of QoS, the status of the system resources shall be suitably monitored and tracked. The provision of QoS entails the correct working of different functionalities. In particular three different mechanisms shall cooperate for the successful QoS provision where different sets of parameters/measurements might be relevant for the three functionalities:
  - CAC;
  - Scheduling;
  - Congestion control.

In FUTURE, the measurement control functionality supports the management of the above functionalities for the S-UMTS case on the air interface (Uu interface). Moreover, they may depend on the considered satellite constellation. This happens not only because of the impact of the propagation delay of different satellite constellations but also because of the very different hand-off rates which may result, for example, or for an LEO and a GEO constellation. So this functionality is implemented by taking the link-specific conditions under certain satellite constellations into account [i.e., in the downlink (forward link) direction and then in the uplink (reverse link) direction for GEO and LEO, respectively].

- *Active set handling:* The active set handling for the satellite segment (S-UMTS) is based on the satellite diversity concept where its procedure for QoS provision clearly depends on the considered satellite constellation (LEO, GEO systems). Unlike T-UMTS, S-UMTS has its own different characteristics through a nonselective satellite fading channel, which preserves the multiplex orthogonality and minimizes intrabeam interference. On the other hand, similar to the T-UMTS case, the signal replies that are sent by the gateway to different noncolocated satellites introduce interbeam interference, producing, in its turn, a capacity loss. This loss could be taken into account and kept within acceptable boundaries.

The difference between the proposed S-UMTS active set handling algorithm and the T-UMTS one lies in the time scale considered for incoming measures and/or in the threshold introduced for different monitored beams. On the other hand, the actions taken into account by these algorithms are the same.

- *MAC scheduling:* In the scope of FUTURE, scenarios are investigated in which different services and/or applications can simultaneously be provided to two mobile users by a unique physical channel (downlink shared channel, DSCH) per satellite beam. The MAC scheduler has as a main function the scheduling of users, connections, and packets. Besides, this module has to provide to the CAC module the information about the available bandwidth in order to accept/reject a new connection, and to ASH power control the information about the TE, which the data packets have to address. This action indicates how the radio resources use can be optimized, in particular, how to maximize the amount of accepted requests and the data flow throughtput, as well as how to satisfy the QoS parameters. In order to respect the radio access QoS subcontract, the MAC scheduling function assigns WCDMA codes and radio frames to data packets to meet the contracted QoS parameter.

7.4.2 Physical Architecture of the QoS Testbeds

Figure 7.31 shows the system architecture with which the QoS experiment was performed in the scope of the project VIRTUOUS.

The system architecture is made up of 5+1 blocks, five Linux-based PCs and one other device [25, 26].

The PHY device is a black box in this context and it represents the physical layer emulator that introduces channel noise, delays, several satellite constellations, user mobility utilization, and so on. In fact, in the framework of the VIRTUOUS/FUTURE projects, the radio link was represented by a suitable module, named ROBMOD, that targeted the emulation of the main features of the radio access section of the real T-S-UMTS integrated network.

Directly connected to the PHY layer are the MAC/RLC device and the application portal server device. An application portal client device is connected to the MAC+RLC device. The application portal server device is equipped with server software, like ftp, Web, or Real Player, in order to test the implemented QoS algorithms with several QoS classes proposed by 3GPP. The application portal client device is equipped with the corresponding software for communication with servers. In the QoS experiment, a set of the above-mentioned applications is triggered simultaneously, and all the traffic generated from these applications is captured from a module named *sniffip* and separated into different connections to the RLC+MAC device, which is treated with the prefixed QoS parameters from the MAC module. The RLC+MAC device is responsible for the packet scheduling that is applied to select data from the parallel channels received by the application portal. In this phase, all the captured data is transformed into a single stream to be sent through the physical emulator. The demultiplexing task runs on the other side, and it is in charge of decoding the single stream, reconstructing the different channels and sending everything to the application portal, where the IP packets are put again on the net. The reverse path is almost the same.

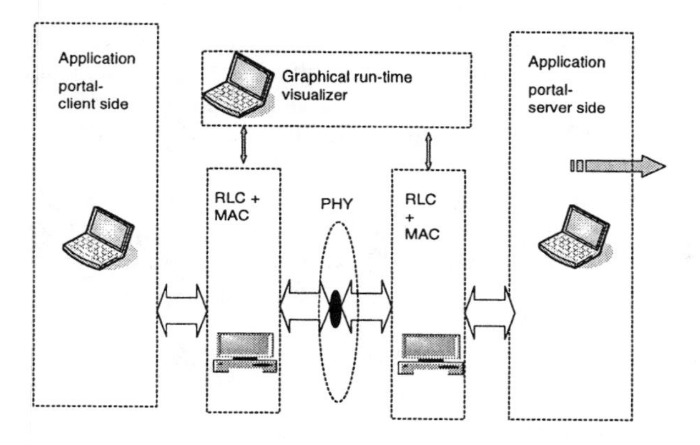

Figure 7.31 System architecture of the VIRTUOUS testbed.

The scheduling algorithm contained in the MAC module is easily configurable. In particular, it is also possible to change the scheduling algorithm in order to compare the performance of different QoS implementations. Three algorithm modules were implemented: the 3GPP proposed one, the fixed length, and an innovative VIRTUOUS algorithm [28]. Comparative logs of the tests can be easily obtained from the fourth (optional) module: the graphical visualizer. This is an external module that can be connected to the MAC block in order to view the transmission performance in real-time and to log all information sent from the MAC during the scheduling task. In this way it is possible to further elaborate the demonstration results.

The FUTURE QoS experiment was extended from the VIRTUOUS one in the following way:

- Representation of a more realistic scenario by including two users;
- Implementation of a call setup: consequently, CAC and the required CN functionality are implemented;
- Implementation of active set handling.

In this respect, special attention was paid to the need for dynamically establishing, modifying, and releasing PDP contexts, as well as to the access method to the physical layer simulation element (PLSE) emulator for both data and measurements.

The modified testbed architecture comprised three main parts: the user part, the network part, and the PLSE.

At the user side, the U0 App Host contained an NAS functionality, an RRC functionality, as well as end user applications, while the U0 NFSE computer encloses modules with the RLC and MAC functionalities.

At the network side, the radio functionality (RRC, RLC, and MAC) was located in a single PC, the GW network function simulator element (NFSE), whereas the CN emulator incorporates NAS functionality and the App Server contains the network side of the end user applications. The PLSE acts as the link between the user side and the network side (note that this link is virtual in the uplink path of user #1), implementing the physical layers of both users and gateway.

It is worth noting that the term NFSE came from the ROBMOD and VIRTUOUS projects, although the modules inserted in the FUTURE NFSE PCs are different from those inserted in the NFSE PCs of the previous projects. In fact, any computer with the required communication means (RS232 ports) to access the PLSE can serve as a FUTURE NFSE PC.

All of the computers in the testbed ran the Linux Operative System.

7.4.3 QoS Experiments

Both QoS experiments investigated the following three points:

- *Feasibility proof:* test the correctness of all the implemented functions and services;
- *Performance measurement:* show if all the QoS constraints required by each established connection are respected. It is one of the ways to verify if the system is optimally designed. To perform this task, many meaningful probes are collected, such as instantaneous bit rate, transfer delay, jitter, and packet error ratio.
- *Maximum system capabilities:* verify the limits of the demonstrator through activation of as many connections as possible.

In order to meet the above-mentioned investigation points, trials of both projects were compared, and the variations were combined and the performances of the implemented concepts were analyzed.

7.4.4 Summary

The described laboratory testbed demonstrates the performance of the investigated QoS provision concepts and furthermore represents a meaningful QoS system architecture for the target UMTS.

7.5 CONCLUSIONS

QoS is a term with many interpretations, but it should reflect how end users perceive the quality of their communication. The requirements are different from application to application. Many parameters influence QoS, for instance: bit/packet error rate, delay, bandwidth, and packet retransmission rate.

The quality of the radio channel is the traditional parameter that impacts QoS whether it is real-time or non-real-time traffic.

Imperfect mobility and handover procedures will also affect QoS. The QoS requirements are strongly linked to the performance and the load of the radio access networks.

A given service and/or network provider offers QoS-based services to its customers. The span of these services is limited to its network boundaries. On the other hand, this service provider is aware that many other service providers, scattered over the Internet also provide QoS-based services to their customers. From a centric view, this service provider wants to benefit from the QoS infrastructure made up with all the QoS-enabled networks to expand its QoS-based services to customers outside its own administrative network.

Therefore, networks should be ready to convey interdomain QoS traffic before customers can initiate end-to-end service level specifications negotiations (e.g., interdomain routing).

References

[1] IST-2000-30106 Project REAL, "Real Time Dynamic Bandwidth Optimization in Satellite Networks," http://cordis.lu/ist/ka4.home.

[2] IST-2000-26298 Project MAMBO, "MultiServices Management Wireless Network with Bandwidth Optimization," http://www.cordis.lu/ist/ka4.home.

[3] IST-2001-37961 Project MESCAL, "Management of End-to-End Quality of Service Across the Internet at Large," http://www.mescal.org/.

[4] IST-1999-10167 Project VIRTUOUS, "Virtual Home UMTS on Satellite," at http://www.cordis.lu/ist/ka4.home.

[5] IST-2000-25355 Project FUTURE, "Functional UMTS Real Emulator," http://www.cordis.lu/ist/ka4.home.

[6] IST-2000-30106 Project REAL, Deliverable 2.1, "Plform Definition," November 2001, http://www.4plus.com/real/.

[7] 3G TS 23.101v301; General UMTS Architecture, 3GPP Organization, April 1999.

[8] IST-2000-30106 Project REAL, Deliverable 4.1.3, "Final Report on Trials Execution," January 2003, http://www.4plus.com/real/.

[9] Benjamin, I. P., "3G Wireless Network Architecture UMTS Versus CDMA2000," ELEN6951, Wireless and Mobile Networking II, Columbia University.

[10] 3GPP TS 23.060 v5.0.0, "General Packet Radio Service (GPRS)," http://www.3gpp.org.

[11] IST-2001-37961 Project MESCAL, D1.2, "Initial Specification of Protocols and Algorithms for Interdomain SLS Management and Traffic Engineering for QoS-based IP Service Delivery and Their Test Requirements," January 2004, http://www.mescal.org/.

[12] IST-2001-37961 Project MESCAL, D1.1, "Specification of Business Models and a Functional Architecture for Interdomain QoS Delivery," June 2003, http://www.mescal.org/.

[13] Van Heuven, P., et al., "D1.4: Final Architecture, Protocol, and Algorithm Specification," TEQUILA, EU IST Project IST-1999-11253, April 30, 2002.

[14] IST-2001-37961 Project MESCAL, D1.4, "Issues in MESCAL Interdomain QoS Delivery: Technologies, Bi-Directionality, Interoperability, and Financial Settlements," January 2004, http://www.mescal.org/.

[15] 3GPP TS 23.207 v5.5.0, "End-to-End QoS Concept and Architecture."

[16] 3GPP TS 23.107 v5.6.0, "QoS Concept and Architecture."

[17] IST Project FUTURE, D03.01 "Description and Functional Design of the FUTURE QoS Procedures for FUTURE UMTS Demonstrator, " http://www.cordis.lu/ist/ka4.home.

[18] ETSI TR 101 865 v1.2.1, "Satellite Component of UMTS/IMT-2000: General Aspects and Principles," http://www.etsi.org.

[19] 3GPP TR 23.846 v1.2.0, "Multimedia Broadcast/Multicast Service (MBMS); Architecture and Functional Description."

[20] Liljebladh, M., and R. Geva, "Integration Problems Between IP Multicast and DiffServ Concepts in Mobile Networks."

[21] Li, Z., and P. Mohapra, "QoS-Aware Multicasting in DiffServ Domains," http://www.ieee.org.

[22] Striegel, A., "DSMCast: A Scalable Approach for DiffServ Multicasting," http://www.ieee.org.

[23] Seoung-Hoon, O., "QoS Guaranteeing Concepts in the VIRTUOUS and FUTURE IST Projects," http://www.comnets.rwth-aachen.de/typo3conf/ext/cn_download/.

[24] Del Sorbo, F., G. Lombardi, and F. Ventrone, "Final Implementation QoS Architecture for the IST Project VIRTUOUS Demonstror," *Proceedings of the IST Mobile Summit 2001*, Barcelona, Spain, September 2001, CD-ROM.

[25] IST Project VIRTUOUS, Deliverable D03.02, "VIRTUOUS Demonstror Architecture and Reference Models," http://www.cordis.lu/ist/ka4.home.

[26] Schultz, D. C., S. -H. Oh, and G. Lombardi, "A QoS Concept for Packet Oriented S-UMTS Services," *Proceedings of the IST Mobile Summit 2002*, Thessaloniki, Greece, June 2002, CD-ROM.

[27] IST Project FUTURE, D03.01, "Description of a Functional Design of the QoS Procedures for the FUTURE UMTS Demonstrator," http://www.cordis.lu/ist/ka4.home.

[28] Delli Priscoli, F., et al., "Joined Transport Form Selection and Scheduling Algorithm in VIRTUOUS," *Proceedings of the IST Mobile Summit 2001*, Barcelona, Spain, September 2001, CD-ROM.

Appendix

List of IST Projects

The following is a list of projects mentioned throughout this book. They are listed alphabetically by project name. A full list of all IST projects can be found at http://www.cordis.lu/ka4/mobile.htm.

IST-2001-37385 6HOP *Protocols for Heterogeneous Multi-Hop Wireless IPv6 Networks*

IST-1999-10731 ADAMAS *Adaptive Multicarrier Access System*

IST-2000-26222 ANTIUM *Advanced Network Radio Identification equipment for Universal Mobile Communications*

IST-2000-25133 ARROWS *Advanced Radio Resource Management for Wireless Services*

IST-2000-31079 BLUE *Bluetooth communications facilities for Electronic Scales*

IST-2001-32686 BROADWAY *The way to broadband access at 60GHz*

IST-1999-10287 CAST *Configurable Radio with Advanced Software Technology*

IST-2001-38229 CAUTION ++ *Capacity and Network Management Platform for Increased Utilization of Wireless Systems of Next Generation++*

IST-2000-25382 CELLO *Cellular Network Optimization Based on Mobile Location*

IST-2000-30066 CODIS *Content Delivery Improvement by Satellite*

IST-2001-33093 CREDO *Composite Radio and Enhanced Service Delivery for the Olympics*

IST-2000-26040 EMILY *European Mobile Integrated Location System*

IST-2001-33182 ESTA *Ethernet Switching at Ten Gigabit and Above*

IST-2001-32449 EVOLUTE *Seamless Multimedia Services over All IP-Based Infrastructures*

IST-2000-30116 FITNESS *Fourth-Generation Intelligent Transparent Networks Enhanced Through Space-Time Systems*

IST-2000-28695 FLEXIMATV *Flexible and Intelligent (S)MATV Systems*

IST-2001-32125 FLOWS *Flexible Convergence of Wireless Standards and Services*

IST-2000-25355 FUTURE *Functional UMTS Real Emulator*

IST-2000-25091 IBIS *Integrated Broadcast Interaction System*

IST-2000-30148 I-METRA *Intelligent Multi-Element Transmit and Receive Antennas*
IST-2000-30155 LOVEUS *Location Aware Visually Enhanced Ubiquitous Services*
IST-2001-32620 MATRICE *Multicarrier CDMA Transmission Techniques for Integrated Broadband Cellular Systems*
IST-2001-37961 *MESCAL Management of End-to-End Quality of Service Across the Internet at Large*
IST-2000-25394 MOBY DICK *Mobility and Differentiated Services in a Future IP Network*
IST-2001-34263 MODIS *Mobile Digital Broadcast Satellites*
IST-2001-34561 MUMOR *Multi Mode Radio*
IST-2000-25390 OBANET *Optically Beam-Formed Antennas for Adaptive Broadband Fixed and Mobile Wireless Access Networks*
IST-2001-36063 OPIUM *Open Platform for Integration of UMTS Middleware*
IST-2001-35125 OVERDRIVE *Spectrum Efficient Uni- and Multicast Services Over Dynamic Multi-Radio Networks in Vehicular Environments*
IST-2001-34157 PACWOMAN *Power Aware Communications for Wireless Optimised Personal Area Network*
IST-2000-26459 PRODEMIS *Promotion and Dissemination of the European Mobile Information Society*
IST-2001-33307 QoSAM *Quality of Services in the Digitised AM Bands*
IST-2001-34692 REPOSIT *Recommendations for Internet Usage on 2.5G and 3G – RIU253*
IST-1999-11316 R-FIELDBUS *High Performance Wireless Fieldbus in Industrial Related Multi-Media Environment*
IST-2001-32549 ROMANTIK *ResOurce Managment and Advanced Transceiver Algorithms for Multihop Networks*
IST-2000-25030 SATIN *Satellite-UMTS IP-based Network*
IST-2001-34344 SATIP6 *Satellite Broadband Multimedia System for IPv6 Access*
IST-2001-34091 SCOUT *Smart User-Centric Communication Environment*
IST-2000-25350 SHAMAN *Secure Heterogeneous Access for Mobile Applications and Networks*
IST-2000-30173 STINGRAY *Space Time Coding for Reconfigurable Wireless Access Systems*
IST-2001-38354 STRIKE *Spectrally Efficient Fixed Wireless Network Based on Dual Standards*
IST-2001-32710 UCAN *Ultra-Wideband Concepts for Ad-Hoc Networks*
IST-2001-35189 ULTRAWAVES *ULTRA Wideband Audio Video Entertainment Systems*
IST-1999-20947 WIN *Wireless Internet Network*
IST-1999-10025 WIND-FLEX *Wireless Indoor Flexible High Bitrate Modem Architecture*

IST-2001-37466 WIRELESSCABIN *Development and Demonstrator of Wireless Access for Multimedia Services in Aircraft Cabins*
IST-1999-21056 WIRELESSINFO *Wireless Supporting Agricultural and Forestry Informations System*
IST-2001-37680 WWRI *Wireless World Research Initiative*
IST-1999-12300 WSI *Wireless Strategic Initiative*

About the Editor

Ramjee Prasad received his B.Sc. (eng.) from the Bihar Institute of Technology, Sindri, India, and his M.Sc. (eng.) and Ph.D. from the Birla Institute of Technology (BIT), Ranchi, India, in 1968, 1970, and 1979, respectively.

Dr. Prasad joined BIT as a senior research fellow in 1970 and became an associate professor in 1980. While Dr. Prasad was with BIT, he supervised a number of research projects in the area of microwave and plasma engineering. From 1983 to 1988, he was with the University of Dar es Salaam (UDSM), Tanzania, where he became a professor of telecommunications in the Department of Electrical Engineering in 1986. At UDSM, Dr. Prasad was responsible for the collaborative project Satellite Communications for Rural Zones with Eindhoven University of Technology, the Netherlands. From February 1988 through May 1999, Dr. Prasad was with the Telecommunications and Traffic Control Systems Group at Delft University of Technology (DUT), where Dr. Prasad was actively involved in the area of wireless personal and multimedia communications (WPMC). Dr. Prasad was the founding head and program director of the Center for Wireless and Personal Communications (CWPC) of International Research Center for Telecommunications–Transmission and Radar (IRCTR). Since June 1999, Dr. Prasad has been with Aalborg University. Initially, Dr. Prasad was the codirector of the Center for PersonKommunikation (CPK) until 2002, and from January 2003 Dr. Prasad became the research director of the Department of Communications Technology. Now, Dr. Prasad holds the chair of wireless information and multimedia communications. Dr. Prasad has been the director of the Center for TeleInfrastruktur (CTIF) at Aalborg University since the center was established on January 28, 2004.

Dr. Prasad has much experience in the European research programs. Dr. Prasad was actively involved in the ACTS program within the project FRAMES (Future Radio Wideband Multiple Access Systems) as a DUT project leader. He was also involved in the ACTS supporting project, ASAP, which published the results of the ACTS mobile and wireless projects in two volumes entitled *Towards a Global 3G System: Advanced Mobile Communications in Europe, Volumes 1 & 2*, available from Artech House. Dr. Prasad continued his European involvement within the Fifth Research Framework in a number of IST projects, such as

PRODEMIS, STRIKE, CELLO, and PACWOMAN, to name a few. Dr. Prasad has much experience as a project leader of several international, industrially funded projects. He is the project coordinator of the European Sixth Research Framework Program integrated project "My Personal Adaptive Global NET (MAGNET)" and its follow-up. Dr. Prasad has published more than 500 technical papers, contributed to several books, and has authored, coauthored, and edited 16 books: *CDMA for Wireless Personal Communications*; *Universal Wireless Personal Communications*; *Wideband CDMA for Third Generation Mobile Communications*; *OFDM for Wireless Multimedia Communications*; *Third Generation Mobile Communication Systems*; *WCDMA: Towards IP Mobility and Mobile Internet*; *Towards a Global 3G System: Advanced Mobile Communications in Europe, Volumes 1 & 2*; *IP/ATM Mobile Satellite Networks*; *Simulation and Software Radio for Mobile Communications*; *Wireless IP and Building the Mobile Internet*; *WLANs and WPANs towards 4G Wireless*; *Technology Trends in Wireless Communications*; *Multicarrier Techniques for 4G Mobile Communications*; *OFDM for Wireless Communication Systems*; and *Applied Satellite Navigation Using GPS, GALILEO, and Augmentation Systems*, all published by Artech House. His current research interests lie in the area of wireless networks, packet communications, multiple-access protocols, advanced radio techniques, and multimedia communications.

Dr. Prasad has served as a member of the advisory and program committees of several IEEE international conferences. He has also presented keynote speeches, and delivered papers and tutorials on WPMC at various universities, technical institutions, and IEEE conferences. He was also a member of the European cooperation in the scientific and technical research (COST-231) project dealing with the evolution of land mobile radio (including personal) communications as an expert for the Netherlands, and he was a member of the COST-259 project. He was the founder and chairman of the IEEE Vehicular Technology/Communications Society Joint Chapter, Benelux Section, and is now the honorary chairman. In addition, Dr. Prasad is the founder of the IEEE Symposium on Communications and Vehicular Technology (SCVT) in the Benelux, and he was the symposium chairman of SCVT'93.

In addition, Dr. Prasad is the coordinating editor and editor-in-chief of *Wireless Personal Communications* and a member of the editorial board of other international journals, including the *IEEE Communications Magazine* and *IEE Electronics Communication Engineering Journal*. Dr. Prasad was the technical program chairman of the PIMRC'94 International Symposium held in The Hague, the Netherlands, September 19–23, 1994, and also of the Third Communication Theory Mini-Conference in Conjunction with GLOBECOM'94, held in San Francisco, California, November 27–30, 1994. He was the conference chairman of the 50th IEEE Vehicular Technology Conference and the steering committee chairman of the second International Symposium WPMC, both held in Amsterdam, the Netherlands, September 19–23, 1999. Dr. Prasad was also the general chairman of WPMC'01, which was held in Aalborg, Denmark, September

9–12, 2001, and was the general chairman for the First International Wireless Summit 2005, held in Aalborg on September 18–21, 2005.

Dr. Prasad is also the founding chairman of the European Center of Excellence in Telecommunications, known as HERMES. He is a fellow of IEE, a fellow of IETE, a senior member of IEEE, a member of the Netherlands Electronics and Radio Society (NERG), and a member of IDA (Engineering Society in Denmark).

Index

The Artech House Universal Personal Communications Series

Ramjee Prasad, Series Editor

4G Roadmap and Emerging Communication Technologies, Young Kyun Kim and Ramjee Prasad

802.11 WLANs and IP Networking: Security, QoS, and Mobility, Anand R. Prasad and Neeli R. Prasad

CDMA for Wireless Personal Communications, Ramjee Prasad

From WPANs to Personal Networks: Technologies and Applications, Ramjee Prasad and Luc Deneire

IP/ATM Mobile Satellite Networks, John Farserotu and Ramjee Prasad

Multicarrier Techniques for 4G Mobile Communications, Shinsuke Hara and Ramjee Prasad

OFDM for Wireless Communications Systems, Ramjee Prasad

OFDM for Wireless Multimedia Communications, Richard van Nee and Ramjee Prasad

Practical Radio Resource Management in Wireless Systems Sofoklis A. Kyriazakos and George T. Karetsos

Radio over Fiber Technologies for Mobile Communications Networks, Hamed Al-Raweshidy and Shozo Komaki, editors

Simulation and Software Radio for Mobile Communications, Hiroshi Harada and Ramjee Prasad

Space-Time Codes and MIMO Systems, Mohinder Jankiraman

TDD-CDMA for Wireless Communications, Riaz Esmailzadeh and Masao Nakagawa

Technology Trends in Wireless Communications, Ramjee Prasad and Marina Ruggieri

Third Generation Mobile Communication Systems, Ramjee Prasad, Werner Mohr, and Walter Konhäuser, editors

Towards a Global 3G System: Advanced Mobile Communications in Europe, Volume 1, Ramjee Prasad, editor

Towards a Global 3G System: Advanced Mobile Communications in Europe, Volume 2, Ramjee Prasad, editor

Towards the Wireless Information Society: Heterogeneous Networks, Ramjee Prasad, editor

For further information on these and other Artech House titles, including previously considered out-of-print books now available through our In-Print-Forever® (IPF®) program, contact:

Artech House
685 Canton Street
Norwood, MA 02062
Phone: 781-769-9750
Fax: 781-769-6334
e-mail: artech@artechhouse.com

Artech House
46 Gillingham Street
London SW1V 1AH UK
Phone: +44 (0)20 7596-8750
Fax: +44 (0)20 7630-0166
e-mail: artech-uk@artechhouse.com

Find us on the World Wide Web at: www.artechhouse.com